Limited Use License Agreement

This is the John Wiley and Sons, Inc. (Wiley) limited use License Agreement, which governs your use of any Wiley proprietary software products (Licensed Program) and User Manual (s) delivered with it.

Your use of the Licensed Program indicates your acceptance of the terms and conditions of this Agreement. If you do not accept or agree with them, you must return the Licensed Program unused within 30 days of receipt or, if purchased, within 30 days, as evidenced by a copy of your receipt, in which case, the purchase price will be fully refunded.

License: Wiley hereby grants you, and you accept, a non-exclusive and non-transferable license, to use the Licensed Program and User Manual (s) on the following terms and conditions only:

a. The Licensed Program and User Manual(s) are for your personal use only.

b. You may use the Licensed Program on a single computer, or on its temporary replacement, or on a subsequent computer only.

c. The Licensed Program may be copied to a single computer hard drive for playing.

d. A backup copy or copies may be made only as provided by the User Manual(s), except as expressly permitted by this Agreement.

e. You may not use the Licensed Program on more than one computer system, make or distribute unauthorized copies of the Licensed Program or User Manual(s), create by decompilation or otherwise the source code of the Licensed Program or use, copy, modify, or transfer the Licensed Program, in whole or in part, or User Manual(s), except as expressly permitted by this Agreement. If you transfer possession of any copy or modification of the Licensed Program to any third party, your license is automatically terminated. Such termination shall be in addition to and not in lieu of any equitable, civil, or other remedies available to Wiley.

Term: This License Agreement is effective until terminated. You may terminate it at any time by destroying the Licensed Program and User Manual together with all copies made (with or without authorization).

This Agreement will also terminate upon the conditions discussed elsewhere in this Agreement, or if you fail to comply with any term or condition of this Agreement. Upon such termination, you agree to destroy the Licensed Program, User Manual(s), and any copies made (with or without authorization) of either.

Wiley's Rights: You acknowledge that all rights (including without limitation, copyrights, patents and trade secrets) in the Licensed Program (including without limitation, the structure, sequence, organization, flow, logic, source code, object code and all means and forms of operation of the Licensed Program) are the sole and exclusive property of Wiley. By accepting this Agreement, you do not become the owner of the Licensed Program, but you do have the right to use it in accordance with the provisions of this Agreement. You agree to protect the Licensed Program from unauthorized use, reproduction, or distribution. You further acknowledge that the Licensed Program contains valuable trade secrets and confidential information belonging to Wiley. You may not disclose any component of the Licensed Program, whether or not in machine readable form, except as expressly provided in this Agreement.

WARRANTY: TO THE ORIGINAL LICENSEE ONLY, WILEY WARRANTS THAT THE MEDIA ON WHICH THE LI-CENSED PROGRAM IS FURNISHED ARE FREE FROM DEFECTS IN THE MATERIAL AND WORKMANSHIP UNDER NORMAL USE FOR A PERIOD OF NINETY (90) DAYS FROM THE DATE OF PURCHASE OR RECEIPT AS EVIDENCED BY A COPY OF YOUR RECEIPT. IF DURING THE 90 DAY PERIOD, A DEFECT IN ANY MEDIA OCCURS, YOU MAY RETURN IT. WILEY WILL REPLACE THE DEFECTIVE MEDIA WITHOUT CHARGE TO YOU. YOUR SOLE AND EX-CLUSIVE REMEDY IN THE EVENT OF A DEFECT IS EXPRESSLY LIMITED TO REPLACEMENT OF THE DEFECTIVE MEDIA AT NO ADDITIONAL CHARGE. THIS WARRANTY DOES NOT APPLY TO DAMAGE OR DEFECTS DUE TO IMPROPER USE OR NEGLIGENCE.

THIS LIMITED WARRANTY IS IN LIEU OF ALL OTHER WARRANTIES, EXPRESSED OR IMPLIED, INCLUD-ING, WITHOUT LIMITATION, ANY WARRANTIES OF MERCHANTABILITY OR FITNESS FOR A PARTICULAR PUR-POSE.

EXCEPT AS SPECIFIED ABOVE, THE LICENSED PROGRAM AND USER MANUAL(S) ARE FURNISHED BY WILEY ON AN "AS IS" BASIS AND WITHOUT WARRANTY AS TO THE PERFORMANCE OR RESULTS YOU MAY OBTAIN BY USING THE LICENSED PROGRAM AND USER MANUAL(S). THE ENTIRE RISK AS TO THE RESULTS OR PERFORMANCE, AND THE COST OF ALL NECESSARY SERVICING, REPAIR, OR CORRECTION OF THE LICENSED PROGRAM AND USER MANUAL(S) IS ASSUMED BY YOU.

IN NO EVENT WILL WILEY OR THE AUTHOR, BE LIABLE TO YOU FOR ANY DAMAGES, INCLUDING LOST PROFITS, LOST SAVINGS, OR OTHER INCIDENTAL OR CONSEQUENTIAL DAMAGES ARISING OUT OF THE USE OR INABILITY TO USE THE LICENSED PROGRAM OR USER MANUAL(S), EVEN IF WILEY OR AN AUTHORIZED WILEY DEALER HAS BEEN ADVISED OF THE POSSIBILITY OF SUCH DAMAGES.

General: This Limited Warranty gives you specific legal rights. You may have others by operation of law which varies from state to state. If any of the provisions of this Agreement are invalid under any applicable statute or rule of law, they are to that extent deemed omitted.

This Agreement represents the entire agreement between us and supersedes any proposals or prior Agreements, oral or written, and any other communication between us relating to the subject matter of this Agreement.

This Agreement will be governed and construed as if wholly entered into and performed within the State of New York. You acknowledge that you have read this Agreement, and agree to be bound by its terms and conditions.

s:/sale/. . . wordfilz/amgt/licagmt3.doc

SPEECH PROCESSING AND SYNTHESIS TOOLBOXES

its numerous uses for modeling and analyzing speech data. Software is provided to calculate the parameters of linear prediction speech models. Aspects of formant speech synthesis are introduced from both the theoretical and practical points of view. Here, the reader learns the importance of the frequency characteristics of speech. The methods of spectral analysis are provided in the software as well as a method for synthesizing speech using the spectral characteristics. There is software to examine procedures for converting the speech of one speaker to sound like that of another speaker, that is, voice conversion. For example, the reader is able to convert the speech of a male speaker to sound like that of a female speaker or a child. Software is provided to analyze and alter the temporal structure of the speech signal. The reader, for example, is able to automatically parse speech into various features, such as voiced segments, unvoiced segments, nasal and non-nasal segments, fricatives, stops, and so forth. The reader can then alter the duration of these segments (e.g., shorten or lengthen the segments, delete or add segments, and so forth.) Finally, the reader is able to synthesize the speech of the altered data file. This software is useful for creating speech with a "high speaking rate" or speeded-up speech. It can also generate speech with a "slow speaking rate." Another application is the creation of speech databases for speech recognition. There is a software model of the vibratory motion of the vocal folds that provides various views of the vocal folds in vibratory motion. The parameters of the vocal fold model can be adjusted to change the vocal fold tension, length, thickness, mass, and so on, so that the reader can observe the effects of these parameters on the vibratory motion of the vocal folds. The vibratory motion of the vocal folds is important because it influences the quality of speech production. The articulatory speech synthesizer included with this text uses a model of the vocal tract to synthesize speech. One objective of the software and theory is to illustrate the effects that speech models and speech analysis procedures have on the quality of synthesized speech. This software allows the user to synthesize speech by generating a vocal tract shape that corresponds to a set of formant frequency characteristics. Simulated annealing is used as the optimization procedure. The reader learns how to design a speech excitation signal that includes the effects of the subglottal system, turbulence noise, and the nasal tract and sinus cavities. Various appendices provide the theoretical bases for the software.

The text also provides a glossary of speech terms; numerous references to the literature; a listing of common standards used in speech processing, synthesis, coding, and recognition; and an appendix that describes the methods and theory for assessing the intelligibility and quality of speech.

The book has ten chapters and thirteen appendices. The material is suitable for a graduate level one-semester course and has been used for such a course at the University of Florida. In this course, each student has a PC at his or her desk. Each class period devotes from 50 to 100 percent of the time on the analysis or synthesis of speech data. The instructor, on a one-on-one basis with the student, monitors the student's work in class. Each class period expands the student's practical experience in speech analysis and synthesis. This practical, experimental exposure to speech data is supplemented by discussions of the theory provided in the appendices. The prerequisites include an understanding of digital signal processing. The reader should be familiar with MATLAB. While a course in random processes is not required, it is useful, such as Childers (1997). Several general references include digital audio (Pohlman, 1991, 1995), an overview of speech production (Denes and Pinson, 1993), and acoustic phonetics (Stevens, 1998).

The material is designed to be taught in sequence, starting with Chapter 1. Each chapter points the reader to the appropriate appendix as needed.

There are two versions of the software. One is a stand-alone version that does not require MATLAB to be installed. The other version does require MATLAB. These versions of the software are discussed in the software installation and introduction.

PREFACE

The purpose of this text is to teach speech analysis and synthesis through user–computer interaction. Most texts in the signal processing field teach only, or mostly, theory. The practice is left for other courses, or is often omitted completely. Speech analysis, synthesis, and recognition are highly dependent on knowledge of the features and properties of the speech signal. However, the individual approaching this field for the first time is offered mostly theory. This book provides a means to study the features and properties of speech as a signal without having to record data and write software to analyze the data. An extensive speech database is provided on the accompanying CD-ROMs along with various software programs to analyze the data. The text also provides the theoretical basis underlying the software algorithms used for speech analysis and synthesis. The goal of this approach is to strike a balance between theory and practice, thereby aiding the student's understanding of the basic concepts, assumptions, and limitations of the theory of speech analysis and synthesis. In other words, the text strives to provide methods for data analysis as well as the theoretical background to comprehend the analysis results. A close coupling of the theory and practice facilitates the understanding of both, and enhances the understanding of the theory.

The text meets this goal by incorporating features available in no other book. First, it provides an extensive database of speech files spoken by numerous speakers. Second, it provides a collection of application software for speech analysis and synthesis that is not available elsewhere. The graphic user interfaces included in the software require only that the user point and click the computer mouse to achieve a desired analysis or synthesis. The chapters of the text teach the reader various methods for speech analysis and synthesis, starting with simple examples and building to sophisticated procedures. The theory behind the software is covered in the various appendices.

The text covers nearly all aspects of speech analysis and synthesis, including data collection and measurement procedures, the theory of speech data processing, and the application of digital signal processing procedures to speech analysis and synthesis. The text does not discuss speech coding or speech recognition. However, much of the material presented is relevant to these topics. The reader learns aspects of speech production, methods for labeling features of the data, and the properties and characteristics of speech data. Some examples of speech analysis techniques include speech waveform analysis, such as the calculation of energy and zero-crossing contours. Other options include editing the data (cutting and pasting), zooming in on the data, as well as scrolling through the data. The reader can also calculate pitch and jitter, the cepstrum, and other characteristics. Various spectral estimation procedures are provided, as well as the calculation of spectrograms. All speech files can be heard through the computer audio system using the play features of the software. There are provisions to allow the user to change the data sampling rate. The estimation of the shape of the vocal tract from speech data is an option available to the reader. Various aspects of speech production are presented, including the classification and labeling of sounds, such as vowels, fricatives, stops, and so on. The database provided with the text is discussed at length, along with the measurement procedures, which include the simultaneous digitization of both the speech signal and an electroglottographic signal that monitors the vibratory motion of the vocal folds. The theory of linear prediction is covered along with

CONTENTS

Acquisitions Editor   *Bill Zobrist*
Marketing Manager   *Katherine Hepburn*
Senior Production Editor   *Robin Factor*
Illustration Editor   *Sigmund Malinowski*
Cover Designer   *Madelyn Lesure*

This book was set in *New Times Roman* by *TechBooks*, Inc. and printed and bound by *Donnelley/Willard*. The cover was printed by *Lehigh Press*.

The book is printed on acid-free paper. ∞

Library of Congress Cataloging-in-Publication Data:

Childers, Donald G.
    Speech processing and synthesis toolboxes / Donald G. Childers.
        p.   cm.
    Includes bibliographical references (p.   ).
    ISBN 0-471-34959-3 (cloth : alk. paper)
    1. Speech processing systems.   2. Speech synthesis.   I. Title.
    TK7882.S65C485   1999
    006.4′5—dc21   99-38270
                                                            CIP

ISBN 0-471-34959-3

Printed in the United States of America

10 9 8 7 6 5 4 3 2 1

SPEECH PROCESSING AND SYNTHESIS TOOLBOXES

D. G. CHILDERS

JOHN WILEY & SONS, INC.

New York / Chichester / Weinheim / Brisbane / Singapore / Toronto

software must have their attribute set to "archive." Thus, the user must change the file attribute after copying the software from the CD-ROM to the user's hard disk. A simple procedure for doing this is as follows: First, copy the speechgui_matlabrt folder from the CD-ROM to the user's hard disk, preferably disk C. Open Windows Explorer. Change directory to speechgui_matlabrt\toolbox\local. From the Windows Explorer menubar select Tools, Find, Files or Folders. Type *.mat in the Named location and verify that the Look in location is speechgui_matlabrt\toolbox\local. Select Find Now. All 71 files with the mat extension will be displayed in the window called Find: Files Named *.mat. In this window select Edit, followed by select All. All files with the mat extension will be highlighted. In the same window select File, Properties. The properties window will appear. Uncheck the "Read-only" attribute and check the "Archive" attribute, select apply, ok. All files will have their attribute changed from "Read-only" to "Archive." Next, using Windows Explorer, change directory to speechgui_matlabrt\toolbox\local\artm\data. Select all files in this data folder. Then select the Properties button in the Windows Explorer toolbar to change the file attribute from "Read-only" to "Archive." Change directory to speechgui_matlabrt\toolbox\local\formant_track\data. Select all files in this data folder and change their attribute from "Read-only" to "Archive." Finally, change directory to speechgui_matlabrt\toolbox\local\formant_track. Select the file formtk_4 and change its attribute from "Read-only" to "Archive."

This same process must be repeated for the speech_toolboxes folder, which must be copied from the CD-ROM to the user's hard disk, preferably disk C. Open Windows Explorer. Change directory to speech_toolboxes. From the Windows Explorer menubar select Tools, Find, Files or Folders. Type *.mat in the Named location and verify that the Look in location is speech_toolboxes. Select Find Now. All 68 files with the mat extension will be displayed in the window called Find: Files Named *.mat. In this window select Edit, followed by select All. All files with the mat extension will be highlighted. In the same window select File, Properties. The properties window will appear. Uncheck the "Read-only" attribute and check the "Archive" attribute, select apply, ok. All files will have their attribute changed from "Read-only" to "Archive." Next, using Windows Explorer, change directory to speech_toolboxes\Chap_10\artm\data. As described previously, select all files in this data folder and change their attribute from "Read-only" to "Archive." Change directory to speech_toolboxes\Chap_10\formant_track\data. Select all files in this data folder and change their attribute from "Read-only" to "Archive." Finally, change directory to speech_toolboxes\formant_track. Select the file formtk_4 and change its attribute from "Read-only" to "Archive."

RUNTIME SERVER VERSION OF THE SOFTWARE

The runtime version is contained in the folder named speechgui_matlabrt, which stands for speech graphic user interface for Matlab runtime server. The file structure contained in this folder is shown in Figure I.1.

Contained within the speechgui_matlabrt folder are the bin folder and the toolbox folder. The latter contains the local, matlab, and signal folders. The contents of the bin folder are shown in the right pane of Figure I.1. Prior to starting the runtime version, the user must copy the speechgui_matlabrt folder from the CD-ROM that accompanies this text to an appropriate disk on the user's computer. Do not place this folder in another folder because this increases the path length, and this can cause application execution problems.

SOFTWARE: INSTALLATION AND INTRODUCTION

INTRODUCTION

There are two versions of the software used in this text. MATLAB powers both versions. One version runs under the MATLAB Runtime Server. This version provides the complete functionality of the regular version of the MATLAB application software, but does not provide a command-line interface to the end user. The MATLAB Runtime Server does not allow the end user to access the MATLAB command window, and does not execute the standard MATLAB M-files. An application that uses the MATLAB Runtime Server can only execute MEX-files and runtime P-files. Furthermore, the application software must supply a graphic user interface (GUI) for the end user. In summary, the major features that distinguish the Runtime Server version from the regular MATLAB version are:

- The command-line window is not active.
- A graphic user interface for the entire software package is provided.
- The error messages that usually are shown in the command-line window are trapped and not displayed.
- Standard M-files are not recognized.
- On startup, the Runtime Server executes the file matlabrt.p instead of matlabrc.m.

The advantage of the Runtime Server version of the software is that it does not require a MATLAB installation. This version of the software is completely self-contained. All necessary files are included as runtime P-files. The publisher makes the Runtime Server version available through a special license with MathWorks, Inc., Natick, MA. The author developed the runtime version under this license using a developer kit. The disadvantage of the runtime version is that it cannot be modified or extended without the use of the Runtime Server kit and the appropriate license.

The regular version of the software must have a version of MATLAB installed. The latest version at the time of this writing is MATLAB version 5.2. Both versions of the software are outlined in this introductory chapter on software installation. In addition, a brief introduction is given to the software.

INSTRUCTIONS FOR FILE ATTRIBUTE CONVERSION

The process that created the CD-ROMs for this text protected all files by changing their attribute to "read-only." Certain files for both the stand-alone and Matlab versions of the

This text is a result of numerous research projects by the author, some of which were funded by the National Science Foundation, others by the National Institutes of Health, and still others by the University of Florida. The author is grateful to many people whose assistance has greatly facilitated the development of this text. The most notable are former doctoral students. I am particularly indebted to the following (listed in alphabetical order): Chieteuk Ahn, Keun Sung Bae, Kwei Chan, Minsoo Hahn, Yung-Sheng "Albert" Hsiao, Yu-Fu Hsieh, Hwai-Tsu Hu, Ajit L. Lalwani, C. K. Lee, Kyosik Lee, Minkyu "MK" Lee, Pedro P. L. Prado, Yean-Jen "James" Shue, Yuan-Tzu Ting, John M. White, Chun-Fan Wong, Changshiann "John" Wu, Ching-Jang "Charles" Wu, Ke Wu, and a recent master's degree student Karthik Narasimhan, who provided assistance with aspects of the software development. These individuals contributed to the completion of this text in various ways. Their specific contributions are noted within the text.

The editorial team of John Wiley and Sons, Inc. provided guidance and encouragement throughout the development and production of the book. I especially appreciate the efforts of the Engineering Editor, Bill Zobrist; Penny Perrotto, Senior Editorial Assistant; Katherine Hepburn, Engineering and Computer Science Marketing Manager and Robin Factor, M. Lesure, Jenny Welter of John Wiley and Sons, Inc. as well as Eleanor Umali of Tech Books, Inc. I am sure there are numerous others unknown to me, without whose help the book would not have been completed.

Finally, I would like to thank my wife, Barbara, who helped proofread various versions of the manuscript. She went through this same process only a few years previously with another book I wrote, and I believe she hopes not to have another such experience. Alas, I have no secretary to thank. I typed the entire manuscript and the numerous revisions, captured the software graphic user interface figures that illustrate the use of the software, and prepared the other figures that supplement the text.

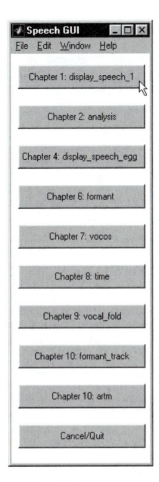

FIGURE I.3   Graphic user interface for the runtime version of the software.

The graphic user interface shown in Figure I.3 allows the user to select and start any of the nine speech software packages supplied with this text. The cancel/quit button terminates the runtime server and clears the screen. To start Chapter 1, display_speech_1, press the button and a figure similar to Figure I.4 appears. Figure I.4 shows the application software after the user has loaded the speech file b.dat, as described in Chapter 1. The reader is

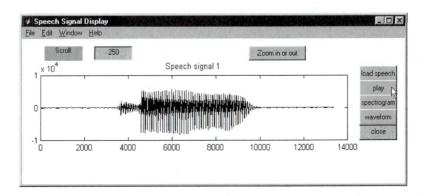

FIGURE I.4   Display_speech_1 application with the speech file b.dat loaded.

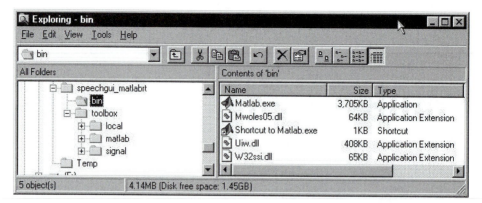

FIGURE I.1 File structure for the runtime server version of the software.

Then change directory, using Windows Explorer, to the bin directory and highlight the shortcut to matlab.exe file. Press the right mouse button, and select properties. Figure I.2 will appear, where the shortcut tab has been selected.

Figure I.2 shows disk E as the location of the speechgui_matlabrt folder. Change the location to the appropriate disk, e.g., Start in: C:\speechgui_matlabrt\toolbox\ local, assuming the user has installed the folder on disk C. For Target: change E to C Select the General tab. Be sure that the option "Read-only" is not checked and that the option "Archive" is checked. Click apply and ok. Now the user can start the runtime version by double clicking the shortcut to matlab.exe icon. (This icon can be moved to the desktop if desired.) Upon starting the runtime version, Figure I.3 appears.

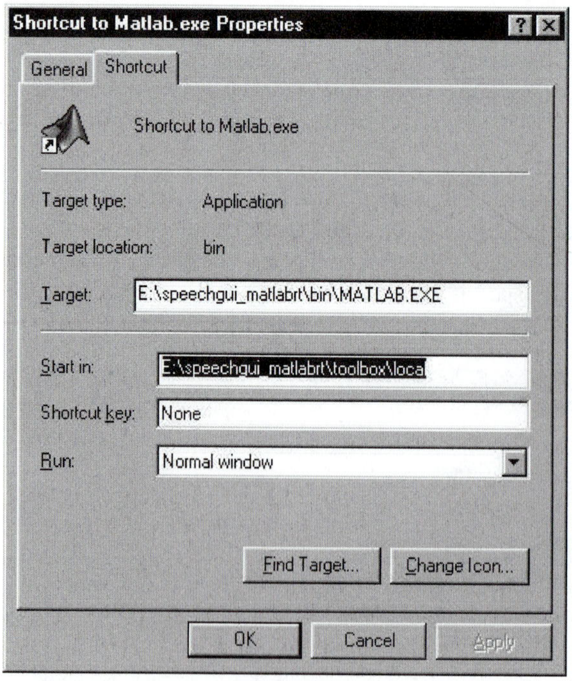

FIGURE I.2 Shortcut to matlab.exe properties.

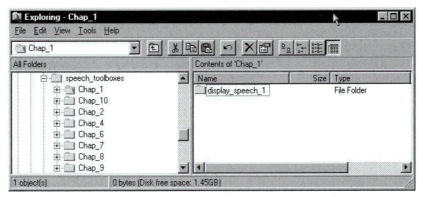

FIGURE I.5 The contents of the speech_toolboxes folder and the Chap_1 folder.

referred to Chapter 1 for additional details on the use of this application. The other chapters describe the use of the remaining software applications.

Contents of the Local Folder

The local folder within the toolbox folder of the speechgui_matlabrt folder contains the complete application software for this text, as well as some additional files that are required by the runtime server. All files are P-files, except for a few M-files that are provided for use by those readers who have a version of MATLAB installed. The use of these M-files will be explained later.

The folders within the local folder are the same folders contained within the speech_toolbox folder shown in Figure I.5. The speech_toolbox folder contains the software to be used with the MATLAB software. To use this software, follow the installation instructions provided in Chapter 1 and the subsequent chapters.

Error Messages Using the Runtime Server

Nearly all error messages that occur using the runtime server version of the speech application software are trapped. However, occasionally a message window does appear, as shown in Figure I.6. This error message occurred when the user did not select a speech file to

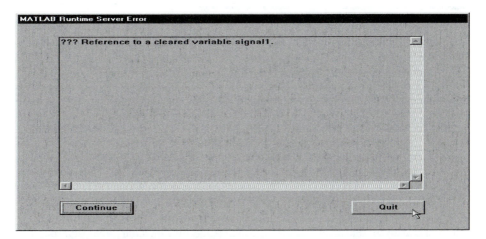

FIGURE I.6 A runtime server error message.

load using the display_speech_1 application load function. In this case, the user can press the continue button and the software will continue to function correctly. Usually, however, when such error messages occur, the user is advised to press the quit button in Figure I.6, followed by a press of the cancel/quit button in Figure I.3. This action terminates both the application software and the runtime server software, and clears all variables and errors. The user then can start the runtime version software again by double clicking the shortcut to matlab.exe icon, as described above.

Additional Comments About the Runtime Server Application Software

The major differences between the runtime and regular MATLAB versions of the application software are:

- The command-line window is not shown in the runtime version.
- The graphic user interface available in the runtime version is not available in the regular version, but can be added by the user.
- Most error messages are trapped in the runtime version and are not shown. There are a few exceptions as shown in Figure I.6.
- A few options do not work in the runtime version, such as the print option in the synthesizer of the articulatory speech synthesizer (artm).
- The background and line color of some of the plots and graphs in the runtime version differ from those in the regular version.
- A few application graphs and plots for the runtime version software differ slightly in appearance from the regular MATLAB version. However, these differences are minor.
- The runtime version of the speech application software appears to run slower than the regular MATLAB version.
- The variables and the paths are not always cleared or reset properly in the runtime version. Consequently, the user may occasionally need to cancel/quit the runtime version and restart the desired application.

THE REGULAR MATLAB SOFTWARE

The software for the regular version of MATLAB is contained in the speech_toolboxes folder contained on the CD-ROM that accompanies this text. The contents of this folder are shown in Figure I.5. This folder can be installed in the toolbox folder of Matlab. However, a less cumbersome installation is to copy each application folder (e.g., display_speech_1, artm, formant, time, etc.) to the MATLAB toolbox folder. This form of the installation is described in Chapters 1 through 10. The user then starts MATLAB, changes directory in the command-line window to the desired application folder, and types the name of the application file within the command-line window. For example, to start the display_speech_1 application, start Matlab, change directory to the display_speech_1 folder, type speech_1_display, and a display similar to Figure I.4 appears, without the speech file b.dat loaded. The use of this software is more fully described in Chapter 1. The use of the other software applications is described in Chapters 2 through 10.

Each software application folder contains a flowchart of the application. For example, the analysis folder for Chapter 2 contains the file flowchart_analysis.doc. This file is a

Microsoft Word document. Two of the applications, Chapter 1 (display_speech_1) and Chapter 4 (display_speech_egg), do not supply flowcharts because of the simplicity of these two applications.

The user can add a graphic user interface like that shown in Figure I.3 for the MATLAB version if desired. The steps required are briefly outlined as follows.

- Copy the main_speechgui M-file and the other *_path M-files from the speechgui_matlabrt\toolbox\local folder to the speech_toolboxes folder and be sure the path structure is properly set.

- Change the names of the main files in each application. For example, change the main file in the artm folder to main_artm, etc. The exceptions that need not be changed are the speech_1_display.m and speech_egg_display.m files.

- Verify that the callback names to these applications are properly named in the main_speechgui M-file.

- Change the main quit/cancel files in each of the applications to reload the main_speechgui file so that the speechgui window will reappear after each application is closed.

OPERATING SYSTEM AND PLATFORM

The runtime version of the software has been tested with both Windows 95 and Windows 98 operating systems on both desktop and laptop PC platforms. It has not been used with Unix or tested on a Macintosh platform. It has not been tested in a classroom environment.

The regular MATLAB version has been tested in a classroom, PC laboratory environment at the University of Florida by graduate students in a speech analysis and synthesis course. Some students also successfully used the software on Sun Microsystems, Inc., Palo Alto, CA, Unix machines. The software was not tested on Macintosh platforms.

SUMMARY

Both versions of the speech software provided with this text function in the same manner. The use of each application is described in detail in Chapters 1 through 10. **The runtime version does not require MATLAB to be installed**. It has a graphic user interface that allows the user to access any application. The regular MATLAB version does require version 5.2 of MATLAB to be installed and does not have the master graphic user interface shown in Figure I.3. However, it does have the other graphic user interfaces shown in Chapters 1 through 10.

CHAPTER 1

INTRODUCTION

1.1 INTRODUCTION

The background assumed for the material covered in this text includes a familiarity with sampling, analog-to-digital and digital-to-analog systems, quantization, discrete linear systems, z-transforms, discrete Fourier transforms, fast Fourier transforms, and digital filters, including finite impulse response (FIR) and infinite impulse response (IIR) filters. A familiarity with MATLAB® is required. However, software packages are provided with the text, so little programming is required. The software is written for MATLAB version 5.2, the full version. It has not been tested with version 4.2c or with the student versions of 4.2c or 5.2.

The purpose of the text is to teach aspects of speech analysis and synthesis using interactive software in a MATLAB environment. To this end, the text provides several interactive software packages as well as speech data that the reader can use to gain extensive experience with speech data. No programming experience is required to use these software packages. The text contains chapters that describe in detail how to use this software. The text also discusses aspects of the theoretical background of the algorithms used in the speech analysis and synthesis software. One goal of the text is to achieve a balance between theory and practice with regard to the processing of speech data. It is the author's belief that a close coupling of theory and practice facilitates the understanding of both. This is particularly so for specialized data sets such as speech. It is difficult to advance the field of speech analysis and synthesis without an understanding of the characteristics, features, and properties of speech data. Thus, while this book may tend to stress the practical over the theory, the hope is that the software will facilitate learning both theory and practice, without having to learn programming.

The text does not cover aspects of audio equipment, except in special cases. Thus, material on microphones such as proximity effect, condenser, and dynamic are not covered in detail. There is no discussion of recording environments such as sound booths or studios, or of other equipment such as amplifiers, speakers, compact disks (CDs), and digital-signal-processing (DSP) boards. There is a brief description of some auxiliary equipment, such as electroglottographic (EGG) devices, sound pressure measures, the Rothenberg mask, and other devices.

Note that the reader may not be familiar with most of the terms used in this introduction. Do not be alarmed. A glossary of selected terms appears in Appendix 1. Furthermore, subsequent chapters describe these terms in more detail. The objective in this chapter is to provide an overview of the material to be presented in the text and to hopefully motivate the reader that there is much to learn.

1.2 APPLICATIONS

Speech analysis, synthesis, and recognition applications include telephone systems, coding, data compression, voice mail, workstations, personal computers, and networks. Speech and

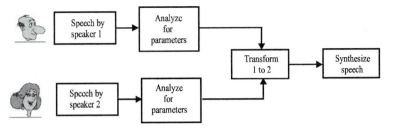

FIGURE 1.1 Block diagram of voice conversion system.

audio coding include statistical models, quantization, and companding. Speech synthesis includes several forms, such as formant, articulatory, linear prediction, miscellaneous synthesizers, and text-to-speech systems. One application of speech synthesis is called voice conversion, where the objective is to synthesize a voice with desired characteristics. For example, one may wish to create a voice that sounds like Mickey Mouse. To accomplish this task, one can try to convert the voice of one speaker to sound like that of another speaker by transforming or converting the parameters of one speaker's speech to those of another speaker's speech, as outlined in Figure 1.1. Text-to-speech synthesis (outlined in Figure 1.2) requires speech synthesis-by-rule, which includes rules for text-to-phoneme conversions, phoneme-to-feature rules, feature-to-parameter rules, and parameter-to-speech rules. Another application is speaker recognition, identification, and/or verification for such applications as banking by voice. Voice prints and forensic applications are a factor in law enforcement. Human–machine communications systems now being developed include methods to accommodate voice input for multiple speaker voice types, and multiple dialects and/or accents. These systems can be speaker dependent or independent systems.

Text–to–Speech Synthesis

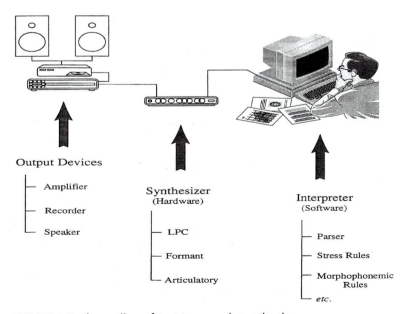

FIGURE 1.2 An outline of text-to-speech synthesis.

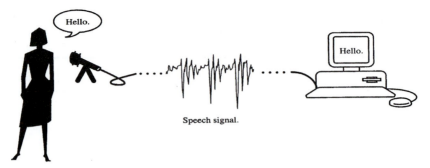

FIGURE 1.3 A simple speech command system.

Applications include voice-operated typewriters, voice messaging, response, and mail, as depicted in a simple manner in Figure 1.3.

Some recent computer applications for email and web applications are examining the use of concatenated speech segments for speech synthesis. This is an old approach that has been revitalized because computational power is now much less expensive than in years past and storage is also more readily available. Applications are appearing now that have animated, speaking "helpers" for word processors, Web applications, and e-mail. These are text-to-speech systems with special features added. A few examples of these modern synthesizers include A T & T, which can be seen at Web site www.att.com. This demonstration provides examples of the synthesis of a child, man, woman, and a singer. Microsoft Agent Beta (Microsoft, Inc., Richmond, WA) is available at Web site www.microsoft.com/intdev/agent. This demonstration works with Microsoft Internet Explorer 4.0 and contains an animated talking "genie."

While speech recognition is not covered in this text, the analysis and synthesis techniques presented here can be helpful for this task. Speech recognition procedures use

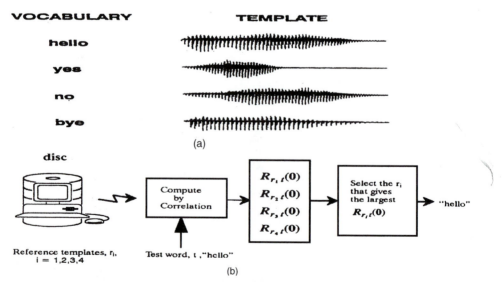

FIGURE 1.4 (a) Templates for speech recognition task. (b) A simple isolated word speech recognition scheme using the templates.

methods for pattern recognition, correlation, clustering, distance measures, and vector quantization. Methods for recognizing isolated words use dynamic time warping to align words spoken at different rates and hidden Markov models to characterize the transitions from one speech segment to another. Methods for recognizing continuous speech include classifying phonemes, segmenting and classifying words, syntactic and semantic analysis. A simple isolated word speech recognition scheme is outlined in Figure 1.4.

There are various speech systems, including communications, coding for efficient data transmission, storage, reduction, and security. Synthesis can be used for singing and music as well as speech. Measurement and analysis techniques are used for the parametric representation of speech as well as scientific modeling, coding, and to study intelligibility, naturalness, and quality. Applications of speech systems include multimedia, teaching aids for the deaf and language training.

EXAMPLE 1.2.1

A numerical example might be useful to indicate the potential for speech analysis and modeling techniques. Consider a speech signal with 5000 Hz bandwidth, sampled at 10 kHz with 12 bits/sample. This gives a 120 kbits/sec transmission rate. Suppose we can model this speech signal with a set of parameters that need updating only every 20 msec. Assume that this model uses 10 parameters, where each parameter can be represented with 12 bits. This gives us 120 bits/20 msec = 6 bits/msec = 6 kbits/sec. This is a saving of a factor of 20. In other words, with this approach we can send 20 speech signals in the same interval as compared with only one speech signal using the original method. The signal is reconstructed at the receiver by reversing the process. ■

1.3 HUMAN SPEECH PRODUCTION FOR AMERICAN ENGLISH

The primary assumption for this text is that the speech production and speech characterization is for American English. We examine aspects of acoustics and sound propagation for the human-speech system. This includes descriptions of the vocal tract, oral tract, and nasal tract. Thus, we discuss the role the pharynx, the velum, and the articulators, including the jaw, lips, palate, tongue, and teeth, play in speech production. Models of the vocal tract can be described using concatenated tubes. The articulators are important because they change the vocal tract shape, thereby, changing the sound produced. The articulators affect the constrictions in the vocal tract and the place of articulation, affecting the production of vowels, consonants, and voiceless sounds, such as fricatives. Constrictions in the vocal tract affect the production of turbulence in the flow of air, thereby, influencing the production of fricatives, sibilants, and other consonants. A schematic of the vocal tract and a simple mechanical model is shown in Figure 1.5. Figure 1.6 presents a digital signal processing model of speech production that is also used for speech synthesis.

The text examines some aspects of sound classifications, including consonants, fricatives, sibilants, stops, plosives, and vowels. Adaptation, assimilation, and coarticulation describe stringing sounds together. The field of phonetics (Stevens, 1998) encompasses the naming or labeling of sounds, which includes phones, allophones, phonemes, syllables, demisyllables, diphones, dyads, and morphs. Other types of labeling include the classification of sounds according to the type of vocal tract excitation, such as voiced, unvoiced, mixed excitation (both voiced and unvoiced), nasalization, and even, silence. Another

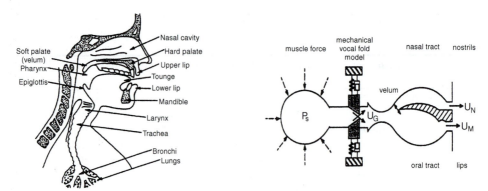

FIGURE 1.5 A schematic of the vocal tract (after Holmes, 1988) and a mechanical model.

characteristic of speech is prosodics, which includes sound timing and duration, pauses, breaks, stress, pitch, loudness, and intonation. Linguistics includes speech syntax and semantics.

The analysis of speech and the design of speech synthesis and recognition systems require various models for speech perception, speech production, and speech waveform and spectra models. These models include acoustic tube models, filter models, pole-zero models, models of the oral and nasal tracts, and speech radiation models.

The sources of data for speech analysis and modeling include photography and videotapes, electroglottography, photoglottography, x-ray technology, magnetic radiation imaging (MRI), and the inverse filtering of speech.

1.4 SPEECH ANALYSIS

A primary objective of speech analysis is to parameterize the speech signal to reduce the bandwidth and to characterize the speech signal with only a few features. Time–domain analysis procedures include data windowing to examine short segments (frames) of data that are presumed to be stationary over the windowed time interval. An example of this is shown by using a software program called speech_1_display provided with the software accompanying this text. To use this program, copy the folder containing the program to a directory in MATLAB. The folder name is display_speech_1. Be sure to keep all of the m-files together in the folder since the main m-file (speech_1_display) calls these other

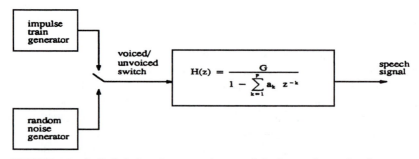

FIGURE 1.6 A digital-signal-processing model of speech production.

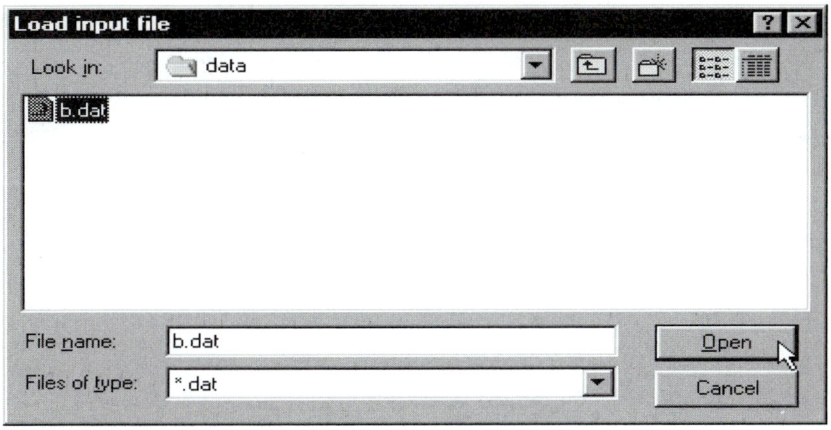

Load input file   ? ×

Look in: data

b.dat

File name: b.dat   Open

Files of type: *.dat   Cancel

FIGURE 1.7   An illustration of the Load data window.

m-files. Start MATLAB. Change directory to this folder and type speech_1_display in the MATLAB command window. To load a speech file, place the mouse cursor on the Load button and press the left mouse button. The activation of the Load button brings up the Load Input File window shown in Figure 1.7. The user can open the data folder or can change directory to another folder that contains the desired data. Assume that the data folder is opened and the ASCII data file called b.dat is loaded, as shown in Figure 1.7. The speech file must be in ASCII format with a dat extension. Figure 1.8 shows the speech file b.dat loaded, which is the word "be" spoken by a male speaker. The user can select one of the various buttons, e.g., play or spectrogram. The play option works only if a sound board is available. Figure 1.9 shows the spectrogram of the data. The waveform button redisplays the speech signal if the spectrogram is plotted. Note that a zoom in or out button is available to zoom the waveform data display. Pressing this button changes the mouse cursor to a cross hair. Move the cross hair to the desired initial x-axis data location to zoom in. Press the left mouse button once. Move the cross hair to the desired end of the zoom in x-axis segment. Press the left mouse button once again. This zooms in one level on the loaded data file. The zoomed-in waveform is plotted in the panel. Repeat these steps to zoom in another level. To stop the zoom-in process, press the right mouse button twice slowly while the cross hair is visible. The cross hair disappears and the standard mouse cursor reappears. To zoom out one level, press the zoom-in or -out button. The cross hair reappears. Press

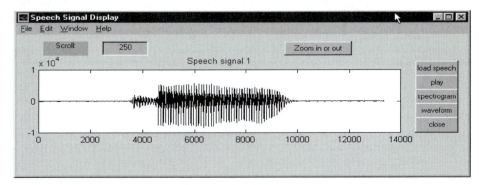

FIGURE 1.8   The data display after loading data.

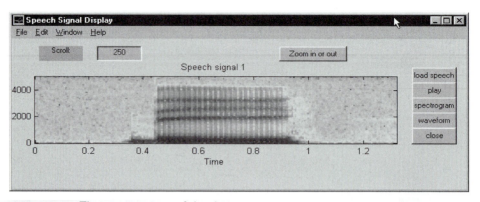

FIGURE 1.9 The spectrogram of the data.

the left mouse button, followed by a press of the right mouse button. The data are zoomed out one level. Each repetition of this sequence zooms out another level, until the original signal level is reached. Press the right mouse button twice slowly to exit the zoom-out option.

The scroll option is to be used after zooming in on the data at least one level. The scroll option becomes available automatically after zooming in. The default scroll value is 250 data points. This value can be changed by highlighting the number using the mouse cursor and typing in a new value in the scroll window panel. Press the left (right) mouse button to scroll the data to the left (right). You can continue scrolling in either direction until the end of the data record. The scroll option can be used at any zoom-in level, as long as the zoom-in level is at least one.

The play option is designed to play the data displayed. Thus, if the data are zoomed in, the data played are the data shown in the panel. Similar remarks apply to the spectrogram option.

The Close button closes out the Speech signal display window.

Figure 1.10 is a plot of the data zoomed in the middle of the vowel "e" of the word "be." The important thing to note is that this vowel is periodic with very few variations in the waveform. This is true over many periods of the data. We shall return to this matter in later chapters.

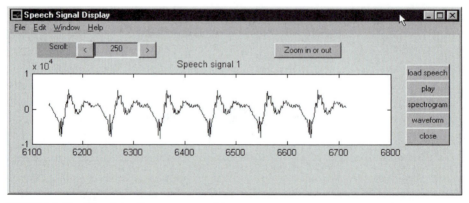

FIGURE 1.10 The data shown zoomed in.

Specific speech waveform features commonly examined are energy, the zero crossing rate, and the autocorrelation function. These features are helpful in segmenting and labeling the speech segments. Labeling can be as simple as voiced or unvoiced, or it can be as complicated as identifying specific phonemes. Frequency–domain analysis procedures include various spectral analysis algorithms and the spectrogram. The spectral analysis algorithms include linear prediction (or autoregression), moving average, and a combination of autoregression and moving average.

One primary objective of speech analysis is to determine the fundamental frequency of voicing, which is often called the pitch. The fundamental frequency of voicing is a physical characteristic that is measured as one would measure the frequency of a sinusoid or a tone. However, pitch is a sensation. In general, when the frequency of a tone is increased, we hear a rise in pitch. And similarly for a decrease in the tone's frequency. Pitch is a psychological phenomenon and is measured only by asking listeners to make judgements about the frequency changes they hear. Nevertheless, the terms fundamental frequency of voicing and pitch are often used interchangeably. Pitch is a critical feature in speech synthesis and recognition. Consequently, there are two basic types of analysis: pitch asynchronous analysis, which uses a fixed frame length of data for analysis; and pitch synchronous analysis, where the data analysis frame varies dynamically with the pitch period. A plot of the pitch frequency (fundamental frequency of voicing) versus time is often called the pitch contour.

Another objective of speech analysis is to measure the resonances of the vocal tract. These resonances are called the formant frequencies or more simply the formants. A plot of the formants versus time is called the formant tracks or formant contours. In a similar manner, there are gain contours for the gain of models of the vocal tract as well as contours for voiced/unvoiced/mixed excitation sounds, nasalization, frication, aspiration, and silence. To accomplish the measurement of such contours, we require methods to detect and label various speech features.

1.5 INVERSE FILTERING

Inverse filtering, or deconvolution, of speech is a special analysis technique that is used to estimate the excitation waveform and the formant frequencies. There are several algorithms used for this analysis procedure.

1.6 ASSESSING SPEECH INTELLIGIBILITY AND QUALITY

At the present time, there are very few quantitative measures of speech intelligibility and quality. Consequently, speech scientists rely on listening tests to assess the quality of speech synthesis or speech coding and reconstruction methods. For speech synthesis systems, one is concerned with the quality, intelligibility, and naturalness of the synthetic speech for various speaker voices, such as male, female, child, or voices that are hoarse, harsh, or breathy. The quality of speech is affected by various speaker attributes that include physical, social, psychological, emotional attributes and the status of the health of the speaker. Other factors of importance include intra-speaker variations, psychological and physiologic factors, speaking rate, and the voice quality of the speaker, such as whether the voice is hoarse or breathy. There are also inter-speaker variations that include dialect and/or accent and vocal tract size. Acoustic factors, such as environment and measurement transducers, also affect quality and intelligibility. Figure 1.11 shows examples of four word

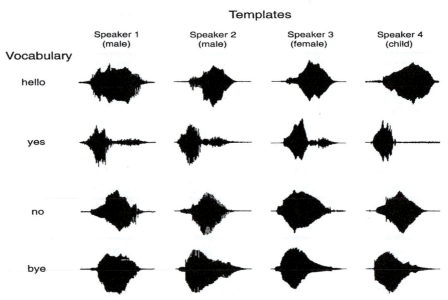

FIGURE 1.11 Examples of four words spoken by four speakers to illustrate variability.

waveforms spoken by four different speakers. Note the variations in shape, amplitude, and duration. These are factors that can be important for models.

1.7 SUMMARY

The goal of this text is to teach aspects of speech analysis and synthesis using interactive software in a MATLAB environment. Theoretical aspects of the algorithms used in the software packages are discussed. However, the emphasis is on learning the features of speech data. Subsequent chapters expand on the material outlined in this chapter.

The outline of the text is as follows. The next chapter describes the first speech toolbox, which is used for speech analysis. While the algorithms for the various programs are not discussed until later chapters, the purpose of introducing the software early is to let the reader become familiar with the analysis of speech data. The experience gained from examining data serves as a motivator for understanding the theory behind the software algorithms. Also, this experience illustrates the difficulty in designing algorithms for tasks such as recognizing word boundaries, especially for various environmental conditions and various speakers. Chapter 3 discusses aspects of speech production, the definitions of phonemes, and their features. Chapter 4 describes various speech measurement procedures and the types of data that speech scientists have used, and are using, to gain insight into speech features, so that new speech analysis algorithms can be developed. We focus on the use of the electroglottograph, since we provide an extensive database of synchronized speech and electroglottographic signals. Chapter 5 develops the theory of linear prediction, which is used extensively in speech analysis, modeling, coding, and synthesis. The second speech toolbox is described in Chapter 6, which is the formant synthesizer. This toolbox is used for speech and voice synthesis. Chapter 7 describes the analysis and synthesis toolbox for voice conversion. A toolbox for the time modification of speech is described in Chapter 8. Next, Chapter 9 presents a toolbox for several models of the vocal folds. These models can be used for studying aspects of vocal fold vibratory motion and generating models of the glottal area function as well as the volume–velocity waveform, and the electroglottographic signal. Chapter 10 discusses the toolbox for articulatory speech synthesis. This toolbox describes a model of the vocal tract and its use for generating synthetic speech.

There are a number of appendices. Appendix 1 provides a glossary of terms that are commonly used in speech. Appendix 2 contains a list of the major references used throughout the text. Appendix 3 provides a list of the major standards that are used for speech analysis, synthesis, and coding. A large speech and electroglottographic data set, provided with the text, is described in Appendix 4. Appendix 5 contains examples of waveforms, spectra, and spectrograms for selected vowels and consonants. An outline of the theory behind the algorithms used in the software introduced in Chapter 2 is discussed in Appendix 6. Two source excitation waveform models, namely the LF model and the polynomial model, are described in Appendix 7. Appendix 8 provides some additional theoretical background for the voice conversion system toolbox presented in Chapter 7. Appendix 9 gives a description of the theory and use of the time modification toolbox described in Chapter 8. The theory for the Chapter 9 toolbox on vocal fold modeling is given in Appendix 10. Appendix 11 presents the theory for the Chapter 10 toolbox on articulatory synthesis. Appendix 12 discusses aspects of procedures for assessing the intelligibility and quality of speech. Finally, Appendix 13 provides a list of the toolboxes that accompany this book.

Note that the original data file and folder names begin with a lower case letter. However, when the files were copied, Microsoft Windows Explorer in Windows 95 changed the first letter of every file to an upper case letter. For example, the file m0125s.dat became M0125s.dat, and so forth. This should not present a problem, since we use an m-file called basename.m to strip the extension from a file during a load operation. The file basename.m also allows the use of a lower or upper case letter as the first letter in a file name. In summary, a data file (x.dat) can use either a lower or upper case letter as the first letter in its file name. If an error should ever occur upon loading a data file (x.dat) that has an upper case first letter, then rename the file using a lower case letter as the first letter in the file name. Since the original data file names and folders all began with a lower case first letter, this notation is used throughout the book. However, the data file and folder names on the CDs that accompany this text begin with an upper case letter.

PROBLEMS

1.1 For the sentence "We were away a year ago" (file m0125s.dat) use the program speech_1_display to locate the word boundaries. Describe how you accomplished this task. (Note: as explained in Section 1.7, the file m0125s.dat is the same as file M0125s.dat.)

1.2 For the sentence "Should we chase those cowboys?" (file m0127s.dat) use the program speech_1_display to locate the word boundaries. Describe how you accomplished this task.

1.3 Describe the differences, if any, between the features or characteristics of the two sentences (files m0125s.dat and m0127s.dat) for Problems 1.1 and 1.2.

1.4 While we have not formally defined the fundamental frequency of voicing in this chapter, how would you define it based on the limited discussion in this chapter? Apply your definition to measure the fundamental frequency of voicing for the sentence "We were away a year ago" (file m0125s.dat).

Describe how you accomplished this measurement task. Does the fundamental frequency of voicing vary throughout the sentence or is it constant?

1.5 Repeat Problem 1.4 for the same sentence, but file f0625s.dat. What is the major difference between your results for this problem and Problem 1.4?

1.6 While we have only defined a formant in very broad terms in this chapter, nevertheless try to determine the number or formants for the sentence "We were away a year ago" (file m0125s.dat). Describe how you accomplished this task.

CHAPTER 2

SPEECH ANALYSIS TOOLBOX

2.1 INTRODUCTION

This chapter discusses and illustrates the use of the software for speech analysis. The first task is to load the software package, called analysis, into a subdirectory of MATLAB version 5.2. This subdirectory contains the folders shown in Figure 2.1, as well as the m-file **main.m**. *The analysis is initiated by the following steps:*

1) *Start MATLAB,*

2) *In the Matlab command window change directory to the analysis subdirectory, and*

3) *Type main in the MATLAB command window.*

The Main menu window appears as shown in Figure 2.2, where the user can select one of the four buttons shown. Almost without exception, the most common selection is the File button, which if pressed brings up the File menu shown in Figure 2.3.

2.2 FILE MENU

At this point, the user has a number of options, which are described, first in overview and then in depth. Upon pressing the Load button, the user can load an ASCII data file, which appears in a window called Input Signal. The Save button lets the user save a data file in ASCII format. Rate conversion allows the conversion of the loaded data file, sampled at one rate, to a new data file, sampled at a new rate. The Waveform generator provides for the generation of a few simple waveforms, including random noise. The .wav to ASCII button is to be used to convert a x.wav file to an ASCII file. And similarly for the ASCII to .wav button.

The use of these options is discussed more fully in the next section. The Play button plays the data displayed in the Input Signal window, provided a sound board is available. The play function is intended to be used with speech data sampled at 10 kHz. Finally, the activation of the Cancel button closes the File window. We now describe these various options in more detail.

2.3 LOAD

The activation of the Load button brings up the Load Input File window shown in Figure 2.4. The user can open the data folder or can change directory to another folder that contains the desired data. Assume that the data folder is opened and the ASCII data file called b.dat is loaded. The data are plotted in Figure 2.5 as the Input Signal with the global variable

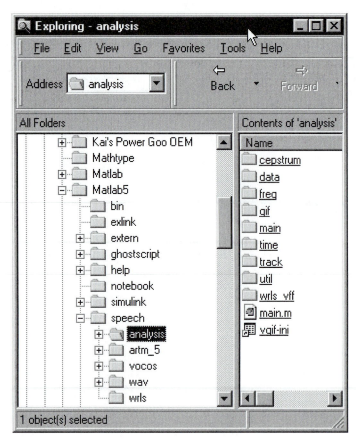

FIGURE 2.1   Contents of analysis directory.

name SPEECH (all upper case letters). The time scale is in number of samples. Since the interval between samples is 10^{-4} seconds, the length of the data record in Figure 2.5 is approximately 1.4 seconds. The software package is designed for use with data sampled at 10 kHz. Although data with other sampling rates can be loaded and analyzed, errors will most certainly occur. So only data sampled at 10 kHz should be loaded.

FIGURE 2.2   Main menu.

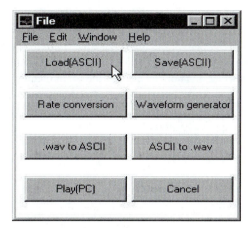

FIGURE 2.3 File menu.

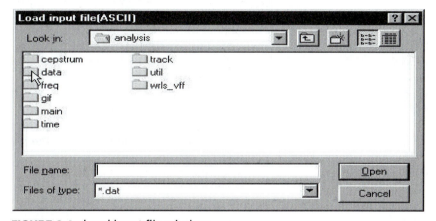

FIGURE 2.4 Load input file window.

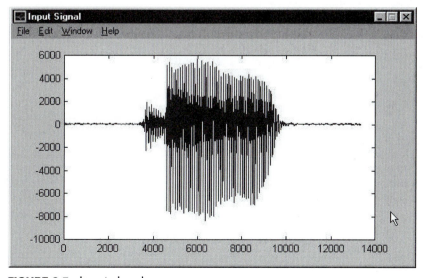

FIGURE 2.5 Input signal.

The Input Signal window should not be closed by pressing the × button at the upper right of the window. Such an action erases the SPEECH data and further data analysis is not possible. Thus, this window should be left open (or put on the task bar) until the quit button is pressed in the Main window. However, other menu windows or data-display windows can be closed (or put on the task bar) as needed. Do not close either the Main menu window or the Input Signal window until you are ready to quit the analysis software package. Selecting the print option under the file pull-down menu at the upper left of the figure, allows the user to print the Input Signal window. This applies to all data-display windows.

2.4   OPTIONS AFTER LOAD

The user has numerous options to select. One of the simplest options is to play the data by pressing the Play button. Provided a sound board is available the Input Signal is heard through the computer speakers. Another option is to activate the Cancel button, which closes the File window. The most typical options after loading a data file are to either analyze the data or to edit the data. Both of these options are described shortly. The Rate Conversion button provides the user with the ability to convert a data file sampled at one sampling frequency to another sampling frequency. Pressing the Rate Conversion button opens the window shown in Figure 2.6. The sampling frequency of the Input Signal is displayed in the upper panel. This display is usually 10,000 Hz, since the software is designed for a 10 kHz sampling frequency. The user must know the sampling frequency for the loaded data file, and change this value if it is not correct. Next, the user enters the desired new sampling frequency, e.g., 8000 Hz. Then press the Convert button and the data are resampled using the appropriate MATLAB functions, and plotted as shown in Figure 2.7. The upper panel displays the resampled data, while the lower panel displays the original Input Signal. The user can play the resampled data using the Play button in the File window. The resampled

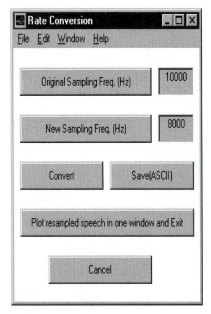

FIGURE 2.6   Rate conversion.

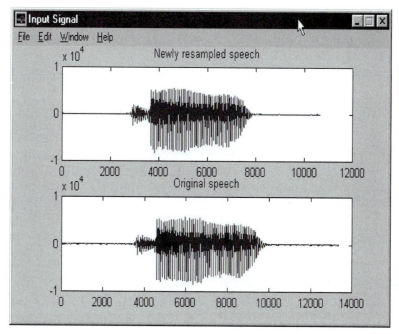

FIGURE 2.7 Input signal after rate conversion. Original data sampled at 10 kHz, newly resampled data sampled at 8 kHz.

data can be saved to a file by activating the Save button in the Rate Conversion window. The newly resampled data are displayed in the upper panel of Figure 2.7 and becomes the Input Signal, which can be analyzed or edited. Both of these options are discussed shortly. The two other buttons in the Rate Conversion window are self-explanatory. If the user elects to replot the resampled data so that it appears as one panel in one window, this data becomes the Input Signal, whether or not it has been saved to a file. Note, however, that if the data being resampled is to be analyzed or edited, the recommended procedure is to save the resampled data as a new file, and then load the new file before conducting further analyses or editing options. The user must remember that the software is designed for data sampled at 10 kHz.

One note is needed about rate conversion and play. The audio quality of the data played via a sound board is dependent on several factors, including the sampling frequencies and the number of bits per sample allowed by the sound board. For example, the 10 kHz speech data provided with this software sounds very good provided that the sound board supports a 10 kHz sampling frequency. However, if the sound board supports an 8 kHz sampling frequency, for example, then the playback of the 10 kHz data sounds "scratchy." If the 10 kHz data are resampled at 8 kHz and then played at 8 kHz [using the MATLAB function *soundsc*(data, 8000, 16)], then the audio quality is quite good.

2.5   wav TO ASCII

This option is provided as a convenience, allowing the user to convert wav files to ASCII for analysis. To use this option, do not load an ASCII file. After opening the File window, press the .wav to ASCII button and the window shown in Figure 2.8 appears, allowing the user

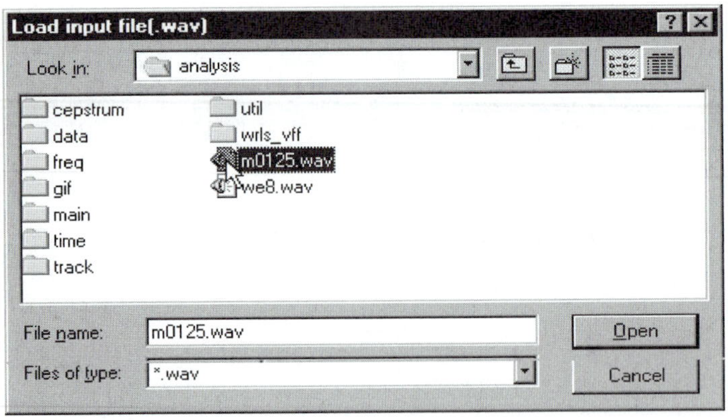

FIGURE 2.8   Load a wav file.

to select and load a wav file. Upon doing so, the wav file is played through the sound board and an Action window appears as shown in Figure 2.9. This latter window allows the user to play the data repeatedly, if desired. The converted data can be saved to an ASCII file or the user can exit this option. Note that upon converting a wav file, a message is printed in the MATLAB Command window, as shown in Figure 2.10, informing the user of the sampling frequency, number of bits per sample, and other items.

This option uses MATLAB functions wavread and wavwrite. There are similar functions for auread and auwrite. However, we have not implemented these in this software package.

2.6   ASCII TO wav

Upon pressing the button for this option, a window appears asking the user to select an ASCII speech file. After the file is selected, the following message appears in the MATLAB Command window

The sample rate of the input file is assumed to be 10000.
Saving data into x.wav (wav format).

FIGURE 2.9   Action window.

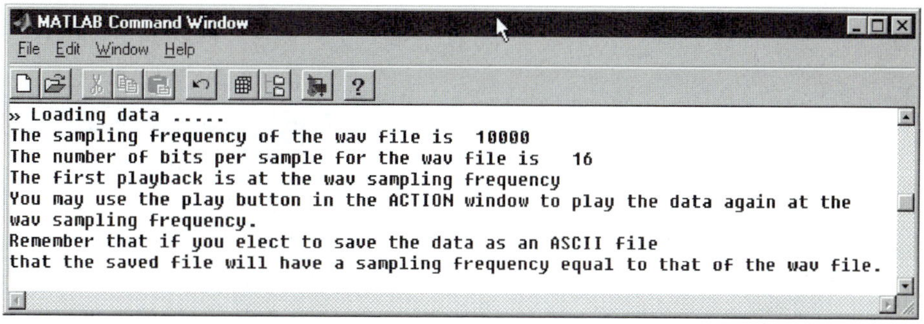

FIGURE 2.10 Matlab Command window showing characteristics of the wav file.

2.7 WAVEFORM GENERATOR

This option allows the generation of a few simple waveforms, including noise, which uses the MATLAB functions rand and randn. The types of waveforms available are shown in the pull-down menu of the upper left panel in Figure 2.11. After selecting a waveform, the user can set the sampling frequency, the length of the desired data record, the frequency of the data record, and the waveform amplitude. The waveform is then generated and plotted in the Input Signal window by pressing the generate button, as shown in Figure 2.12. The data can be saved to an ASCII file, if desired. If a random waveform is to be generated (uniform or normal), the user can select the desired seed value. For further information on random waveform generation, consult MATLAB help or the book, *Probability and Random Processes Using MATLAB*, McGraw-Hill, 1997 by D.G. Childers. The waveforms generated using this option can be analyzed further using the options described in the following sections.

2.8 EDIT

The Edit function in the Main window allows the user to zoom in on the data; scroll through the data; cut, insert, and save segments of the data; as well as play the data through a sound board. The Edit window is shown in Figure 2.13. The Edit function should be activated

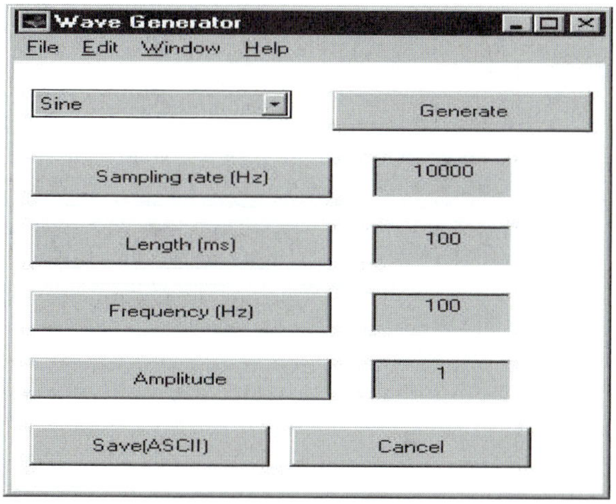

FIGURE 2.11 Waveform generation options.

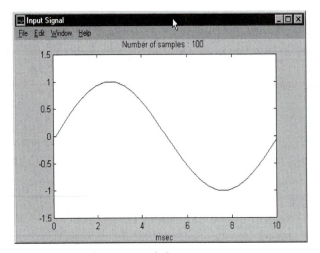

FIGURE 2.12   A generated sinewave.

only after an ASCII data file is loaded using the Load option in the File window. First, we illustrate the zoom feature on the b.dat file. The user presses the Zoom button and slides the mouse cursor to the Input Signal window. This changes the cursor to a cross hair. Suppose we zoom in on a segment of the b.dat file starting at about 3200. Move the cross hair to this location and press the left mouse button once. Next, select the end of the desired data segment to be about 5000 by moving the cross hair to that location and press the left mouse button once again. This zooms in one level. The zoomed-in waveform is plotted in the upper panel of the Input Signal window as shown in Figure 2.14. The lower panel shows the original Input Signal prior to zooming. Repeat these steps to zoom in another level. To stop the zoom-in process, press the right mouse button twice slowly, while the cross hair is in the zoomed data panel. The cross hair disappears and the standard cursor arrow reappears. To zoom out one level, press the Zoom button, slide the cursor to the upper panel

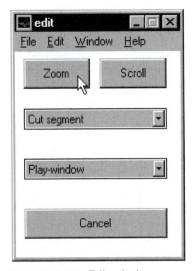

FIGURE 2.13   Edit window.

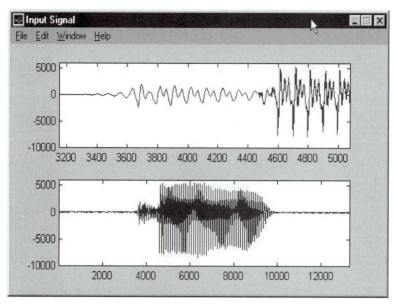

FIGURE 2.14 Input signal zoomed in one level, shown in upper panel.

of the Input Signal window, and observe that the cross hair reappears. Press the left mouse button, followed by a press of the right mouse button, to zoom out one level on the data. Each repetition of this sequence zooms out another level, until the original signal level is reached. At this point, the user can press the right mouse button twice slowly, and the data are replotted as one waveform in one panel in the Input Signal window.

The scroll feature is to be used after zooming in on the data at least one level. To activate the scroll option, press the Scroll button, move the cursor to the Input Signal upper panel, press the left (right) mouse button to scroll the data to the left (right). You can continue scrolling in either direction until the end of the data record, at which point, the data will scroll no further. To exit scroll, press both mouse buttons simultaneously. This can be difficult on the first few attempts, but it can be accomplished with some patience and practice.

The data shown in the upper panel under zoom and scroll represents the Input Signal and can be analyzed without saving the data to a file. However, if further analysis is desired, it is recommended that the data shown in the upper panel be saved as an ASCII file and then reloaded as a new file.

The pull-down menu shown next to the cut segment in the Edit window provides three options: cut segment, insert segment, and save segment. The cut segment option is destructive, discarding the data that are selected to be cut. The save segment option does not cut the segment; rather this option saves a copy of the selected segment to an ASCII file. The insert segment option lets the user insert an ASCII data file at a selected position in the Input Signal window. Again, the data displayed in the Input Signal window is the Input Signal and can be analyzed. However, it is recommended that the data be saved as an ASCII file and then reloaded prior to further analysis. An example illustrating the cutting of the beginning segment of the b.dat file is shown in Figure 2.15, while Figure 2.16 shows this same segment inserted at the end of the data record. To accomplish this task, we first click on the save segment option, move the cursor to the Input Signal window, whereupon the cursor changes to a cross hair. We then select the beginning and ending points of the segment to be saved by pressing the left mouse button as described under zooming. This

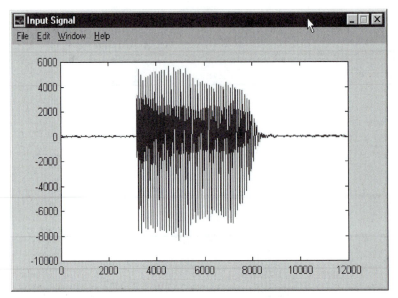

FIGURE 2.15   Input signal after saving and cutting the leading segment.

segment is saved to an ASCII file. This procedure is then repeated using the cut segment option. Then the insert segment option is selected and the cross hair is moved to the end of the data record and the left mouse button is pressed. A window opens, allowing the user to select the desired file to be inserted. Once the file name is selected, it is inserted at the location selected, as shown in Figure 2.16. Using various combinations of cut, insert, and save the user can synthesize words and sentences by concatenating phonemes and words.

The play-cursor pull-down menu provides the options to play the data in the Input Signal window in two ways: between two selected points of the data or the entire data

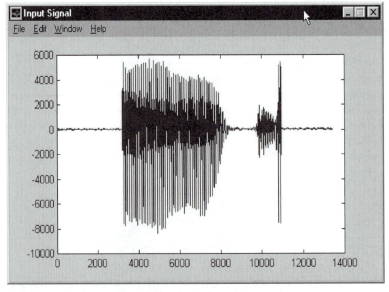

FIGURE 2.16   Input signal after inserting the saved segment.

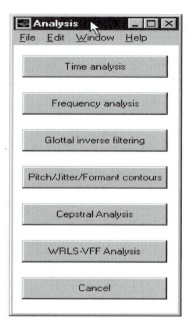

FIGURE 2.17 Analysis menu.

window. The play-cursor option works by selecting this option, moving the cursor to the Input Signal window where it becomes a cross hair. Next, press the left mouse button to select the beginning point to the desired segment, move the cross hair to the desired end point of the segment, and press the left mouse button again. The data between the two marked points are played through the sound board. This option can be repeated for other data segments. The play window option plays the entire data record displayed in the Input Signal window. The cancel button closes the edit window, but not the Input Signal window.

2.9 ANALYSIS

Various analysis methods are available, as shown in Figure 2.17, including time–domain analysis; frequency–domain analysis (spectral analysis); glottal inverse filtering; pitch, jitter, and formant contour calculations; cepstral analysis; and weighted recursive least squares analysis with a variable forgetting factor (WRLS-VFF). The use of these techniques is described in the remainder of this chapter. Prior to selecting one of the analysis methods in Figure 2.17, be sure to load a data file as described previously.

2.10 TIME–DOMAIN ANALYSIS

The options available under time–domain analysis appear in Figure 2.18, and include a demonstration of the process of windowing a data record, energy and zero crossing (ZCR) calculations, and calculation of the biased autocorrelation function. The Save button is a pull down that allows the user to save the energy, zero crossing, and autocorrelation functions. The options for the Property button appear in Figure 2.19. The Window length, Overlap, and Window type buttons apply to the Windowing, Energy and ZCR, and Autocorrelation

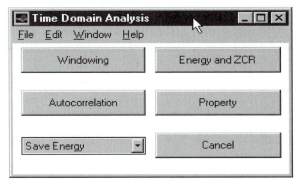

FIGURE 2.18 Time–domain analysis window.

options. The Threshold button applies to only the Energy and ZCR option. The select starting point for autocorrelation applies to only the autocorrelation option.

2.11 PROPERTY WINDOW: TIME–DOMAIN ANALYSIS

First, we describe the use of this window for the Windowing option under time domain analysis. Select the desired window length in data samples (points) by moving the slider or by clicking the mouse cursor on the number and typing in the desired value. The range is 4 to 2048. Next, select the desired window overlap as a percentage of window length. Skip the threshold value, since this does not apply for the Windowing option. Select the desired window type from the pull-down menu. The options available include hamming, hanning, kaiser, triangular, bartlett, blackman, rectangular, and chebyshev, all of which are functions in MATLAB. Once these selections have been made, press the Apply button. Then, press the Window button in the Time Domain Analysis window. An animated data analysis

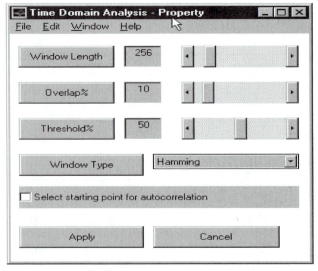

FIGURE 2.19 Property window for time–domain analysis.

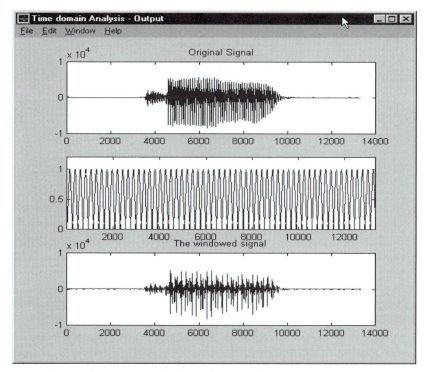

FIGURE 2.20 Illustration of windowing of a data record.

window appears as shown in Figure 2.20. The Input Signal data is displayed in the upper panel, the window is shown moving across the data in the middle panel, and the windowed data is replotted in the lower panel. After the windowing process is completed, the static display is shown in Figure 2.20 for the default parameter values shown in Figure 2.19. The primary purpose of this option is to illustrate the data windowing process. No save option is provided. One note of caution, do not move and click the mouse cursor in another window while the window analysis calculations are underway. This can cause the data to be plotted in an incorrect window.

2.12 ENERGY AND ZCR

For this example, accept the default property settings, except set the Threshold to 10% of the Input Signal amplitude level. Press the Energy and ZCR button. The calculated energy and zero crossing rate are shown in Figure 2.21. While the calculations are being performed for this option, the results are displayed in an animated-like manner. The completed results appear in Figure 2.21. The results vary with the window length, overlap, threshold, and window type. These results can be saved to a file.

2.13 AUTOCORRELATION

The biased autocorrelation function is calculated as follows. In the Property window, select the desired window length, overlap, window type, and check the select starting point for

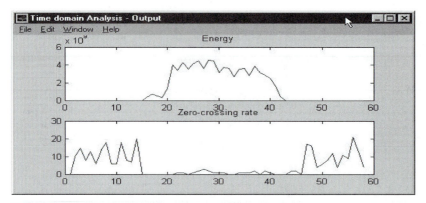

FIGURE 2.21   Illustration of energy and ZCR calculations.

autocorrelation. The threshold option does not apply for this option. Press the Apply button and then press the Autocorrelation button in the Time Domain Analysis window. Slide the mouse to the Input Signal window. The mouse cursor changes to a cross hair. Move the cross hair to the desired starting location and press the left mouse button. The biased autocorrelation function for the windowed data is calculated and displayed as shown in Figure 2.22, where for this example the cross hair is placed at the beginning of the b in the word b.dat. This operation can be repeated at various locations in the Input Signal window. To exit this option, press the right mouse button, the cross hair vanishes, and the cursor returns. The autocorrelation function can be saved to a file for spectral analysis. If the selected starting point is not checked, the cross hair does not appear, and the autocorrelation is calculated only at the beginning of the data record.

2.14 FREQUENCY–DOMAIN ANALYSIS

The available frequency domain analysis methods are shown in Figure 2.23. We start with the Spectrogram. Upon pressing this button, the window shown in Figure 2.24 appears. Select the Property button and the window shown in Figure 2.25 comes up. The user can select the Frame (window) length for data analysis, as well as the percentage of overlap,

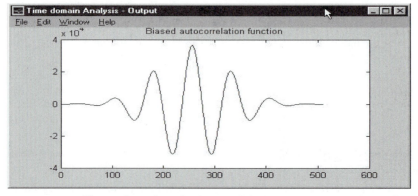

FIGURE 2.22   Illustration of a calculation of the biased autocorrelation function.

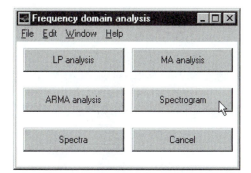

FIGURE 2.23   Frequency–domain analysis window.

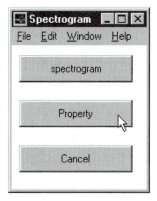

FIGURE 2.24   Spectrogram window.

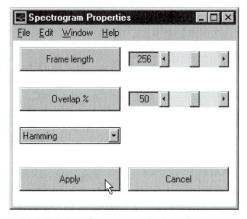

FIGURE 2.25   Property window for spectrogram.

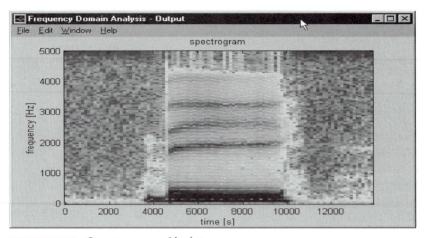

FIGURE 2.26 Spectrogram of b.dat.

and the type of window in a manner similar to that used for time–domain analysis. Press the Apply button and then press the Spectrogram button in the Spectrogram window. The spectrogram for the data in the Input Signal window is calculated and shown in Figure 2.26. The results vary with the frame length, overlap, and type of window.

2.15 SPECTRA

The Spectra option in Figure 2.27 provides the user with various spectral estimation options, including FFT, Periodogram, Blackman-Tukey, Music, and Esprit. First, open the Property window, shown in Figure 2.28 and select the desired window frame length and type of window for any of the options. The number of poles is for Music and Esprit only. The user can decide whether or not to select the starting point for the data analysis of the Input Signal. Figure 2.29 shows the FFT calculated using the select starting point option with the cross hair set at the beginning of the b in b.dat. The select starting point option is canceled by pressing the right mouse button, as mentioned previously. These steps can be repeated using the FFT option or any one of the other options. The successive calculations are plotted in the same window as superimposed waveforms. If this superimposition is not desired, then close out the display window after each calculation by pressing the × in the upper right corner. Then, the calculated spectrum is displayed in a new window each time. The superimposition feature was implemented so that the results for the various

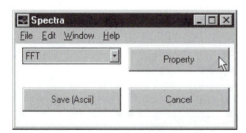

FIGURE 2.27 Spectra window.

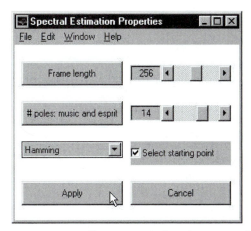

FIGURE 2.28 Property window for spectral estimation.

spectral analysis algorithms could be compared in the same window. Note that one can also compare spectral calculations using the same algorithm. In this case, the user selects the desired algorithm and then selects data from various locations within the Input Signal waveform. Note that the frequency scale is in Hz, since we assume the signal is sampled at 10 kHz.

2.16 LP ANALYSIS

Linear prediction (LP) spectral analysis is available with this option. Pressing this button brings up the window shown in Figure 2.30. The Property window appears in Figure 2.31, where the user can select the window frame length, number of poles for the LP model, the type of window, as well as the starting point. We discuss the Order selection later. Press the Apply button and then select the type of algorithm for calculating the LP model, which includes autocorrelation, covariance, modified covariance, burg,

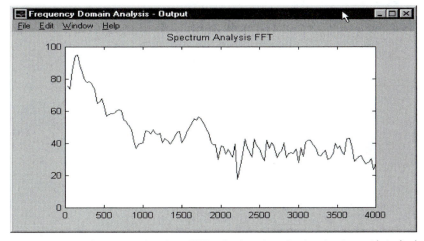

FIGURE 2.29 An example of an FFT calculated at the beginning of b in b.dat.

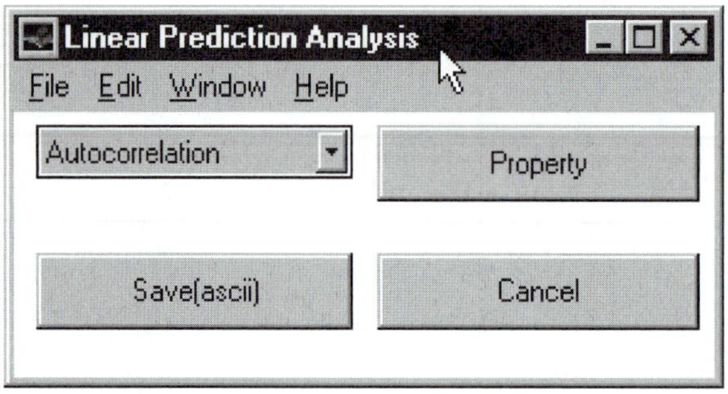

FIGURE 2.30 Linear prediction ⟨LP⟩ window.

recursive maximum likelihood, or perceptual linear prediction. Figure 2.32 shows the result for the autocorrelation method with the starting point placed at the beginning of the e in b.dat. This figure contains a plot of the FFT (blue solid) and the LP model (red dashed) superimposed in the left panel, while the right panel shows the pole-zero plot for the LP model, which in this case, is 14 poles. The user can uncheck the pole-zero plot option, if desired. The LP spectrum can be saved, but not the pole-zero plot or the FFT spectrum. As before, the cross hair can be moved to another location in the Input Signal and a new spectral estimate is plotted. However, there is no superimposition since this would make the plots too difficult to interpret. The other algorithms give similar results using the same procedures.

The Order selection option button in the Property window calculates the model "error criterion" versus the model order for the data in the Input Signal window using three criteria: final prediction error (FPE), Akaike information criterion (AIC), and criterion autoregressive transfer function (CAT). An example is shown in Figure 2.33 for the b.dat file.

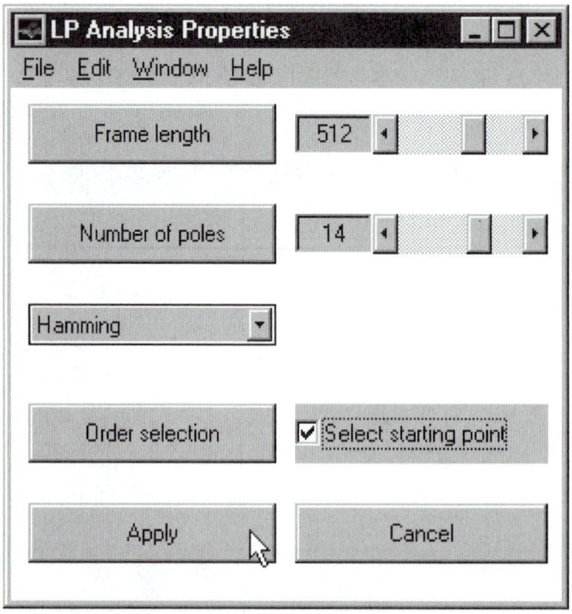

FIGURE 2.31 Property window for LP analysis.

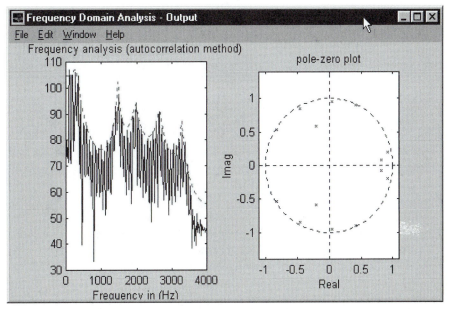

FIGURE 2.32 Illustration of an LP spectrum, FFT spectrum, and LP pole-zero plot.

2.17 ARMA ANALYSIS

An autoregressive-moving average (ARMA) spectral estimation model is calculated in this option. The ARMA window appears in Figure 2.34, while the Property window is shown

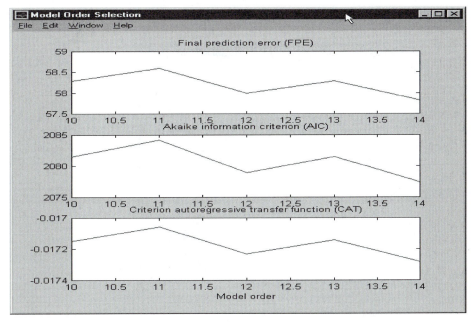

FIGURE 2.33 Order selection calculation for the Input Signal data.

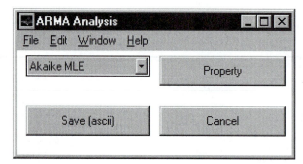

FIGURE 2.34 ARMA analysis window.

in Figure 2.35. The property options available are similar to those discussed previously, with the exception that for the ARMA model, the user can also select the number of desired zeros. An example of an ARMA model spectrum and the corresponding pole-zero plot is given in Figure 2.36 with the starting point selected at the beginning of e in b.dat.

The ARMA spectrum can be saved to a file in a manner similar to that used for the previous techniques. The algorithms available for ARMA modeling include Akaike maximum likelihood estimate (Akaike MLE), modified Yule–Walker likelihood estimate (MYLE), least squared MYLE (LS-MYLE), and the Mayne–Firoozan method.

2.18 MA ANALYSIS

The moving average (MA) analysis window appears in Figure 2.37, and the corresponding Property window in Figure 2.38. An example of the spectrum calculated by this method is shown in Figure 2.39 for the starting point selected near the e in b.dat. The MA spectrum can be saved to a file. There is only one algorithm (Durbin's) implemented for this method.

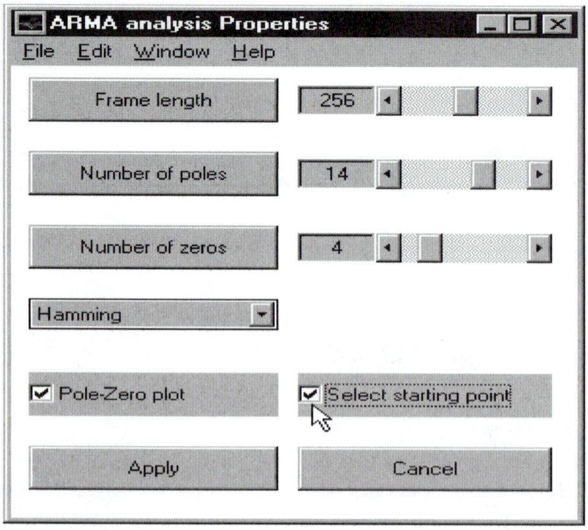

FIGURE 2.35 Property window for ARMA.

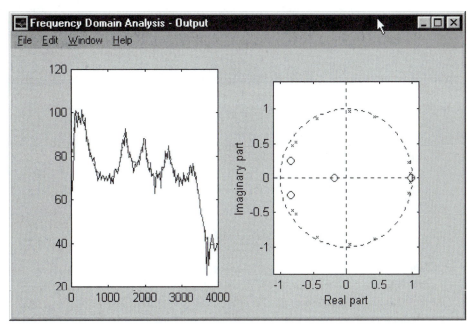

FIGURE 2.36 Illustration of an ARMA spectrum, FFT spectrum, and the pole-zero plot.

2.19 GLOTTAL INVERSE FILTERING

This option requires some effort on the part of the user. However with practice, the user can become skilled and obtain very good results. The Glottal Inverse Filtering window is shown in Figure 2.40, and its Property window is illustrated in Figure 2.41, where the user can select the window frame length, the window overlap, and the number of poles and zeros for the vocal tract model. The entire data record in the Input Signal can be analyzed or just the data between cross hair marks. The data are pre-emphasized and windowed by the hamming window within the program. Figure 2.41 shows the options selected for this example. The Apply button is pressed and the mouse cursor is moved to the Input Signal, where the cursor becomes a cross hair. For this example, the data beginning just before the b of b.dat file is selected as the beginning point, and the ending point is just over

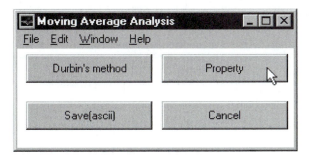

FIGURE 2.37 Moving average window.

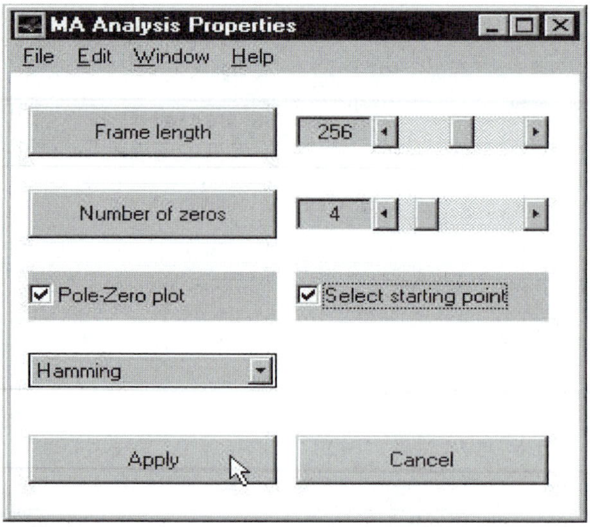

FIGURE 2.38 Property window for MA spectral analysis.

6000. To start the analysis, the user must select one of the algorithms for inverse filtering from the pull-down menu, which provides four selections: pitch synchronous (manual), pitch asynchronous (manual), asynchronous analysis, and iterative analysis. The first two require user interaction. The latter two are automatic. Figures 2.42 through 2.45 show the windows that appear if the pitch synchronous (manual) option is selected. To process the data, the user must now make selections from those in Figure 2.42. The user examines the data displayed in Figure 2.43, particularly the top waveform, which displays the frequency response (transfer function) for the residual of the vocal tract model. If the inverse filtering model is a good one, then this frequency response should be flat (or white). The middle

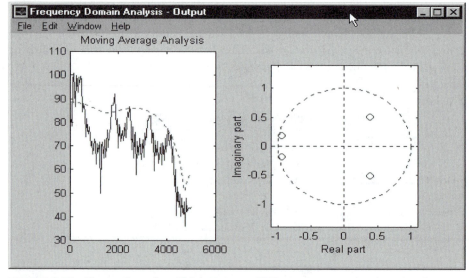

FIGURE 2.39 Illustration of an MA spectrum, FFT spectrum, and pole-zero plot.

FIGURE 2.40 Glottal inverse filtering.

panel shows the transform of the differentiated glottal volume velocity (dgvv) waveform, and the lower panel shows the transform of the glottal volume velocity (gvv) waveform. Figure 2.42 shows the values for the formants and the corresponding bandwidths for the vocal tract model. These values can be adjusted by the user to make the frequency response in the upper panel flatter. Each time a change is made to one or more of these values, press the Go button to observe the effect these changes have on frequency responses and the dgvv and gvv waveforms. When the user is satisfied with the changes, then press the Save button. Continue these steps until the data are completely analyzed. After each save operation, a new set of waveforms is plotted, as shown in Figure 2.44, which shows the time–domain waveforms for the residue, the differentiated glottal volume velocity (dgvv), and the glottal volume velocity (gvv). The time scale is in data points analyzed. If the entire Input Signal is being analyzed, then the time scale agrees with the Input Signal time base, otherwise the time scale is rescaled to the number of data points selected for analysis. Upon completion of the analysis, the frequency response window in Figure 2.43 is changed to that shown

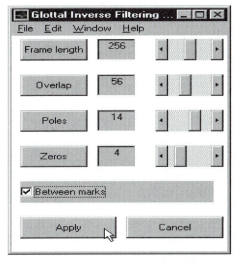

FIGURE 2.41 Property window for glottal inverse filtering.

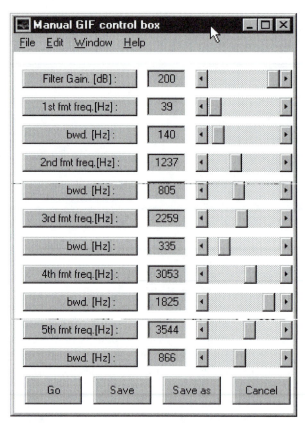

FIGURE 2.42 Formant frequencies and bandwidths.

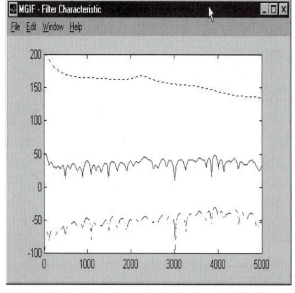

FIGURE 2.43 Filter characteristics for vocal tract model.

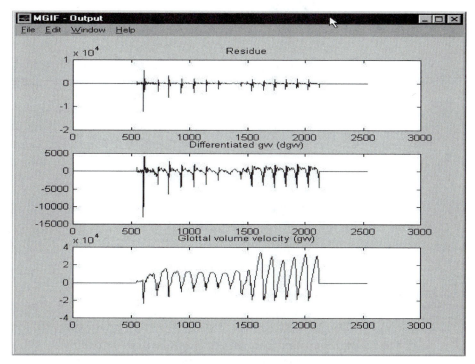

FIGURE 2.44 Inverse filtering waveforms.

in Figure 2.45, namely the pitch (lowest contour) and the first five formant frequency contours for the data analyzed. The time waveforms for the residue, dgvv, and gvv are plotted as shown in Figure 2.44. The data can be saved to a file. The figures can be printed out. Note that a window also appears telling the user that the analysis has been completed. This window contains a Cancel button. Do not press this Cancel button until the data are saved (if desired) and the data are printed (if desired). Otherwise, the data displays are erased.

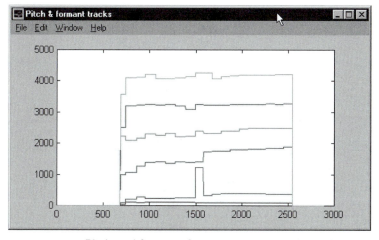

FIGURE 2.45 Pitch and formant frequency contours (tracts).

If the option between marks is not selected, then the entire data record in the Input Signal is analyzed in the same manner as that described. This can take some time. Consequently, the user may want to experiment with selected data segments at first.

The pitch asynchronous option works in a manner somewhat differently from the option described previously. For this option, the data selected between marks (or the entire data record) are analyzed and plotted as described. The user can modify the results by changing the formant frequencies and their bandwidths, as shown in Figure 2.42. Pressing the Go button displays the changes in the frequency responses and the waveforms. To accept the changes, press the Save button as described. The pitch and formant contours are not plotted.

The asynchronous and iterative analysis methods are automatic and do not involve user interaction. The frequency responses are not plotted, only the dgvv and gvv waveforms. Only the dgvv waveform can be saved. The gvv waveform can be derived by integrating the dgvv waveform, if desired.

2.20 PITCH, JITTER, FORMANT CONTOURS

This option is automatic, with the results shown in Figure 2.46 for the b.dat as the Input Signal. The upper panel is the pitch contour, the next panel down is the jitter (perturbation, order 1), the third panel from the top shows the formant contours, and the bottom panel replots the Input Signal for ease of comparison. The pitch contour is smoothed with a median filter of length 5. The horizontal scale for all three panels is the number of samples. The vertical scale for the pitch contour is the pitch in Hz. The vertical scale for the perturbation panel is in Hz. The jitter in percent would be the average (or maximum) value divided by the average pitch period times 100. The vertical scale for the formant contour is Hz.

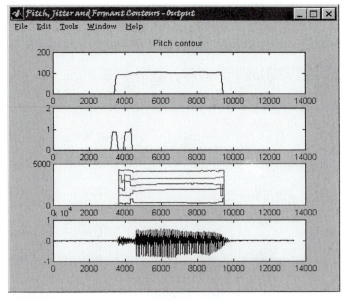

FIGURE 2.46   Pitch, jitter, formant contours.

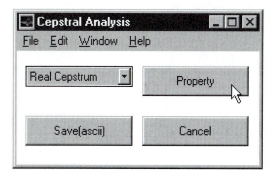

FIGURE 2.47 Cepstral analysis window.

2.21 CEPSTRAL ANALYSIS

The Cepstral analysis window appears in Figure 2.47, with the corresponding Property window displayed in Figure 2.48, which has the usual options, including window frame length, overlap, window type, and an option to select the starting point. After the desired options are selected, as shown, for example, in Figure 2.48, then press the Apply button. Next, select either the real cepstrum or the complex cepstrum to be calculated. Move the mouse cursor to the Input Signal and select the desired starting point. The real and complex cepstra are shown in Figures 2.49 and 2.50, respectively, for the starting point being selected at the beginning point of e in b.dat. The horizontal frequency scale is in number of samples. The user can move the cross hair to another location on the Input Signal to calculate another cepstrum. Exit by pressing the right mouse button. The results can be saved to a file.

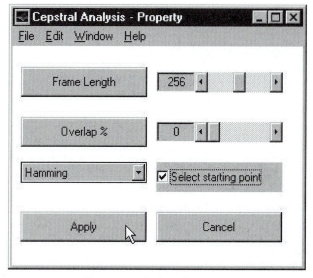

FIGURE 2.48 Property window for cepstral analysis.

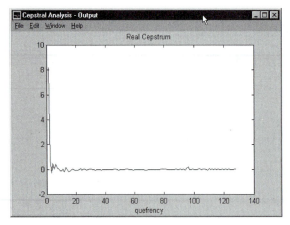

FIGURE 2.49 Real cepstrum.

2.22 WRLS-VFF ANALYSIS

The weighted recursive least squared, variable forgetting factor (WRLS-VFF) analysis window is shown in Figure 2.51. The user can select the number of poles and zeros for the model, the minimum VFF (lambda), and the desired error. However, lambda and the error are adjusted automatically as the number of poles and/or zeros are changed. So the user can choose to use these model "default" values. Upon pressing the Go button, the calculations begin for the entire data record in the Input Signal. This may take some time. Upon completion of the calculations, the first window to be plotted is shown in Figure 2.52, which plots lambda (VFF) in the uppermost panel, the estimated input excitation waveform in the second panel, the error in the third panel down, and the speech Input Signal in the bottom panel. Finally, the waterfall plot of the spectrum of the vocal tract model is plotted in Figure 2.53. This plot can take some time, so be patient. Do not move the mouse and click it in an open window during this time, since this causes the data to be plotted incorrectly in that window.

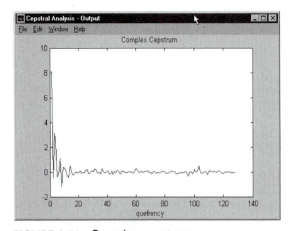

FIGURE 2.50 Complex cepstrum.

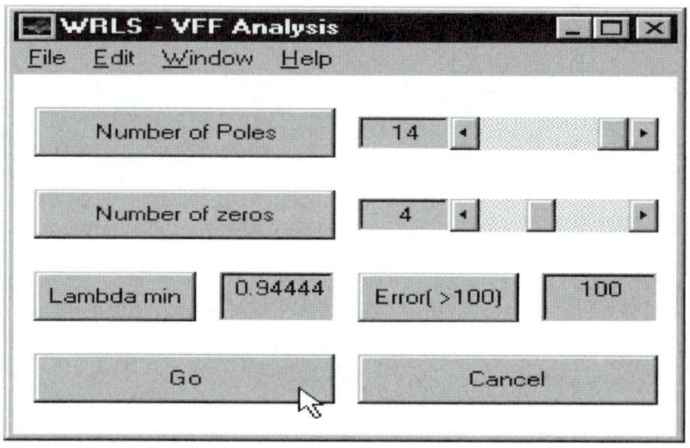

WRLS - VFF Analysis

File   Edit   Window   Help

| Number of Poles | 14 |
| Number of zeros | 4 |

| Lambda min | 0.94444 | Error(>100) | 100 |

| Go | Cancel |

FIGURE 2.51   WRLS-VFF analysis window.

It has been observed that the value of the Error depends on several factors, and can vary over a wide range of values. It is also highly data dependent. It is suggested that the user vary this parameter in steps of 10 until an acceptable value is found for the given data record. In general, the more predictable the data, the lower the value of the error. It is presumed that the error is smaller for synthesized speech than for real, digitized speech. The goal of the program is to have lambda make a transition from 1 to lambda_min only once per glottal excitation pulse period. If the error is too small, lambda makes many transitions per glottal cycle. If the error is too large, lambda will remain constant ($= 1$) and will not make a transition. Thus, the purpose of the software is to have the graph of lambda reflect an estimate of the glottal closure instants, with one transition per glottal pulse period.

The algorithm calculates the model frequency response once per frame. The default value for the number of samples per frame is 10, which corresponds to the frequency response being calculated once every 1 msec.

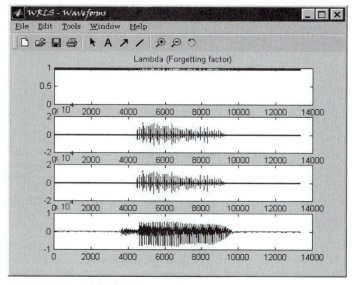

FIGURE 2.52   WRLS-VFF waveforms.

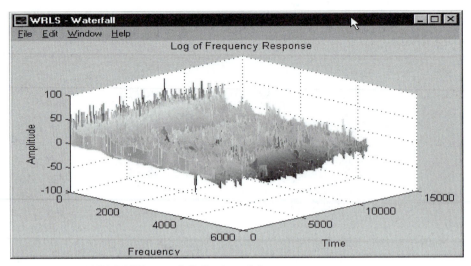

FIGURE 2.53 WRLS-VFF waterfall display of the spectrum.

2.23 MISCELLANEOUS MATTERS

If the user replots saved files, note that the plot shape is correct. However, the time or frequency scale, as the case may be, will not reflect the original scale, since the saved data is a matrix (vector) of the number of data points and does not contain the scaling information. Therefore, the user should make notes of the original scale and plot the data accordingly.

Sometimes the user will make an error in the use of the software, perhaps making an incorrect selection, or clicking the wrong option or the wrong mouse button. When this happens, a software variable may be set incorrectly, leading to an incorrect software calculation or data plot. A beginning user may not notice such errors at first. However, as the user becomes more skilled, such errors are apparent. The best way to reset the software is to quit the analysis program and restart, loading the data file and selecting the desired analysis option again. This is necessary only occasionally. The glottal inverse pitch synchronous (manual) analysis option is prone to such errors because it requires extensive user interaction. Errors rarely occur in the other analysis options.

Do not move and click the mouse button while the software is making calculations, since this can cause the data to be plotted in an incorrect window.

Appendix 6 discusses aspects of the theory for the various algorithms described previously. This appendix also contains two papers, one describing the WRLS-VFF algorithm in detail with some results, and the other describing an algorithm for silent, voiced/unvoiced/mixed (four-way) classification of speech. The latter algorithm uses the speech and EGG data files. However, the algorithm can be modified to use only the speech signal.

Recall from Chapter 1 that the original data file and folder names began with a lower case letter. However, these data file names now begin with an upper case letter. Thus, file m0125s.dat is the same as file M0125s.dat.

PROBLEMS

2.1 For the sentence, "We were away a year ago" (file m0125s.dat), use the software described in this chapter to parse the sentence into words. To accomplish this task use the edit option, which contains the zoom, scroll, cut, save, insert, and play features. Load the data file and then isolate the individual words using a combination of the options within the edit window. Verify your results by playing back the individual words. Once you have accomplished this task, examine each word to determine if there are certain common features about the word boundaries (the beginning and the end of the word). Next synthesize the nonsense sentence "A year away we ago were." Describe the quality of this synthesized sentence. If the sentence does not sound "right," what do you think is wrong? Save your results so that you can use them in the next few problems.

2.2   Repeat Problem 2.1 for the sentence "Should we chase those cowboys?" (file m0127s.dat). In this problem, synthesize the nonsense sentence "Cowboys chase those we should." Save your results so that you can use them in the next few problems.

2.3   Using the results from Problems 2.1 and 2.2, synthesize the sentence "A year ago cowboys were away we should chase." Describe the quality of this synthesized sentence. If the sentence does not sound "right," what do you think is wrong?

2.4   The purpose of this problem is to compare the parsing of the two sentences in Problems 2.1 and 2.2 with the results obtained using the Energy and ZCR options under the Time Domain Analysis window. Set the window length to 256, the overlap to 10%, the threshold to 10%, and the window type to hamming. Calculate the Energy and ZCR contours using the software. Compare these contours with your results from Problems 2.1 and 2.2. Do these contours have features that agree with your parsing of the sentences, i.e., do the beginning and ending word boundaries that you found in Problems 2.1 and 2.2 agree with the results calculated using the Energy and ZCR contours? Describe in words an algorithm that could make use of these contours to automatically parse the sentences into words. You do not have to implement the algorithm, only describe it.

2.5   Repeat Problem 2.4 but now vary the parameters as follows.

> Word length: 64, 128, 256, 512
> Overlap: 10%, 20%, 30%, 50%
> Threshold: 0%, 10%, 20%, 40%
> Window type: hamming, rectangular

There are (4) (4) (4) (2) = 128 possible combinations. Which combination of word length, overlap, threshold, and window type appears to give the best word parsing of the two sentences? Are there major differences for the various combinations of parameter values?

2.6   Load the speech file b.dat. Zoom in on the e portion of the data until there are about five repetitions of the waveform. Using the hamming window, calculate the autocorrelation function of the zoomed-in data. The autocorrelation function should be similar to that shown in Figure 2.22. Save the autocorrelation function to a file. How is the interval between successive peaks in the autocorrelation function related to the zoomed-in data? Does this result change significantly if the data window is the bartlett or the rectangular window? What do you think this interval is called? Design an algorithm for automatically calculating this interval. You do not have to write an m-file, only put the algorithm into words in a step-by step form.

2.7   Calculate the FFT of the autocorrelation function you saved for Problem 2.6. Describe how the peaks in the FFT are related to certain features of the autocorrelation function.

2.8   Calculate the spectrogram for the b.dat file. The result should be the same as that shown in Figure 2.26. What do the dark bands represent? Can you identify the interval for the b in the spectrogram?

2.9   Determine 5 successive FFTs for the b.dat file starting at the beginning of the b, i.e., starting at about point 3000 (0.3 sec). Let the interval between each FFT be about 1000 samples (0.1 sec). Compare the successive FFTs. Is it possible to use this spectral information to determine the boundary between the b and e sounds in the b.dat file?

2.10   Repeat Problem 2.9 using the LP spectrum option and compare the successive LP pole-zero plots. (We will define the LP spectrum in later chapters, but for now think of the LP spectrum as another algorithm to calculate the spectrum). Is it possible to use this spectral information to determine the boundary between the b and e sounds in the b.dat file? We will discuss this further in the following chapters.

2.11   Load the file m0118s.dat (the s sound). Calculate the spectrogram. Calculate the autocorrelation function for a short segment of this file using the hamming window. How does this spectrogram and autocorrelation function compare to the spectrogram and autocorrelation function for the e sound in the b.dat file?

2.12   Repeat Problem 2.11 for the file m0122s.dat (the z sound). What do you think the differences between the s and z sounds are due to?

CHAPTER 3

SPEECH PRODUCTION, LABELING, AND CHARACTERISTICS

3.1 LANGUAGE

The American Heritage dictionary defines language as "the aspect of human behavior that involves the use of vocal sounds in meaningful patterns and, when they exist, corresponding written symbols to form, express, and communicate thoughts and feelings." An alternate definition might be "any method of communicating ideas as by a system of signs, symbols, gestures, or the like; the language of algebra or computer languages such as C or FORTRAN." Thus, one might study a language in one or more fields.

3.2 LINGUISTICS

Linguistics is the study of the rules by which speech sounds are assembled in a language. Linguists study the units of language and how they function.

3.3 PHONETICS

Phonetics is the study and the classification of the sounds of speech. The discipline of phonetics includes:

- The physiology of speech production and the movements of the vocal organs (articulators) to produce sounds and words.
- Acoustic phonetics, which involves the acoustic description of speech sounds.

3.4 SOUND CLASSIFICATIONS

A basic unit of linguistics is the phone, which is an articulated sound. Articulation is the process of producing sounds by manipulating the articulators, which changes the configuration of the vocal tract. The vocal tract includes the pharynx and the oral cavity to the lips. Producing sounds by clapping one's hands or stamping one's feet is not a form of articulation, although it is a means of communication. A collection of phones is called phonemes. Phones are sounds. Phonemes are the elements of speech. Phonemes may be combined into larger units called syllables, a definition of which is not always agreed upon by linguists. A common definition of a syllable is that it has a central vowel that is preceded and succeeded by one or more consonants, which form a subset of the language of phonemes. The definition of a phoneme is such that if one phoneme is replaced by another, then the meaning of the word may be changed. There may be restrictions in a language concerning the usage of phonemes. For example, note that English never begins a syllable with the phoneme "ng."

A combination of syllables forms a word, which is usually a combination of two to five phonemes. However, there are words with as few as one phoneme, for example, a and I, and as many as ten or more, for example, synthesizers, and Mississippi.

The ten most frequently used English words are I, the, a, it, to, you, of, and, in, and he. We usually select words from a vocabulary of 5000 to 10,000 words; however, 3000 words are adequate for conversation. Later, we will specify the ten most common phonemes of English.

Sentences are larger linguistic units, which require a grammar, or rules that determine the manner by which sequences of words are joined together. For example, the following sentence is allowed.

- The bear's ear is blue.

However, the sentence

- Bear's ear blue the is.

is not allowed. Grammar is not the only determinant of sentence word order, for example, the sentence must make "sense." To illustrate this, note that "The fox jumped the river." is both grammatically correct and sensible. However, the sentence "The idea jumped the river." does not make sense, although it is grammatically accurate.

As we progress higher in the structure of language we come to semantics, which is the study of the meaning of words. The manner by which we order words is influenced by the grammar of the language and the meaning we wish to communicate.

3.5 PROSODY

Prosodic features of speech convey information about the long-term variations of pitch, intensity, and timing. We perceive prosodic features as stress and intonation, which are difficult to measure, model, and simulate in speech synthesis. For example, we express the distinction between a question and statement and doubt by changes in intonation or changes in our fundamental frequency of voicing. A speaker's emotional state, gender, health, and other factors may also be conveyed by variations in intonation. Stress is used in conversation to indicate the importance of words. For example, the statement

- **I** will be the judge of that.

as contrasted with

- I will be the judge of **that**.

have two different meanings to both the speaker and the listener when spoken.

There is no good way to indicate stress and intonation in writing; these two factors are used nearly exclusively in speech. Lexical stress is used by a speaker to emphasize a syllable in a word, e.g., CAMpus. Stress and intonation are very effective communication techniques and have a connection to meaning, since variations in these factors can convey information about a speaker's emotional state, gender, health, and other factors.

Stress and intonation carry information important to the quality of the speech signal and are, therefore, important for speech synthesis. Intonation is perceived as changes in the fundamental frequency of voicing, F_0. A plot of the time variation of F_0, or its reciprocal, the period of voicing, is often called a pitch contour. In actuality, as mentioned in Chapter 1, pitch is the perception of F_0. The terms pitch and F_0 are often incorrectly used interchangeably. Stress is perceived as changes in increased speaking effort as well as changes in intensity, pitch, and duration of a sound.

Stress and intonation represent two suprasegmental features, another being duration. The term suprasegmental feature refers to speech characteristics that extend beyond a short segment. For example, a suprasegmental feature, such as pitch, may extend over an entire sentence, while a vowel feature is localized within a syllable within a word. Speech sounds vary in duration. For example, some vowels are longer than other vowels, and some consonants are longer than other consonants. It is even possible for the same vowel to vary in duration depending upon the location of the vowel within a word. For example, word-ending vowels tend to be longer than word-beginning vowels.

3.6 ORTHOGRAPHY

Orthography is a system of spelling that follows a set of rules and/or common usage. Spelling provides us with a visual means of representing spoken words with written letters. However, spelling is fallible, as we know. For example, how do you pronounce "read"? It is sometimes "red" and sometimes "reed," depending on the context. The words "way" and "weigh" are pronounced the same but have different meanings.

3.7 PHONEMIC TRANSCRIPTION OR LABELING

This is a procedure to convey pronunciation information as unambiguously as possible. It is a technique to provide an allophonic transcription that is more accurate than spelling rules.

3.8 ADAPTATION, ASSIMILATION, AND COARTICULATION

The sounds one produces are influenced and altered by the neighboring sounds. One type of such influence is called phonetic adaptation. Such adaptations are the result of variations in the manner by which a speaker moves the articulators, thereby causing changes in the vocal cavities. The manner and extent to which these cavities are changed are affected by the past, present, and following phonemes in a particular utterance. The positions of the articulators and the shape of the cavities for one phoneme influence the movements of the articulators for producing neighboring phonemes. One example of adaptation comes about as our speaking rate increases. The faster we speak, the less likely the tongue will reach specific target positions for specific phonemes. In summary, adaptation is the process of varying, or changing, a phoneme due to the influence of the shape of the vocal tract for neighboring phonemes. Adaptation is the result of one articulator modifying its movement due to phonemic context. Thus, the production of a phoneme is influenced by its neighboring vocal tract shapes. However, the acoustic manifestation of the phoneme does not change.

Assimilation is an excessive form of adaptation. If a phoneme changes sufficiently to become more like its neighboring phonemes, then such a change in the phoneme is called assimilation. One example can be seen in the "sh" sound at the beginning of the word "should," due to the influence of the voiced phoneme following "sh." The influence can either be anticipatory of the next phoneme or a carry over from the previous phoneme. Assimilation results in an actual change in the sound of the phoneme.

Coarticulation is defined as the movement of two articulators at the same time for different phonemes. Coarticulation can occur with or without a change in sound production. One example is for the word "two." We can pronounce this word by moving the tongue for the phoneme "t" while we simultaneously round our lips for "u." However, we can also pronounce this word with no lip rounding. On the other hand, coarticulation can result in a "smearing" of segmental boundaries between phonemes, as in assimilation, which can modify the characteristics of the phoneme. For example, for the same phoneme "k" we have two different pronunciation in the words "keel" and "cool." These two phonemes are called allophones of the phoneme "k." Another example is the phrase "keep them," which is often pronounced as "keep em." And another example is the sentence "Up and at them," which we often say as "Up an at em." The term coarticulation is often used to mean that assimilation is also occurring.

3.9 ELEMENTARY ASPECTS OF SPEECH PRODUCTION

A simplistic representation of speech production is depicted in Figure 3.1, which is from a paper by Dennis Fry (1979) and drawn by L. S. Johnson. An idea originates in the brain, which can be expressed in words, which are stored in memory. Control is communicated to the articulators, including the lungs, jaw, lips, tongue, velum, and vocal folds. In Figure 3.1 the speaker is in the process of speaking the words, "The rain in Spain falls mainly in the plain," from Pygmalion by Shaw, which later became the play, My Fair Lady, by Lerner and Loewe.

Figure 3.2 is a more scientific view of speech production, although still simplistic. Yet, this is the beginning of a model for speech production based on physiology. The acoustic signal radiated from the lips is perceived by the listener. In American English, there are on the order of 41 phonemes, more or less. Most languages have 30 to 50 phonemes.

Speech is not an orderly, precise string of phonemes. It is more like a series of phonemes that have onset and offset slopes in amplitude, which contribute to the transitions between phonemes. Thus, speech is a sequence of "blurred" phonemes that are not usually pronounced precisely, thereby making phoneme identification from waveforms or spectrograms difficult.

While it is beyond the scope of this book to provide a detailed description of the anatomy and physiology of human speech production, we do provide a brief introduction to the vocal system. The principal laryngeal function is to provide a protective closure for the respiratory system. The larynx, supported by ligaments and controlled by muscles activated by numerous nerves, is composed of soft tissues encased in a cartilaginous skeleton. A mucous membrane lines the larynx, as it does the trachea. There are two pairs of folds: the vocal folds, or cords, and the ventricular (false vocal) folds; both pairs are membranes that extend into the air passageway.

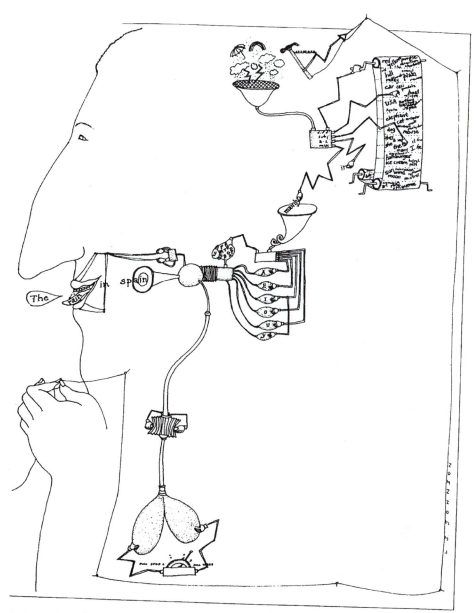

FIGURE 3.1 A representation of speech production.

Figure 3.3 shows a sketch of an individual with a laryngeal mirror (a dental mirror) located at the back of the throat, allowing a view of the vocal folds from the open mouth. Figure 3.4 is a x-ray photograph of a side view of the head of a man. This x-ray shows a laryngeal mirror located at the back of the throat, as in Figure 3.3.

Chapter 1 presented the human vocal system. The lungs are typically considered as air reservoirs, which are capable of expelling air up the trachea to the vocal folds. For voiced sounds (such as vowels), the air pressure increases until the folds are pushed apart forming a slit known as the glottis. A puff of air passes through this glottal opening, setting the vocal

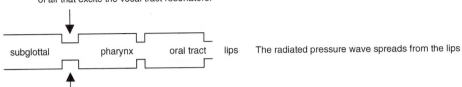

The vocal folds vibrate to modulate the air, creating puffs
of air that excite the vocal tract resonators.

subglottal   pharynx   oral tract   lips   The radiated pressure wave spreads from the lips

Area of opening between the vocal folds is the glottis.

FIGURE 3.2   A model of speech production.

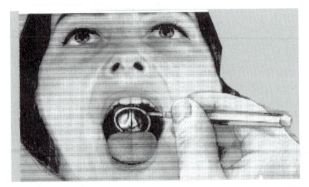

FIGURE 3.3   A view of the vocal folds using a laryngeal mirror.

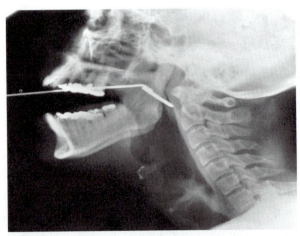

FIGURE 3.4   An x-ray of the mouth and throat with laryngeal mirror.

pressure reduced, velocity increased

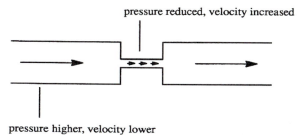

pressure higher, velocity lower

FIGURE 3.5 Bernoulli effect.

folds into vibratory motion. The myeloelastic-aerodynamic theory of phonation proposes that this vibratory motion is the result of the interplay of two forces. One force is the subglottal pressure, which causes air to push the adducted folds apart, releasing a puff of air. The other force is the Bernoulli effect (the linear combination of pressure and velocity squared is constant), which is a suction phenomenon that pulls the cords together (adducts the glottis) when the air velocity through the glottis is relatively large. If the glottis is initially abducted (opened), then the Bernoulli effect will cause the glottis to close.

The myoelastic–aerodynamic theory originated with Helmholtz and Muller in the 19th century and was later expanded by van den Berg in the 1950s (Borden and Harris, 1980, pg. 75). The myoelastic part of the theory arises from the fact that the muscles (myo) change the elasticity and tension of the vocal folds to bring about changes in the frequency of vibration of the vocal folds. The mass of the vocal folds also affects the vocal fold vibratory frequency. As the vocal folds become shorter and thicker, they become more massive, and the frequency of vibration is lowered. As the folds become longer and thinner, they become less massive and can vibrate at a higher frequency. Elastic folds vibrate faster because they are able to "bounce" back at a more rapid rate. Tense folds vibrate faster than slack folds. The folds are stretched to make them more tense. The muscles regulate the thickness and tension of the vocal folds. The aerodynamic part of the theory says that the driving force for vibration is airflow. The air expelled from the lungs activates the vibration of the vocal folds. The Bernoulli effect, shown in Figure 3.5, is one factor that affects the vibration of the vocal folds; another is the recoil force of the vocal folds.

The length of the vocal folds for males is 17 to 24 mm, and 13 to 17 mm for females. The folds can stretch from 3 to 4 mm. For a male voice phonating at a fundamental frequency of $F_0 = 120$ Hz, the glottis is approximately 10 mm long and 4.25 mm wide, giving a glottal area of approximately 42.5 mm^2. The actual glottal area is less because the glottis is not rectangular, but oval. A truer glottal area would be about 28 mm^2. At $F_0 = 340$ Hz, the glottis is approximately 11.5 mm long and 1.65 mm wide, giving a glottal area of approximately 19 mm^2. Since the glottal area is more nearly oval than rectangular, a more accurate glottal area would be 13 mm^2. The size of the glottis in these cases can be compared to that of a dime, which is 18 mm (0.7 inches) in diameter, with an area of 254 mm^2 (see Figure 3.6).

Let this represent a dime: circle with dia 4 mm, area 12.56 mm^2
dia. 18 mm, area 254 mm^2

FIGURE 3.6 Comparison of the glottis with a dime.

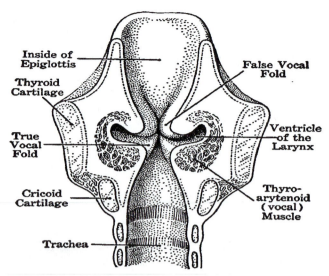

FIGURE 3.7   The vocal folds.

An artist's rendition of a frontal section of the larynx, the cartilages, and connective tissues that encloses the vocal folds, is shown in Figure 3.7 (after V. A. Anderson, 1977). Figure 3.8 is a sequence of 23 ultra high-speed film frames of the vocal folds during vibration. This sequence is to be viewed from right to left, top to bottom. The sequence begins with the vocal folds nearly closed in frame 1 and fully open at frame 4, then closed at frame 7, and so on. Since the filming was done using a laryngeal mirror, anterior is at the top and posterior is at the bottom of each frame, and the right vocal fold is at the left. Later chapters discuss this further and provide models of the vibratory motion of the vocal folds. A digitized sequence of frames of one vibratory cycle of the vocal folds along with a movie of this sequence is contained in a folder on the accompanying CDs. See the README file.

As the vocal folds vibrate, they modulate the flow of air from the lungs, creating pulses of air that are known as the volume–velocity of air flow through the folds and vocal tract. In physics courses, one usually studies the particle motion of air. For speech, we are

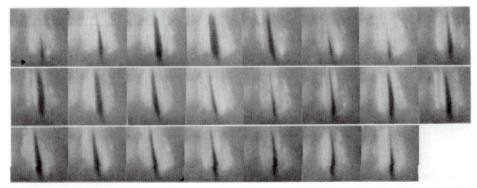

FIGURE 3.8   A sequence of ultra high-speed film frames of the vibratory motion of the vocal folds.

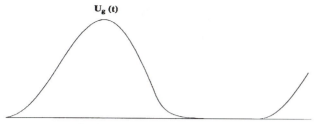

$U_g(t)$

FIGURE 3.9   A glottal volume–velocity pulse.

concerned with the movement of a volume of air through the vocal tract. A sketch of a volume–velocity waveform (pulse) of air versus time is shown in Figure 3.9.

The succession of air pulses generated as result of the vibratory motion of the vocal cords sets up an acoustic field that continues to travel up the vocal tract. The phonation or sound generated has an auditory correlate (or pitch) that is directly related to the frequency of vibration and a loudness that is determined by the amplitude of the acoustic pressure wave. The frequency of oscillation of the folds is determined by their mass, length, thickness, elasticity, and compliance, as well as by the subglottal pressure.

The volume–velocity (rate of air flow), as it passes through the glottis, modified by the pharynx, mouth, and nasal cavities, is finally radiated from the lips and nostrils as voiced sounds. All voiced sounds such as all vowels and voiced consonants (e.g., b, z, and g), originate at the vocal cords. The glottal volume–velocity is considered the acoustic source waveform for voiced sounds. The other sound is called "unvoiced." In this case, the vocal cords are held apart so that air expelled from the lungs up the trachea is unaffected by glottal vibrations. There are two fundamental types of unvoiced sounds: fricative and plosive. The former is typified by the sound s, which is produced by expelling air through a constriction such as the teeth to produce turbulent air flow. The latter unvoiced sound, typified by p as in puff, is created by the rapid release of air pressure built up behind a closure such as the lips. Since fricatives and plosives can also be voiced, as in b, d, g, and z, one should be cautious of oversimplifications.

The vocal folds are held open (abducted) for unvoiced sounds. While for voiced sounds, the vocal folds are brought together (adducted). To bring the vocal folds into a vibratory mode, the subglottal pressure is greater than the supraglottal pressure. There are two major types of source excitation, voiced, where the vocal folds vibrate, and unvoiced, where the vocal folds are held open. A third type of excitation is sometimes called mixed voiced and unvoiced excitation or more simply mixed excitation. This type of excitation is a combination of voiced and unvoiced excitation, for example, for the phoneme "z," as in zebra. In this latter case, the vocal folds vibrate and there is also a turbulent sound produced at the tip of the tongue near the teeth.

Primarily the lung pressure, the tension of the folds, and their mass, affects the fundamental frequency of vibration of the vocal folds. Consequently, the average, minimum, and maximum fundamental frequency for men, women, and children differs. Table 3.1 summarizes these differences.

Primarily, the source or the shape of the volume–velocity waveform affects the quality of the voice, while the movement of the articulators affects the intelligibility of speech. This is illustrated in Figure 3.10.

Figure 3.11 shows several glottal pulse waveforms. On the left, the pulse becomes more skewed to the right for pulses 4, 3, 2, 1. The shape of pulse 4 is typical of a breathy

TABLE 3.1. Average, Minimum, and Maximum Fundamental Frequencies for Men, Women, and Children

	F_0 average (Hz)	F_0 minimum (Hz)	F_0 maximum (Hz)
Men	125	80	200
Women	225	150	350
Children	300	200	500

voice, shape 3 for a falsetto voice, shape 2 for a modal (normal) voice, and shape 1 for a vocal fry (creaky) voice. On the right in Figure 3.11, the abruptness of closure is the characteristic that is important. Pulse 1 is typical of modal voice, pulse 2 of vocal fry, pulse 3 of falsetto, and pulse 4 of breathy voice. One can see that the excitation of a breathy voice is one that has almost a sinusoidal type of volume–velocity waveform, indicating that the vocal folds never achieve full or complete closure, with some air leakage at all times. For modal voice, the excitation pulse typically occupies approximately 65% of the pitch period, while for vocal fry, the pulse is about 25% of the pitch period. We discuss voice types again later.

The vocal tract is usually considered to be the entire cavity or passageway from the glottis to the lips. The basic assumption of nearly all speech production models is that the source (glottis) is independent of the vocal tract. The vocal tract may be modeled in various ways, and the model may, in turn, be excited by numerous source waveforms. However, the assumption of independence between source and vocal tract is not always valid, for example, as in the production of certain transient sounds such as p in pot.

The primary element controlling the vocal tract is the tongue, which divides the vocal tract into two resonant cavities, the pharynx and mouth, which, in turn, determine the transmission characteristic of the vocal tract. This characteristic can also be modified by coupling into the nasal cavity under the control of the soft palate or velum. As shown in Figure 1.5 in Chapter 1, the nasal tract begins at the velum and ends at the nostrils. The length of the nasal tract is about 12 cm.

The vocal tract transmission characteristic is such that it causes certain frequency components of the laryngeal signal to pass with less attenuation than others. Examples follow later. The resonances, or peaks, in the spectrum are referred to as formants, the center frequencies of which are designated in their ascending order of appearance as F_1, F_2, and so forth. Each formant also has a bandwidth. Studies indicate that the first two or three formants generally suffice for the perceptual characterization of most voiced English vowels

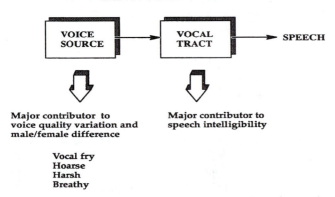

FIGURE 3.10 Quality of voice and intelligibility of speech.

FIGURE 3.11 Glottal pulse waveforms for various voice types.

and consonants and, therefore, of speech. The higher formants are important for natural sounding or good quality speech. A similar formant structure is also observed for unvoiced sounds. The most significant frequency range for speech is approximately 250 to 3000 Hz.

The formants occur at the natural frequencies of the human vocal tract, which is actually a nonuniform acoustical tube. To a first-order approximation, we can model the vocal tract as a uniform acoustical tube closed at one end and open at the other. Recall from physics that the natural frequencies (standing waves) of a tube closed at one end and open at the other are determined by the wavelength, which in this case is $\lambda = 4L$. Thus, the natural frequencies, F_n, of this model are

$$F_n = (2n + 1)c/4L, \quad \text{for } n = 0, 1, 2, 3, 4, 5, \ldots \tag{3.9.1}$$

where c is the velocity of sound in air ($\cong 34{,}000$ cm/sec) and L is the length of the tube, which for the adult male is approximately 17 cm. Making these substitutions, we have $F_n = (2n + 1)500$ Hz. Thus, for this idealized model in the range from 0 to 5 kHz, there are five equally spaced formants. Figure 3.12 shows the first three formants from 500 Hz to 2.5 kHz. In actuality, the vocal tract is not uniform, and the formant locations as well as their bandwidths depend on the particular configuration of the vocal tract. For continuous speech, the speaker is continuously changing the vocal tract as well as the glottal excitation.

3.10 VOICE VERSUS SPEECH

The situation depicted in Figure 3.10 is overly simplistic. While it is true that the shape of the source excitation waveform does affect the quality and type of vocal characteristics, it is not true that this is exclusively the case. The type of voice is also affected by the movement and placement of the articulators, by the rate of speaking, as well as by other factors. So the configuration of the vocal tract does affect vocal characteristics. As an example, New England and New York dialects are often called clipped because native

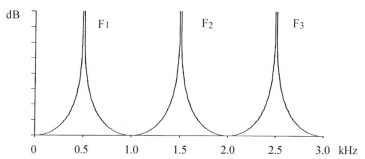

FIGURE 3.12 An idealization of the formants for a tube closed at one end and open at the other.

New Englanders speak rapidly, often truncating some phonemes. They often habitually make phoneme substitutions, and/or insertions, and/or omissions. On the other hand, native residents of southern states in the United States often speak slowly, with drawn-out vowels. These speaking styles come about through habit and are due to variations in speaking rate and in articulatory movements. Another example is accents. A specific sentence can be spoken in standard American English as well as with a Spanish or German or other accent. In this case, the perceived differences can be due again to differences in the movement of the articulators, with perhaps some differences in the shape of the excitation waveform as well. Perhaps the major point to remember is that voice or vocal characteristics are different from speech. This can be illustrated by thinking of voice and speech synthesis. The primary goal of speech synthesis is to produce speech that is highly intelligible with good vocal qualities that sounds like either a "standard" male or female voice. Speech synthesis for most commercial applications does not require the mimicking of a specific vocal characteristic or voice type. The primary goal of voice synthesis is to reproduce a specific voice, such as that of former President John F. Kennedy, or former actor John Wayne, or Mickey Mouse; or one may want to create a new voice. Mel Blanc was known as the man of a thousand voices because he created the voices of Bugs Bunny, Porky Pig, Daffey Duck, Tweety Bird, Sylvester the Cat, and many others. One goal of voice synthesis is to be able to create new vocal characteristics much like a composer may create a new musical composition. In summary, voice and speech synthesis are different, even though most engineering authors do not make this distinction, although they should. Thus, the characteristics of voice are not the same as the characteristics of speech, although they may be related. The parameters we use to model a voice differs from those we use to model speech.

3.11 SOURCE-FILTER MODEL

The source-filter model, due to Fant (1960), is shown in Figure 3.13. This model of sound production is linear and assumes superposition holds. The source provides the excitation,

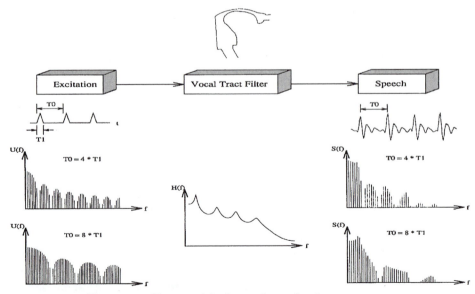

FIGURE 3.13 The source-filter model of speech production.

which is shaped spectrally by the vocal tract filter. Since the source is periodic for vowels, its spectrum consists of the fundamental and its harmonics, or in other words, a line spectrum. This line spectrum is filtered to produce speech. The filter changes with time to simulate the effect of changes in the vocal tract shape. So the vocal tract filter is the primary factor in producing various sounds. We will elaborate on this model in succeeding chapters.

3.12 CLASSIFICATION OF PHONEMES

For a spoken language, the acoustic signal radiated from the lips consists of a string of sound elements called phonemes. There are 30 to 50 phonemes for most languages, while for American English there are about 41 phonemes. There are various ways to classify phonemes.

3.12.1 Acoustic

One classification is according to the type of excitation, for example, voiced and unvoiced.

3.12.2 Articulation

Another classification is by the manner and place of articulation. The typical places of articulation include labial (lips), dental (teeth), labiodental (lips and teeth), alveolar (gums), palatal (palate), velar (soft palate), and glottal (glottis).

3.12.3 Phonemes

Yet another classification is made among vowels and consonants. These phoneme classes contain various subclasses. Vowels tend to be steady, at least over a relatively long time interval, while consonants tend to be brief. There are several schemes for labeling the phonemes. The International Phonetic Alphabet (IPA) is one standard of labels or symbols. An abbreviated IPA list is shown in Figure 3.14. Also shown in this figure is an abbreviated list of the

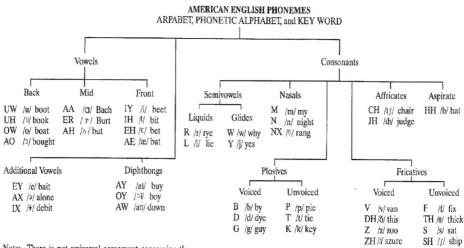

Note: There is not universal agreement concerning the classification of vowels and diphthongs. In addition some authors include additional allophones of various phonemes.

FIGURE 3.14 A list of phonemes using the IPA and upper case ARPAbet alphabets.

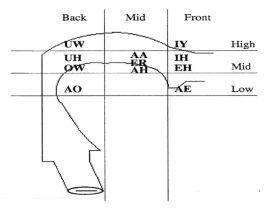

FIGURE 3.15   Some examples of vowel place of articulation.

upper case ARPAbet, or the ARPA (Advanced Research Projects Agency) alphabet, which we shall use, since it is easily typed and can therefore be electronically communicated. Figure 3.14 contains 41 phonemes. It is common practice to bracket phonetic transcriptions with forward slashes, for example, /AA/ as in Bach.

3.13   VOWELS (14)

For our purposes, we consider 14 vowels, which are all voiced. The place of articulation for these vowels varies as shown in Figures 3.14 and 3.15. One example is the vowel /IY/, as in beet, which is a high, front, unrounded (no lip protrusion) vowel. /AO/, as in bought, is a low, back vowel, while /UW/, as in boot, is a high, back, rounded vowel.

3.13.1   Schwa Vowel

A schwa is often called a degenerate vowel, which is a vowel to which other vowels tend to when one articulates rapidly or carelessly during fluent speech. The initial vowel in the word "ahead" is a schwa. A schwa tends to occur when the tongue hump does not have time to move into a correct position, but instead assumes a neutral position in the vocal tract, thereby causing the vocal tract to approximate a uniform tube. The schwa vowel tends to be short in duration and weak in amplitude.

3.13.2   Tense and Lax Vowels

Some vowels and diphthongs (in American English) are intrinsically longer than others and are made by the tongue reaching a more extreme position. The vowels with more extreme tongue adjustments and longer duration are termed tense vowels, more for their function in the language than for their method of production. Some examples appear in open syllables such as see, saw, lah, sew, Sue, say, sow, soy, cue.

Shorter vowels, which can appear in closed syllables (syllables ending with a consonant), but not in open syllables, are called lax vowels because they are produced with less extreme movements. Some examples appear in closed syllables such as sing, strength, sang, song, sung.

3.13.3 Diphthongs (3)

A diphthong is a gliding monosyllabic phoneme starting at one articulatory position for one vowel and moving toward the position of another vowel. There is also a change in vowel resonance.

3.13.4 Jitter and Shimmer

While vowels and diphthongs are considered to be steady, the nature of their production is such that there is a variation between successive periods, which is called jitter, as well as a change in the amplitude, which is called shimmer. These characteristics are illustrated in Chapter 1.

3.13.5 Semivowels: Liquids (2) and Glides (2)

Liquids move the tongue toward the alveolar ridge. Glides are similar to diphthongs, but a glide has a faster transition, with the tongue tip raised and the oral cavity is constricted.

3.14 CONSONANTS (17)

Consonants consist of several subclasses. For example, using the manner of articulation, there are **plosives or stops** (6), which are generated by blocking the air flow in the oral cavity to build up pressure, then the pressure is suddenly released. There are three voiced and three unvoiced plosives. The **fricatives** (8) are often classified by the type of excitation, that is, voiced and unvoiced. There are four voiced and four unvoiced fricatives, the latter form a constriction in the oral tract to produce turbulence (friction). The **nasals** (3) are generated by lowering the velum to connect the nasal cavity to the pharynx and the oral tract. The mouth is also closed. Nasals have a resonance in the nasal cavity, which introduces zeros in the overall system transfer function. The acoustics are such that a nasal murmur is within the 200 to 300 Hz range for males. The resonance or formant is a bit lower for /M/ and /N/ than for /NX/ because there is a progressively decreased volume of the oral cavity as the closure moves back in the mouth. Sometimes, diphthongs and semivowels (liquids, and glides) are grouped with consonants. Figure 3.16 shows the place of articulation for the plosives, fricatives, nasals, and semivowels.

3.15 AFFRICATES (2)

An affricate is a stop with a fricative release. One is voiced and one is unvoiced.

3.16 ASPIRATE (1)

An aspirate is produced by a steady airflow through the open vocal folds.

3.17 CONTINUANTS

These phonemes have a fixed, non-time varying vocal tract, which is excited by a source. Vowels, fricatives (voiced and unvoiced), semivowels, aspirates, and nasals, are all sustained, unlike stops.

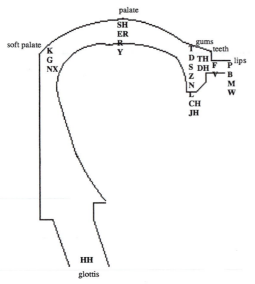

FIGURE 3.16 Place of articulation for plosives, fricatives, nasals, and semivowels.

3.18 NONCONTINUANTS

These phonemes have a changing vocal tract, for example, diphthongs, stops, and affricates.

3.19 SIBILANTS

Sibilants are high-pitched sounds with an obvious hiss as in sigh and shy.

3.20 NONSIBILANTS

Nonsibilants are the other fricatives, not classified as sibilants.

3.21 SONORANTS

These include the nasals and semivowels (liquids and glides) and are produced with less constricted airflow than the obsruents.

3.22 OBSTRUENTS

These include the stops and fricatives. An alternate phoneme classification to Figure 3.14 is shown in Figure 3.17.

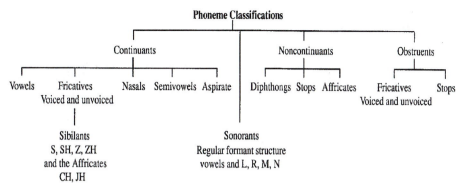

FIGURE 3.17 Summary of phoneme classifications.

3.23 VOICE BAR

A voice bar is a very low-frequency formant, typically near 150 Hz. A voice bar occurs when the vocal folds are vibrating and exciting an occluded nasal and oral tract. Voiced fricatives often contain a voice bar, e.g., /Z/.

3.24 ALLOPHONES

Allophones are a variant of a phoneme, as for /K/, as in keel and cool.

3.25 STATISTICS

The eleven most frequently occurring phonemes of English are (Denes, 1963):

/AX/, as in about, 9.04%	/L/, as in let, 3.69%
/T/, as in ten, 8.40%	/M/, as in met, 3.29%
/IH/, as in bit, 8.25%	/DH/, as in that, 2.99%
/N/, as in net, 7.08%	/K/, as in kit, 2.90%
/S/, as in sat, 5.09%	/AY/, as in buy, 2%
/D/, as in debt, 4.18%	

3.26 WHISPER

A whisper is produced like an aspirate, by a steady flow of air through the glottis.

3.27 SPECTRAL CHARACTERISTICS

Each phoneme has certain spectral characteristics. One example of the waveform, spectrum, and spectrogram for the phoneme /IY/ is shown in Figure 3.18. Additional examples for the vowels and some consonants appear in Appendix 5. Figure 3.19 shows a typical vocal tract shape, waveform, cross-sectional vocal tract area, and spectrum for the vowel /IY/

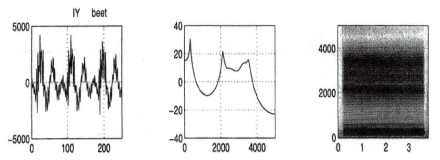

FIGURE 3.18   The waveform, spectrum, and spectrogram for the vowel /IY/.

calculated using the articulatory speech synthesizer described in a later chapter. Additional examples for the vowels appear in Appendix 5.

The formant features for the vowel /IY/ for both a male and female speaker appear in Figure 3.20. The formant frequencies are shifted upward by about 20% for the female speaker relative to the male speaker.

Figure 3.21 illustrates the "phoneme" labeling of the all voiced sentence "We were away a year ago," spoken by a male speaker. In this figure, the labeling is done by indicating the spelling of the words. This is not the conventional procedure, but it serves as an aid to mark the location of the various phonemes and to indicate how the phonemes vary in duration for this sentence. Figure 3.22 shows the pitch contours for both a male and female speaker for the sentence, "We were away a year ago."

It is often useful to know the formant frequencies and bandwidths for the various vowels. Such data were collected some years ago by Peterson and Barney (1952), and recalculated by Childers and Wu (1991) using digitized data and more modern analysis techniques. The latter data are presented in Figures 3.23 and 3.24 for both male and female speakers. These data can be used for speech and speaker recognition tasks. The vowel

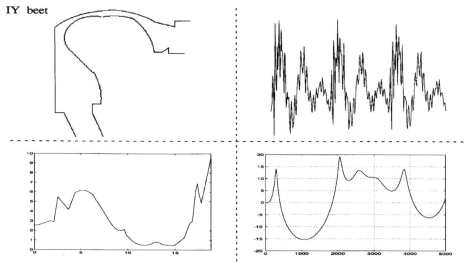

FIGURE 3.19   The vocal tract shape, waveform, cross-sectional area, and spectrum for the vowel /IY/.

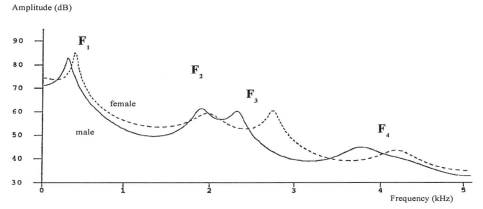

Amplitude (dB)

FIGURE 3.20 The formants for the vowel /IY/ for a male and female speaker.

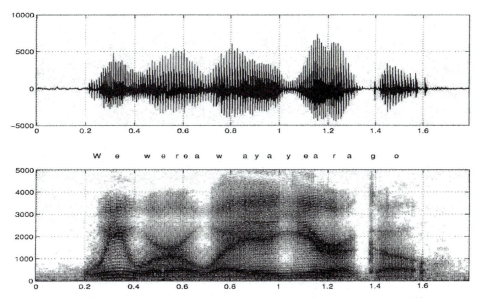

FIGURE 3.21 The labeled waveform and spectrogram for the sentence, "We were away a year ago."

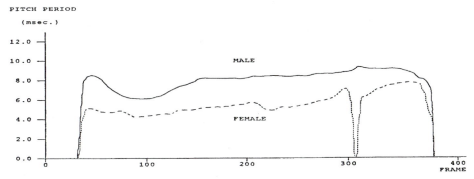

FIGURE 3.22 The pitch contours for both a male and female speaker for the sentence, "We were away a year ago."

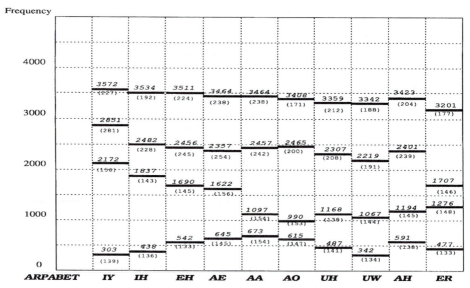

FIGURE 3.23 Average formant frequencies and bandwidths for ten vowels for male speakers.

triangle is obtained by plotting the second formant, F_2, versus the first formant, F_1, as shown in Figure 3.25 for male speakers. This data gives one an idea of how the first two formants vary for the vowels.

Table 3.2 contains a summary of minimum and inherent durations of some of the phonemes. This data was compiled by Klatt and appears in Allen, Hunnicutt, and Klatt (1987). The data is intended for use in speech synthesis. However, it is also useful as a phoneme feature and similar data has been used in isolated word recognition tasks (Gupta

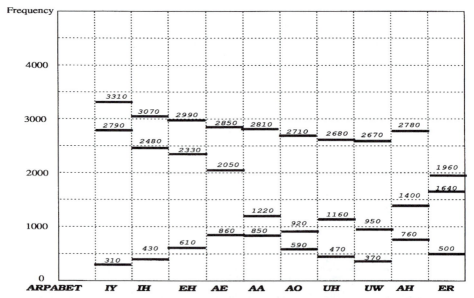

FIGURE 3.24 Average formant frequencies and bandwidths for ten vowels for female speakers.

TABLE 3.2. A Summary of Minimum and Inherent Phoneme Durations (msec)

Vowels

AA	100	240	AE	80	230	AH	60	140
AO	100	240	AW	100	260	AX	60	120
ER	80	180	AY	150	250	EH	70	150
IH	40	135	IX	60	110	UW	70	210
IY	55	155	OW	80	220			
OY	150	280	UH	60	160			

Sonorant consonants

L	40	80	HH	20	80	R	30	80
Y	40	80	W	60	70			

Nasals

M	60	70	N	50	60	NX	60	95

Fricatives

DH	30	50	F	80	100	S	60	105
SH	80	105	TH	60	90	V	40	60
Z	40	75	ZH	40	70			

Plosives

B	60	85	D	50	75	T	50	75
G	60	80	P	50	90			
K	40	80						

Affricates

CH	50	70	JH	50	70

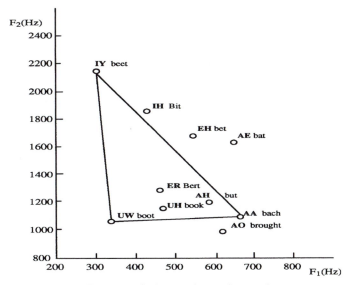

FIGURE 3.25 The vowel triangle for male speakers.

et al., 1992). According to Klatt, the inherent duration has no special significance, other than as a starting point for rule application. It is roughly the duration of the phoneme in non-sense consonant–vowel–consonant or consonant–consonant–consonant context, for example, b–vowel–b or c–consonant–b.

3.28 SUMMARY

Since this text attempts to stress the examination of aspects of speech using toolboxes, we outline certain features of speech data here to assist the reader in understanding characteristics of speech that can be useful in speech synthesis and recognition.

The excitation for speech can be voiced/unvoiced/mixed (V/U/M). Unvoiced excitation tends to be turbulent noise, caused by constrictions in the vocal tract. Unvoiced excitation tends to be high frequency with a high number of zero-crossings in the waveform and low amplitude (low energy). Voiced excitation is generally periodic in nature and high amplitude (high energy).

The vocal and nasal tracts introduce resonances causing the formants in the speech signal. The resonances can be modeled with an all-pole transfer function, as we will see in Chapter 5. When the velum is open, coupling the nasal tract to the oral tract, we have two all-pole models in parallel; thereby introducing zeros in the overall vocal tract model. So nasal sounds tend to introduce antiresonances caused by zeros. Nasal sounds tend to have formants with larger bandwidths than vowels because of the antiresonances.

Some features of the phonemes are summarized in Table 3.3. This summary is only indicative of the features and is not necessarily valid for all situations.

Additional features become apparent in certain applications. For example in word boundary identification tasks, it is difficult to determine when a word begins or ends with

- Weak fricatives (/F/, /TH/, /HH/).
- Weak plosives (/P/, /T/, /K/).
- Nasals (at the end of words).
- Voiced fricatives that die out (or become devoiced) at end of word.
- Trailing off of vowels at the end of a word.

In most speech analysis, we also assume we know the background noise level. This knowledge is useful for setting threshold levels for algorithm decision making.

Be sure that you read Appendices 4 and 6.

Additional references for the material presented in this chapter include Denes and Pinson (1993); Edwards (1992); Fant (1973); Olive, Greenwood, and Coleman (1993); Potter, Kopp, and Kopp (1966); Stevens (1998).

TABLE 3.3. A Summary of Some Phoneme Features

Sound	Energy	Excitation	Zero-Crossing	Formants	Duration
Vowels	High	V	Low	Yes	Long
Stops	Low	U and M	High	No	Short
Fricatives	Low	U and M	High	No	Short
Nasals	Medium	V	Low	Yes	Long

PROBLEMS

3.1   Read Appendix 4, which describes the data set provided with this book. Note that the data includes speech and electroglottographic files for speakers with normal larynges, a set of data for patients with vocal disorders, a set of data that mimics various voice types, an other data folder, and an additional data folder. The purpose of this problem is to examine the data for speakers m01 and f06 in the normal data folder. Calculate the first three formants for the twelve vowels in the data set, i.e., files m0103s.dat through m0114s.dat using the spectrogram and the Pitch/Jitter/Formant options. Repeat this task for subject f06. Do the results generally agree with the data in Figures 3.23 through 3.25? Do the results agree with the data in Appendix 5? Does the pitch for these two subjects agree with the data in Table 3.1?

3.2   Repeat Problem 3.1 for the fricatives, files m0115s.dat through m0123s.dat and files f0615s.dat through f0623s.dat. Only the spectrogram can be used, since for fricatives there is no voicing or it is weak so that the Pitch/Jitter/Formant software gives an error. Do the features you determine generally agree with those summarized in Table 3.3 and in Appendix 5 for the fricatives? Is it generally true that the fricatives have no (or very weak) formants?

3.3   Since the vowels and fricatives for files m0103s.dat through m0123s.dat and f0603s.dat through f0623s.dat are steady waveforms with little variation, we can also examine the FFTs. Calculate an FFT for each file using a centrally located section of the file. Use a hamming window of duration 512. Do the results generally agree with the data in Figures 3.23 through 3.25? Do the results agree with the data in Appendix 5? Does the pitch for these two subjects agree with the data in Table 3.1? Compare your results from Problems 3.1 and 3.2. How can you estimate the pitch from the FFTs?

3.4   Use the Cepstrum option to determine the pitch contour for files m0125s.dat and f0625s.dat. You will have to record the pitch for each segment and plot the data. Use a hamming window length of 512 with 10% overlap. What is the average pitch for each speaker? Does this result agree with Table 3.1? Now use the Pitch/Jitter/Formant option and calculate the pitch contour again. Do the two pitch estimates generally agree?

3.5   For the beginning of the file m0125s.dat, i.e., the "We were" segment, use the autocorrelation option to calculate the pitch contour.

3.6   Design a pitch period estimation algorithm using the autocorrelation function. You do not have to implement this algorithm in MATLAB. However, write the algorithm in some detail and discuss those sections that might introduce errors in the computation.

3.7   The purpose of this problem is to compare the waveforms and spectrograms for the same subject but for different vocal conditions, i.e., normal (modal) voice (file m0125s.dat), vocal fry (creaky voice) (file m0305ms.dat), breathy voice (file m0405ms.dat), hoarse voice (file m0505ms.dat), computer voice (file m0605ms.dat), and rough voice (file m0705ms.dat). The latter five files are in the Mimic folder. Describe the major differences between the features of the waveforms and the spectrograms for these six different voice types. What is the major contributor to the differences between these voice types?

3.8   The purpose of this problem is to compare the jitter for the six voice types considered in Problem 3.7. The files to be compared are as follows: normal (modal) voice (file m0125s.dat), vocal fry (creaky voice) (file m0305ms.dat), breathy voice (file m0405ms.dat), hoarse voice (file m0505ms.dat), computer voice (file m0605ms.dat), and rough voice (file m0705ms.dat). The latter five files are in the Mimic folder. Calculate and compare the pitch/jitter/formant contours for these files. What are the jitter values for all of these voices? Which voice type has the least jitter and which the largest jitter? An average jitter greater than 5% to 7% is often considered a deviant voice by clinicians. Would any of these voices be considered deviant? Note that the jitter is plotted in Hz on the vertical scale just as is the pitch contour. To calculate the percent jitter, calculate the average jitter, divide by the average pitch, and multiply by 100.

3.9   The purpose of this problem is to illustrate the difficulty of identifying and labeling word boundaries for sentences with different phoneme content. The three files to be compared are m0125s.dat,

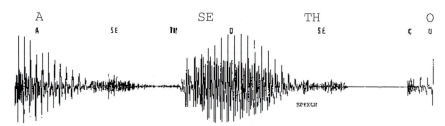

FIGURE P3.10 An illustration of the zoom option for phoneme labeling.

m01zs.dat (in the Other folder), and spe9_1.dat (in the Additional folder). The three sentences are "We were away a year ago" (all voiced), "That zany van is azure" (voiced fricatives), and "We saw the ten pink fish" (unvoiced plosives and fricatives), respectively. Use the Energy and ZCR option with a hamming window of length 256 and 10% overlap to make your first estimate of the word boundaries. Next calculate the spectrogram for each file and label the boundaries by hand. Finally, use the Zoom in and Play options to locate the word boundaries. Compare and discuss your results.

3.10 The purpose of this problem is to practice phoneme labeling for five sentences: "We were away a year ago" (all voiced) (m0125s.dat), "Early one morning a man and a woman ambled along a one mile lane" (many nasals) (m0126s.dat), "Should we chase those cowboys" (fricatives) (m0127s.dat), "That zany van is azure" (voiced fricatives) (m01zs.dat; Other folder), and "We saw the ten pink fish" (unvoiced plosives and fricatives) (spe9_1.dat; Additional folder). Use plots of the waveforms and spectrograms to identify and label the phoneme boundaries. The Zoom option and the Energy and ZCR and the pitch and formant contours may also be helpful. Do not use the Play option. Do this problem without the help of your ears. Make a list of the major features that you found useful for identifying and labeling the phonemes. When you have finished, compare your results with the results in Appendix 5. Discuss the possibility of designing an automatic phoneme recognition system. An illustration of how the zoom option can help phoneme labeling is shown in Figure P3.10. The segment shown is taken from the sentence "Should we chase those cowboys?"

CHAPTER **4**

DATA AND MEASUREMENTS

4.1 MEASUREMENTS OF LARYNGEAL FUNCTION

The larynx, just over the tongue and below the epiglottis, is so near yet so far away. We can feel it with our fingers as the "Adam's apple." Perhaps it is because it is so "near" that we feel particularly exasperated when we are frustrated in our attempts to observe the vibratory motion of the vocal folds during normal speech. Many methods have been used to study the normal and diseased larynx. Here we provide an overview of major techniques used to evaluate and measure laryngeal function. Our major categories (outlined in Table 4.1) are arbitrary, but functional. The first, a collection of methods aimed at displaying the movement of the vocal folds, is observational; these methods employ various wavelengths across the spectrum, that is, ultrasonic, infrared, visible, electron beam, and x-ray, to achieve this goal. The next category is glottography, which includes techniques that indirectly measure aspects of the glottal opening by either electrical or optical methods. The third and final category encompasses techniques that attempt to extract parameters related to laryngeal function by processing the acoustic or speech waveform, by measuring or evaluating aural parameters, or by determining the airflow and pressure at various points in the vocal system. See Childers (1977) for additional references to the literature in this area. Our coverage is neither exhaustive nor all inclusive; we do not discuss various biochemical laryngeal evaluative methods.

4.2 OBSERVATIONAL

4.2.1 Direct and Indirect Laryngoscopy

The larynx has been directly viewed through a puncture in the throat, but this is extremely rare for scientific studies. Laryngeal function is primarily observed indirectly with the aid of a laryngoscope or mirror. This was first attempted in 1807 by Bozzini, the inventor of the laryngoscope, a two-channel tube similar to a periscope (Childers, 1977, pg. 381). In Bozzini's day, one channel allowed light from a candle to enter and be reflected from a mirror onto the vocal folds. The adjacent channel was designed to provide a pathway for the reflected light to return to the viewer. However, the mechanism was not successful, probably because of inadequate illumination. By 1860, physicians world-wide were attempting to use laryngoscopy and to improve on the procedure; however, the principal factor in the success of indirect laryngoscopy in clinical practice has been the evolution of improved lighting conditions.

Indirect laryngoscopy is simple to use, inexpensive, and requires no elaborate equipment or data processing. Unfortunately, the procedure provides no permanent record useful

75

TABLE 4.1.   Laryngeal Evaluative Methods

4.2 Observational
  4.2.1. Direct and indirect laryngoscopy
  4.2.2. Microscopy using light and electron beam
  4.2.3. Still photography, stroboscophy, and cinematography:
  black and white, color, and infrared
  4.2.4. Fiberoptics
  4.2.5. Ultra high-speed cinematography
  4.2.6. Radiography: x-ray, stroboscopic laminography, tomography,
  cinefluourography, and magnetic resonance imaging
  4.2.7. Ultrasound
4.3 Glottography: optical and electrical
  4.3.1. Photoelectric (or optical) glottography
  4.3.2. Electroglottography, laryngography
  4.3.3. Electromyography
4.4 Acoustic, aural, and airflow
  4.4.1. Inverse filtering
  4.4.2. Acoustical measurements
  4.4.3. Spectrograms
  4.4.4. Perturbations: jitter, shimmer
  4.4.5. Fundamental frequency
  4.4.6. Intensity
  4.4.7. Vocal quality
  4.4.8 Glottal airflow and subglottal pressure

for medical history or therapy. Furthermore, when the laryngoscope or mirror is used in a subject who can tolerate the instrument, the subject is able to phonate only a very limited range of sounds, usually sustained vowels, since the vocal folds are otherwise hidden by the tongue or epiglottis. In addition, the mirror may interfere with normal vocal activity and distort the sound being phonated. Since the vocal folds vibrate rapidly, the viewer is not able to follow their detailed motion by use of laryngoscopy alone.

4.2.2 Laryngeal Microscopy

This procedure has used both light and electron beams to study laryngeal functioning, but is not widely used, even in scientific studies.

4.2.3 Still Photography, Stroboscopy, and Cinematography

In order to overcome some of the disadvantages of indirect laryngoscopy, investigators have used photography in various forms. The developed film provides a permanent record that can be analyzed for scientific details. Photography offers the investigator a means by which he or she can "stop" the vibratory motion of the vocal folds.

The first known attempt to photograph the larynx was in 1860. This method is still used today, primarily to preserve images obtained by indirect laryngoscopy, but the method is also useful as a teaching tool, for documentation, and as a basis for comparison of pre- and post-treatment conditions for various laryngeal diseases. The record is easily obtained, relatively inexpensive, and available to educators and clinicians alike. The technique, how-ever, is dependent upon indirect laryngoscopy and its accompanying inherent problems. Finally, the image on each frame of the film is a blurred composite of the many vocal cord vibrations that occurred during the exposure of each successive film frame.

Stroboscopic laryngoscopy illuminates the vibrating vocal folds via the laryngeal mirror with an intermittent or flashing light providing the illusion that the rapid vibratory motion of the vocal folds has been stopped or slowed. The stroboscope was first suggested as a method for investigating the movement of the vocal folds in 1866, but was not actually tried until 1878. Despite improvements in stroboscopic laryngoscopy, its primary limitation is that the investigator sees only brief "snapshots" of consecutive cycles of the vibratory pat-tern; he or she never sees a complete vibratory cycle. The image on the film is a composite of the vocal folds illuminated by the strobe at consecutive phases or cycles in the vibratory pat-tern of the vocal folds. Because of this, the composite image may contain artifacts not easily recognized. For example, the vocal folds may vary slightly in periodicity or their opening or closing pattern may vary over successive illuminations by the strobe. The image on the film represents a composite of these variations. Thus, measurements taken from the film may be an incorrect representation of the actual activity that took place during the vibratory cycle. Since the vibratory motion of the vocal folds is usually aperiodic, stroboscopy is not a particularly effective means for analyzing vocal cord function; nevertheless, it has been, and is still, used in a clinical setting despite the fact that it possesses the general limitations of indirect laryngoscopy. This technique will probably not be used in the future since laryngeal research has progressed to a much higher level of sophistication. Current investigative proce-dures require considerably more quantifiable detail concerning the motion of the individual cycles of the vibratory pattern of the vocal folds. Stroboscopy cannot supply these data.

Cinematography or regular-speed motion picture photography, using cameras with a film rate of approximately 24 frames per second, was first employed to film the larynx in 1913. Motion pictures, combining the principles of stroboscopy with those of laryngeal photography, have also been taken of the larynx.

Regular-speed cinematography also provides a permanent record of the gross function and pathologies of laryngeal structures, and the films have proven useful in both diagno-sis and follow-up treatment in a clinical environment. The procedure is relatively simple to use and is not particularly expensive. The primary disadvantage of this technique has already been described, that is, there is no quantifiable data concerning glottal dimensions or vibratory movements are available from the films.

The three types of film that have been used in the photography techniques are black and white, color, and infrared. Color film provides an almost exact representation of the condition of the tissue; however, the speed at which color film can be exposed is generally less than that for black and white. Infrared is not used frequently, but it does tend to enhance those areas on the vocal folds where an excessive inflammation or concentration of blood may be present, which, in turn, raises the temperature of the surrounding tissue.

4.2.4 Fiberoptics

This subcategory is not unique, but rather a result of a recent development that provides an alternate method for illuminating, viewing, and photographing laryngeal function. The primary advantage that fiberoptics offer is that the fiberoptic bundle can be inserted through the nasal passages, allowing the subject or patient to articulate normal speech. The insertion of the bundle can, however, be painful. The vibratory pattern of the vocal folds is filmed through the eye piece of the fiberoptic bundle with film rates from 24 to 64 frames per second having been achieved.

A special fiberoptic laryngoscope has also been designed for use with high-speed stroboscopic light and a high-speed camera. Fiberoptics have also been used in microlaryngoscopy and microsurgery, for supraglottal illumination during glottography in conjunction with the standard laryngeal mirror and for illumination with videotaping.

As mentioned, the unique advantage of fiberoptics in illumination and photography is that the larynx can be observed during the production of normal connected speech. The primary disadvantage appears to be that, at the present time, only 64 frames per second has been achieved in cinematography, but research is underway to improve the light source so that eventually ultra high-speed photography should be possible with fiberoptic systems.

4.2.5 Ultra High-Speed Cinematography

While regular-speed cameras operate at 24 frames per second, ultra high-speed motion picture photography is capable of filming speeds from 4000 to 10,000 (or higher) frames per second. Such rates are required to achieve a detailed slow-motion study of the vibratory pattern of the vocal folds. The first use of ultra high-speed cinematography in laryngeal research was reported in 1937 at Bell Telephone Laboratories and by Moore (1936), in a PhD dissertation, using an apparatus similar to that illustrated in Figure 4.1. The analysis of these films launched numerous subsequent investigations. The Bell Telephone Laboratories' (Murray Hill, NJ) films achieved good photographs at a rate of 4000 frames per second.

The significance of these rates can be illustrated as follows. The ratio of the rate of exposure to the rate of projection is the factor at which the vibratory motion is slowed, for example, if the rate of exposure is 4000 frames per second and the rate of projection is 16, then the motion is slowed by $4000/16 = 250$. Thus at these rates, if the vocal folds are performing 125 vibrations per second, they then appear to make one-half vibration per second, or one vibration every 2 seconds when the film is projected onto the screen. With the exception of the work using fiberoptic bundles, ultra high-speed cinematography uses indirect laryngoscopy, typically employing a laryngeal mirror. Both black and white and color film is used.

The advantages of ultra high-speed cinematography are that the vibratory pattern of the vocal folds can be viewed in slow motion, and that a detailed frame-by-frame analyses

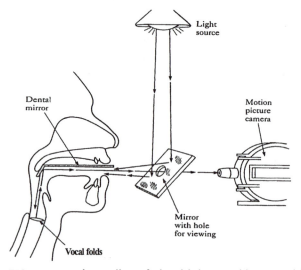

FIGURE 4.1   An outline of ultra high-speed laryngeal filming technique.

can be made. The vocal folds are typically filmed while the subject or patient is phonating a sustained vowel, but it is possible to film nonlinguistic phenomena such as a cough, laughter, or clearing of the throat.

4.2.6 Radiography

This procedure employs x-rays as an agent to expose a film plate that can be developed to yield an image of the larynx and the surrounding structure. Since the vocal folds are soft tissue, the laryngeal images are generally of poor quality, due to the low radiographic contrast. The cervical vertebrae often obscure the laryngeal image in a frontal or anteroposterior view. These techniques have found their most use recording the motion of the articulators. The outlining of the soft tissue of the tongue and velum is facilitated by gluing lead pellets to the tongue and palate. These pellets are quite visible in the films.

The first x-rays of the larynx were attempted in 1913 from a lateral view. The procedure has undergone many changes since that time. Frontal and lateral views have been taken along with laminographic or tomographic (body sectioning) procedures, as well as cineradiographic or cinefluouragraphy. Radiography will not become a major investigative tool in laryngeal research. It is expensive, offers a risk to the subject due to the cumulative radiation dosage, and relevant data are difficult to obtain. The procedure is not able to provide data about the glottal area or the vibratory frequency, two parameters presently considered essential in laryngeal research.

Magnetic resonance imaging has been used in recent years because it offers less risk than x-ray exposure and it can readily capture the motion of the articulators, particularly when the pellets mentioned above are used.

4.2.7 Ultrasound

Ultrasonic systems have been used in a manner analogous to x-ray systems to obtain images of the skeletal structure within the body. However, only limited success has been achieved in laryngeal research due to the soft tissue in the vocal folds. In the hope that greater success might be achieved in another way, investigators have used ultrasonic systems to measure the velocity of vocal cord motion during speech by detecting the doppler shift caused by the reflection of the ultrasonic wave from the surface of the vocal fold.

Glottal closure has also been measured by using a source on one side of the neck and a transducer on the other. When the folds are open, the air reflects the ultrasonic wave; when the folds are closed, the wave passes through the tissue from the source to the transducer.

The procedure is easy to apply and the patient experiences no discomfort and can speak normally; thus, the tongue and epiglottis do not interfere with the data-collection process. It has been noted that most commercially available equipment does not generate ultrasound frequencies sufficiently high enough to provide adequate resolution of vocal cord motion. It is unlikely that ultrasound will be used extensively in laryngeal research since the data obtained by this procedure are limited to information about glottal closure.

4.3 GLOTTOGRAPHY: OPTICAL AND ELECTRICAL

The techniques to be discussed in this section are indirect methods for monitoring laryngeal function. The data obtained are generally related to glottal area or closure. These procedures have, and are being, used because they are simple to apply and overcome many of the disadvantages of the observational techniques discussed previously. However, as will be pointed out, these indirect procedures also have their limitations.

4.3.1 Optical Glottography

Perhaps the major impetus for the interest in optical or photoelectric glottography is the hypothesis that the volume–velocity of airflow at the glottis is proportional to the glottal area. Under this hypothesis, one aspect of laryngeal function may be measured using a simpler procedure than photographic techniques. For example, optical glottography attempts to determine glottal function by measuring the amount of light that passes through the glottis. The light source may either be above the vocal folds, introduced through the mouth or nasal passages, or introduced through the neck below the larynx. In the former case, the light that passes through the glottis is usually monitored by a photocell next to the neck below the larynx. In the latter case, a curved light-conducting rod or fiberoptic bundle might be introduced through the oral cavity with the photocell attached to the end of the rod or bundle outside the lips. As the vocal folds vibrate, the light is modulated and picked up by the photocell whose electrical output may be displayed on an oscilloscope or a strip chart recorder or recorded on magnetic tape.

The results of this technique have sometimes agreed with and sometimes differed from data measured by ultra high-speed films on a frame-by-frame basis. A possible cause for this is that the light can be reflected in irregular and unpredictable ways within the throat and from the mucosal surfaces of the vocal folds. Even the placement of the photocell may affect the monitored waveform.

This method cannot account for vocal fold movements away from or toward the light source, but the ease with which this method can be applied makes the procedure very attractive. It is relatively inexpensive and can be used in a clinical setting. If the artifacts can be isolated and the data substantiated through the simultaneous application of other procedures, such as ultra high-speed cinematography, then optical glottography may be increasingly used in both research and clinical environments.

4.3.2 Electroglottography

Introduced in 1958 by Fabre, this method measures the variation in electrical impedance across the neck as the vocal folds vibrate (Childers, 1977, p. 385). The impedance increases as the glottis becomes enlarged and decreases as the glottis closes. Thus, the variations in glottal area are related to the variations in impedance.

Electroglottography has been used by several investigators and has been compared to optical glottography and to ultra high-speed photography. The laryngograph is an electroglottographic instrument that depends on the amount of vocal fold contact rather than the width of the glottis. This instrument uses modified electrodes that are applied in a manner different from those previously used.

Electroglottography and laryngography do not measure the glottal area or other glottal-related parameters. The procedure is apparently limited to frequency measurements. The technique is inexpensive, comfortable, easily used, portable, and does not interfere with normal speech. Despite its limitations, it is expected that this procedure will be increasingly applied because it can provide a quick and permanent record in a clinical setting. Figure 4.2 illustrates the procedure for recording the electroglottographic signal.

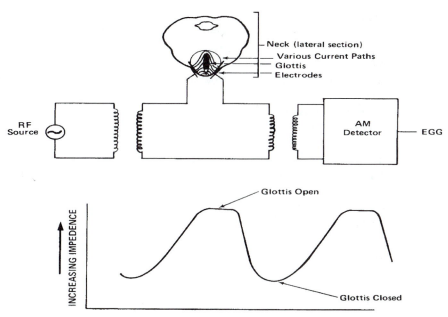

FIGURE 4.2 Electroglottography and the electroglottographic waveform.

4.3.3 Electromyography

Here, the laryngeal muscle activity is monitored. Therefore, this procedure is not a glottographic technique; however, since it is related, the method has been so categorized. The activity of the laryngeal muscles is monitored using needle electrodes inserted into the muscles.

4.4 ACOUSTIC, AURAL, AND AIRFLOW

This category is a collection of loosely related topics. The first and second subcategories cover procedures that attempt to extract parameters related to laryngeal function by processing the acoustic or speech waveform.

4.4.1 Inverse Filtering

We shall shortly discuss this topic at length as part of linear prediction. The theoretical foundation for this procedure is founded on the premise that the glottal volume–velocity of air, modified by the resonance and damping characteristics of the vocal tract, is radiated from the lips and nose as speech. Therefore, if a model of the vocal tract could be constructed, it would be possible to process the speech or acoustic signal in a reverse (or inverse) manner to obtain an estimate of the glottal volume–velocity. In 1959, Miller at Bell Telephone Laboratories (Murray Hill, NJ), was apparently the first to obtain such an estimate. This work spurred additional effort, which attempted to verify that the glottal volume–velocity waveform obtained by inverse filtering was a true representation of the airflow at the vocal folds. Thus, both optical and electroglottograms were taken and shown to be quite similar to the waveforms obtained by inverse filtering.

Rothenberg in 1973 modified the inverse filtering method to obtain an estimate of the volume–velocity of air at the mouth by using a pneumotachographic mask to measure the pressure differential across the mask's screen. The signal Rothenberg obtained was inverse filtered to yield an estimate of the glottal volume–velocity. Prior to the introduction of Rothenberg's method, the acoustic pressure wave was inverse filtered. However, Rothenberg's procedure yielded a signal that was not as susceptible to low frequency noise, was more accurate at low frequencies (even down to zero frequency), and was easily calibrated in amplitude.

Another modification to inverse filtering was introduced by Sondhi in 1975, who pointed out that the glottal volume–velocity waveform could be measured as follows. The subject inserts a hard-walled uniform tube into his mouth; the vocal tract and tube cross-sections are identical, and the tube termination is made reflectionless. The glottal source, therefore, feeds into a uniform lossless tube that is effectively infinite in extent. A probe microphone inserted anywhere in the tube will pick up a waveform identical in shape to the glottal volume–velocity waveform, but slightly delayed. This method has not yet been used extensively by others. The tube diameter must be matched to the vocal tract cross-section of the subject. Furthermore, this procedure cannot be applied to recorded speech. The subject must be present.

The concept of inverse filtering, as introduced by Miller, is receiving considerable attention, since it can generally be applied to recorded speech. It is simple and safe to use, the subject experiences no discomfort, and the speech may be continuous.

4.4.2   Acoustic Measurements

In acoustic measurements, intensity levels are often measured with respect to the reference intensity of 10^{-12} watts/m^2 or 0.0002 dyne/cm^2, which is the threshold of hearing. This is sometimes called the zero-reference level for acoustic sound-pressure level (SPL) measurements, which is the lowest SPL we can perceive.

$$\text{intensity level} = 10 \log(P_1/P_2)\, \text{dB} \qquad (4.4.2.1)$$

The threshold of feeling, the loudest level before discomfort begins is 120 dB SPL. Table 4.2 provides a summary of various SPLs.

Microphones are typically used for SPL measurements. They typically have a dynamic range of 125 dB. A carbon microphone is one that has found frequent use in telephone handsets. A simple diagram is shown in Figure 4.3. The diaphragm compresses a loose pack

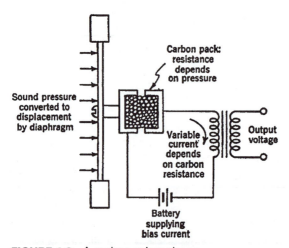

FIGURE 4.3   A carbon microphone.

TABLE 4.2. Sound-Pressure Levels for Various Conditions and Events (dB SPL)

12-inch Cannon at muzzle	220
	210
Rocket engines	200
	190
	180
	170
	160
Jet engine	150
Threshold of pain	140
	130
Threshold of feeling	120
Thunder	110
Niagara Falls	100
Subway	90
Factory	80
Busy street	70
	60
Office	50
Audience noise	40
Quiet home	30
Recording studios	20
	10
Threshold of hearing	0

of carbon granules, thereby increasing their area of contact and decreasing the electrical resistance of the pack. Thus, sound-pressure waves cause the resistance to fluctuate. If a constant current is passed through the carbon granules, the voltage drop across the pack is inversely related to the instantaneous sound pressure.

Another common microphone is a moving coil microphone, illustrated in Figure 4.4. Here a coil is moved in a magnetic field by the sound pressure to induce a voltage.

A third microphone is the condenser microphone shown in Figure 4.5. Here the sound pressure acts on a diaphragm to vary a capacitance (condenser). This type of microphone requires a current source. It is considered an excellent microphone and is often used in speech, singing, and music research studies.

4.4.3 Spectrograms

The spectrogram provides a time history of the spectrum, as shown in Figure 4.6, for the sentence "We were away a year ago," as spoken by a male. The amplitude of the spectrum is plotted in relative terms as variations in the gray level. This time history provides a time–frequency picture of the variations of the spectral resonances, called the formants. Thus, we can view changes in the bandwidths of the formants and changes in the formants themselves as the spoken words change. Present-day procedures for calculating the spectrogram use FFT techniques. MATLAB provides a function specgram that we use. For each data segment or frame, the amplitude of the spectrum is quantized and mapped to an assigned gray level.

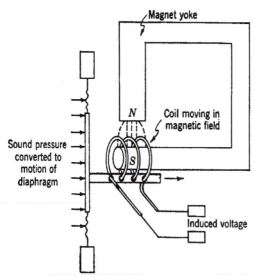

FIGURE 4.4 A moving coil microphone.

The data segment is updated and the process repeated. The spectral estimates are then plotted versus time, as shown in Figure 4.6.

The historical reason for this type of display is that some years ago, before digital techniques, the spectra and spectrogram were estimated using analog filters. For example, the spectrogram was estimated by repeatedly passing a tape recording of the signal through a bandpass filter, whose center frequency was shifted after each pass of the data. The tape recording of the data was made into a loop, which could be played over and over. The sweep of the bandpass filter was synchronized with the tape loop. The energy of the filter output for each pass of the tape loop controlled the heat of a stylus, which in turn burned a special paper. The more energy at the filter output, the darker the burn on the paper,

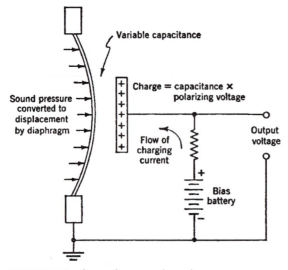

FIGURE 4.5 A condenser microphone.

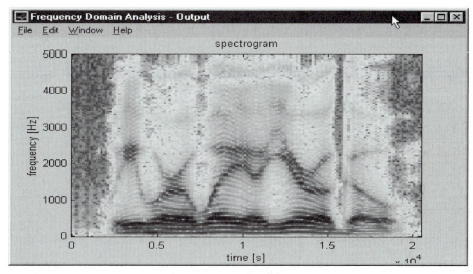

FIGURE 4.6 A spectrogram of the sentence, "We were away a year ago."

thereby creating a gray level display of the energy of the signal in the filter band at that particular center frequency. The filter center frequency then shifted up in frequency and the taped data started a new loop. Figure 4.7 illustrates this process. It took approximately 5 minutes or so to plot the spectrogram of a 1 or 2 second record of speech. But this display has proven to be useful to speech researchers and is the reason it is still used today with digital techniques. See Olive et al. (1993) and Potter et al. (1966) for spectrograms of speech data.

4.4.4 Perturbations: Jitter and Shimmer

Perturbations in the pitch period are called jitter, while similar perturbations in pitch amplitude are known as shimmer. Such perturbations occur naturally during continuous speech, but measurements from acoustic waveforms have demonstrated that the perturbations for pathologic and normal speakers differ. Speech synthesis has been used to verify that the

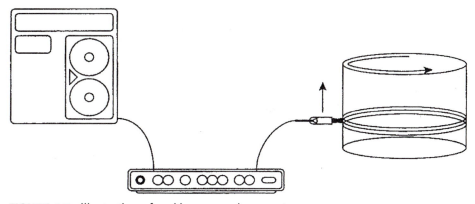

FIGURE 4.7 Illustration of making an analog spectrogram.

proper manipulation of jitter and shimmer does, in fact, produce speech that is indicative of either normal or pathologic voice quality.

4.4.5 Fundamental Frequency

The compliance, mass, length, and elasticity of the vocal folds affect the speaker's fundamental frequency. This is a parameter that can be measured quantitatively from the acoustic signal, the glottal volume–velocity, and the electroglottographic waveform. Deviations in the fundamental frequency may be indicative of a functional disorder or an existing pathology.

4.4.6 Intensity

This aural parameter, sometimes referred to as amplitude, can also be measured quantitatively or assessed by a trained listener. It, too, can serve as an indicator of a laryngeal disorder. A typical example is a weak voice, which may be due to pathology such as a paralysis of respiratory muscles causing insufficient subglottal pressure. A laryngeal paralysis can produce a similar effect.

Excessive intensity may occur simultaneously with an increase or a decrease in fundamental frequency, with the former causing a shrill voice, the latter leading to hoarseness. The clinician and speech pathologist learn to listen for deviations in both intensity and fundamental frequency as elementary signs of a laryngeal disorder.

4.4.7 Vocal Quality

This parameter is difficult to assess and is usually associated with nasality and hoarseness, but other descriptive terms such as breathy, husky, harsh, throaty, and strained are also used. No quantitative measure exists for these terms, and there may be considerable disagreement among researchers with respect to the relative importance of this variable. However, some attempts at quantification of quality have been made, for example, jitter and shimmer are now routinely used. The condition of the vocal folds as well as that of the vocal tract affects the quality of the voice.

4.4.8 Glottal Airflow and Subglottal Pressure

While many measurements in this area have been made, it is significant that the glottal volume–velocity has apparently never been measured directly. The average subglottal pressure has been measured directly with a pressure transducer introduced into the subglottal region through a tracheal puncture and can be measured by lowering a transducer through the glottis into the trachea.

4.5 MORE ON ELECTROGLOTTOGRAPHY

A mathematical model of the electroglottographic (EGG) waveform as a function of time is

$$\text{EGG(t)} = \frac{k}{A(t) + C} \qquad (4.5.1)$$

Idealized Descriptive EGG Waveform

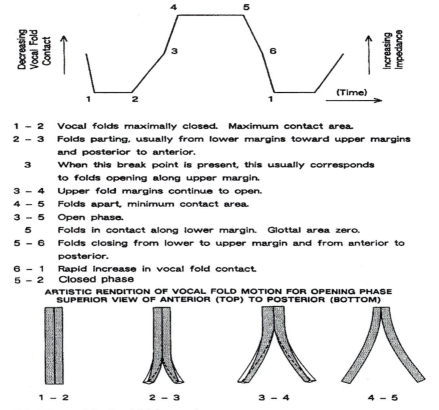

1 – 2	Vocal folds maximally closed. Maximum contact area.
2 – 3	Folds parting, usually from lower margins toward upper margins and posterior to anterior.
3	When this break point is present, this usually corresponds to folds opening along upper margin.
3 – 4	Upper fold margins continue to open.
4 – 5	Folds apart, minimum contact area.
3 – 5	Open phase.
5	Folds in contact along lower margin. Glottal area zero.
5 – 6	Folds closing from lower to upper margin and from anterior to posterior.
6 – 1	Rapid increase in vocal fold contact.
5 – 2	Closed phase

**ARTISTIC RENDITION OF VOCAL FOLD MOTION FOR OPENING PHASE
SUPERIOR VIEW OF ANTERIOR (TOP) TO POSTERIOR (BOTTOM)**

FIGURE 4.8 Idealized EGG waveform.

where A(t) is the vocal fold contact area, k is a scaling constant, and C is a constant proportional to the shunt impedance specified for the situation when A(t) = 0. This mathematical model, and its relationship to features of the EGG waveform depicted in Figure 4.8, has been partially verified by Childers and co-workers (1986; 1990). The vocal fold events are labeled on the EGG waveform and correspondingly on the artistic rendition of the vocal fold motion. The vocal folds are stylized and depict only the anterior one-third segment of the folds. The upper and lower vocal fold margins are out of phase. A vocal fold model has been developed using MATLAB and will be discussed in another chapter. A comparison of this EGG waveform model to actual data can be made via Figure 4.9 where, from top to bottom, the waveforms are speech, EGG, differentiated EGG, and glottal area. All waveforms are aligned. These data are for a male subject with a normal larynx with F_0 approximately 120 Hz at a low intensity.

Several vocal fold vibratory (glottal area) events are related to corresponding events (segments or peaks) of the EGG or the DEGG waveforms. For example, the instant of the opening of the glottis and the instant of the positive peak in the DEGG are coincident, as are the instant of the closure of the glottis and the instant of the negative peak in the DEGG. Similarly, the instant of the maximum glottal area and the instant of the maximum positive peak in the normalized EGG waveform are coincident. The open quotient (OQ) measured

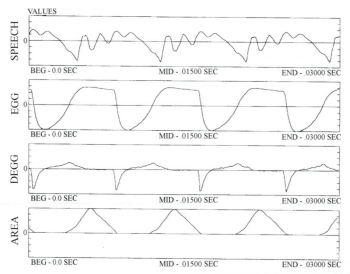

FIGURE 4.9 Data waveforms: speech, EGG, DEGG, and glottal area.

from the glottal area and the OQ measured from the EGG are nearly the same, where

$$OQ = \frac{\text{duration of glottal open phase}}{\text{duration of glottal cycle}} \qquad (4.5.2)$$

Also the average perturbation measured from the glottal area and that measured from the EGG are nearly the same.

The results of such studies have shown that the EGG and DEGG waveforms are useful for extracting information about the vibratory motion of the vocal folds and that such measurements are noninvasive and easy to accomplish. A major use for such measurements is to assist in the validation of measurements made from the speech waveform only. In addition, the EGG waveform can be used as a second "channel" for processing the speech waveform for speech analysis and recognition tasks. One example of this appears in the paper in Appendix 6.

In order to facilitate the comparison of the EGG and DEGG waveforms with the speech waveform, we have developed a display program described next.

4.6 DATA-APPENDIX 4

The data described in Appendix 4 are contained on two CD-ROMs. One CD-ROM contains the Normal folders, while the other CD-ROM contains the Disorder folder, the Mimic folder, the Other folder, the Additional folder, and the Area folder. The files generally include a sychronized EGG file as well as a speech file. Be sure to read this appendix for additional details.

4.7 DISPLAY OF SPEECH, EGG, AND DEGG

The m-file called speech_egg_display.m in the display_speech_egg folder can be used to load, play, and compare speech, EGG, and DEGG ASCII data files. Change directory to

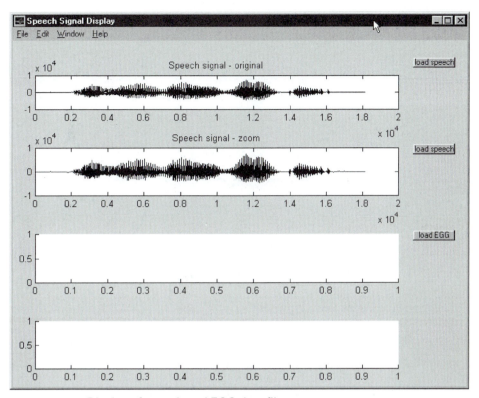

FIGURE 4.10   Display of speech and EGG data files.

the display_speech_egg folder and type speech_egg_display in the MATLAB Command window to initiate the program. The window will be similar to that shown in Figure 4.10, except that only the top load speech button will appear. No other buttons will be present. The user is to load a speech file, which will be plotted in the top panel. Repeat the load for the same speech file for the next panel. After completing the loading of the speech file twice, the results appear as shown in Figure 4.10. Next load the corresponding EGG file. There is a slight pause (30 to 60 sec) after loading the EGG file, since the program calculates the DEGG. Once this calculation is completed, the EGG and DEGG waveforms are plotted and the remaining buttons for the window are displayed. This is shown in Figure 4.11. At this point, the user can play the files. Also, the waveforms or spectrograms can be compared. The purpose for loading the speech file twice is that the top panel displays the speech file in its original form, while the lower three panels can display the data in a zoom-in manner.

In summary, to use the speech_egg_display.m file, follow these steps: first load the desired speech file in the top panel, then load the same speech file in the second panel from the top, next load the corresponding EGG file in the third panel. After the DEGG is calculated, the EGG and DEGG waveforms are plotted along with the remaining window buttons, as shown in Figure 4.11.

The user can select one of the various buttons, e.g., Play or Spectrogram, for each of the displayed signals. The Waveform button redisplays the signal if the spectrogram is plotted. Note that a zoom-in or-out button is available to zoom the waveform data display. Pressing this button changes the mouse cursor to a cross hair. Move the cross hair to the desired initial x-axis data location to zoom in. The cross hair can be place at any desired

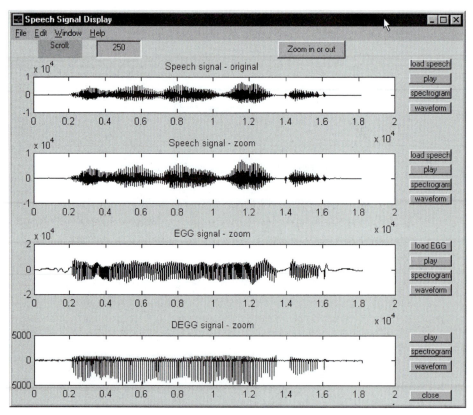

FIGURE 4.11 Display of speech, EGG, and calculated DEGG signals.

location on any of the waveforms. Press the left mouse button once. Move the cross hair to the desired end of the zoom in x-axis segment. Press the left mouse button once again. This zooms in one level on the signals in the three lower panels, as shown in Figure 4.12, which shows the zoomed-in waveforms plotted in each panel. Repeat these steps to zoom in another level. To stop the zoom-in process, press the right mouse button twice slowly while the cross hair is visible. The cross hair disappears and the standard mouse cursor reappears. To zoom out one level, press the zoom-in or-out button. The cross hair reappears. Press the left mouse button, followed by a press of the right mouse button. The data are zoomed out one level. Each repetition of this sequence zooms out another level, until the original signal level is reached. At any point in the zoom-out process, press the right mouse button twice slowly to exit the zoom-out option. Note that the scale of the zoomed-in data show the x-axis locations of the selected beginning and ending points of the desired zoomed in data segment. Thus, the user can refer to the top panel to keep track of the location of the zoomed-in data segment.

The scroll option is to be used after zooming in on the data at least one level. The scroll option becomes available automatically after zooming in. When the data are zoomed in, the scroll display is updated by adding the < and > scroll buttons to the figure, as shown in Figure 4.12. The default scroll value is 250 data points. This value can be changed by highlighting the number in the scroll window with the mouse cursor and typing in a new value. Press the left (right) mouse button on the < (>) button to scroll the data to the left

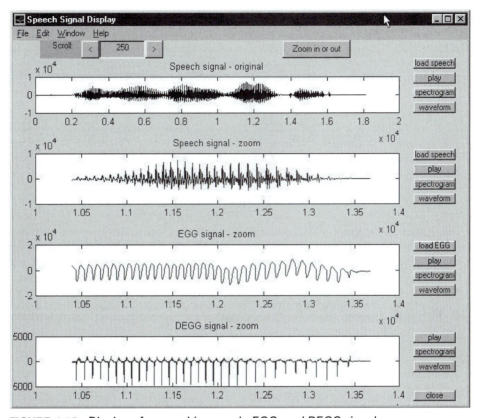

FIGURE 4.12 Display of zoomed-in speech, EGG, and DEGG signals.

(right). You can continue scrolling in either direction until the end of the data record. The scroll option can be used at any zoom-in level, as long as the zoom-in level is at least one.

The play option is designed to play the data displayed. Thus, if the data are zoomed in, the data shown in the panel are played. Similar remarks apply to the spectrogram option. The Close button closes out the Speech signal display window.

4.8 SUMMARY

The purpose of this chapter is to introduce the reader to many of the measurement procedures that are used in speech science. However, usually only the speech signal is used in most, if not all, speech-engineering applications, including speech recognition and speech synthesis. Nevertheless, some of the measurement techniques are useful to assist in the validation of algorithms that have been developed for speech only analysis. One such technique is the EGG. An extensive data set of synchronized EGG and speech waveforms is included with this text. Also, one algorithm that uses both the EGG and speech waveforms for silent and voiced/unvoiced/mixed (four-way) classification of speech is discussed in Appendix 6.

The following problems are a mixture of theoretical and experimental. Their purpose is to examine the theory behind some existing algorithms and to illustrate the development of additional algorithms for speech analysis. After this chapter, we move on to linear prediction and its use in speech analysis and synthesis.

PROBLEMS

4.1   The average magnitude difference function is useful for pitch estimation and offers less computational complexity than the autocorrelation function. Suppose that a function x(t) is periodic with period T. Then a function

$$d(n) = x(n) - x(n - k)$$

will be zero for k = 0, ±T, ±2T, ... For short segments of voiced speech one would expect that d(n) would be nearly zero, although not exactly so, at multiples of the period. The short time average magnitude of d(n) as a function of k will be small whenever k is close to the period. Thus, the short time average magnitude difference function (AMDF) is defined as

$$AMDF(k) = \sum_{m=-\infty}^{\infty} |x(n + m)w(m) - x(n + m - k)w(m - k)|$$

When x(n) is close to periodic in the interval spanned by the window, w(n), then AMDF(k) will have sharp nulls for k = T, 2T, ... It is not uncommon for the window to be the rectangular window. The AMDF requires fewer multiplications than the autocorrelation function. Consequently, it has been used in real-time speech processing systems. Implement this algorithm in MATLAB and test its use for pitch estimation for the sentence "We were away a year ago" file m0125s.dat. Compare this result with that obtained using the pitch algorithm available in the analysis software. Next, test the algorithm on the sentence "Should we chase those cowboys?" file m0127s.dat, and compare your results with that obtained using the pitch algorithm available in the analysis software. You may want to use the zoom option so that the results are more easily seen. Save your results for use in another problem below.

4.2   Clip the speech waveform using a center clipper, i.e., delete the speech waveform between ±C. This can reduce the amount of the data that needs to be processed for the autocorrelation estimation of the pitch. The threshold in the Pitch/Jitter/Formant option can be used for this purpose; however, the waveforms are not displayed. Therefore, implement in MATLAB a center clipper for speech. Let C be a variable that the user can set. Then center clip the sentence "We were away a year ago" file m0125s.dat with C = 70% of the maximum of the data. Next calculate the short time autocorrelation function using a hamming window of length 256. Plot the original data, the center clipped data, and the short time autocorrelation function. Estimate the pitch contour from the short time autocorrelation function. Play the original data as well as the center clipped data. Discuss your impressions of this auditory evaluation. Repeat the problem with C = 40% and C = 90%. Discuss your results. Save your results for use in another problem below.

4.3   The pitch algorithm in the Pitch/Jitter/Formant option uses median smoothing to eliminate sharp discontinuities in the contour due to algorithm errors. Design a median filter of length 3 and one of length 5. Apply each filter to the pitch contours estimated in Problems 4.1 and 4.2. MATLAB function, medfilt1, can be helpful for this problem as well as the median filter implementation in the pitch/jitter/formant m-files. Does a median filter give a more reasonable contour? Why?

4.4   The cepstrum can be used for pitch estimation as well as formant estimation. In this problem, estimate the pitch contour using the cepstrum (real or complex) for the sentence "We were away a year ago" (file m0125s.dat). Compare your result with the results from Problems 4.1 and 4.2 and the pitch/jitter/formant algorithm.

4.5   Implement in MATLAB a formant estimation algorithm using the Cepstrum. Analyze the sentence "We were away a year ago" (file m0125s.dat) using your implementation and compare your results with that using the pitch/jitter/formant algorithm.

4.6   Use the software described in this chapter to compare the speech, EGG, and DEGG for the sentences "We were away a year ago" and "Should we chase those cowboys?" (files m0125s.dat and m0125e.dat and m0127s.dat and m0127e.dat, respectively). For the first sentence note that the EGG and DEGG are not of much use for estimating the voiced/unvoiced segments. Why? However, for the second sentence the EGG and DEGG are helpful. Design an algorithm for classifying the speech into voiced/unvoiced segments using both the speech and EGG (DEGG) waveforms. You do not have to implement the algorithm. Compare your algorithm with that in Appendix 6. What is a major problem with such algorithms?

4.7   If $\hat{x}(n)$ is the complex cepstrum of $x(n)$, then match the first column of possible properties of $\hat{x}(n)$ with the second column of properties to be considered for $x(n)$. For each property in the first column determine the corresponding property in the second column. In all cases, assume $x(n)$ is real. Any property in the second column can be used only once. (After a problem in Oppenheim and Schafer, 1975)

1. $x(n) = -x(-n)$
2. $x(n) = x(-n)$
3. $x(n)$ real
4. $x(n) = 0, \ n < 0$

(a) $\hat{x}(n)$ real
(b) $\hat{x}(n) = -\hat{x}(-n)$
(c) $\hat{x}(n) = 0, \ n < 0$

5. $\displaystyle\sum_{n=-\infty}^{\infty} x^2(n) = 1$

6. $\displaystyle\sum_{n=-\infty}^{\infty} |x(n)| = \frac{1}{\sqrt{2\pi}}$

4.8   Let $x_1(n)$ and $x_2(n)$ be two sequences and $\hat{x}_1(n)$ and $\hat{x}_2(n)$ their complex cepstra. If $x_1(n) * x_2(n) = \delta(n)$, determine the relationship between $\hat{x}_1(n)$ and $\hat{x}_2(n)$.

4.9   Suppose we have two sequences $x_1(n)$ and $x_2(n)$, with transforms $X_1(z)$ and $X_2(z)$, respectively. Suppose $x_1(n)$ is minimum phase and $x_2(n)$ is maximum phase. If $|X_1(z)| = |X_2(z)|$ for z on the unit circle, determine the relationship between $x_1(n)$ and $x_2(n)$.

4.10   The purpose of this problem is to compare the speech, EGG, and DEGG waveforms and spectrograms for the same speaker, but for different vocal conditions, i.e., normal (modal) voice (file m0125s.dat), vocal fry (creaky voice) (file m0305ms.dat), breathy voice (file m0405ms.dat), hoarse voice (file m0505ms.dat), computer voice (file m0605ms.dat), and rough voice (file m0705ms.dat). The latter five files are in the Mimic folder along with the corresponding EGG files. Describe the major differences between the features of the waveforms and the spectrograms for these six different voice types. What is the major contributor to the differences between these voice types? Do the EGG and DEGG waveforms offer any features that might be helpful in voiced/unvoiced classification?

4.11   The purpose of this problem is to compare the jitter for the six voice types considered in Problem 4.10. The files to be compared are as follows: normal (modal) voice (file m0125s.dat), vocal fry (creaky voice) (file m0305ms.dat), breathy voice (file m0405ms.dat), hoarse voice (file m0505ms.dat), computer voice (file m0605ms.dat), and rough voice (file m0705ms.dat). The latter five files are in the Mimic folder along with their corresponding EGG files. Calculate and compare the Pitch/Jitter/Formant contours for these files. Do this for both the speech and EGG files. What are the jitter values for all of these voices? Compare the results for the speech analysis with that for the EGG analysis. Which voice type has the least jitter and which the largest jitter? An average jitter greater than 5% to 7% is often considered a deviant voice by clinicians. Would any of these voices be considered deviant? How do the results for the speech analysis compare with that for the EGG analysis?

4.12   The purpose of this problem is to illustrate the difficulty of identifying and labeling word boundaries for sentences with different phoneme content. The three files to be compared are m0125s.dat, m01zs.dat (in Other folder), spe9_1.dat (in Additional folder). The three sentences are "We were away a year ago" (all voiced), "That zany van is azure" (voiced fricatives), and "We saw the ten pink fish" (unvoiced plosives and fricatives). Use the Energy and ZCR option with a hamming window of length 256 and 10% overlap to make your first estimate of the word boundaries using the speech files only. Repeat this using only the EGG files. Next calculate the spectrogram for each file and label the boundaries by hand. Finally, use the Zoom in and Play options to locate the word boundaries. Compare and discuss your results. Could the EGG data be useful?

4.13 The purpose of this problem is to practice phoneme labeling for five sentences: "We were away a year ago" (all voiced) (m0125s.dat), "Early one morning a man and a woman ambled along a one mile lane" (many nasals) (m0126s,dat), "Should we chase those cowboys?" (fricatives) (m0127s.dat), "That zany van is azure" (voiced fricatives) (m01zs,dat; Other folder), and "We saw the ten pink fish" (unvoiced plosives and fricatives) (spe9_1.dat; Additional folder). Use plots of the speech and EGG waveforms and spectrograms to identify and label the phoneme boundaries. The Zoom option and the Energy and ZCR and the Pitch and Formant contours may also be helpful. Do not use the Play option. Do this problem without the help of your ears. Make a list of the major features that you found useful for identifying and labeling the phonemes. When you have finished, compare your results with the results in Appendix 5. Discuss the possibility of designing an automatic phoneme recognition system. Could the EGG data be useful?

CHAPTER 5

LINEAR PREDICTION

5.1 INTRODUCTION

We saw in Chapter 3 that a tube closed at one end and open at the other supports standing waves that are related to the length of the tube. This concept can be generalized to a concatenation of tubes of differing widths and lengths. For modeling purposes, the vocal tract can be considered as straight (not curved) and, thus, approximated as a variable area tube. The linear wave equation applies, and wave motion in the tract is planar to a good approximation. We can illustrate that wave motion is planar as follows. The wavelength for sound at 4000 Hz is

$$\lambda = \frac{c}{f} = \frac{350 \text{ m/s}}{4000 \text{ Hz}} \cong 8.7 \text{ cm} \tag{5.1.1}$$

Note that 8.7 cm is greater than the average diameter of the vocal tract, which is typically of the order of 2 to 3 cm. Thus, for all practical purposes, we can consider sound propagation in the vocal tract to be planar. Consequently, the vocal tract shape can be approximated by the cross-sectional area as a function of position along the tract.

For vowels, the air flow can reach 700 cm^3/sec. However, the opening of the tract is usually less than 1 cm^2. This means that the Mach number is less than 0.02, which is low, thereby implying that the linear wave equation is a good approximation. Nonlinear effects only become important for the flow through the vocal folds and through narrow constrictions in the vocal tract, as in the production of fricatives.

For an ideal, lossless, rigid tube with a cross-sectional area A(x), one can use the approximate equations of motion and continuity to describe wave propagation in the vocal tract. If u(x, t) is the volume–velocity and p(x, t) is the pressure, then

$$\begin{aligned} \frac{\partial p}{\partial x} &= -\frac{p}{A} \frac{\partial u}{\partial t} \quad &\text{equation of motion} \\ \frac{\partial u}{\partial x} &= -\frac{A}{\rho c^2} \frac{\partial p}{\partial t} \quad &\text{equation of continuity} \end{aligned} \tag{5.1.2}$$

where ρ is the density of air and c is the velocity of sound. Note that these equations are the same as those for an RLC transmission line, and, thus, comparisons can be made between parameters associated with a tube and those with a transmission line, as illustrated in Figure 5.1.

For a tube of length L with a complex exponential excitation, one can use Equation (5.1.2) to calculate the frequency response and impedance of the tube. Upon doing so, we find that the frequency response is

$$\text{frequency response} = \frac{1}{\cos(2\pi f L/c)} \tag{5.1.3}$$

This equation has poles at

$$f_n = \frac{2n + 1}{4L} c, \quad n = 0, \pm 1, \pm 2, \ldots \tag{5.1.4}$$

Acoustical parameter	Electrical parameter
p – Pressure	v – Voltage
u – Volume velocity	i – Current
ϱ/A – Air mass inertia (acoustic inductance)	L – Inductance
$A/(\varrho c^2)$ – Air compressibility (acoustic capacitance)	C – Capacitance
Viscous loss	R – Series resistance
Heat conduction loss	G – Shunt resistance
Yielding wall	Z_w – Shunt impedance

Transmission-Line Circuit Network

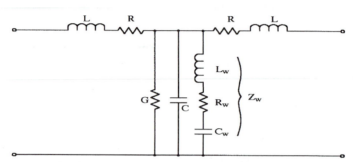

FIGURE 5.1 Analogies between acoustic and electrical transmission line parameters.

which is the result we mentioned in Chapter 3. For $c = 34,000$ cm/sec and $L = 17$, $f_n = (2n + 1)/4L \cong (2n + 1)500$ Hz, which are the odd multiples of 500 Hz. These are the natural frequencies, or eigenfrequencies, of the system model and they represent the formants of the vocal tract. Such formants occur regardless of the vocal tract shape. For example, one can concatenate two tubes of different widths and lengths and show that resonances (formants) appear in the system-transfer function just as they do for a single tube. However, the location of the resonances varies with the shape of the tubes. Thus, the formants change as the shape of the vocal tract changes.

In a nonideal tube, there are losses due to viscous friction, thermal conduction, and yielding walls. The effect of yielding walls is to slightly shift the low-frequency formants upward in frequency as well as increase their bandwidth. There is not much effect on the high-frequency formants. For example, the formant at 500 Hz can shift up to 504 Hz and have a bandwidth of 53 Hz. The effect is on the order of 0.8%.

The effect of radiation at the lips and nostrils is that of a high-pass filter approximated by differentiation in the time domain. This effect tends to "lift-up" the amplitude of the formants at the high frequencies.

Another factor is turbulence generated at a constriction. Turbulence occurs when the Reynolds number exceeds a critical value. The definition of the Reynolds number is

$$R_e = \frac{\rho w}{\mu} \frac{u}{A} \tag{5.1.5}$$

where ρ is the density of air, μ is the coefficient of viscosity, and w is the width of the constriction. For the glottis, w is the separation between the vocal folds. For a tract constriction, w is the average radius of the constriction. Noise is generated at the constriction only if the Reynolds number is greater than a critical value R_c. When this occurs, the root mean square (rms) value of the generated noise is proportional to the following

$$N \propto \left(R_e^2 - R_c^2\right) \tag{5.1.6}$$

At the glottis, the noise is usually considered a series pressure (voltage) source, while at a tract constriction, the noise can be modeled better as a shunt velocity (current) source.

For synthesis there are generally two excitation sources, a periodic excitation source for voiced sounds and a noise generator for unvoiced sounds. There are various waveform models for the periodic excitation. We will discuss these in a later chapter. We shall also consider a detailed model of the vocal tract in a later chapter, where we show how the transfer function is calculated and how the model can be used for speech synthesis.

In summary, the vocal tract can be modeled by a set of resonances (formants or poles) that depend on the vocal tract shape. The bandwidths of the lowest two formants depend primarily on wall losses. The bandwidths of the higher formants depend primarily on viscous friction losses, thermal losses, and losses due to radiation. The nasal tract also has poles. The nasal tract can be considered as parallel with the oral tract. Because the mouth is closed for nasal production, the oral tract can trap frequencies, thereby eliminating these frequencies in the radiated sound. This trapping of frequencies is modeled with zeros in the overall system-transfer function. Furthermore, the excitation is modeled as two sources: one periodic wave-form generator (approximated with a two pole transfer function) for voiced sounds and one noise generator for unvoiced sounds. There is also radiation at the lips, which is modeled as a single zero, which is equivalent to differentiation in the time domain. The complete model is summarized in Figure 5.2, where A_V and A_N are gain terms for the voiced and noise source models, respectively. The major point of this model, from the point of view of this chapter, is that the vocal tract model $V(z)$ can be an all pole model. This is why linear prediction modeling is so important in speech analysis, synthesis, and recognition. Therefore, in this chapter we develop an all-pole model for the vocal tract. This model is also used for spectral analysis.

5.2 ESTIMATES OF THE AUTOCORRELATION FUNCTION

Various applications call for autocorrelation estimates, including data modeling such as linear prediction and power spectral estimation, both of which we discuss in this chapter. There are two major estimators for the autocorrelation function that are used on a regular basis, both of which are implemented in MATLAB with the function xcorr.

Definition 5.1. The first estimate is called the **unbiased autocorrelation estimate**, and is expressed as

$$\hat{R}_{XX}(k) = \frac{1}{N - |k|} \sum_{i=0}^{N-1-|k|} X(i + |k|)X(i), \quad \text{for } |k| < N \tag{5.2.1}$$

The summation is sometimes expressed as from 1 to $N - |k|$, but the total number of terms

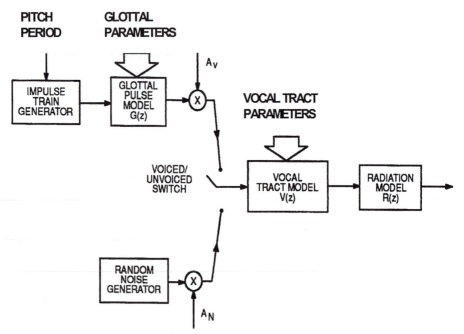

FIGURE 5.2 A model of vocal tract sound production.

is the same, namely, N, which denotes the number of data points available. The choice depends solely on the manner by which the data is indexed, that is, whether the initial value of the data starts at $i = 0$ or $i = 1$. Note that when we calculate Equation (5.2.1), we use an observed data record. Another comment is that we use upper case to denote that the process is a random process. In this case, the random process is a random sequence, since we are dealing with discrete data. Thus, $X(n)$ is a random sequence, while $x(n)$ is a particular member function (or sample function) of the random sequence. Also, the hat over the R denotes an estimate.

This estimate is unbiased, since the expectation of the estimate is the true autocorrelation

$$E[\hat{R}_{XX}(k)] = \frac{1}{N - |k|} \sum_{i=0}^{N-1-|k|} E[X(i + |k|)X(i)] = \frac{1}{N - |k|} \sum_{i=0}^{N-1-|k|} R_{XX}(k) = R_{XX}(k)$$

(5.2.2)

where E denotes the expected value. When the random variables are zero mean Gaussian, the variance of this estimate is approximately

$$VAR[\hat{R}_{XX}(k)] = \frac{1}{N - |k|} \sum_{i=-(N-1-|k|)}^{N-1-|k|} \left\{ \left(1 - \frac{|i|}{N - |k|} \right) \left(R_{XX}^2(i) \right. \right.$$
$$\left. \left. + R_{XX}(i + |k|)R_{XX}(i - |k|) \right) \right\}$$

(5.2.3)

Provided k is fixed, this estimate of the autocorrelation function approaches zero as N approaches infinity, making this estimate consistent. We discuss this more fully shortly. One problem with this estimate is that its largest value does not always occur at the origin.

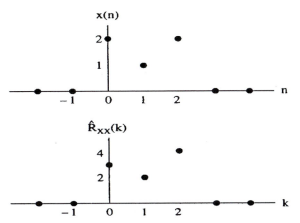

FIGURE 5.3 An example of an autocorrelation estimate that is not positive definite.

This is illustrated in the following example. For correlation matrices, this means that the estimate is not always positive definite.

EXAMPLE 5.1

Consider the data sequence, $x(n)$, shown in Figure 5.3. The autocorrelation estimate is calculated as follows.

$$\hat{R}_{XX}(0) = \frac{1}{3}[2^2 + 1^2 + 2^2] = 3$$

$$\hat{R}_{XX}(1) = \frac{1}{2}[(2)(1) + (1)(2)] = 2$$

$$\hat{R}_{XX}(2) = \frac{1}{1}[(2)(2)] = 4$$

■

The last example shows that this autocorrelation estimate does not have its largest value at the origin.

Definition 5.2.   Another estimate is called the **biased autocorrelation estimate**, and is given as

$$\hat{R}_{XX}(k) = \frac{1}{N} \sum_{i=0}^{N-1-|k|} X(i + |k|)X(i), \quad \text{for } |k| < N \tag{5.2.4}$$

where $X(i)$ is a zero mean Gaussian random sequence. Although this estimate is biased, it is asymptotically unbiased as N increases for k fixed. And the variance is approximately

$$\text{VAR}[\hat{R}_{XX}(k)] = \frac{1}{N^2} \sum_{i=-(N-1-|k|)}^{N-1-|k|} \left\{ (N - |i| - |k|)\left(R_{XX}^2(i) + R_{XX}(i + |k|)R_{XX}(i - |k|)\right) \right\} \tag{5.2.5}$$

This estimate is also consistent as N increases, provided k is fixed. It can also be shown that this estimate does have its largest value at the origin, which in correlation matrix terms means that this estimate is positive definite.

Given the above two possible autocorrelation estimates, which is the "best?" Suppose $|k|$ approaches N; then the variance of the unbiased estimate becomes large, because k, the shift in the data, is large. This shift k is often called the number of lags. When this shift or number of lags is large, then the estimate of the autocorrelation function is unreliable, since there is very little overlap of the data sequence with its shifted version. Consequently, this has lead to the rule of thumb that the number of lags should not be greater than 10% of the length of the data sequence or N/10. On the other hand, the variance of the biased estimate does not become large as k becomes large. In addition, the expected value of this estimate goes to zero, not to the true autocorrelation function. Since the bias is as large as the autocorrelation function we are estimating, we do not obtain a good estimate.

Now let k be fixed and let N increase. Then the variance of the biased estimate decreases as does the bias. The variance of the unbiased estimate also decreases. So we remain perplexed about which estimate is the "best." It is conjectured that the mean square error of the biased estimate is less than that for the unbiased estimate. Thus, it is recommended that the biased autocorrelation estimate be used when possible.

Calculating the autocorrelation estimate involves several practical matters. First, is the data sequence. This is typically of finite length and can be thought of as being observed through a window, where a typical window is rectangular. This process sets the data outside the window to zero. Thus, as we calculate the autocorrelation estimate for increasing k, we have less and less data contributing to the estimate and more and more zeros. This is one reason why the estimate becomes unreliable as the number of lags increases. Figure 5.4 illustrates this point using a sliding rectangular window to represent the data. Note that the total number of data points is N, where the range is zero to $N - 1$.

Sometimes the data may be periodically extended outside the observation window. Here the idea is that this periodic extension is "better" in some sense than using zeros outside the window. In either case, the experimenter must decide where such assumptions are justified.

A rectangular window may be replaced by a window with less sharp leading and trailing edges in many applications. The reason for this is that the fast rise and fall time of the leading and trailing edges of the rectangular window introduce large spectral sidelobes in the spectral estimate, as well as less high frequencies due to the window alone. This type of windowing distorts the spectral estimate. We will discuss this more later.

Both of the autocorrelation estimates can be calculated using the function xcorr in the MATLAB Signal Processing Toolbox. However, they can also be calculated using vector

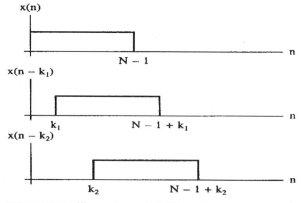

FIGURE 5.4   Illustration of shifting the data when calculating the autocorrelation estimate.

notation as follows. Let the data sequence be represented in vector form as

$$\vec{x}(t)^T|_0 = [x(0)\ x(1)\ x(2) \cdots x(N-1)\ \underbrace{0\ 0\ 0 \cdots 0]}_{K-1\ \text{zeros}}$$

$$\vec{x}(t)^T|_1 = [0\ x(0)\ x(1)\ x(2) \cdots x(N-1)\ \underbrace{0\ 0\ 0 \cdots 0]}_{K-2\ \text{zeros}}$$

$$\vec{x}(t)^T|_2 = [0\ 0\ x(0)\ x(1)\ x(2) \cdots x(N-1)\ \underbrace{0\ 0\ 0 \cdots 0]}_{K-3\ \text{zeros}} \qquad (5.2.6)$$

$$\vec{x}(t)^T|_{k-1} = [\underbrace{0\ 0 \cdots 0}_{K-1\ \text{zeros}}\ x(0)\ x(1)\ x(2) \cdots x(N-1)]$$

Note that Equation (5.2.6) could use either upper or lower case notation for the vectors and their elements. Since the discussion is primarily directed toward formulating data vectors, we adopted the lower case notation. Thus, we have K data vectors, numbered from zero to $K-1$, where $K < N$. The $K-1$ zeros appended to the end of the first data sequence and to the beginning of the last data sequence represents the largest value of the lag to be calculated in the autocorrelation estimate, that is, the number of lags ranges from 0 to $K-1$. The subscript on the vector represents the number of lags, k, in the data sequence. To calculate the biased estimate of the autocorrelation function we do the following.

$$\hat{R}_{XX}(0) = \vec{x}(t)^T|_0 \vec{x}(t)|_0 = [x(0)\ x(1) \cdots x(N-1)\ 0\ 0 \cdots 0] \begin{bmatrix} x(0) \\ x(1) \\ \cdot \\ \cdot \\ x(N-1) \\ 0 \\ 0 \\ \cdot \\ \cdot \\ 0 \end{bmatrix}$$

In general, we calculate the autocorrelation estimate in the following manner. Note that N is the length of the data, not the length of the data plus the number of appended zeros.

$$\hat{R}_{XX}(0) = \frac{1}{N}\vec{x}(t)^T|_0 \vec{x}(t)|_0$$

$$\hat{R}_{XX}(1) = \frac{1}{N}\vec{x}(t)^T|_1 \vec{x}(t)|_0 \qquad (5.2.7)$$

$$\hat{R}_{XX}(k) = \frac{1}{N}\vec{x}(t)^T|_k \vec{x}(t)|_0$$

One can also create a data matrix, [X], of the data vectors. The matrix has $(N+K-1)$ rows and (K) columns. The autocorrelation matrix, $[\hat{R}_{XX}(k)]$, is calculated by multiply the transpose of the data matrix by the data matrix as follows.

$$[X] = \begin{bmatrix} x(0) & 0 & 0 & \cdots & 0 \\ x(1) & x(0) & 0 & \cdots & 0 \\ x(2) & x(1) & x(0) & \cdots & 0 \\ \vdots & \vdots & \vdots & \cdots & 0 \\ 0 & x(N-1) & & \cdots & x(0) \\ 0 & 0 & x(N-1) & \cdots & x(1) \\ \vdots & \vdots & \vdots & \cdots & \vdots \\ 0 & 0 & 0 & \cdots & x(N-1) \end{bmatrix} \tag{5.2.8}$$

Then

$$[\hat{R}_{XX}(k)] = \frac{1}{N}[X]^T[X] \tag{5.2.9}$$

This matrix calculation yields a K × K Toeplitz matrix, that is, a matrix that is symmetric about the main diagonal and one that has equal elements on the principal diagonal. This autocorrelation matrix is also positive definite. However, note that the matrix multiplication requires more computations than are necessary since the autocorrelation estimate is symmetric about the origin. In a similar manner we can calculate the unbiased autocorrelation estimate. However, if we do this by matrix multiplication, then the resulting autocorrelation matrix is not Toeplitz, and, furthermore, the matrix may not be positive definite. This is another reason why the unbiased estimate is not usually used.

EXAMPLE 5.2

Consider the data shown in Example 5.1. The biased autocorrelation matrix for K = 3 (K − 1 = 2) is

$$[X] = \begin{bmatrix} 2 & 0 & 0 \\ 1 & 2 & 0 \\ 2 & 1 & 2 \\ 0 & 2 & 1 \\ 0 & 0 & 2 \end{bmatrix} \quad [X]^T = \begin{bmatrix} 2 & 1 & 2 & 0 & 0 \\ 0 & 2 & 1 & 2 & 0 \\ 0 & 0 & 2 & 1 & 2 \end{bmatrix}$$

$$[\hat{R}_{XX}(k)] = \frac{1}{3}\begin{bmatrix} 2 & 1 & 2 & 0 & 0 \\ 0 & 2 & 1 & 2 & 0 \\ 0 & 0 & 2 & 1 & 2 \end{bmatrix}\begin{bmatrix} 2 & 0 & 0 \\ 1 & 2 & 0 \\ 2 & 1 & 2 \\ 0 & 2 & 1 \\ 0 & 0 & 2 \end{bmatrix} = \frac{1}{3}\begin{bmatrix} 9 & 4 & 4 \\ 4 & 9 & 4 \\ 4 & 4 & 9 \end{bmatrix}$$

Thus, the number of lags, k, ranges from 0 to 2 with autocorrelation estimate values of 3, 4/3, and 4/3, respectively. If we had let K = 4, then the value of the autocorrelation estimate for lag 3 would be zero. ■

5.3 ESTIMATES OF THE POWER SPECTRUM

5.3.1 Classical Estimation

Many problems in electrical engineering require a knowledge of the distribution of the power of a random process in the frequency domain, for example, the design of filters to remove noise, to cancel signal echoes, or to represent features of a signal for pattern recognition. If the data records are long, then reliable power spectrum estimates can be obtained using standard fast Fourier transform (FFT) techniques. However, if the data records are short, as they usually are, then the task of obtaining reliable, that is, small bias and small variance, power spectrum estimates becomes difficult. Spectral estimators are usually classified as parametric and nonparametric. The latter generally make no assumptions about the statistics of the data other than that it is wide sense stationary or, perhaps, ergodic. The parametric spectral estimators typically use models of the data. In this case, the data may be modeled as a moving average process or as an autoregressive (linear prediction) process. The parametric approach to spectral estimation typically results in estimates with a smaller bias and variance for a given data record length than nonparametric methods. There is no one "best" spectral estimator at this time. One reason for this is that the spectral estimate is data dependent. Therefore, one estimator is good for one type of data, but not for another type of data. If we know *a priori* the type of data we are to analyze, then the choice of the type of spectral estimator to be used can be narrowed. However, we usually do not have such knowledge. Consequently, we often use a "bouquet" of spectral estimators and use our experience to make a judgment as to which estimator is the most accurate for the given data.

The power spectrum is usually defined using expectation, that is, the expected value of the magnitude squared of the Fourier transform of the data, or another definition, is the Fourier transform of the autocorrelation function determined by expectation. In practical situations, we usually have limited knowledge of the data as well as limited data records. So we need estimates of the autocorrelation function, as defined previously.

Perhaps the oldest and most commonly used spectral estimator is the periodogram, which is one of the classical spectral estimators.

Definition 5.3. The **periodogram** is defined as

$$\hat{S}_{XX}(f) = \frac{1}{N}|X_N(f)|^2 = \frac{1}{N}\left|\sum_{n=0}^{N-1} X(n)e^{-j2\pi fn}\right|^2 \tag{5.3.1.1}$$

where $X(n)$ is a random process.

The periodogram is related to the transform of the biased estimate of the autocorrelation function as follows. Suppose we window the data sequence so that

$$X_N(n) = w(n)X(n)$$

where $w(n)$ is the window sequence and is zero for $n < 0$ and $n \geq N$. Then

$$X_N(z) = \sum_{n=-\infty}^{\infty} X_N(n)z^{-n} = \sum_{n=0}^{N-1} w(n)X(n)z^{-n}$$

However, we have

$$\hat{R}_{XX}(k) = \frac{1}{N}\sum_{n=0}^{N-1-|k|} X_N(n+|k|)X_N(n) = \frac{1}{N}[X_N(k) * X_N(-k)]$$

This estimate is zero for $|n| \geq N$ due to the window. The spectral estimate is

$$\hat{S}_{XX}(f) = \sum_{k=-(N-1)}^{N-1} \hat{R}_{XX}(k)z^{-k}\big|_{z=e^{-j2\pi f}} = \sum_{k=-(N-1)}^{N-1}\left[\frac{1}{N}[X_N(k) * X_N(-k)]\right]z^{-k}\big|_{z=e^{-j2\pi f}}$$

$$= \frac{1}{N}X_N(z)X_N(z^{-1})\big|_{z=e^{-j2\pi f}} = \frac{1}{N}|X_N(f)|^2$$

While the periodogram seems to be a reasonable spectral estimator, it has some faults. First, the estimator is biased. If we denote the ensemble average autocorrelation function of the window as $R_{ww}(k)$ and the spectrum of the window as the Fourier transform of this autocorrelation function as $S_{ww}(f)$, then the expected value of the periodogram turns out to be the true spectrum of the data convolved with the power spectrum of the window. Thus, if the window is short, then the bias of the periodogram is large. As the window length in the time domain increases, the bias decreases and approaches the true power spectrum of the data. Furthermore, the variance of the periodogram does not decrease as N increases. This is because no averaging of the estimate is done either by using the expectation operator or by time averaging. So the periodogram is not a good spectral estimator, even for long data records. Nevertheless, it is commonly used.

Several proposals have been made to improve the power spectral estimate of the periodogram. One such proposal suggests that the data record be segmented into M independent data records each with L points, as shown in Figure 5.5. These data records can overlap if desired, as shown. The periodogram is then calculated for each segmented data record, and the periodogram for the entire data record is the average of the periodograms for each segment. However, the bias of this spectral estimate is larger than the periodogram for the original data record, although the variance of the estimate is reduced by the factor M over that of the periodogram of a single L point segment. Note that this approach has fewer points in the spectral estimate due to the segmentation of the data. Consequently, the spacing between spectral line components is increased, thereby, decreasing the spectral resolution of the estimate.

Since the variance of the spectral estimate of the segmental averaging approach is reduced, this implies that "smoothing" the periodogram of the original (unsegmented) data record may give an improved spectral estimate. Thus, another approach is to segment the autocorrelation estimate and average the segmented autocorrelation estimates, as shown in Figure 5.6. This approach yields a smoothed spectral estimate of the periodogram in that the periodogram is convolved with the power spectrum of a triangular window. However,

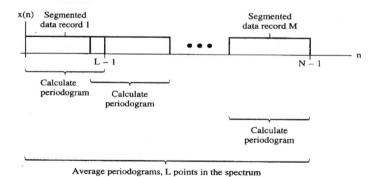

FIGURE 5.5 Illustration of segmenting the data record to improve the statistical stability of the power spectral estimate.

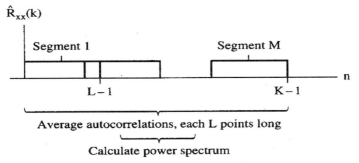

FIGURE 5.6 Illustration of segmenting the autocorrelation function to improve the statistical stability of the power spectral estimate.

since K is 10% of N, and L is less than K, we have even fewer points in the spectral estimate than averaging the periodograms. But, we again achieve improved statistical stability (a reduction in the variance of the estimate), at the cost of reduced spectral resolution.

Appending M zeros to extend the data record from length N to N + M does not result in increased resolution in the spectral estimate. The spectral lines in the transform are closer, thus assisting the interpretation of the spectral envelope. But no new information is added by appending zeros. This process is basically one of interpolation.

Windowing the data record will reduce the sidelobes of the periodogram. However, the bandwidth of the main lobe is increased, which reduces the spectral resolution of the estimate. There are a number of data windows available in the MATLAB Signal Processing Toolbox. The effect of these windows on speech data is easily explored using the software provided in Chapter 2. If the autocorrelation function estimate is windowed before transforming, then the spectral estimate may have negative values. Thus, the transform of the window should be nonnegative.

Another classical spectral estimator is the Blackman–Tukey estimate.

Definition 5.4.   The **Blackman–Tukey spectral estimate** is

$$\hat{S}_{XX}(f) = \sum_{k=-M}^{M} w(k)\hat{R}_{XX}(k)e^{-j2\pi fk} \qquad (5.3.1.2)$$

where $N - 1 \geq M$, and $w(k)$ is a window applied to the autocorrelation estimate, which is zero for $|k| > M$. The value of M is usually restricted so that $N/10 \geq M$. The window is usually normalized such that $w(0) = 1$ and $1 \geq w(k) \geq 0$ for $k \neq 0$. The windows in MATLAB as well as others can be used. Typically, the biased autocorrelation estimate is used. The Blackman–Tukey spectral estimate is biased, and for certain conditions is asymptotically consistent, since the variance of the estimate does decrease as N increases.

EXAMPLE 5.3

In this example, we compare the periodogram and the Blackman–Tukey spectral estimates for a speech data record of the sustained vowel /IY/ as in see. The data record is shown in Figure 5.7 along with the spectral estimates. All spectra are plotted on the same dB scale. The frequency axis is scaled such that 0.5 is 5000 Hz. The windowed data contain approximately three pitch periods of data, representing about 25 msec of data sampled at 10 kHz. For the periodogram, we calculate the spectrum using a rectangular window and a hamming window

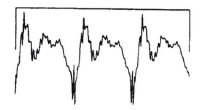

(a) Speech signal with rectangular window.

(b) Speech signal with hamming window.

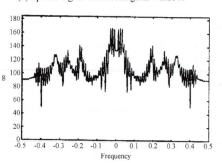

(c) Periodogram of speech signal with rectangular window.

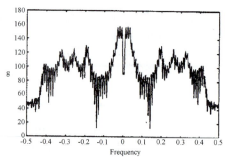

(d) Periodogram of speech signal with hamming window.

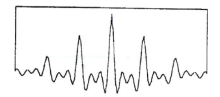

(e) Autocorrelation function with rectangular window.

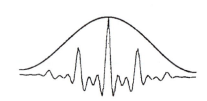

(f) Autocorrelation function with hamming window.

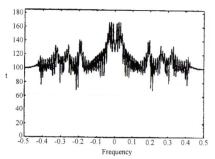

(g) Blackman–Tukey spectrum with rectangular window.

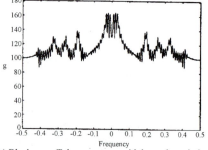

(h) Blackman–Tukey spectrum with hamming window.

FIGURE 5.7 An example of power spectra for the vowel /IY/.

applied to the data record. For the Blackman–Tukey spectral estimate, we first calculate the biased autocorrelation estimate and then apply the rectangular and hamming windows to the autocorrelation function prior to calculating the power spectrum. The hamming window tends to enhance the peaks and the nulls of the periodogram over that for the rectangular window. The power spectrum of the rectangular windowed autocorrelation function is similar to that for the corresponding periodogram, as one might expect. The Blackman–Tukey power spectrum with the hamming window is the "smoothest" of all four, with perhaps the best defined peaks. We shall return to this example again after we discuss the parametric spectrum methods, at which point, it will be seen that the methods shown here do not provide good estimates of the spectral peaks (formants) around 300, 2200, and 3000 Hz. This example can be done using the software supplied with this book in Chapter 2. ■

5.3.2 Parametric Spectral Estimation

The parametric spectral estimation procedures are relatively new, having appeared within the last 30 years. Their attraction is that they obtain good spectral estimates for short data records. The methods tend to be data adaptive in that they adjust themselves to be least disturbed by power at frequencies other than the one being estimated.

One parametric spectral estimator is called the maximum entropy estimator. The idea for this estimator is that instead of appending zeros to increase the length of the estimated autocorrelation function estimate, the estimated autocorrelation function is to be extrapolated (predicted) beyond the data limited range. The principle used for this extrapolation process is that the spectral estimate must be the most random or have the maximum entropy of any power spectrum that is consistent with the sample values of the calculated autocorrelation function. The objective is to add no information as a result of the prediction process, but yet make an improvement over that obtained by just appending zeros. This spectral estimate is equivalent to linear prediction or autoregression and is the same as the maximum likelihood method for Gaussian data.

Perhaps the simplest way to introduce the parametric spectral estimation procedure is with data models. One way to generate a data model is to excite a filter and measure the filter output. For example, Figure 5.8 shows a filter with both poles and zeros [an autoregressive-moving average (ARMA) model] such that the output is given as

$$X(n) = -\sum_{i=1}^{p} a_i X(n-i) + \sum_{i=0}^{q} b_i E(n-i) \tag{5.3.2.1}$$

Note that we still denote a random process with upper case notation and a particular member function of the random process with lower case notation. Thus, Equation (5.3.2.1) denotes the difference equation for an ARMA random process data model, which we can also write as

$$X(z) = -\sum_{i=1}^{p} a_i z^{-i} X(z) + \sum_{i=0}^{q} b_i z^{-i} E(z) \tag{5.3.2.2}$$

The minus sign in front of the a_i coefficients is one convention that is also used in MATLAB. The input $E(n)$ to the filter is the driving function, which is typically white noise, while the output $X(n)$ is the random process to be modeled. This model is called an

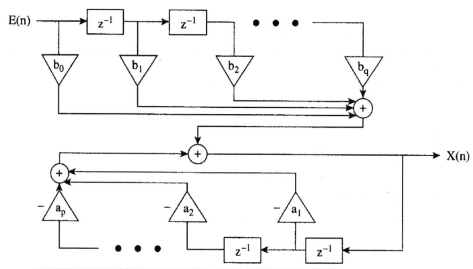

FIGURE 5.8 An ARMA filter model.

ARMA(p, q) process. The transfer function for the ARMA filter is

$$H(z) = \frac{X(z)}{E(z)} = \frac{\displaystyle\sum_{i=0}^{q} b_i z^{-i}}{\displaystyle\sum_{i=0}^{p} a_i z^{-i}} = \frac{B(z)}{A(z)} \tag{5.3.2.3}$$

where $B(z)$ represents the moving average (MA) branch of the filter, while $A(z)$ represents the autoregressive (AR) branch, which is also called the linear prediction (LP) branch. Note that there are q zeros in the MA branch of the filter, and p poles in the AR branch.

Since

$$X(z) = H(z)E(z) \tag{5.3.2.4}$$

The power spectrum is expressed as

$$S_{XX}(f) = S_{EE}(f)|H(f)|^2 \tag{5.3.2.5}$$

This is not an estimate, since this last step can be done using ensemble averaging. However, we will soon address the estimation problem.

If all of the $a_i = 0$, except for the unity coefficient in $A(z)$, which is usually called $a_0 = 1$, then we have an all zero, MA, model of order q, that is

$$X(n) = \sum_{i=0}^{q} b_i E(n-i) \tag{5.3.2.6}$$

which is shown in Figure 5.9.

The MA(q) power spectrum for a white noise excitation with variance σ_{EE}^2 is

$$S_{MA}(f) = S_{XX}(f) = \sigma_{EE}^2 |B(f)|^2 \tag{5.3.2.7}$$

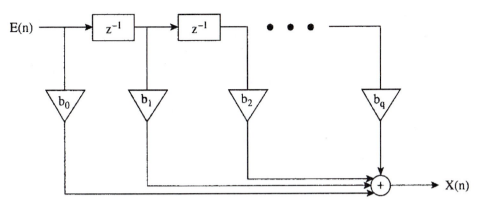

FIGURE 5.9 A MA(q) filter model.

If all the $b_i = 0$ except for $b_0 = 1$, then

$$X(n) = -\sum_{i=1}^{p} a_i X(n - i) + E(n) \tag{5.3.2.8}$$

which is an all pole, AR(p), process, also called a LP(p) process, shown in Figure 5.10, and the power spectrum is

$$S_{AR}(f) = S_{XX}(f) = \frac{\sigma_{EE}^2}{|A(f)|^2} \tag{5.3.2.9}$$

The estimation process requires a procedure for estimating the a_i and b_i coefficients given a data record. There are several theorems that tell us more about these data models. One theorem is called the Wold decomposition theorem, which says that a wide sense stationary random process can be decomposed into a component that is random and one that is deterministic. Another theorem says that any ARMA or AR process can be represented by a unique MA process of infinite order, i.e., MA(∞). The third theorem says that any ARMA or MA process can be represented by an AR(∞) process. These theorems tell us that if we should select the wrong model to represent the data, we can still obtain an adequate approximation, although not an "optimum" approximation, if we use a high enough model

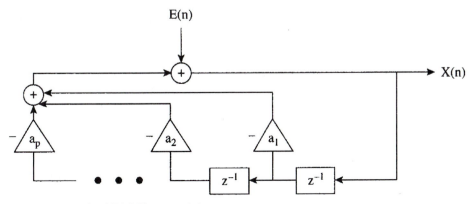

FIGURE 5.10 An AR(p) filter model.

order. In practical terms this is not satisfactory, since if we have to improve the model by increasing its order toward infinity, then we also have to consider a larger data set from which to estimate reliably the parameters of the model. And one purpose of data modeling for spectral analysis is to obtain a good spectral estimate with as few model parameters as possible. So from a practical point of view, it is not a good idea to think that if the result from a low-order model is not satisfactory in some sense, that the model order can be increased until a "good" result is obtained. A better idea is to try another data model.

EXAMPLE 5.4

Consider a first order AR model, such that

$$X(n) = -a_1 X(n-1) + E(n)$$

Then

$$X(z) = -a_1 z^{-1} X(z) + E(z)$$

which can be expressed as

$$X(z) = \frac{E(z)}{1 + a_1 z^{-1}} = z \frac{E(z)}{z + a_1}$$

where $|a_1| < 1$. This process has a simple pole at $z = -a_1$, and depending on the value of a_1, the process will be lowpass or highpass, as shown in Figure 5.11. The process cannot be bandpass unless the process has more than one pole. The power spectrum for a white noise input with variance σ_{EE}^2 is given below and is sketched in Figure 5.11, where the sampling frequency is f_s. Note that we use f_s and F_s interchangeably.

$$S_{XX}(f) = \frac{\sigma_{EE}^2}{|1 + a_1 e^{-j2\pi f}|^2}$$ ■

Further discussion of linear prediction (LP) and linear prediction spectral estimation follows. We also discuss ARMA and MA spectral estimates that are used in the software provided with this book. The reader can also consult Appendix 6 as well as Kay (1988).

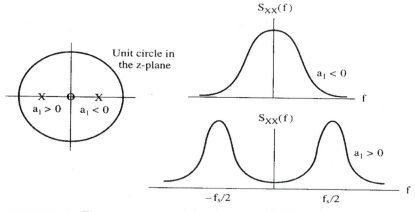

FIGURE 5.11　The power spectral density of an AR(1) model.

5.3.3 Least Squares Estimation

The method we will use in this chapter for estimating the autoregression (AR) parameters is the method of least squares. We will focus on AR estimation only. The method of least squares estimation may be introduced with the following simple situation. We conduct an experiment by measuring specific data values of a function, $x(n)$. We wish to design a filter that will estimate the data at time n using one previous data sample at time $n - 1$, i.e.,

$$\hat{x}(n) = -\hat{a}x(n - 1) \tag{5.3.3.1}$$

where $\hat{x}(n)$ and $x(n)$ are real discrete sequences, and $\hat{a}$ is a parameter to be estimated. The term $\hat{x}(n)$ is an estimate of the true value $x(n)$. Consequently, we have an error, defined as

$$e(n) = x(n) - \hat{x}(n) = x(n) + \hat{a}x(n - 1) \tag{5.3.3.2}$$

Note that this model is predicting the true value of the data at time n using a weighted value of the data at time $n - 1$. This model is called linear prediction (LP) or autoregression (AR). The method of least squares says that the constant $\hat{a}$ can be chosen so that we can minimize the sum of the errors over some interval from $n = 0$ to $n = N - 1$ (or 1 to N), i.e., we minimize the total squared error.

$$E_T^2 = \sum_{n=0}^{N-1} e^2(n) = \sum_{n=0}^{N-1} (x(n) - \hat{x}(n))^2$$

$$= \sum_{n=0}^{N-1} (x(n) + \hat{a}x(n - 1))^2 \tag{5.3.3.3}$$

The total squared error is sometimes normalized by the number of data points, N. However, this has no effect on the solution, as we show below. The total squared error is often called the prediction error or the prediction error power.

A word about notation is appropriate here. The method of least squares assumes that we are dealing with data and does not use statistics. Consequently, the lower case notation is used for functions. It is possible to consider $X(n)$ to be a random process and solve for the parameter $\hat{a}$ using expectation. This procedure minimizes the mean square error and gives the true value (not an estimate) for the parameter.

A necessary condition for a relative minimum is that the partial derivative with respect to $\hat{a}$ is zero. Then

$$\frac{\partial E_T^2}{\partial \hat{a}} = 0 = 2 \sum_{n=0}^{N-1} (x(n) + \hat{a}x(n - 1))x(n - 1) \tag{5.3.3.4}$$

which gives

$$\hat{a} \sum_{n=0}^{N-1} x(n - 1)x(n - 1) = - \sum_{n=0}^{N-1} x(n)x(n - 1) \tag{5.3.3.5}$$

Note that if there had been a multiplicative factor of $\frac{1}{N}$ in (5.3.3.3), then this would have no effect on the solution to (5.3.3.4). Equation (5.3.3.5) can be expressed in terms of autocorrelation estimates as follows.

$$\hat{a} = \frac{-\dfrac{1}{N}\displaystyle\sum_{n=0}^{N-1} x(n)x(n - 1)}{\dfrac{1}{N}\displaystyle\sum_{n=0}^{N-1} x(n - 1)x(n - 1)} = -\frac{\hat{R}_{XX}(1)}{\hat{R}_{XX}(0)} \tag{5.3.3.6}$$

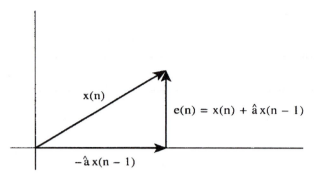

FIGURE 5.12 The principal of orthogonality. The data is orthogonal to the error.

where we will continue to use $\hat{R}_{XX}(k)$ to denote an estimate of the autocorrelation function. Equation (5.3.3.6) gives the least squares estimate for the parameter $\hat{a}$. This estimate gives the best linear prediction of $x(n)$ using only one weighted past data sample. The solution is best in the sense that the total error, that is, the sum of the squared error terms, is minimized.

We will soon generalize the above results for p $\hat{a}_i$ coefficients. However, first we can observe that (5.3.3.4) provides us with the result that the error is orthogonal to the data in a least squares sense, that is

$$\sum_{n=0}^{N-1}(x(n) + \hat{a}x(n-1))x(n-1) = \sum_{n=0}^{N-1}e(n)x(n-1) = 0 \qquad (5.3.3.7)$$

This **principle of orthogonality** is illustrated in Figure 5.12.

Next we note that Equation (5.3.3.5) is called the **normal equation**, which becomes multiple equations when we consider p $\hat{a}_i$ coefficients. The method of least squares makes no assumption about the statistics of the random sequence. However, Gauss proved a theorem, named after him years ago, that says that if the errors are uncorrelated with zero mean and the same variance, then the optimum estimate for the coefficient is the value that minimizes the square of the error between the observed data and the true value. The estimate is optimum in the sense that any linear function can be estimated with the minimum mean square error using the estimator.

We now form a more general linear prediction model using $p\hat{a}_i$ coefficients, and p past values of the data to predict the value of the data at time n. The data are given, and may or may not be from an AR process. This knowledge is often unknown. Nevertheless, we are modeling the data as if it were an AR data model. Therefore,

$$\hat{x}(n) = -\sum_{k=1}^{p}\hat{a}_k x(n-k) \qquad (5.3.3.8)$$

Then, we have

$$x(n) = -\sum_{k=1}^{p}\hat{a}_k x(n-k) + e(n) = \hat{x}(n) + e(n) \qquad (5.3.3.9)$$

which we illustrate in Figure 5.13.

The z-transform of Equation (5.3.3.9) is

$$X(z) = -\left[\sum_{k=1}^{p}\hat{a}_k z^{-k}\right]X(z) + E(z) = \hat{X}(z) + E(z) \qquad (5.3.3.10)$$

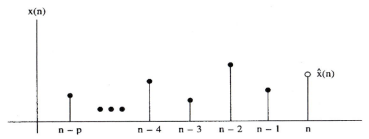

FIGURE 5.13 Linear prediction model.

which can be expressed as

$$X(z) = \frac{E(z)}{1 + \left[\sum_{k=1}^{p} \hat{a}_k z^{-k} \right]} = \frac{E(z)}{A(z)} \tag{5.3.3.11}$$

where

$$A(z) = 1 + \sum_{k=1}^{p} \hat{a}_k z^{-k} \tag{5.3.3.12}$$

Equation (5.3.3.11) is implemented as a filter in Figure 5.14. This form is often called the forward filter model, since $E(z)$, the input excitation, is filtered by $\frac{1}{A(z)}$ to produce $X(z)$, the LP data.

We can also write Equation (5.3.3.10) in the following form.

$$E(z) = \left[1 + \sum_{k=1}^{p} \hat{a}_k z^{-k} \right] X(z) = A(z)X(z) \tag{5.3.3.13}$$

This is the form used in Figure 5.15, where the data, $X(z)$, is "inverse" filtered by $A(z)$ to yield the excitation, $E(z)$. Consequently, this form is called the inverse filter model.

The method of least squares determines the coefficients $\hat{a}_i$ such that the sum of the errors over some interval is minimized. The interval over which we observe the data is what distinguishes the two methods that are used to solve for the $\hat{a}_i$ coefficients. For the **autocorrelation** method, it is assumed that the data are windowed such that the data are

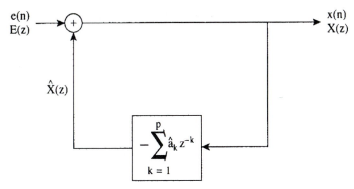

FIGURE 5.14 The forward filter model.

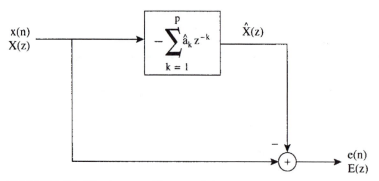

FIGURE 5.15 The inverse filter model.

zero outside the window. Note that in this case the error is likely to be large at both the beginning and ending of the estimation interval, since we are trying to predict the true data using zeros because we have windowed the data. This is why the data window is typically tapered at both ends. This causes the data to gradually increase at the beginning of the data record and similarly gradually decrease at the end, thereby, reducing the error for these segments. The total squared error (total prediction error, or total prediction power) over this interval is

$$E_T^2 = \sum_{n=0}^{N-1} e^2(n) = \sum_{n=0}^{N-1} (x(n) - \hat{x}(n))^2 = \sum_{n=0}^{N-1} \left(x(n) + \sum_{k=1}^{p} \hat{a}_k x(n-k) \right)^2 \quad (5.3.3.14)$$

where $x(n)$ is the windowed data. The minimization is done by

$$\frac{\partial E_T^2}{\partial \hat{a}_i} = 0, \quad i = 1, 2, \ldots, p$$

which gives

$$\sum_{k=1}^{p} \hat{a}_k \sum_{n=0}^{N-1} x(n-k)x(n-i) = -\sum_{n=0}^{N-1} x(n)x(n-i), \quad \text{for } i = 1, 2, \ldots, p \quad (5.3.3.15)$$

When these equations are derived using expectation, they are known as the **normal equations**, and are also sometimes referred to as the **Yule–Walker equations** and the **Wiener–Hopf equations**. We will refer to them by the same names. The equations can be expressed in a more compact form using the autocorrelation estimates as follows.

$$\sum_{k=1}^{p} \hat{a}_k \hat{R}_{XX}(i-k) = -\hat{R}_{XX}(i), \quad i = 1, 2, \ldots, p \quad (5.3.3.16)$$

where

$$\hat{R}_{XX}(k) = \frac{1}{N} \sum_{n=0}^{N-1-|k|} x(n)x(n+|k|) \quad (5.3.3.17)$$

Note that the $\frac{1}{N}$ term is not needed as we mentioned previously. This is only a scale factor and does not affect the solution of the equations for the $\hat{a}_i$. Equation (5.3.3.16) is a set of p equations with p unknowns and can be solved using matrix techniques as follows:

$$
\begin{bmatrix}
\hat{R}_{XX}(0) & \hat{R}_{XX}(-1) & \cdots & \hat{R}_{XX}(-(p-1)) \\
\hat{R}_{XX}(1) & \hat{R}_{XX}(0) & \cdots & \hat{R}_{XX}(-(p-2)) \\
\vdots & \vdots & \cdots & \vdots \\
\hat{R}_{XX}(p-1) & \hat{R}_{XX}(p-2) & \cdots & \hat{R}_{XX}(0)
\end{bmatrix}
\begin{bmatrix}
\hat{a}_1 \\ \hat{a}_2 \\ \vdots \\ \hat{a}_p
\end{bmatrix}
= -
\begin{bmatrix}
\hat{R}_{XX}(1) \\ \hat{R}_{XX}(2) \\ \vdots \\ \hat{R}_{XX}(p)
\end{bmatrix}
\tag{5.3.3.18}
$$

These equations can be solved by matrix inversion, but this is not as efficient as the Levinson algorithm, which solves the equations for the $\hat{a}_i$ in a recursive manner. Algorithms for doing this are available in MATLAB and take advantage of the fact that the autocorrelation matrix is Toeplitz. MATLAB has a function called **lpc** that returns the $\hat{a}_i$ coefficients given the data. Note, that since the autocorrelation function is real and even, then the terms above the main diagonal in the correlation matrix in Equation (5.3.3.18) can be replaced with terms with a positive argument.

The autocorrelation method is guaranteed to give a stable LP filter, that is, one such that the poles of $\frac{1}{A(z)}$ are all within the unit circle in the z-plane. Since we are assuming real data, the $\hat{a}_i$ coefficients are real, implying that the poles of $\frac{1}{A(z)}$ appear in complex conjugate pairs. Thus, for example, if p is even, then we will have an even number of coefficients and an even number of poles, with one-half of the poles being in the upper half of the unit circle in the z-plane and one-half of the poles being in the lower half. Of course, one or more pairs of poles could fall on the real line in the z-plane. If p is odd, then at least one root will be on the real line.

An additional equation is often added to the above equations for the $\hat{a}_i$, namely, the equation for the total squared error (prediction error), which from Equations, (5.3.3.13), (5.3.3.15), and (5.3.3.17) is

$$
E_T^2 = \hat{R}_{XX}(0)a_0 + \sum_{k=1}^{p} \hat{a}_k \hat{R}_{XX}(-k)
\tag{5.3.3.19}
$$

where $a_0 = 1$. Again the estimates with negative arguments in Equation (5.3.3.19) can be replaced with the corresponding estimates with positive arguments. This error will vary with the order p. The reason for introducing a_0 is that this is a convenience when the matrix equations are augmented for the additional unknown, E_T^2, as follows.

$$
\begin{bmatrix}
\hat{R}_{XX}(0) & \hat{R}_{XX}(-1) & \cdots & \hat{R}_{XX}(-p) \\
\hat{R}_{XX}(1) & \hat{R}_{XX}(0) & \cdots & \hat{R}_{XX}(-(p-1)) \\
\vdots & \vdots & \cdots & \vdots \\
\hat{R}_{XX}(p) & \hat{R}_{XX}(p-1) & \cdots & \hat{R}_{XX}(0)
\end{bmatrix}
\begin{bmatrix}
a_0 \\ \hat{a}_1 \\ \hat{a}_2 \\ \vdots \\ \hat{a}_p
\end{bmatrix}
=
\begin{bmatrix}
E_T^2 \\ 0 \\ \vdots \\ 0
\end{bmatrix}
\tag{5.3.3.20}
$$

These equations are also often called the **extended** or **augmented normal equations**.

The **covariance method** for solving for the coefficients is similar to the autocorrelation method. The major difference is that the covariance method does not assume that the data are windowed, and that while N samples of the data are available, the error is windowed such that $N - p$ samples of the error are available. This affects the calculation of the "autocorrelation" values, since we now have

$$
E_T^2 = \sum_{n=p}^{N-1} e^2(n) = \sum_{n=p}^{N-1} (x(n) - \hat{x}(n))^2 = \sum_{n=p}^{N-1} \left(x(n) + \sum_{k=1}^{p} \hat{a}_k x(n-k) \right)^2
\tag{5.3.3.21}
$$

The normal equations have the same form, namely

$$\sum_{k=1}^{p} \hat{a}_k \hat{R}_{XX}(i-k) = -\hat{R}_{XX}(i), \quad i = 1, 2, \ldots, p \tag{5.3.3.22}$$

However, the autocorrelation estimates are different, so a different notation is used to distinguish the two methods, that is,

$$\hat{C}_{XX}(i, k) = \frac{1}{N-p} \sum_{n=p}^{N-1} x(n-i)x(n-k) \tag{5.3.3.23}$$

and

$$\hat{C}_{XX}(i, 0) = \frac{1}{N-p} \sum_{n=p}^{N-1} x(n-i)x(n) \tag{5.3.3.24}$$

where $1 \le i \le p, 0 \le k \le p$, and the data record is N long. No window is applied to the data because we assume we have sufficient data to calculate the desired correlation values. The matrix equations have the same form, namely

$$\begin{bmatrix} \hat{C}_{XX}(1, 1) & \hat{C}_{XX}(1, 2) & \cdots & \hat{C}_{XX}(1, p) \\ \hat{C}_{XX}(2, 1) & \hat{C}_{XX}(2, 2) & \cdots & \hat{C}_{XX}(2, p) \\ \vdots & \vdots & \cdots & \vdots \\ \hat{C}_{XX}(p, 1) & \hat{C}_{XX}(p, 2) & \cdots & \hat{C}_{XX}(p, p) \end{bmatrix} \begin{bmatrix} \hat{a}_1 \\ \hat{a}_2 \\ \vdots \\ \hat{a}_p \end{bmatrix} = - \begin{bmatrix} \hat{C}_{XX}(1, 0) \\ \hat{C}_{XX}(2, 0) \\ \vdots \\ \hat{C}_{XX}(p, 0) \end{bmatrix} \tag{5.3.3.25}$$

The correlation matrix is no longer Toeplitz, but has the properties of a covariance matrix, from which the name for this method is derived. The solution for this method does not use the Levinson algorithm, but instead uses Cholesky decomposition (or other methods), which are also available in MATLAB.

The covariance method may yield estimates of the $\hat{a}_i$ coefficients such that the LP filter, $\frac{1}{A(z)}$, does not have all of its poles inside the unit circle, therefore, the filter may be unstable.

Return to Equation (5.3.3.20) and note that all of the equations are prediction equations for the autocorrelation function. In particular, the last equation in the set of equations represented by (5.3.3.20) can be written as

$$\hat{R}_{XX}(p) + \hat{a}_1 \hat{R}_{XX}(p-1) + \cdots + \hat{a}_p \hat{R}_{XX}(0) = 0 \tag{5.3.3.26}$$

This equation can be considered a prediction equation for $\hat{R}_{XX}(p)$ using the past values of the autocorrelation function and the prediction coefficients. It can be shown that for lag values greater than p, that

$$\hat{R}_{XX}(k) + \hat{a}_1 \hat{R}_{XX}(k-1) + \cdots + \hat{a}_p \hat{R}_{XX}(k-p) = 0 \tag{5.3.3.27}$$

for $k = p + 1, p + 2, \ldots$. Therefore, the linear prediction equations lead to a correlation prediction equation, whereby the correlation function can be extended. This is the basis for the maximum entropy spectral estimator mentioned earlier.

5.3.4 Power Spectral Estimation With Linear Prediction

Linear prediction is a data model. Consequently, it is frequently used to estimate the spectrum of data that are modeled using linear prediction techniques.

Definition 5.5. The **LP power spectral estimate** is the reciprocal of the square of the absolute value of the transform of the LP coefficients.

$$\hat{S}_{XX}(f) = \frac{1}{|A(z)|^2} = \frac{1}{\left|1 + \sum_{i=1}^{p} \hat{a}_i z^{-i}\right|^2}, \quad \text{for } z = e^{j2\pi f} \tag{5.3.4.1}$$

where the $\hat{a}_i$ coefficients are the estimates obtained using the least squares method above, and the number of coefficients, p, is the same as the number of poles of $\frac{1}{A(z)}$. Since $A(z)$ is a polynomial in z, the roots of this polynomial are the zeros of $A(z)$, which in turn are the poles of $\frac{1}{A(z)}$.

Why is Equation (5.3.4.1) considered a good spectral estimate? We explain this as follows. The error can be expressed as

$$E(z) = A(z)X(z) \tag{5.3.4.2}$$

where the $A(z)$ in this equation represents the z-transform of the estimated coefficients. Then

$$|E(z)|^2 = |A(z)|^2|X(z)|^2 \tag{5.3.4.3}$$

and the true power spectrum is

$$S_{XX}(f) = |X(z)|^2, \quad \text{for } z = e^{j2\pi f} \tag{5.3.4.4}$$

The total squared error can be expressed as

$$E_T^2 = \frac{T}{2\pi} \int_{-\pi/T}^{\pi/T} |E(e^{j2\pi f})|^2 \, d\omega = \frac{T}{2\pi} \int_{-\pi/T}^{\pi/T} \frac{S_{XX}(f)}{\hat{S}_{XX}(f)} \, d\omega \tag{5.3.4.5}$$

where T is the sampling interval or $\frac{1}{T}$ is F_s, the sampling frequency. Now note that minimizing the error is equivalent to minimizing the integral of the ratio of the true power spectrum to the estimate. This ratio is positive. Both global and local errors are minimized by minimizing the integral. A local error may be thought of as an error at a particular frequency. Globally, the total error is determined by how well the true spectrum and the estimated spectrum match over the entire frequency range, regardless of the shape of the spectrum. Local errors are a measure of the difference between these two spectra at a given frequency. If the true spectrum is greater than the estimate (model), then the total error is larger than if the opposite is true.

Thus, after minimizing E_T^2, we would expect the estimate to match the true spectrum better in those regions where the true spectrum is greater than the estimate. Thus, the estimate, or model spectrum, should follow the true spectrum in the vicinity of the peaks, or local maxima. However, this is not necessarily the case in regions where minima (notches) occur, since the local error is already small and the emphasis is on reducing the largest errors. Thus, the estimate should be a good estimate of the spectral envelope of the true spectrum. Since it is usually the peaks in the spectrum that we want to estimate, the LP estimate is often used.

Spectral flatness is defined as the spectrum of the prediction error

$$E_f = \frac{\exp\left\{\dfrac{T}{2\pi}\displaystyle\int_{-\pi/T}^{\pi/T} \log |E(\omega)|^2 \, d\omega\right\}}{\dfrac{T}{2\pi}\displaystyle\int_{-\pi/T}^{\pi/T} |E(\omega)|^2 \, d\omega} = \frac{\exp\left\{\dfrac{T}{2\pi}\displaystyle\int_{-\pi/T}^{\pi/T} \log |E(\omega)|^2 \, d\omega\right\}}{E_T^2} \qquad (5.3.4.6)$$

where $|E(\omega)|^2$ is the spectrum of the prediction error. The spectral flatness is the ratio of the geometric mean to the arithmetic mean of the spectrum and can assume values between 0 and 1. It is 1 for a constant spectrum. Note that the spectral flatness is inversely proportional to the mean squared prediction error. Thus, minimizing the mean squared prediction error in the time domain is equivalent to finding an all pole filter such that the spectral flatness of the prediction error is maximized. The idea of maximizing the spectral flatness is used in the manual inverse filtering procedures in the software in Chapter 2. Maximizing the spectral flatness of the error in glottal inverse filtering to eliminate (or at least reduce) the formant frequencies reduces the error in the estimates of the linear prediction filter coefficients. This procedure is difficult to achieve automatically, so it is implemented with user manual interaction.

One problem with the LP method is the estimation of the number of coefficients, p, to be used in the model. While there are criteria for this, it is still debated which is the best. One such criterion is the Akaike information criterion (AIC), which is defined as $AIC(p) = (N)\ln(E_T^2) + 2p$, where p is the model order, N is the number of data points available, ln is the natural log, E_T^2 is the total squared error for the pth order model given in Equation (5.3.3.19). The value of AIC can be plotted versus p to determine the "best" model order. The AIC procedure is implemented in the software for this book (see Chapter 2 and Appendix 6). A trial and error procedure is also often used, whereby, the choice of p is determined by the value that gives the smallest error for reasonable variations in p. This is done by varying p and calculating the error for each value of p. Then, we select the value of p that gives the least error. Another problem with the LP method is that it does not give good estimates of sinusoids. However, there are other methods for doing this, such as subspace methods, which use eigenvector decomposition techniques. These methods are covered in other texts (see Kay, 1988; Therrien, 1992).

A major usage of linear prediction, besides spectral estimation, is for speech coding, where the method is called linear predictive coding (LPC). For example, suppose a 10th order model is used to code (represent) 100 samples of speech data. The coefficients are transmitted over a communication channel and the speech is reconstructed (synthesized) using the LPC coefficients in a filter that is excited by a model of the speech excitation waveform. The excitation model may be a discrete multipulse random sequence distributed over a pitch period. This technique can produce very high-quality speech, with a reduction in data of ten to one or so. Similar applications are used in some speaking toys and talking calculators, as well as in speech applications software for personal computers.

The spectral estimate using the LP coefficients determined by the covariance method is usually better (less bias and less variance) than the spectral estimate using the coefficients determined by the autocorrelation method. This is because no data window is used with the covariance method.

While we do not cover the MA or ARMA parameter estimation techniques in depth, an outline of the procedures is briefly presented. The MA method models the data with q zeros. Consequently, the MA power spectral estimation method is useful for estimating broad peaks or sharp nulls in the spectrum, while poles are used to estimate sharp peaks. Durbin's algorithm is often used to estimate the coefficients of the MA model. The idea of this algorithm is to first fit a large order, say L, AR model to the data, where $q \ll L \ll N$. Then let the AR parameters serve as a data sequence to estimate the MA parameters. Thus, the algorithm calculates the autocorrelation matrix using the AR parameters as data and then solves the normal equations for the MA parameters, which are the unknowns. Thus, we have an equation similar to Equation (5.3.3.18), but one that involves the $\hat{b}_i$ coefficients. The idea for this algorithm comes from the fact that an MA model is equivalent to an infinite order AR model. So that $B(z) = \frac{1}{A(z)}$, or $B(z)A(z) = 1$. By taking inverse transforms, we obtain an equation involving the AR coefficients and the MA coefficients that can be arranged into a form like the AR (or LP) equations.

The ARMA model is used to estimate spectra where there are both poles and zeros, which calls for a technique to estimate both the peaks and the nulls in the spectrum. Estimating the ARMA parameters is more difficult and involves nonlinear equations. There are no optimal solution methods available, so suboptimal methods are used. One procedure first estimates an AR(p) model using the data. The data is then filtered by A(z) to produce an approximate MA model. This data is then used to estimate the MA(q) parameters. Another method solves a set of nonlinear equations.

The software introduced in Chapter 2 has algorithms for both MA and ARMA spectral estimates; see Appendix 6 as well.

EXAMPLE 5.5

In this example, we compare in Figure 5.16 the magnitude squared of the FFT (periodogram) of a data record of the vowel /IY/ with the LP spectral estimate with $p = 11$. The LP spectrum matches the envelope of the periodogram well, but not the notches, as expected. Furthermore, the LP spectrum is smoother than the other spectral estimates we have examined. The dB and the frequency scales are the same as for Figure 5.7. Also shown are the pole locations in the z-plane. ■

EXAMPLE 5.6

Consider the following autocorrelation matrix and determine the LP coefficients using MATLAB.

$$\begin{bmatrix} 1.0 & 0.3 & 0.09 \\ 0.3 & 1.0 & 0.3 \\ 0.09 & 0.3 & 1.0 \end{bmatrix} \begin{bmatrix} \hat{a}_1 \\ \hat{a}_2 \\ \hat{a}_3 \end{bmatrix} = - \begin{bmatrix} 0.3 \\ 0.09 \\ 0.027 \end{bmatrix}$$

The $\hat{a}_i$ are -0.3, 0, and 0 for $i = 1, 2, 3$. From Equation (5.3.3.19), the total error is $1 + \hat{a}_1(0.3) + \hat{a}_2(0.09) + \hat{a}_3(0.027)$, which is 0.91. The spectral estimate is found from Equation (5.3.4.1), and is plotted in Figure 5.17. It might be surprising that only coefficient $\hat{a}_1$ is nonzero. However, the autocorrelation function for a first order LP process is proportional to $(-a)^{|k|}$, and such a process has only one pole. In this example $\hat{a}_1 = a = -0.3$, so the autocorrelation function is proportional to $(0.3)^{|k|}$, which agrees with the data given. ■

EXAMPLE 5.7

This example uses the function lpc in MATLAB to determine the power spectral estimate of the data shown in Figure 5.18, where the spectrum is also plotted. This algorithm uses the estimate for the biased autocorrelation function, which is calculated using the data. The same results are obtained if one calculates the autocorrelation function, then calculates the LP coefficients, and then calculates the spectrum. So this MATLAB function is quite convenient to use. The reader is encouraged to examine this function carefully. It is used in the software in Chapter 2. ■

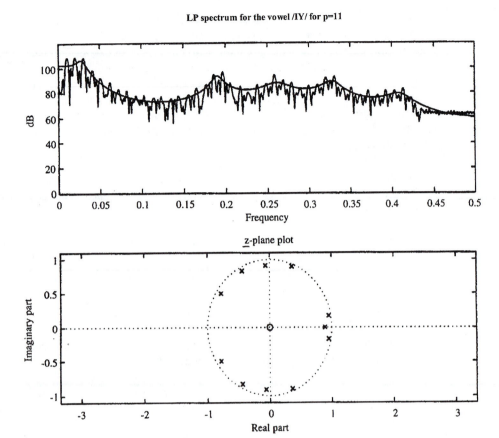

FIGURE 5.16 Comparison of the periodogram and the LP spectrum for the vowel /IY/.

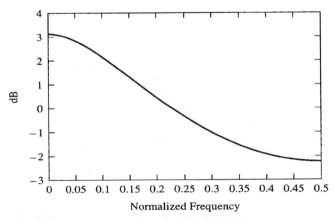

FIGURE 5.17 The spectrum for Example 5.6.

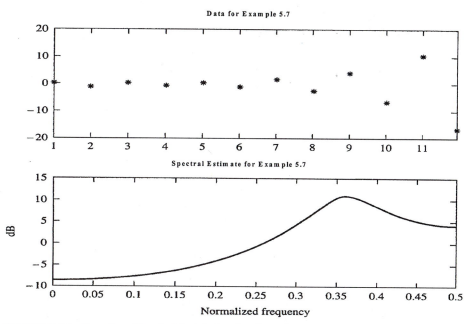

FIGURE 5.18 Results of lpc in Matlab for the data shown.

5.4 ALGORITHMS

We mentioned previously that the Levinson algorithm is used to solve the normal equations. What does this algorithm do? Suppose it is our task to find the pth order prediction filter coefficients for

$$\hat{x}(n) = \sum_{i=1}^{p} \hat{a}_k x(n - k)$$

The Levinson algorithm finds recursively the coefficients for the first-order filter, which we denote as $\hat{a}_1^1$, where the superscript denotes the filter order and the subscript denotes the coefficient number for the order. Next, the algorithm finds the coefficients for the second-order filter, $\hat{a}_1^2$ and $\hat{a}_2^2$, and so on, until the pth order filter, $\hat{a}_1^p, \hat{a}_2^p, \ldots, \hat{a}_p^p$. Note that there are several notations for the filter coefficients besides the one used here. Two others are $\hat{a}_p[1], \hat{a}_p[2], \ldots, \hat{a}_p[p]$, which is used in Kay (1988) and $\hat{a}_{p,1}, \hat{a}_{p,2}, \ldots, \hat{a}_{p,p}$. Thus, the Levinson algorithm gives all the lower order filter coefficients recursively, and it is terminated upon reaching the desired pth order filter. This algorithm requires fewer multiplications and additions than matrix inversion and is thus faster. However, the algorithm does require that the matrix of equations be Toeplitz. Levinson's algorithm is given as follows.

Initialize recursion

$$\hat{a}_1^1 = -\frac{\hat{R}_{XX}(1)}{\hat{R}_{XX}(0)}$$

$$\left[E_T^2\right]^1 = \text{prediction error power for first order filter} = \left(1 - \left(\hat{a}_1^1\right)^2\right)\hat{R}_{XX}(0)$$

Increase order to i = 2, Levinson recursion

$$\hat{a}_i^i = -\frac{\hat{R}_{XX}(i) + \sum_{j=1}^{i-1} \hat{a}_j^{i-1} \hat{R}_{XX}(i-j)}{(E_T^2)^{i-1}} = k_i = \text{reflection coefficient}$$

$$\hat{a}_j^i = \hat{a}_j^{i-1} + \hat{a}_i^i \hat{a}_{i-j}^{i-1}, \quad \text{for } j = 1, 2, \ldots, i-1$$

$$(E_T^2)^i = \left(1 - (\hat{a}_i^i)^2\right)(E_T^2)^{i-1}$$

Increase order by one
if order $= p + 1$ exit

The reflection coefficient is discussed in a later chapter. However, it is determined by having a concatenation of tubes for a vocal tract model. The reflection coefficient determines the amount of volume–velocity reflected at a junction of two tubes. The ith reflection coefficient is given as

$$k_i = \frac{A_{i+1} - A_i}{A_{i+1} + A_i}$$

where A_i is the cross-sectional area of the ith uniform tube. The reflection coefficient is $-1 \le k_i \le 1$. Thus, the reflection coefficient and, therefore, the ith coefficient of the ith order linear prediction filter determine the amount of volume–velocity reflected at a lossless tube junction. The reflection coefficient is also denoted as r_i. The reflection coefficients are also known as the partial correlation coefficients or PARCOR coefficients or the Itakura–Saito reflection coefficients. In the lattice filter problem the computation of the linear prediction coefficients becomes a problem of computing the PARCOR coefficients, which is the correlation coefficient between the forward and backward prediction errors.

The Cholesky decomposition algorithm is used to solve a positive definite matrix, [A], that is not Toeplitz. The idea for this algorithm is to decompose [A] such that [A] = [L][D][L]^H, where [L] is a lower triangular matrix with ones along the main diagonal, [D] is a diagonal matrix with real and positive elements on the main diagonal, and H denotes Hermitian transpose. The set of equations [A]$\vec{x} = \vec{b}$ is solved by two stages of back substitution. Again this algorithm is more efficient than matrix inversion techniques.

Burg's spectral estimator was derived by minimizing the sum of the forward and backward prediction errors in the linear prediction problem. In the material above, we derived the linear prediction filter coefficients by considering only the forward prediction error. However, one can define a backward prediction error in a similar manner. Burg's algorithm derives a set of reflection coefficients. However, these coefficients differ from those estimated by the PARCOR method. The advantage of Burg's approach is that the algorithm finds the reflection coefficients directly from the data, without having to calculate the autocorrelation coefficients. Furthermore, this approach does not need to window the data. The Burg spectral estimator is implemented in the software in Chapter 2.

5.5 ADDITIONAL NOTES ON LP

In the LP model, the error is orthogonal to the prediction term (or data). This is not always valid with real data. For example, voiced speech is a situation where the error is not orthogonal to the prediction term. However, if the speaker's pitch is low, then the error is nearly orthogonal to the prediction term, that is, the cross-correlation function between the speech signal and the error is approximately a delta function. Another way of interpreting this is that the impulse response of the filter has sufficient time to decay (or damp out) before the next glottal excitation occurs. For male voices, this situation is approximately true. However for female and children's voices, this is not the case. The condition of orthogonality is met for the unvoiced case.

An accepted rule for selecting the order of the LP model is due to Markel and Gray (1976).

$$p = F_s + (4 \text{ or } 5), \quad \text{for voiced speech}$$

$$p = F_s, \quad \text{for unvoiced speech}$$

where F_s is the sampling frequency in kHz. Thus, there is one pole pair per kHz of the Nyquist bandwidth, which assumes that there is one formant per kHz. The extra 4 or 5 poles provide "spectral balance" as well as some freedom to adjust the spectrum for spectral roll-off due to the source spectrum.

In speech analysis, especially inverse filtering, the speech signal is often pre-emphasized prior to inverse filtering. There are two major reasons for this procedure. The first is due to the model of the speech production process. Recall that we have assumed that the source can be modeled with two poles and the radiation is modeled as a zero. Since the system is linear, the product of the source-transfer function with the radiation-transfer function leaves one pole, provided we assume one pole of the source approximately cancels the radiation zero. If we now pre-emphasize the speech signal by multiplying it with a transfer function that has a zero at the remaining pole location of the source, then we are left with only the poles of the model that represents the vocal tract filter. A second reason is based on preventing numerical instability. In the autocorrelation method, if the speech signal contains primarily low frequencies, then it is highly predictable. Thus, a large order LP model will result in an ill-conditioned autocorrelation matrix. Thus, pre-emphasis tends to whiten the spectrum, making it flatter and removing some of the spectral tilt due to the source or other causes.

Since the coefficients of A(z) are real, then the roots of A(z) appear in complex conjugate pairs with one pair denoted as $z_i = m_i e^{j\theta_i}$ and $z_i^* = m_i e^{-j\theta_i}$. Then an estimate of the formant frequency is

$$f_i = \frac{\omega_i}{2\pi} = \frac{1}{2\pi}\frac{1}{T}[\arg(z_i)] \text{ Hz} \tag{5.5.1}$$

and the formant bandwidth estimate is

$$b_i = \frac{1}{\pi}\frac{1}{T}|\log(m_i)| \text{ Hz} \tag{5.5.2}$$

When the order of the filter, p, is twice the number of formants in the frequency range 0 to $\frac{1}{2T}$, where T is the sampling interval, the roots of A(z) in the equation are almost exactly the formants. If p is larger than the number of formants, the roots include not only the formants but also some spurious poles corresponding to small peaks in the spectrum and real poles of zero or $\frac{1}{2T}$ Hz. The real poles represent the slope of the spectral envelope. If p is less than the number of formants, some adjacent low-level formants, which very often appear in the high-frequency range, are combined together and approximated with one pole of wide bandwidth. For a telephone voice line, the frequency range is up about 3 to 4 kHz, and usually, p is 8 to 10, but for a high-fidelity voice band that covers 5 to 8 kHz, p is 12 to 16.

Generally, a spurious formant has a relatively wider bandwidth than a true formant. Real poles often appear at 0 or $\frac{1}{2T}$ Hz. They represent the overall slope of the spectral envelope. In all cases, careful selection of the poles is necessary to extract the true formants from the roots of A(z) in order to maintain continuity of formant frequency movement in a frame to frame analysis sequence.

The log area coefficients are given as

$$g_i = \log\left[\frac{A_{i+1}}{A_i}\right] = \log\left[\frac{1 - k_i}{1 + k_i}\right], \quad 1 \le i \le p \tag{5.5.3}$$

We can express the PARCOR coefficients as

$$k_i = \frac{1 - e^{g_i}}{1 + e^{g_i}} \tag{5.5.4}$$

The log area coefficients have found use in speaker recognition systems.

5.6 SUMMARY

Two estimators for the autocorrelation function were presented and discussed. We concluded that the biased estimator is the best.

There are two major power spectral estimators: parametric and non-parametric. The non-parametric spectral estimators include the periodogram and the Blackman–Tukey methods. These methods are often included in a class of classical estimators because they are among the first to be used. The parametric spectral estimators use data modeling methods and are generally considered superior to the classical methods. The parametric procedures require a means to estimate the parameters of the data model that is selected. We used the method of least square, which requires no assumptions about the statistics of the data.

This chapter basically provides a theoretical discussion of some of the software features introduced in Chapter 2 and discussed in Appendix 6.

Consult Furui (1989) and Furui and Sondhi (1992) for additional details.

PROBLEMS

5.1 Show that Equation (5.3.3.20) is equivalent to Equation (5.3.3.18) plus Equation (5.3.3.19).

5.2 This problem is to be completed using analytical methods so that you will gain some insight into the appropriate equations. Suppose you have a sample from a wide sense stationary random sequence as follows.

$$0, 1, 0, -1, 2$$

Determine, $\hat{R}(0)$, $\hat{R}(1)$, $\hat{R}(2)$ using the biased autocorrelation estimate. Next determine the correlation matrix for the autocorrelation method. Finally, determine the sample mean.

5.3 A random process, X(n), is to be approximated by a straight line using the estimate, $\hat{X}(n) = a + bn$. Determine the least squares estimates for a and b if N samples of the error and X(n) are available. Next, suppose you are given the data sequence

$$0, 1, 0, -1, 2, -3, 5, 0, -7, 8$$

Determine the least square straight line fit to the data.

5.4 Suppose a zero mean random sequence, X(n), has an autocorrelation function $R_{XX}(k) = c^{|k|}$. An estimate of X(n) is

$$\hat{X}(2) = -\hat{a}_1 X(1) - \hat{a}_2 X(0)$$

What is the linear prediction estimate for X(2)? What is the total squared error?

5.5 A simple moving average filter processes a zero mean, unit variance random process $X(n)$ such that the filter output is $Y(n) = (\frac{1}{2})[X(n) + X(n-1)]$ for $n = 0, \ldots, N-1$, where $X(-1) = 0$ and $X(N) = 0$ and the $X(n)$ are independent, identically distributed random variables. Find the covariance (autocorrelation) matrix of $Y(n)$. Now let the variance for $X(n)$ be σ_{XX}^2; find the autocorrelation function of $Y(n)$ and the power spectral density.

5.6 Let $R_{XX}(k) = (0.3)^{|k|}$ for $k = 0, 1, 2$. If we model the data with an LP model, we obtain $\hat{a}_1 = -0.3$, $\hat{a}_2 = 0$, $\hat{a}_3 = 0$. Consider the autocorrelation prediction Equation (5.3.3.27) and calculate $\hat{R}_{XX}(3)$. Discuss the implications of this result.

5.7 Derive the following impulse response for the first order $\frac{1}{A(z)}$ prediction filter.

$$h(n) = \delta(n) - \hat{a}_1 h(n-1)$$

where $\delta(n)$ is the unit pulse. Let $h(-1) = 0$. Let $\hat{a}_1 = -0.3$. Analytically generate 10 samples of $h(n)$. Now suppose you have a data sequence

$$0.2, 0.5, 1.1, 0.7, -0.1$$

Determine the error sequence. Calculate the correlation matrix using $\hat{a}_1$ only. (There are two unknowns, and one equation. What can you do?) Repeat the problem for a second-order filter, where

$$h(n) = \delta(n) - \hat{a}_1 h(n-1) - \hat{a}_2 h(n-2)$$

where $h(-1) = h(-2) = 0$ and $\hat{a}_1 = -0.625$ and $\hat{a}_2 = 0.25$.

5.8 The purpose of this problem is to estimate the filter coefficients for a second-order filter, i.e., $p = 2$. Assume you are given $\hat{R}_{XX}(0)$, $\hat{R}_{XX}(1)$, and $\hat{R}_{XX}(2)$. Determine the filter coefficients by matrix inversion. Repeat the estimation using Levinson's algorithm.

5.9 Calculate and plot the periodogram for each of the windows in MATLAB, Bartlett, triangle, kaiser, hanning, hamming, chebyshev, boxcar (rectangular), and blackman. Which window has the lowest sidelobes? Which window has the narrowest main lobe? Which window has the widest main lobe?

5.10 Generate a 100 point data sequence using randn in MATLAB. Calculate the total squared error for a linear prediction model of the data for $p = 1, 2, \ldots, 10$. Plot the error versus p. What value of p has the smallest error? Compare your results with the Akaike information criterion.

5.11 Calculate the spectrograms for the files m0125s.dat and m0127s. Identify and label the vowel regions.

5.12 Generate a 100 point random sequence using rand in MATLAB. Use a first-order LP filter to filter this random sequence. The filter is

$$A(z) = \frac{1}{1 + a_1 z^{-1}}$$

Let $a_1 = -0.1, -0.3$, and -0.7. Calculate and compare the autocorrelation function for the output of the three filters. Do the results agree with the theory? Repeat this problem using randn in MATLAB. Are there any significant differences in your results?

5.13 Suppose we have the following correlation matrix

$$\begin{bmatrix} 1.0 & 0.3 \\ 0.3 & 1.0 \end{bmatrix}$$

Determine the first order LP model. Calculate $\hat{R}_{XX}(2)$ using the autocorrelation prediction Equation (5.3.3.27). Now suppose you are told the correlation matrix is

$$\begin{bmatrix} 1.0 & 0.3 & (0.3)^4 \\ 0.3 & 1.0 & 0.3 \\ (0.3)^4 & 0.3 & 1.0 \end{bmatrix}$$

Compare $\hat{R}_{XX}(2)$ to the true value. Now determine the second order LP model. Calculate $\hat{R}_{XX}(3)$ using Equation (5.3.3.27). Plot and compare the true values of the autocorrelation function with the predicted values.

5.14   Analyze a segment of the vowel /IY/ in the data file m0103s.dat. Use LP models with p = 10, 11, 12, 13, and 14. Window the data with a hamming window prior to analysis. Calculate and plot the LP spectra and compare with the periodogram. Calculate and plot the total squared error versus the order, p. The first three resonances for this vowel (/IY/ as in see) occur at approximately 270, 2290, and 3010 Hz. Which model order appears to give the best results? Does this agree with the periodogram results? Does this agree with the error versus p results? Now that you have the "best" LP model for this data, predict and plot the unwindowed data beyond the last data value in the file, m0103s.dat. Comment on this prediction. For how many samples does the prediction appear to work well? Repeat this problem for the vowel /AA/ (as in Bach) in file m0113s.dat. The first three resonances for this vowel occur at approximately 570, 840, and 2410 Hz. Note that this problem can be done using the software in Chapter 2.

5.15   Apply a hamming window to the speech file m0103s.dat. Find the LP filter $\frac{1}{A(z)}$. Now inverse filter the speech signal and plot the data sequence, e(n). Discuss your result including why the sequence has peaks where it does. Hint: compare the location of the peaks in e(n) with the speech data. This problem can be done using the software in Chapter 2.

5.16   The purpose of this problem is to compare the various spectral analysis techniques for the vowel /IY/, file m0103s.dat. Use the software in Chapter 2. Load the file. Calculate and plot the following: spectra: FFT, periodogram, Blackman–Tukey. Use the following parameter settings: Frame length: 128, Hamming window, Select starting point (which is to be near the middle of the data file). Repeat using frame lengths of 256, 512, and 1024. Compare and discuss your results. Does frame length have an effect?

5.17   Repeat Problem 5.16 but change the hamming window to a rectangular window. Compare and discuss your results. Next compare and discuss the results of Problem 5.16 with the results obtained in this problem. What do you conclude about frame length and window type?

5.18   The purpose of this problem is to compare the autocorrelation and covariance algorithms for LP spectral analysis for the vowel /IY/, file m0103s.dat. Use the Chapter 2 software. Load the file. Calculate the following: LP analysis: autocorrelation and covariance. Use the following parameter settings: frame length: 256, number of poles: 10, 11, 12, 13, 14, 15, 16, select starting point (which is to be near the middle of the data file). Compare and discuss the results for the two algorithms and the seven different settings for the number of poles. Which algorithm appears to be "best"? What value of number of poles appears to be "best"?

5.19   The purpose of this problem is to calculate the model order selection for LP analysis. Use the vowel /IY/, file m0103s.dat. Use the Chapter 2 software. Load the file. Select the LP analysis property window. The settings in the property window are not important. Simply press the Order Selection button. The analyses for the three criteria are calculated and displayed for p = 10 to 14. What value of p gives the least error? Compare these results with the results of Problem 5.18.

5.20   The purpose of this problem is to compare the results for two ARMA spectral analysis techniques with the LP results for Problem 5.18. Use the vowel /IY/, file m0103s.dat. Use the Chapter 2 software. Load the file. Calculate the following: ARMA: Akaike MLE and MYWE. Use the following parameter settings: Frame length: 256, Number of poles: 10, 12, 14, Number of zeros: 2, 4, 6, Hamming window, Select starting point (which is to be the middle of the data file). Compare and discuss the results for this problem with those for Problem 5.18. Does the ARMA analysis improve the estimation of the spectral envelope over that using LP analysis?

5.21   The purpose of this problem is to compare the results for MA analysis with the LP for Problem 5.18. Use the vowel /IY/, file m0103s.dat. Use the Chapter 2 software. Load the file. Calculate the MA spectrum. Use the following parameter settings: Frame length: 256, Number of zeros: 4, 6, 8, Hamming window, Select starting point (which is to be near the middle of the data file). Compare and discuss the results for this problem with those for Problem 5.18. Is LP analysis superior to MA analysis for speech?

5.22   The purpose of this problem is to compare the results for inverse filtering using three algorithms: pitch synchronous (manual), pitch asynchronous (manual), and asynchronous analysis. Use the vowel /IY/, file m0103s.dat. Use the Chapter 2 software. Load the file. Calculate the inverse filtered waveforms and the pitch and formant frequency contours. Follow the instructions in Chapter 2. Use the following parameter settings: Property window: Frame length: 256, Overlap: 56, Poles: 14, Zeros: 0 and 4, Between marks: (select a segment near the middle of the data file of about 2000 to 3000 samples in length). Compare and discuss your results. Does the inverse filtered waveform appear reasonable (ignore any dc offset)? How about the pitch and formant contours? Compare the formant contours with the spectrogram for the data analyzed.

5.23   The purpose of this problem is to compare the Burg spectral analysis algorithm with the LP spectral analysis algorithm for the vowel /IY/, file m0103s.dat. Use the Chapter 2 software. Load the file. Calculate the Burg analysis using the following parameter settings: Frame length: 256, Number of poles: 10, 11, 12, 13, 14, 15, 16, Select starting point (which is to be near the middle of the data file). Compare and discuss the results for the Burg algorithm and the seven different settings for the number of poles with those for Problem 5.18. Which algorithm appears to be "best"? What value of number of poles appears to be "best"?

RESEARCH PROBLEMS

R5.1   Read the paper in Appendix 6 on WRLS-VFF analysis. Conduct a WRLS-VFF analysis of the data used in Problem 5.22. Compare the results with those of Problem 5.22.

R5.2   Obtain a copy of the Hermansky paper on perceptual linear prediction (PLP). Read the paper. Repeat Problem 5.18 using the PLP method. The algorithm is available in the Chapter 2 software. Compare the results of this problem with those for Problem 5.18.

R5.3   Derive the Levinson algorithm. Help is available in several texts.

CHAPTER 6

SPEECH SYNTHESIS AND A FORMANT SPEECH SYNTHESIS TOOLBOX

6.1 INTRODUCTION

Speech synthesis can be defined as the process of creating a synthetic, acoustic replica of a specified speech signal or of a typed text. One of the first, if not the first, attempts to synthesize speech was by Wolfgang Von Kempelen (1791) (Paget, 1930) who invented a mechanical talking machine that could speak whole phrases in French and Italian. During the 19th century, a number of studies were made to investigate the generation of vowel sounds by using mechanical devices (Linggard, 1985; Paget, 1930). Following 1930, researchers, with the aid of electronic instrumentation such as the oscilloscope, began to better understand speech acoustics. This knowledge resulted in the construction of elementary circuit-based speech synthesizers. Dudley's (1936) 10-channel voice coder (vocoder) was the first electronic speech synthesizer. Due to its flexibility, the electronic speech synthesizer outperformed conventional mechanical synthesizers of the time. Dunn (1950) improved the quality of the vocoder. In 1960, the availability of the digital computer made it possible to adopt software programs to perform speech synthesis. The computer allowed speech scientists to implement and evaluate a variety of speech synthesizer designs and encouraged applications of speech synthesis. Several reviews of speech synthesis appear in the following references (Bailly and Benoit, 1992; Bristow, 1984; Carlson, 1993; Cater, 1983; Flanagan 1982; Flanagan, 1972a; Flanagan, 1972b; Flanagan and Rabiner, 1973; Flanagan, et al., 1970; Klatt, 1987; Linggard, 1985; Morgan, 1984; Paget, 1930; Schroeder, 1993; van Santen et al., 1997).

Speech synthesis can be classified by methodology, for example, articulatory, formant, linear prediction, sinusoidal, and so on. Text-to-speech synthesis may use any of these techniques to generate an acoustic signal. However, text-to-speech systems have a number of other procedures that must be implemented as well (Allen et al., 1987; Klatt, 1987; Moulines and Charpentier, 1990).

6.1.1 Articulatory Synthesis

Models of the vocal folds describe the dynamic vibrations of the vocal folds (Flanagan and Ishizaka, 1978; Sondhi, 1975; Titze, 1982; C. J. Wu, 1996). We present in Chapter 9 several such software models. The parameters of these complicated source models are useful for assisting the interpretation of the physiologic and pathologic phenomena of the vibratory motion of the vocal folds. The articulatory model in recent years is based on modeling the movement of the articulators (Allen and Strong, 1985; Bavegard, 1996; Coker, 1976; Coker, 1967; Coker and Fujimura, 1966; Mermelstein, 1973; Parthasarathy and Coker, 1992;

Sondhi and Schroeter, 1986, 1987). The vocal tract cross-sectional area function, $A(x)$, is often specified at 20 to 60 separate points along the oral tube. The area function is usually estimated by acoustic optimization procedures (Atal et al., 1978; Fant, 1960; Gopinath and Sondhi, 1970; Hogden et al., 1996; Hsieh, 1994; Kaburagi and Honda, 1996; Levinson and Schmidt, 1983; McGowan and Lee, 1996; Wu, 1996). However, a small amount of data is available from such sources as x-ray photography and magnetic resonance imaging to validate the cross-sectional area functions estimated using acoustic analysis techniques (Baer et al., 1991; Fant, 1960; Story and Titse, 1996). In a discrete realization, the cross-sectional area function, $A(x)$, is updated every 20 to 50 msec. The interpolation of the area function in both time and space is crucial for synthesizing high-quality speech (Gupta and Schroeter, 1993; Linggard, 1985; Schroeter et al., 1988; Schroeter et al., 1987).

Articulatory synthesis can be classified into several categories, each of which simulates the basic building blocks (excitation source, vocal tract, and radiation filter). These include the concatenation of acoustic tubes (Kelly and Lochbaum, 1962; Maeda, 1982a; Maeda, 1982b; Maeda, 1977; Rubin et al, 1981; Strube, 1982). A second approach models the acoustic properties of the glottis and vocal tract with a set of differential equations (Bocchieri and Childers, 1984; Childers and Ding, 1991; Flanagan and Cherry, 1968; Flanagan and Ishizaka, 1978; Flanagan and Ishizaka, 1976; Flanagan and Landgraf, 1968; Flanagan et al., 1980; Koizumi et al., 1985; Rothenberg, 1981). A third approach is a hybrid time–frequency domain method (Allen and Strong, 1985; Lin, 1992; Sondhi and Schroeter, 1987). Yet a fourth method is a wave digital filter approach (Fettweis and Meerkötter, 1975; Lawson and Mirzai, 1990; Liljencrants, 1985; Meyer and Strube, 1984; Meyer et al., 1989). The basic structure for articulatory speech synthesis is depicted in Figure 6.1.

For the articulatory model, the vocal tract is divided into many small sections and the corresponding cross-sectional areas are used as parameters to represent the vocal tract characteristics. In the acoustic model, each cross-sectional area is approximated by an electrical analog transmission line. To simulate the movement of the vocal tract, the area functions must change with time. Each sound is designated in terms of a target configuration and the movement of the vocal tract is specified by the motion of the articulators.

At the present time, the complexity of articulatory synthesis is partially due to the analysis procedure, which usually requires an "articulatory-to-acoustic inverse transformation" from the speech signal, that is, speech inverse filtering. The complexity of the relationship between articulatory gestures and the acoustic signal makes it very difficult

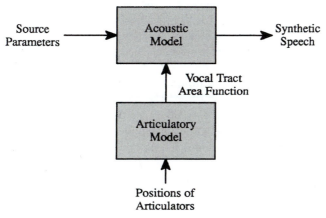

FIGURE 6.1 Basic structure for articulatory speech synthesis.

to automatically generate the details of articulatory control needed to produce a synthetic copy of a given sample of human speech. Despite such drawbacks, articulatory speech synthesis has several advantages. The model has a direct relation to the human speech production process. Consequently, it is conjectured that articulatory synthesis may lead to a simple and elegant synthesis by rule, e.g., text-to-speech applications (Parthasarathy and Coker, 1990; 1992) and articulation-based speech recognition systems (Erler and Deng, 1993). The articulatory parameters in the human voice production system vary slowly. Consequently, researchers have suggested that these parameters are potential candidates for efficient coding, for example, low bit-rate speech communication (Flanagan et al., 1980). To the extent that we can accurately represent the speech gestures (articulatory movements or trajectories), articulatory synthesizers may be valuable for research scientists and physicians, since such synthesizers can be used to study linguistic theories, to provide a feedback mechanism for teaching speech production, and to explore the effects of vocal tract surgical techniques on speech production prior to surgical intervention (Childers, 1991). They hold out the ultimate promise of high quality, natural-sounding speech with a simple control scheme (Klatt, 1987). A properly constructed articulatory synthesizer is capable of reproducing all the naturally relevant effects for the generation of fricatives and plosives, modeling coarticulation transitions, as well as source-tract interaction in a manner that resembles the physical process that occurs in real speech production. Articulatory synthesizers will continue to be of great importance for research purposes and to provide insights into various acoustic features of human speech. Thus, an articulatory synthesizer may provide both an efficient description of natural speech and a means for synthesizing natural-sounding speech. However, a major problem with the articulatory synthesizer is the lack of a means to derive articulatory configurations from the speech signal using speech inverse filtering. We present in Chapter 10 a software implementation of an articulatory speech synthesizer.

6.1.2 Formant Synthesis

A simplified approximation of the speech production mechanism in the acoustic domain was proposed in the late 1950s and called the source-tract or source-filter model (Fant, 1960). In this model, the speech production system is divided into two parts: (1) the excitation source; and (2) the resonant tract. These two parts are assumed to be non-interactive and linearly connected. The formants are the resonances of the vocal tract. The formant synthesizer reproduces the formant structure of speech. The history of the formant synthesizer is closely related to the evolution of electronic technology. In the late 1930s, "terminal-analog" synthesizers were built using analog electrical networks. These analog networks were serial or parallel combinations of second-order resonators. A series of impulse-like waveforms and white noise was applied to the resonators in order to generate vowels and fricative sounds, respectively.

In the 1960s, discrete realizations of formant synthesizers appeared (Flanagan et al., 1962; Gold and Rabiner, 1968; Rabiner, 1968). The resonators for the formant synthesizer were arranged in either a cascade or parallel manner (Flanagan, 1957; Holmes, 1983; Holmes, 1987; Holmes, 1988; Klatt, 1980; Rye and Holmes, 1982). Flanagan (1957) concluded that the serial form is a better model for non-nasal voiced sounds, while the parallel structure is superior for nasal and unvoiced sounds. The reason is that the vocal tract is considered as an all-pole filter for non-nasal voiced sounds and as a pole-zero system for other phonations. Thus, it is quite simple to use the cascade structure to simulate an all-pole system and the parallel form to implement a pole-zero system. Klatt's 1980 system combined the cascade and the parallel structures. Anti-resonators were added to the cascade branch to enhance the ability of the cascade configuration to model nasal and unvoiced sounds. By properly specifying the synthesis variables and using the correct configuration, this synthesizer is capable of synthesizing high quality, intelligible speech (Pinto et al., 1989).

A major factor in achieving high-quality synthetic speech is the extraction of synthesis parameters from the speech signal via accurate analysis procedures. Most of these procedures use the acoustic speech signal as the source for determining the formants (Alku, 1992; Childers and Lee, 1991; Klatt and Klatt, 1990; Markel and Gray, 1974; McCandless, 1974; Olive, 1971). Unfortunately, almost all of these procedures have only addressed the extraction process for vowel-like sounds. Another important factor for achieving high-quality speech synthesis is the design of the excitation waveform (Childers, 1995; Childers and Ahn, 1995; Childers and Hu, 1994; Childers and Lee, 1991; Childers and Wu, 1990). Formant synthesis software is available (e.g., Sensimetrics, Corp., Cambridge, MA) and a few commercial applications are available, such as DECtalk from Digital Equipment Corporation (Maynard, MA). This chapter and the next also describe a software implementation for formant synthesis.

In summary, at present, there are two commonly used formant synthesis models, the cascade/parallel formant synthesizer, pictured in Figure 6.2, developed by Klatt (1980) and Klatt and Klatt (1990) and the versatile parallel formant synthesizer, shown in Figure 6.3, developed by Rye and Holmes (1982). Although there has been dissent about which of the two systems is the better (Holmes, 1983), it is generally agreed that the Klatt model is favored for text-to-speech synthesis, while the Holmes model tends to be used for synthesis-by-analysis systems. The reasons for this are probably related more to the way in which the different synthesis models are controlled, rather than the inherent capabilities of the synthesizers themselves. The formant frequencies, amplitudes, and bandwidths can be

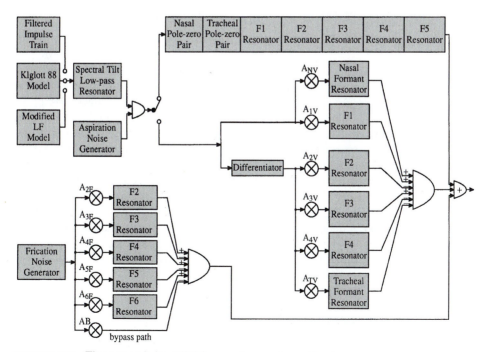

FIGURE 6.2 The cascade/parallel formant speech synthesizer.

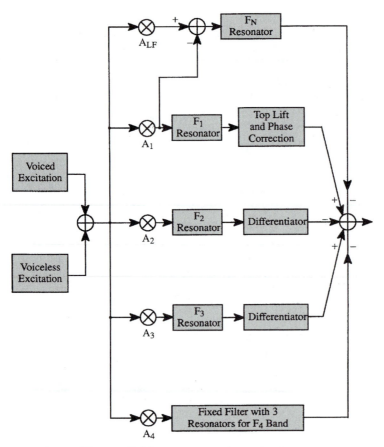

FIGURE 6.3   The parallel formant speech synthesizer.

implemented in the form of a digital filter. The excitation source and the spectral shaping network (filter) that make up a formant synthesizer must be varied dynamically to mimic the changes that occur in the source characteristics and the vocal tract shape during speech production. Since these changes occur relatively slowly, it is possible to use a set of synthesizer parameters (as control signals for the formant synthesizer) to specify a short segment (frame) of the speech signal. These parameters can be used to reduce the amount of data needed to represent the speech signal so that the data rate is approximately 3 kbits/sec. It has been demonstrated that high-quality speech can be generated with such synthesizers (Klatt, 1980; Klatt and Klatt, 1990; Pinto et al., 1989). However, the control tables are complicated. Clearly, such a model has no simple relationship to an articulatory specification of the vocal tract. Although it cannot properly represent the effects of varying glottal impedance or subglottal coupling, or the subtleties of vocal fold motion, many successful speech-synthesis systems are based on formant synthesis since it is possible to make a functional approximation to these effects.

6.1.3 LP Synthesis

The linear predictive (LP) synthesizer is a mathematical all-pole realization of the linear source-tract model (Atal and Hanauer, 1971). Two types of excitation sources are switched at the input of the all-pole system to generate voiced or unvoiced sounds. For voiced sounds,

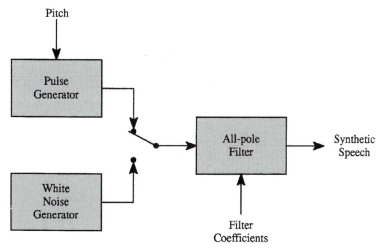

FIGURE 6.4 Basic structure for LP speech synthesis.

the excitation is in the form of a periodic train of pulses sounds (Fujisaki and Ljungqvist, 1986). For unvoiced sounds, the excitation source is generated by random noise, which is controlled by a gain parameter and a spectral parameter. The vocal tract transfer function is characterized by the LP coefficients. The LP model is quite good for speech synthesis and is used extensively in speech coding applications (Atal and Remde, 1982; Childers and Hu, 1994; Childers and Wu, 1990; Singhal and Atal, 1989; Rose and Barnwell, 1990). The basic structure for the LP model is discussed in Chapter 5 and is shown again in Figure 6.4.

Olive (1992) has proposed a mixed spectral representation (formant and LP) to make use of the benefits of both the formant and the LP synthesis schemes. Basically, Olive's strategy is to use the high-order LP scheme for synthesizing unvoiced sounds and the formant scheme for voiced phonations. By carefully considering the discontinuity problem that might arise at the boundary of voiced and unvoiced sounds, Olive (1992) claims that the synthesizer can be used for analysis-synthesis applications and produces high-quality synthesized speech.

6.1.4 Miscellaneous Items

In 1986, McAulay and Quatieri (1986) developed a sinusoidal model for speech analysis and synthesis. This method has found use for speech transformations, such a time-scale and pitch-scale modifications. Moulines and Charpentier (1990) suggested the pitch-synchronous overlap-add (PSOLA) approach for text-to-speech applications. This approach can modify the prosody of the speech and is able to concatenate speech waveforms. The speech is modified in either the time domain or the frequency domain. One application of speech synthesis is called voice conversion or transformation (Childers, 1995; Childers et al., 1989; Iwahashi and Sagisaka, 1995; Kuwabara and Sagisaka, 1995; Mizuno and Abe, 1995; Moulines and Laroche, 1995; Narendranath et al., 1995). Some recent computer applications for e-mail and Web applications are examining the use of concatenated speech segments for speech synthesis. This is an old approach that has been revitalized because computational power is now much less expensive than in years past and storage is also more readily available. Applications are appearing now that have animated, speaking "helpers" for word processors, Web applications, and e-mail. These are text-to-speech systems with special features added. A few examples of these modern synthesizers include:

AT&T, http://www.att.com (this is a synthesis demo of a child, man, woman, and singer), Microsoft Agent beta, http://www.microsoft.com/intdev/agent (this works with Microsoft Internet Explorer 4.0 and contains an animated talking "genie").

6.1.5 Speech Analysis-By-Synthesis

There are several methods for estimating the speech production model parameters from its input and/or output. If both the input and output of the model are given, then the problem is reduced to a system identification problem whereby the design of the model and the estimation of its parameters are the major concern. However, when only the output is known, the problem is an optimization problem, which is the case for speech production modeling. For speech research, the analysis-by-synthesis method often gives good results for estimating the model parameters. The analysis-by-synthesis method for the estimation of the articulatory parameters from the speech signal is called the speech inverse problem. This is illustrated in Figure 6.5. In the speech inverse problem, the excitation, due to the glottal source, is to be estimated. This excitation is either a quasi-periodic pulse train for voiced speech or random noise for unvoiced speech. For the articulatory model, the parameters for speech production are the position vectors of the articulators, which determine the shape of the vocal tract. For the linear prediction model, the parameters for speech production are the LP coefficients. In order to estimate the input excitation waveform and the model parameters, initial estimates of the input waveform and articulatory (or LP) model parameters are made and used to synthesize speech. This synthesized speech is compared with the target (actual or other synthetic) speech and an error is calculated. Then the initial input waveform and model parameters are adjusted iteratively to reduce the error to a predefined threshold value. The error between the synthesized speech and the target speech can be defined in numerous ways. However, there is no guarantee that the parameters obtained by such a procedure are optimal, since the optimization procedure might be the result of a local error minima. Many algorithms have been suggested to solve this problem. Another difficult problem is that the mapping from the model parameters to the speech signal is not a one-to-one mapping. This means that there can be more than one (inverse) solution for one target speech (Atal et al., 1978). The nonunique property of the speech inverse problem can be solved by using proper initial estimates of the input and model parameters (Sorokin, 1994).

Most speech analysis and synthesis procedures have focused on voiced speech because such data is well approximated as a quasi-stationary, which is relatively easy to analyze. Numerous algorithms have been proposed and many of them give successful results for voiced speech. The vocal tract transfer function can be estimated using various spectral estimation algorithms and the glottal source waveform can be extracted by glottal inverse filtering algorithms. The inverse filter is usually obtained from an estimate of the vocal tract transfer function. On the other hand, research on unvoiced speech has not received much attention, mainly because adequate models are not available (Lee, 1996). In addition, the length of unvoiced speech segments is usually short, making it more difficult to analyze than the voiced speech. Thus, estimating the articulatory (or LP) parameters for the unvoiced speech has not been as successful as that for voiced speech.

6.2 CHARACTERISTICS OF VOICE

For speech synthesis several factors become important for representing or replicating the characteristics of a particular voice. One factor is the overall vocal tract dimensions, which determines the formant frequencies and their bandwidths. Another factor is the vibratory

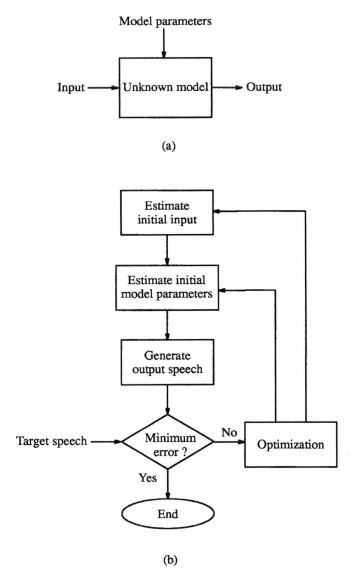

Model parameters

Input → Unknown model → Output

(a)

Estimate initial input

Estimate initial model parameters

Generate output speech

Target speech → Minimum error ? — No → Optimization

Yes

End

(b)

FIGURE 6.5 An outline of an analysis-by-synthesis procedure. (*a*) Overall problem. (*b*) An algorithm.

patterns of the vocal folds, which is affected by the mass and tension of folds. The fundamental frequency of the vocal folds can be estimated using the formula for a vibrating string,

$$F_0 = \frac{1}{2L}\sqrt{\frac{\sigma}{\rho}} \qquad (6.2.1)$$

where L is the length of the vocal folds, σ is the longitudinal stress in the vocal fold tissue, and ρ is the tissue density. Thus, the fundamental frequency is inversely proportional to the vocal fold length and directly proportional to the square root of tissue stress. In addition, the variations in the dimensions of the subglottal apparatus affect the glottal pulse width, pulse skewness, abruptness of closure, and the spectral tilt of the glottal pulse (Ishizaka and

Flanagan, 1972). A third factor is related to the dynamics of speech production, which is affected by the speaker's articulatory skills and speaking habits, which in turn affect accents and dialects and can influence prosodic variations such as intonation, stress, and duration of speech.

Speech synthesis and voice conversion systems should account for these factors. Several such systems include Kuwabara (1984), which used a linear prediction model for speech synthesis. Another system was developed by Childers and co-workers (1989), which focused on male-to-female and female-to-male voice conversion. Abe and colleagues (1988) also investigated voice conversion using codebooks to represent various speaker's vocal characteristics. A similar approach was taken by Savic and Nam (1991) using neural nets. Valbret and co-workers (1992) replaced the excitation source of an LPC vocoder with a residue type waveform.

In the software used in this book, we use three sets of speech parameters: one for the vocal tract, one for the voicing source, and one for prosodic features. We synthesize the desired signal using these three sets of parameters. For voice conversion, discussed in the next chapter, we map (convert) the parameters from the acoustic space for one speaker (source) to the acoustic space of another speaker (target).

6.3 SPEECH ANALYSIS AND SYNTHESIS

The speech analysis and synthesis system introduced in this chapter and expanded on in the following chapter is an outgrowth of Klatt (1980), Olive (1992), and Childers and Hu (1994). Here we describe the general features of the system for voice conversion used in Chapter 7. However, the formant synthesizer described at the end of this chapter is a simplified version, which is intended as an introduction to formant speech synthesis.

For the more general system, the user can select either a formant or an LP synthesizer for voiced sounds, while the LP synthesizer is used for unvoiced sounds. For the voiced excitation source, we use either the LF model or the polynomial model (see Appendix 7). Also included in the speech synthesizer is a control model to control voiced/unvoiced classification and the pitch and gain contours. This feature assists the user in mimicking the speaking style used by various speakers. The speech parameters are obtained by an analysis-by-synthesis procedure. The speech parameters are grouped into three categories: excitation control parameters, excitation source parameters, and vocal tract parameters. During synthesis, the latter two groups of parameters are updated at the beginning of each pitch period, while the first group controls the rate of updating. Figure 6.6 illustrates the speech synthesizer process. The sets of parameters can be altered or modified independently in the parameter modifier portion of the speech synthesis or voice conversion system.

6.3.1 Excitation Source Model

For the speech synthesizer, we need two excitation sources: one for voiced speech and one for unvoiced speech. For unvoiced speech we use a stochastic codebook, which codes the residue using a Gaussian noise generator (Childers and Hu, 1994). Two models are used for voiced excitation: the LF model (Fant et al., 1985) and a polynomial model (Childers and Hu, 1994; Milenkovic, 1993). The latter two models are described in Appendix 7. Other models for voiced excitation include the Rosenberg (1971), Rothenberg and colleagues (1973), Rosenberg and co-workers (1975), Rothenberg (1981), Fant and co-workers (1985), and Klatt and Klatt (1990) models.

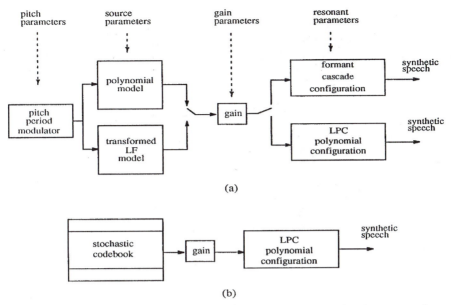

FIGURE 6.6 The speech synthesizer for speech synthesis and voice conversion. (*a*) Voiced sounds. (*b*) Unvoiced sounds.

6.3.2 Control Parameters

There are three types of parameters that control the excitation functions.

1. Voiced/unvoiced classification, v_c. The voiced/unvoiced classification determines which synthesis scheme (voiced or unvoiced) is adopted for each synthesis frame.
2. Gain parameter, g. The human aural system is sensitive to the intensity of speech, the gain parameter (either voiced or unvoiced) is used to control the intensity of synthesized speech.
3. Pitch parameters, p_p. The pitch (period) parameter, p_c, is the parameter that determines the length of the glottal excitation waveform. For unvoiced sounds, the pitch period is fixed at 5 msec.

6.3.3 Excitation Parameters

Either the polynomial model parameters, $c_0, c_1, c_2, c_3, c_4, c_5, c_6$, for the 6th-order polynomial model, or the LF model parameters, specify the shape of the glottal waveform. Note that only one source model is employed in speech synthesis. The user selects the model to be used.

6.3.4 Resonant Tract Parameters

For synthesizing voiced sounds, the formant frequency and bandwidth parameters, $f_1, f_2, f_3, f_4, f_5, b_1, b_2, b_2, b_4, b_5$, determine the resonant frequencies and bandwidths in Hz for the first five resonators of the vocal tract for the formant configuration. The parameters, $a_1, a_2, a_3, a_4, a_5, a_6, a_7, a_8, a_9, a_{10}, a_{11}, a_{12}, a_{13}$, represent the LP coefficients for the LP configuration.

6.3.5 LP Analysis

A frame based LP analysis is used. The speech signal is normalized by the maximum amplitude and segmented into 25 msec frames with a 5 msec overlap. If the final frame is less than 25 msec, then a random noise (30 dB below the peak amplitude) is appended to the frame to fill it out to 25 msec. The speech signal is next filtered with a zero phase filter to remove any low-frequency drift. A 13th order LP model is used for analysis and an orthogonal covariance method is used to calculate the LP coefficients (Ning and Whiting, 1990). The residue in the overlap area is obtained by weighting the forward and backward overlapped sequences. Next, each frame is classified as voiced or unvoiced using the residue signal by simply setting a threshold. Then the pitch period and glottal closure instants (GCIs) are calculated. The GCIs are determined by peak picking the voiced residue signal. The pitch period is estimated using the cepstrum of the residue signal. This data can be used to segment the speech pitch synchronously. And the LP parameters can be estimated pitch synchronously as well. For a formant synthesizer the first five formants can be estimated from the LP polynomial.

The excitation waveform is estimated as follows. For unvoiced sounds, the residue signal is used to find the optimal codeword in the stochastic codebook. For voiced sounds, glottal inverse filtering is used. Two methods are used, one for the formant synthesizer and one for the LP synthesizer.

During speech synthesis, only one excitation waveform is used to excite the vocal tract filter for each pitch period. This can result in large discontinuities in the glottal phase characteristic at frame boundaries. This is alleviated by using an infinite impulse response (IIR) filter to smooth rapid changes in the source parameters.

Vocal noise, which is important for the naturalness of synthesized speech, is added to the voiced excitation. The amplitude of the noise is adjusted to achieve a signal-to-noise ratio of 25 dB. The noise is produced by modulating uniformly distributed white noise with a Gaussian window.

The gain parameter is a function that modulates the power of the excitation waveform and is an important factor affecting the quality of synthesized speech. For an all-pole LP filter, the output can be considered to be composed of two components: one due to the input data sequence and one due to the filter memory. To adjust the gain we use two synthesis filters: one holding the previous LP coefficients to account for the filter memory and one for processing the current excitation using the current LP coefficients. For speech synthesis we use only one vocal tract filter for each pitch period. However, the length of the excitation pulse is extended to twice the pitch period by appending zeros. Consequently, the output filter is twice as long as the pitch period. The first part of the output is due to the current excitation, while the second part of the output accounts for the filter memory. The gain for the excitation is determined by subtracting the memory contribution from the total power. In summary, the synthesized speech for the present pitch period is the current filter output plus the memory contribution from the previous excitation.

6.3.6 Summary of Analysis and Synthesis

In summary, depending on the classification of the speech frame as voiced/unvoiced, one of two excitations is created using the excitation source. If the classification is voiced, the control model determines the length of the pitch period, the starting time of each excitation pulse, as well as the excitation gain. The shape of the excitation pulse is controlled by the source model. For unvoiced sounds, the stochastic codebook supplies the excitation. For voiced sounds, the waveform model generates the excitation pulse according to the source parameters.

There are two configurations for the resonant (vocal) tract that models the slow-varying frequency response of the vocal tract. The vocal tract filter for unvoiced sounds is obtained from a 13th order LP analysis. For voiced sounds, the filter is either derived from LP analysis (LP configuration) or a polynomial expansion process (formant configuration). The latter cascades five second-order polynomials. Each second-order polynomial is associated with a specific set of formant frequencies and bandwidths. Thus, there are two types of resonant tract configurations for our system. The user can select either a formant or an LP configuration for voiced sounds, while the LP representation is used for unvoiced sounds. The user can also select either the polynomial model or the transformed LF model for the excitation source.

6.4 FORMANT SYNTHESIZER TOOLBOX

The formant synthesizer, described below, is implemented as a cascade all-pole filter (formant filter) that can be excited by two excitation sources, a LF glottal waveform generator and an excitation noise generator. For voiced excitation, both excitation generators can be used in combination. For unvoiced excitation, only the noise waveform generator is used. A block diagram of the system and an overview of its operation are shown in Figure 6.7. See Appendix 7 for a description of the LF model and the noise generator. Note that the polynomial excitation model is not available for this formant synthesizer. Also see Appendix 8 for additional detail on the synthesizers outlined here and used in the next chapter on voice conversion.

As with the other software in this text, copy the folder named formant to a directory of your choice. Start MATLAB. Change directory to the formant folder in the Matlab command window. Type main. The main Graphic User Interface (GUI) opens as shown in Figure 6.8. This main window consists of three function buttons that let the user specify the parameters for synthesizing speech, and one button to quit/close the window. The sequence of operations is described as follows.

Pressing the Source Specification button, opens the window shown in Figure 6.9. The purpose of this window is to generate the source (excitation) waveform for the formant synthesizer. To generate a new excitation waveform, press the Specify New Excitation button. This opens the window shown in Figure 6.10, which is the new excitation window. The only option here is to specify the number of frames to be synthesized. The sampling frequency is assumed to be 10 kHz. The default number of frames is 20, which appears in the upper box. The slider controls the specification of the new value, or the user can highlight

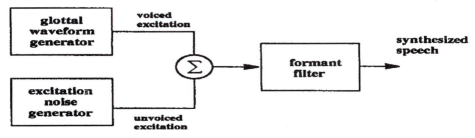

FIGURE 6.7 A block diagram of the formant synthesizer.

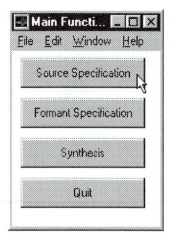

FIGURE 6.8 The main function window.

FIGURE 6.9 The source menu window.

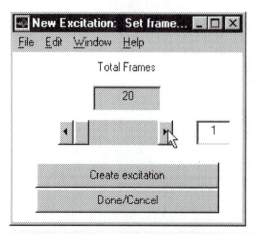

FIGURE 6.10 New excitation window to specify the number of frames.

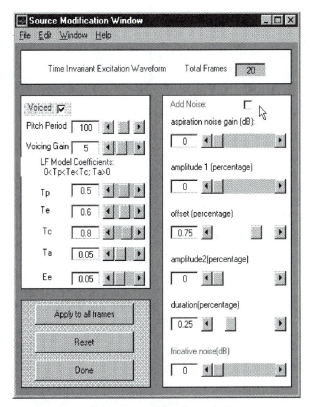

FIGURE 6.11 Source modification window.

the number in the box to the right of the slider and type in a new value. The smallest number of frames is 1 and the largest is 200. These values can be changed in the appropriate m-file. Once the number of frames is specified, the user presses the Create button. Pressing the Done button closes the window, whereby a message window appears stating that a default source file has been created and the user can modify this file. It is not necessary to press the Done button, since the window will be closed after the user closes the modify source window, which is discussed next. In its present configuration, it is not possible to synthesize words with this formant synthesizer. However, a slight change would allow this. However, this seems unnecessary, since the synthesizers in Chapters 7 and 8 do provide this capability.

Pressing the Modify Specified Source button in Figure 6.9 opens the window shown in Figure 6.11. The purpose of this window is to specify the source parameters for the excitation waveform. There are two excitation source generators. One is based on the LF model and uses periodic pulses. The other is for noise generation. This window shows the default values for the various parameters. At the top of the figure, the number of frames to be synthesized is shown. This value cannot be changed here or in the formant specification window, which is discussed later. The window shown is for the voiced option. The unvoiced option is discussed later. For the voiced option, the synthesis is pitch synchronous and the pitch period and the frame length are the same. Thus, the product of the pitch period and the number of frames is the length of the synthesized excitation file. The default pitch period in number of samples is 100 (minimum is 10 and maximum is 200), the voicing gain is 5 (minimum is 0 and maximum is 20), and the default parameters for the LF model are as shown. The LF source model

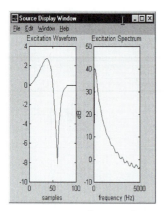

FiGURE 6.12 Source display window.

and the noise model are discussed in Appendix 7. These LF parameter values can be altered by the user. However, some caution is required since there is a delicate relationship between the parameter values and the excitation waveform generated. An error message is printed in the MATLAB command window if certain parameter values are not accepted. Generally speaking, it is best to make only very slight changes in the LF parameter values, being sure to satisfy the relationship that $0 < t_p < t_e < t_c$ and $t_a > 0$. If the user also desires a noise source to be added to the voicing source, then check the Add Noise box. To add noise, the aspiration gain must be set to a value greater than zero. This is necessary to alter the values of the next four parameter settings for the noise model, namely, amplitude 1, offset, amplitude 2, and duration (consult Appendix 7). The sum of offset and duration cannot exceed one. The fricative noise parameter is independent of the other noise parameter settings. As the user alters any of the parameters, the changes in the waveform and spectrum of the excitation are displayed in Figure 6.12. To view these changes, the user must press the apply to all Frames button at the bottom of the window. If the user wishes to return to the default parameter settings, press the reset button. Once the desired parameter values have been selected, press the Apply to All Frames button, followed by a press of the Done button. This action closes the source modification and new excitation windows and stores the excitation waveform and its parameter values in a file, which can be saved to a file on hard disk.

Another available option is to set the voicing gain to zero. Under this option, an LF model excitation waveform is not generated. The user must select the add noise option and set the aspiration noise to a value greater than zero. Then the other noise parameters can be set as desired. This option uses only noise as the excitation. However, it is pitch synchronous and periodic and is not unvoiced excitation. Thus, the excitation waveform becomes a periodic "noise" waveform.

A designed source excitation file can be saved as a waveform as either a mat or dat file. These files are not pitch synchronous, that is, the pitch must be estimated when loaded. This also applies to a saved unvoiced excitation file. This is discussed under the load file options. The other save format is also as a mat file. However, this option saves the parameter vectors, such as pitch, and is therefore pitch synchronous. The save file options are shown in Figure 6.13. This window appears as a result of pressing the Save Source button in Figure 6.9.

An existing excitation file can be loaded. This option is available in the Source window in Figure 6.9. Pressing the Load File popup button provides the options shown in Figure 6.14. The load mat and dat options allow the user to load data files (not parameter vector files)

FIGURE 6.13 Save excitation data.

created either by this program or by another program, such as by inverse filtering. A file (iy_ex_model.dat) created by another program is available in the data folder. This file is a neural-net model excitation trained using the /IY/ portion of the b.dat file. Upon loading this file, a message window appears informing the user that the file is being analyzed for the glottal closure instants. This analysis is pitch asynchronous. It is frame based using a frame length of 200 samples. Once this analysis is complete, the user can press the Modify Specified Source button in Figure 6.9, whereupon a message window appears, as shown in Figure 6.15, informing the user that the number of frames is 25 and that the first frame is unvoiced. The excitation waveform and its spectrum are shown in Figure 6.16. Both of these windows are cleared in about 15 seconds. Note that there are slightly more than two excitation waveforms shown in the window and that the excitation is clearly voiced. Thus, the analysis that the first frame is unvoiced is incorrect. This is due to the fact that

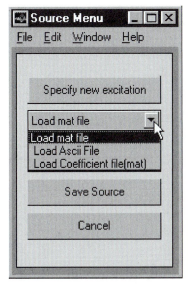

FIGURE 6.14 Load excitation option.

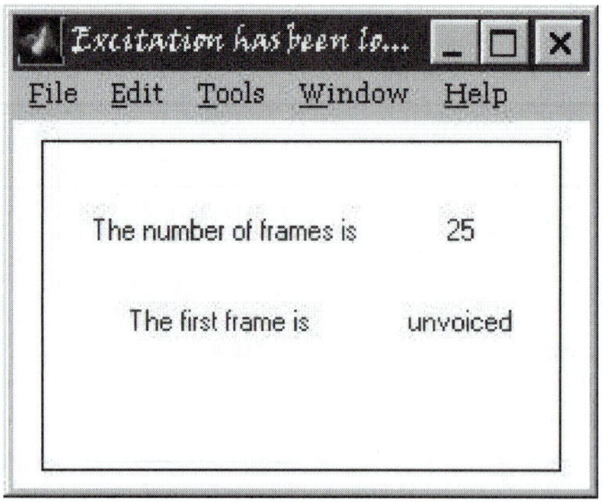

FIGURE 6.15 Message.

the analysis is frame based. Subsequent frames are analyzed correctly as voiced. The user cannot modify any of the parameters of a loaded data file (mat or dat). However, if the user loads a mat file created by this software that stored the parameter vectors (coefficient file), then the user can modify the parameter values just as though the file is being created in the first place. The number of frames cannot be changed.

The comments above apply to a loaded unvoiced excitation file as well. It, too, is analyzed using frame based analysis and errors can occur in this case, just as discussed above.

If the user checks the Voiced box in Figure 6.11, then the window shown in Figure 6.17 appears. The user can return to the voiced excitation window by checking the Voiced box again if desired. For unvoiced excitation, the user cannot set the pitch period or any of the LF model parameters. Only the aspiration noise source can be specified. The changes in the excitation waveform can be viewed in Figure 6.18 upon pressing the Apply to all frames button. If the user wishes to return to the default settings, then press the Reset button. Once the desired values have been specified, press the Apply to All Frames button, followed by a

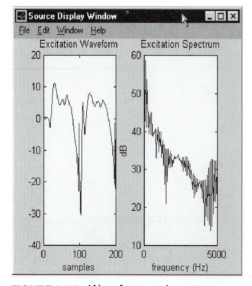

FIGURE 6.16 Waveform and spectrum.

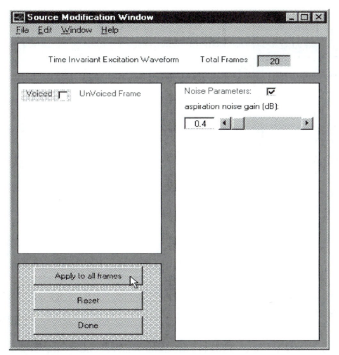

FIGURE 6.17 Unvoiced source modification window.

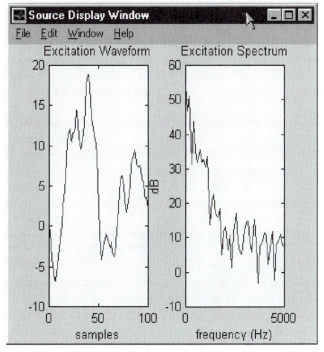

FIGURE 6.18 Unvoiced source display window.

FIGURE 6.19 Formant specification window.

press of the Done button. This action closes the source modification and new excitation windows and stores the excitation waveform as a file. The file does not contain any pitch information or other such data and cannot be stored as mat coefficient file. It can be stored only as a mat or dat data file. The unvoiced excitation file is one continuous file and contains no pitch information.

Once the source has been specified, either as a newly created file or as a loaded file, then the user proceeds to the formant specification. Press the Formant Specification button in the Main window shown in Figure 6.8. This action opens the window shown in Figure 6.19. The user now presses the Specify New Formants button to create a new formant file in a manner similar to that for creating an excitation file. This action causes a message window to appear stating that default formants have been created and that the user can modify these values.

Next, the user presses the Modify Specified Formants button, which opens the window shown in Figure 6.20. The purpose of this window is to specify the formant frequencies and bandwidths to simulate the desired vocal tract filter. The default formants and bandwidths are for the vowel /IY/, with some modifications to the bandwidths. There are six sliders each for setting the formant frequencies and the formant bandwidths. The pole-zero plot and the frequency response of the resulting vocal tract filter are displayed in Figures 6.21 and 6.22. The number of frames cannot be changed. If the specified excitation is voiced (unvoiced), then the user is reminded of this by the message at the top of the left panel. The pole-zero and the vocal tract frequency response plots reflect the changes made by the user in the formant frequencies and bandwidths. This is accomplished by the user pressing the Apply to All button. Press the Reset button to reset the default values. Once the desired values are set, press the Apply All button followed by a press of the Done button.

Once excitation and formant files are specified, then a speech file may be synthesized by pressing the Synthesis button in the Main window in Figure 6.8. The speech is synthesized and displayed in Figure 6.23 along with the excitation waveform. Shown here is the voiced excitation and the synthesized /IY/. There is a small change in the waveform for the first

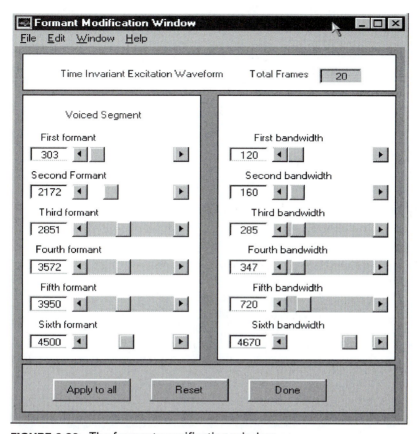

FIGURE 6.20   The formant specification window.

pitch period as the synthesizer initializes. This file can be played. The spectrogram can be viewed as in previous speech toolboxes.

If the excitation is unvoiced, then the synthesized display appears as shown in Figure 6.24. For an unvoiced excitation the excitation is not pitch synchronous, rather it is one

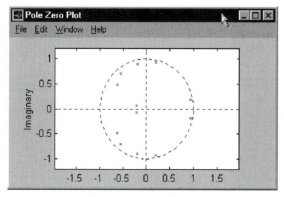

FIGURE 6.21   The vocal tract filter pole-zero plot.

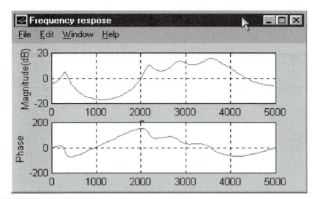

FIGURE 6.22 The vocal tract frequency response.

continuous noise excitation file, which is filtered by the vocal tract filter to generate synthesized speech. Speech synthesized in this manner sounds very much like whispered speech.

When the synthesized speech is displayed, as in Figures 6.23 and 6.24, a window, shown in Figure 6.25, appears. The options provided by this window allow the user to clear the data display and/or save the synthesized speech file as a dat file. The saved speech file can be loaded and played using the analysis toolbox or the speech_display toolbox. An example of a saved synthesis file is included in the data folder. This file

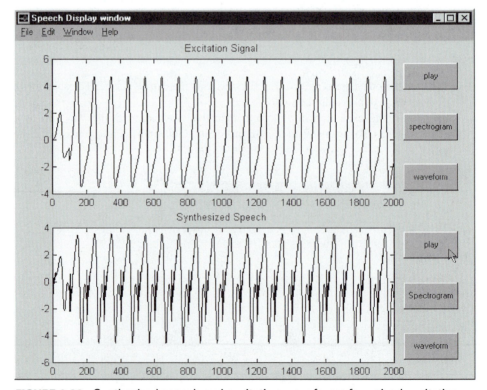

FIGURE 6.23 Synthesized speech and excitation waveforms for voiced excitation.

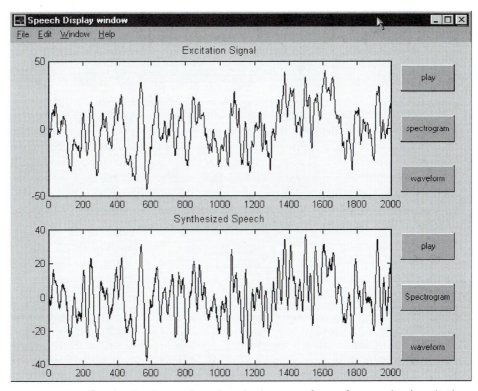

FIGURE 6.24 Synthesized speech and excitation waveforms for unvoiced excitation.

(syn_speech_iy_ex_model.dat) is the speech synthesized using the iy_ex_model.dat file.

Once excitation and formant files are created and a speech file synthesized, the user can start again and create new files. To do this, start by pressing the Specify New Excitation button and proceed as described previously.

Note that jitter cannot be specified in the excitation file and cannot be incorporated into the synthesized speech. The synthesizers in Chapters 7 and 8 do include jitter.

FIGURE 6.25 Cancel and/or save synthesized speech.

6.5 SUMMARY

This chapter has provided an overview of speech synthesis and its use. A description of speech analysis-by-synthesis was outlined. The chapter concluded with a description of a speech synthesis system that is used in Chapter 7 for voice conversion. Also included in this chapter is a simple formant speech synthesizer that contains only a few of the more sophisticated features of the system used in Chapter 7. Appendix 7 should be consulted for details on two excitation models: the LF and the polynomial models, as well as a description of the use of these models for creating various voice types. Also included in Appendix 7 is a description of the noise model used in the formant synthesizer. Appendix 8 contains additional details on the voice conversion system used in Chapter 7, which was outlined in this chapter. Additional details on both the formant synthesizer described in this chapter and the synthesizer described in the next chapter can be found in the following references: Hu (1993), Shue (1995), and Hsiao (1996).

PROBLEMS

6.1   The purpose of this problem is to synthesize the vowel /IY/ using the formant synthesizer described in this chapter. Use the default settings for the LF source, noise excitation, and the formants. Increase the number of frames so that you can assess the synthesized vowel more easily. This synthesized vowel is for that of a modal, male voice. Evaluate your result by a listening test by comparing the data and the spectrogram to the /IY/ in the file m0103s.dat.

6.2   Repeat Problem 6.1. However, set the formant frequencies and bandwidths to the values given in Figure 3.19 (a female voice). Also lower the pitch period.

6.3   Repeat Problem 6.1 for the vowel /AA/.

6.4   The purpose of this problem is to synthesize the vowel /IY/ using the formant synthesizer described in this chapter. Your task is to create a vowel that sounds like a vocal fry. Use the model values given in Appendix 7 for both the LF model and the noise model. Be sure to specify a sufficient number of frames, e.g., 100, so that you can evaluate your result via a listening test. Evaluate your result by a listening test and comparing the spectrogram to the /IY/ in the file m0305s.dat in the mimic folder. This file is the sentence, "We were away a year ago," so you will have to isolate an appropriate segment for comparison.

6.5   The purpose of this problem is to synthesize the vowel /IY/ using the formant synthesizer described in this chapter. Your task is to create a vowel that sounds like a breathy voice. Use the model values given in Appendix 7 for both the LF model and the noise model. Be sure to specify a sufficient number of frames, e.g., 100, so that you can evaluate your result via a listening test. Evaluate your result by a listening test and comparing the spectrogram to the /IY/ in the file m0405s.dat in the mimic folder. This file is the sentence "We were away a year ago," so you will have to isolate an appropriate segment for comparison.

6.6   The purpose of this problem is to synthesize the vowel /IY/ using the formant synthesizer described in this chapter. Your task is to create a vowel that sounds like a harsh voice. Use the model values given in Appendix 7 for both the LF model and the noise model. Be sure to specify a sufficient number of frames, e.g., 100, so that you can evaluate your result via a listening test. Evaluate your result by a listening test and comparing the spectrogram to the /IY/ in the file m0705s.dat (and file m0505s.dat) in the mimic folder. This file is the sentence "We were away a year ago," so you will have to isolate an appropriate segment for comparison.

6.7   Repeat Problem 6.6 for a falsetto voice. There is no data file to compare your result with.

6.8 The purpose of this problem is to synthesize the vowel /IY/ using the formant synthesizer described in this chapter. Your task is to create a vowel that sounds like a whispered voice. It is recommended that you use an unvoiced excitation. Be sure to specify a sufficient number of frames, e.g., 100, so that you can evaluate your result via a listening test. Evaluate your result by a listening test and examining the spectrogram. There is no data in the files to compare to. However, your results should be similar to the results for a breathy voice, but perhaps sound even better.

6.9 This problem is difficult. Your task is to synthesize the vowel /IY/ (or another of your choice). However, the excitation file is to be obtained by inverse filtering the file m0103s.dat (or a file of your choice). Compare your result with the data in file m0103s.dat (or the file you selected) using a listening test and spectrograms.

CHAPTER 7

VOCOS—A VOICE CONVERSION TOOLBOX

7.1 INTRODUCTION

This chapter describes software that can be used for speech analysis, speech synthesis, and voice conversion, also known as voice transformation or voice manipulation. The software provides graphical user interfaces (GUIs) with various features to assist the user in speech analysis and synthesis tasks. For example, the user can select analysis algorithms, inspect and correct analysis derived parameters, align the acoustic parameters of two speakers with different speaking rates using dynamic time warping (DTW), and synthesize speech. Speech synthesis can mimic the voice of a speaker or convert the voice of one speaker to sound like that of another. Data displays are provided to assist the user in judging the correctness of the analysis results. The primary acoustic features that are measured include voiced/unvoiced (voice type) classification, pitch and gain contours, formant frequencies and bandwidths, and the shape and type of glottal excitation waveform. An overview of the major characteristics of the software is shown in Figure 7.1. The software system is called VOCOS for Voice Conversion System. The software is to be installed in a subdirectory in MATLAB in a manner similar to that described for the analysis system. Additional details are provided in Chapter 6, Appendices 7 and 8, and Hsiao (1996).

7.2 MAIN FUNCTION

The analysis algorithms are based on a fixed-frame linear prediction (LP) analysis method. To start the software change directory to the VOCOS directory and type main in the MATLAB command window. The Main Function window shown in Figure 7.2 appears. The Analysis button provides algorithms for the analysis of speech data. The Correction button lets the user correct an analyzed speech data file. The Modification button allows the user to modify parameter values of an analyzed speech file. The Synthesis button provides a method for synthesizing speech using the parameters obtained during speech analysis. There are two types of synthesis: linear prediction (LP) and formant. The selection of the type of synthesizer is automatic and determined by the type of vocal tract model selected by the user during the analysis phase. The Close button closes out the Main Function window.

7.3 SPEECH ANALYSIS—LINEAR PREDICTION (LP)

Upon pressing the Analysis button, the window shown in Figure 7.3 appears. The topmost button, Specification, is used to specify the desired analysis parameters, which are available in the Specification window, shown in Figure 7.4. This window contains a set of default

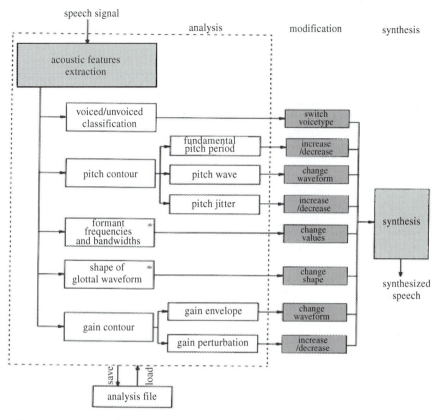

FIGURE 7.1 Voice conversion software system. The blocks with an asterisk are not available for all analysis parameter settings.

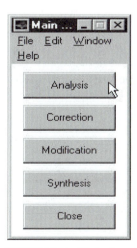

FIGURE 7.2 Main function window.

FIGURE 7.3 Analysis window.

values that can be reset using the Default button. There are two source models: polynomial and simplified LF. The number of formants can be selected as well as the window frame length and the window overlap. The two vocal tract models are linear prediction and formant. The order of the LP model can be selected. After the desired parameters have been set, press the Apply button, otherwise reset the values to the default by pressing the Default button. Finally, press the Return button to close out the Specification window and return to the Analysis window for the next selection.

Next, load a data file for analysis by pressing the Load Speech File button. This brings up the window shown in Figure 7.5, which shows the contents of the `spedata` subdirectory. This directory is designed to contain speech data files for analysis or voice conversion. If the desired data file is not in this directory, then change directory to the desired location. In Figure 7.5 the `m11.dat` file is selected and loaded. This file is displayed as the Input Signal in Figure 7.6. This is the sentence, "We were away a year ago," spoken by a male speaker. This sentence is to be analyzed. Remember, do not close the Input Signal window or the Main Function window until the analysis is complete, since all data will be lost. To proceed, the user presses the Execution button in Figure 7.3. Note that the visual correction option for analysis is checked as the default. This is almost always the desired option. On occasion, the user can uncheck this option. For example, when an analysis of familiar data is being performed and the user knows that visual correction is not necessary or is to be omitted.

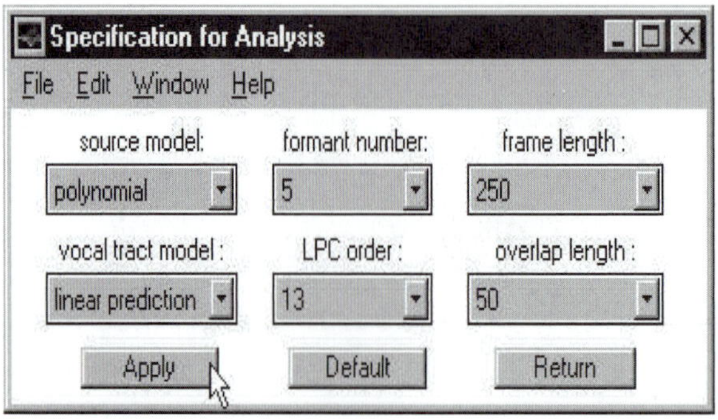

FIGURE 7.4 Specification window for analysis.

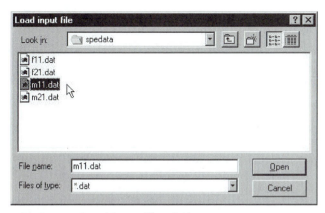

FIGURE 7.5 Load input file window.

The data files stored in the spedata directory are dat speech files. The data are sampled at 10 kHz. The prefix f/m refers to female/male speaker. The next digit is the speaker identification number, and the following digit is the sentence identification number. Thus, m11.dat is the speech file for male speaker 1, sentence 1, which is, "We were away a year ago." The file names have been shortened for this chapter only as a convenience, the reasons for which will become apparent later.

After starting the execution of the software, a message window appears stating that the program is working. When the first stage of calculations is completed, a second message window appears stating that the voice types (voiced/unvoiced) are classified and that the user can inspect the results. In addition, the action window shown in Figure 7.7 appears, allowing the user to view the results of the first stage of calculations or to continue. Upon selecting the Voice Type Inspection option, the results shown in Figures 7.8 and 7.9 appear. These results are obtained via an automatic voice type classification algorithm that can make mistakes, thereby, introducing unwanted errors. Figure 7.8 is an action window that allows the user to scroll through the data to view the results of the voice type classification algorithm. Press the > (<) button to scroll the data to the right (left). This figure shows the voice type classification at the top of the data as either V or U for voiced or unvoiced, respectively. The

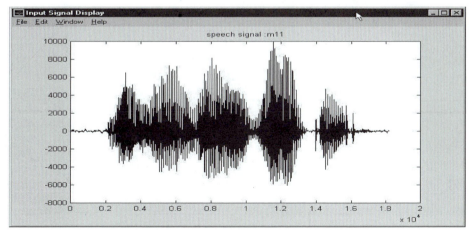

FIGURE 7.6 Input signal for analysis.

FIGURE 7.7 Voice type inspection window.

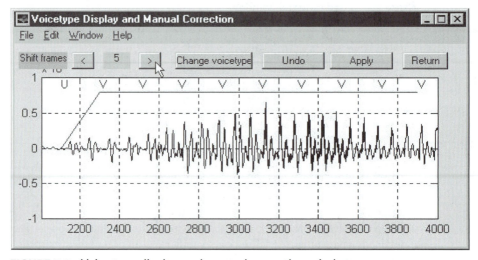

FIGURE 7.8 Voice type display and manual correction window.

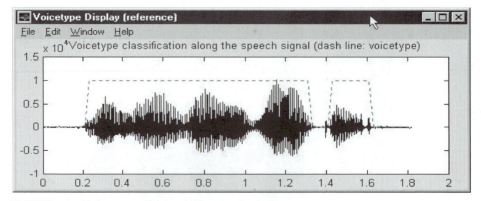

FIGURE 7.9 Voice type display (reference) window.

FIGURE 7.10   GCI inspection window.

inclined, solid line indicates the onset of voicing and remains constant as long as voicing is present. An overview of the V/U classification for the complete data file is shown in Figure 7.9, where the dashed line superimposed over the data indicates the voiced data regions. As the user scrolls through the data in Figure 7.8, errors can be found and corrected as follows. If an error is found, say a frame is marked by the algorithm as voiced (V) and it should be unvoiced (U), then press the Change Voice Type button. The mouse cursor changes to a cross hair. Move the cross hair to the V and press the left mouse button. The V changes to a U and the solid line changes to zero, indicating unvoiced. A V > U indication is printed under the V to remind the user that a change has been made. The cross hair changes back to the usual cursor. To undo this operation, press the Undo button. To retain the change, press the Apply button. Perhaps the best procedure to follow is to make all the desired changes before the Apply button is pressed. However, changes can be made after the Apply button is pressed by following the above steps. Once the Return button is pressed and the analysis continues, changes cannot be made until the analysis is completed. This will be discussed later.

The analysis is continued by pressing the Continue button, whereupon a message window appears informing the user that the glottal closure instants (GCIs) are located and that the results can be inspected, in a manner as described previously. Upon selecting the GCIs Inspection button in Figure 7.10, Figures 7.11 and 7.12 are displayed.

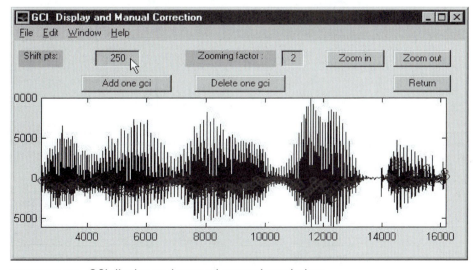

FIGURE 7.11   GCI display and manual correction window.

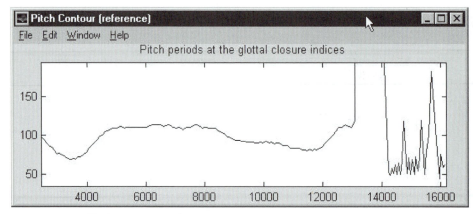

FIGURE 7.12   Pitch contour (reference).

The action window in Figure 7.11 allows the user to correct the algorithm detection errors of the glottal closure instants (GCIs). At the upper left is a panel where the user can set the number of points that the data are shifted each time. The zooming factor default setting is 2, but this can also be changed. The user can zoom in or out. The buttons on the next row are used to add or delete a GCI one at a time. Figure 7.12 shows the pitch contour calculated using the GCIs. This contour appears reasonable, except beyond point 14,000, which is the beginning of the word "go." We can see sharp changes in the pitch contour in the region 14,000 to 16,000, which should not be present, indicating some errors in the detection of the GCIs in this region. To correct these errors, we zoom in on Figure 7.11, as shown in Figure 7.13. To accomplish this, proceed as follows. In Figure 7.11, move the mouse cursor to the zoom factor panel and press the Zoom in Button. The mouse cursor changes to a cross hair. Move the cross hair to a location near 14,000 on the data and press the left mouse button. The data zooms in a factor of 2 and is replotted. Repeat this step one or more times. Note that once the zoom-in action is activated, that markers < and > appear on either side

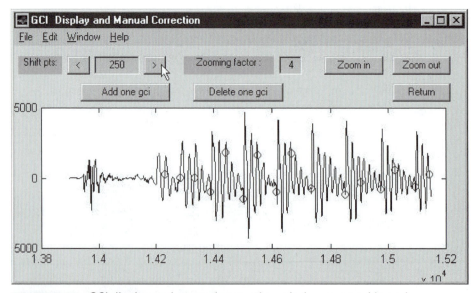

FIGURE 7.13   GCI display and manual correction window zoomed in and scrolled to 14,000.

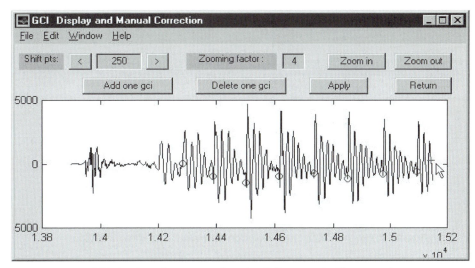

FIGURE 7.14   GCI display after correcting the GCI errors in Figure 7.13.

of the shift points panel. Scroll to the right until you reach the point 1.4×10^4 as shown in Figure 7.13. Observe that there are errors in the GCIs detection at almost every other open circle on the data. We correct these errors by removing the unwanted GCIs by pressing the Delete One GCI button, whereupon the mouse cursor becomes a cross hair. Move the cross hair to the circle to be deleted and press the left mouse button. Repeat this process until all errors are corrected. An example of the correction process is shown in Figure 7.14, where the unwanted GCIs shown in Figure 7.13 are deleted. If a GCI is to be added, follow a similar procedure. Each time a GCI is added or deleted, the Pitch contour is updated. Scroll through the data as needed. Upon making the necessary corrections, the final Pitch contour appears as shown in Figure 7.15. Note that the contour is now smoother in the vicinity of 14,000 to 16,000. To save the corrections, press the Apply button in Figure 7.14, followed by a press of the Return button. Then press the Continue button in Figure 7.10.

A message window appears stating that calculations are being performed. Another message window appears after the calculations are complete, and states that the results can be saved. The Analysis window changes to that shown in Figure 7.16, where a Save Analysis

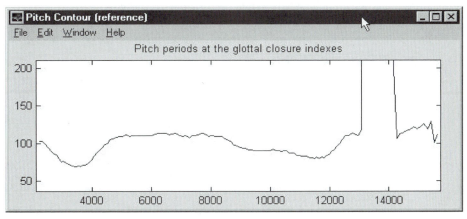

FIGURE 7.15   Pitch contour after GCI corrections.

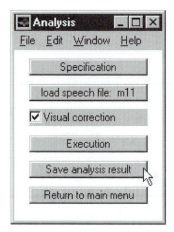

FIGURE 7.16 Analysis window upon completion of data analysis.

Result button appears. The user can save the results to a file. However, if the save option is not selected, the results of the analysis are retained in memory and the user can now do additional corrections, modifications, or synthesis. Pressing the Save File button brings up the Save window shown in Figure 7.17. The file name appears as m11.mat because we analyzed a data file named m11.dat. The file name is shown changed to m11_poly_lp_anal.mat to illustrate that the user can change the name as desired. The extension to this file is mat rather than dat, because a mat file is designed for storing a number of parameter vectors. Note, however, that the prefix, m11, is not important because the analysis files store the speech data as a vector along with the parameter vectors calculated via the analysis software. Press the Apply button in Figure 7.17. The results of the analysis are saved to the desired file. A message window appears stating that the results are saved and that the user can return to the Main window. If the user presses the Return button, both the Analysis menu window and the Input Signal window close. The user can analyze a new data file, make corrections or modifications to the data just analyzed, or synthesize speech using the parameters in the file m11_poly_lp_anal.mat. Suppose we synthesize a speech file.

7.4 SYNTHESIZE

Speech synthesis attempts to mimic the voice of the original speaker using the parameters measured in the analysis phase. Press the Synthesis button in the Main window. The

FIGURE 7.17 Save window.

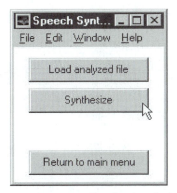

FIGURE 7.18 Synthesis window.

Synthesis window shown in Figure 7.18 appears. The available options are to Load a previously analyzed data file, such as m11_poly_lp_anal.mat, or Synthesize, or Return. In this case, the user can press the Synthesize button, since the analyzed data are still in memory. On doing so, a message window appears stating that the synthesis process is underway. After the calculations for synthesis are completed a message window appears stating that the synthesis is completed and that the synthesis file can be saved. In addition, the synthesized and original data waveforms are plotted in Figure 7.19. Recall that the original data file is m11.dat. Figure 7.19 is an action window that lets the user play the data files, plot the spectrograms of the data as in Figure 7.20, or replot the waveforms. The Return button closes the figure and returns to the Synthesis window. The synthesized data can be saved to a file by pressing the Save Synthesized Speech button in Figure 7.21, which appears after the synthesis process is completed. This save option is similar to that described previously, except that the file saved is a dat file. For example, in this case the saved data file name is m11_poly_lp_syn.dat.

Note that the software stores the analyzed parameters, e.g., source model (polynomial or LF), vocal tract model (LP or formant), and so forth, as well as the speech data loaded as the original data file. The type of speech synthesis is selected automatically as either linear prediction (LP) or formant depending on whether the analysis phase used the LP or formant vocal tract model, respectively.

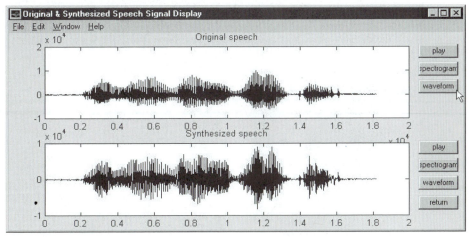

FIGURE 7.19 Original and synthesized speech signal display window showing waveforms.

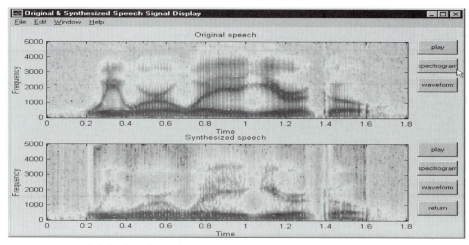

FIGURE 7.20 Original and synthesized speech signal display window showing spectrograms.

7.5 SPEECH ANALYSIS—FORMANT

The previous analysis procedure used the polynomial model as the source and the vocal tract linear prediction model with the other parameters set to the default values. In this case, the formant frequencies are not calculated. To illustrate this aspect of the software, return to the Main menu window, select Analysis, and set the source model to simplified LF and the vocal tract model to formant. Leave the other parameters set to their default values. The source model does not affect the vocal tract model calculations, so it could have remained set at polynomial. Load the m11.dat file again for analysis. Press the Execution button. The voice type and GCI analyses are performed as described previously. The third and last analysis in this case is the formant analysis, which can be inspected in Figures 7.22 and 7.23. These two figures are similar to those for the voice type and GCI analysis results. Figure 7.22 is an action figure that allows the user to alter the circled formant frequency values as follows. First, Zoom In on the data by pressing the Zoom In button. The mouse cursor changes to a cross hair, which is to be moved to the desired location for a correction. Click the left mouse button, and the data are zoomed in. While corrections can be made to the formant values without zooming in, it is easier to observe the results in the zoom-in mode. The user can scroll the data by pressing the < or > buttons. Next, press the Select Formant to Correct

FIGURE 7.21 Synthesis window showing save option.

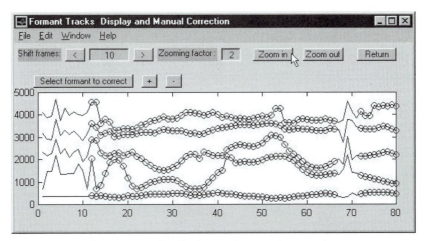

FIGURE 7.22   Formant tracks display and manual correction.

button. The mouse cursor changes to a cross hair. Move the cross hair to the circle to be corrected and press the left mouse button. An × appears within the circle and the cross hair reverts to the normal mouse cursor. If the circled formant frequency is to be reduced in value, then place the mouse cursor on the minus sign and press the left mouse button repeatedly until the circle moves down to the desired location. An example is shown in Figure 7.24. If the circled value is to be increased, then press the + button. Continue with any additional corrections as needed. To save the corrections, press the Apply button, followed by a press of the Return button. This completes the analysis and the results can be saved as described previously. The results for this analysis are saved to the file m11_lf_for_anal.mat.

7.6   CORRECTION

Pressing the Correction button in the Main menu window brings up the window shown in Figure 7.25. Note that there is no load option. The corrections are to be made to the file in memory. Assume this file is m11_poly_lp_anal.mat. Pressing any of the buttons activates the respective option. Proceed as described previously. Note that no formant information is available for this file, since we selected the linear prediction model for the vocal tract

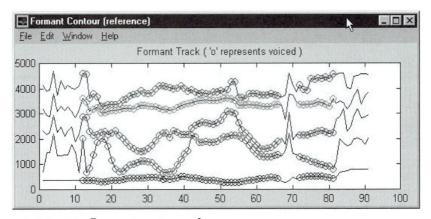

FIGURE 7.23   Formant contour reference.

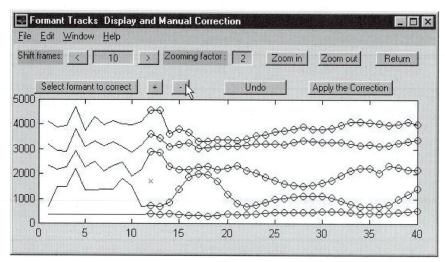

FIGURE 7.24 An example of correcting the first value of the second formant contour.

model. If we had selected the formant model, then formant information would be available. We discuss this option later. After making any necessary corrections, save the corrected file before returning to the Main window.

7.7 MODIFICATION

The Modification option window appears in Figure 7.26. The available options are described next. The Load button lets the user load a previously analyzed mat file, such as m11_poly_lp_anal.mat. However, if the file is already in memory, then the load option is not necessary.

FIGURE 7.25 Correction window.

FIGURE 7.26 Modification menu window.

7.8 VOICE TYPE

Pressing the Voice Type button opens the window shown in Figure 7.27, which shows the voiced and unvoiced segments of the analyzed data as a solid line. Modifications can be made by first pressing the > marker at the upper left once, thereby moving the frame location to 2. The display automatically changes with + signs appearing on the figure, and the +

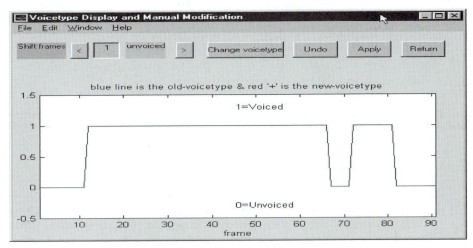

FIGURE 7.27 Voice type modification window.

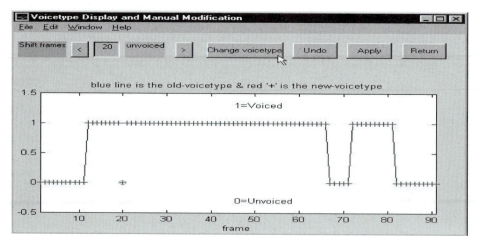

FIGURE 7.28 Voice type modification window. Frame shifted to 20 and changed from voiced to unvoiced.

sign at 2 is circled. To illustrate this, suppose we shift the location to 20 by either pressing > repeatedly or by typing in the number 20 in the panel and pressing >. Next, press the Change Voice Type button. The circled + moves from voiced to unvoiced, as shown in Figure 7.28. Pressing the Undo button can reverse this action. After all changes are made, press the Apply button followed by a press of the Return button.

7.9 PITCH CONTOUR

After pressing the Pitch contour button, three windows appear allowing the modification of the pitch contour. The left window (Figure 7.29) supplies the action buttons and sliders for altering the several factors. The upper right window (Figure 7.30) shows the pitch wave and the pitch jitter. The lower right window (Figure 7.31) displays the fundamental pitch period and the pitch contour. The pitch contour consists of three factors that can be altered separately. The fundamental pitch period is controlled by the topmost slider. Click the left side button of the slider to decrease the value, or the right side button to increase the value. The blue waveform moves down or up, respectively The jitter waveform is controlled by the lower slider and can be changed in a similar manner. It may appear that the pitch wave is altered when altering the jitter. However, this is due to a vertical scale change on the figure. The pitch wave is modified as described next.

In the middle of the left window, there are two menus and one button for modeling the pitch wave. The add knob | delete knob popup menu is used to mark or delete a critical point on the pitch wave with a knob that controls the shape of the modeled waveform. The placement of a knob on the waveform segments the waveform. The line_fit | parabola-fit(1) | parabola-fit(2) | cubic-fit | default popup menu provides five models for each segment of the pitch wave. The line_fit model draws a straight line between the two successive knobs. See MATLAB Help for the parabola-fit and cubic-fit models (help parafit and help cubfit). Briefly, the parabola-fit models determine a second-order polynomial for each segment of the waveform. The parabola-fit(1) model is an upward polynomial and the parabola-fit(2) model is a downward polynomial. The cubic-fit model determines a cubic polynomial for the pitch wave. The default model is the data waveform with no model fit to the data. The modeled wave is drawn and the knobs are marked by + signs. The draw wave button enables

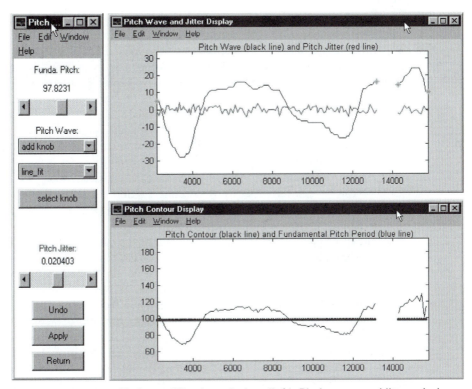

FIGURE 7.29–7.31 Pitch modification window (left). Pitch wave and jitter window (upper right). Pitch contour and fundamental pitch period window (lower right).

the knob to be moved in the vertical direction to a desired location that does not have to be on the pitch wave. As an example, let us add a knob at the first negative peak of the waveform and another knob on the waveform at about 5000 on the horizontal axis. These two knobs appear in Figure 7.32. Select the line_fit option. The mouse cursor changes to a cross hair on the waveform. Press the left mouse button and straight line models are drawn between the knobs, as shown in Figure 7.32. Pressing the Undo button can reverse this action. The other line drawing models work in the same manner. Once the user places the desired knobs

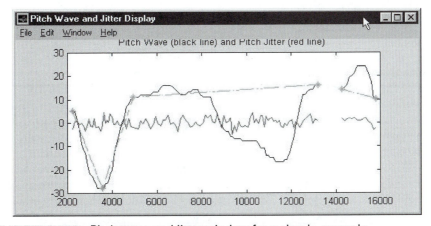

FIGURE 7.32 Pitch wave and jitter window for a simple example.

on the waveform, the various line drawing models can be tried in succession. For example, the straight line fit can be made as described previously. To change the straight line fit to the parabola-fit(2), for example, select the parabola-fit(2) model. The mouse cursor changes to a cross hair. Move the cross hair to a location between knob 1 and knob 2. Click the left mouse and a parabola is fit between these two successive knobs. Repeat the entire process again for the next two successive knobs, until parabola-fit(2) is redrawn as a model fit to the waveform. Of course, an alternate procedure is to undo the line_fit model, place new knobs on the waveform, and select the parabola-fit(2) model.

The parabola-fit models attempt to avoid discontinuities at knobs and also attempt to fit the models to the data. As a consequence, sometimes the fit by a parabola model between the knobs may be an upward or downward parabola, or even a straight line.

A knob can be removed (deleted) by selecting the delete knob option. The mouse cursor changes to a cross hair on the figure. Move the cross hair to the desired knob and click the left mouse button. The knob will be circled. The mouse cursor returns. Next click the Undo button and the knob will be deleted. A knob can be moved by pressing the Select Knob button. The mouse cursor changes to a cross hair on the figure. Select the desired knob to be moved with the cross hair and press the left mouse button. The mouse cursor returns. Press the Up or Down button repeatedly to move the knob accordingly. Select the desired line drawing model. Move the cross hair to a location left of the moved knob and press the left mouse button. Note that the drawing retains a memory of the location of the previous knob location for the segment to the right of the knob. Repeat the previous operations for the segment to the right of the knob. A line model is drawn for this segment. Remember that at any time, the Undo button can be pressed to erase the line drawing models. Only press the Apply button after you are sure that the desired model is the one you want. If you should do this and find you want to cancel the operation, then you must start over by loading again the desired x.mat file, for example file m11_poly_lp_anal.mat. Press the Return button to close these three windows and return to the main modification function window.

7.10   GAIN CONTOUR

The three windows for modifying the gain contour are similar to those shown for the Pitch contour. Figure 7.33 is the action window, while Figure 7.34 shows the gain envelope and the gain perturbation. Figure 7.35 depicts the gain contour. Note that the voiced (V) and unvoiced (U) regions are labeled on Figures 7.34 and 7.35. The gain envelope and perturbation are altered separately as follows. The perturbation is controlled by a slider. Press the left side of the slider to decrease the value, or the right side to increase the value. The procedures for modifying the gain envelope are similar to those for modifying the pitch wave as discussed.

A knob can be added or deleted as described for the Pitch contour. And the Gain contour can be modeled with the same line drawing options.

The label grossly option is the default and provides a plot of the data with two knobs at either end. With this option, the user can add knobs as desired and select the desired line drawing option to model the gain contour. Only voiced speech should be modeled. The label finely option shows the actual data values on the gain contour waveform as + signs. With this option the user can zoom in and out on the data. Knobs cannot be added or deleted with this option. A knob can be selected and moved, however. Line drawing options can be selected.

To cancel a modification, click the Undo button. Press the Apply button when all modifications are completed. Press the Return button to close these three windows and return to the main Modification function window.

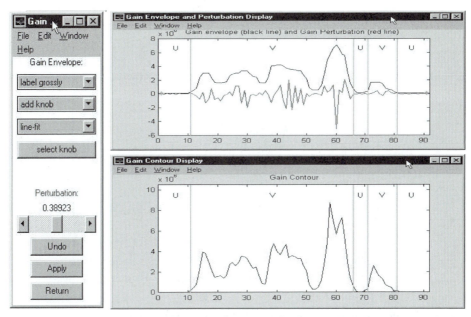

FIGURE 7.33–7.35 Gain modification window (left). Gain envelope and perturbation window (upper right). Gain contour window (lower right).

7.11 GLOTTAL WAVEFORM

This modification option is not available for the polynomial source model. It can be used only for the LF source modeling. Figures 7.36 and 7.37 show the windows for modifying the differentiated LF glottal flow model for the file m11_lf_for_anal.mat. Recall that this file used the LF source model and the formant vocal tract model. Figure 7.36 shows the function buttons and sliders for altering the LF timing parameters. The user can also select the frame of interest using either the < down or up > buttons or the slider. Figure 7.37 displays the

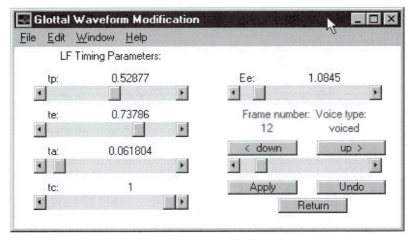

FIGURE 7.36 Modification window for the differentiated LF glottal waveform model.

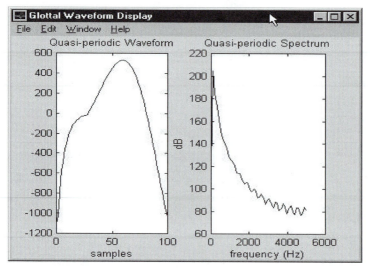

FIGURE 7.37 Differentiated LF glottal waveform model and its spectrum.

LF differentiated glottal waveform model and its spectrum for the selected frame. The user should be careful about setting the LF timing parameters. An error message occurs if the setting is not a valid setting for a model parameter, for example, if T_p is larger than T_e.

7.12 SPECTROGRAM (FORMANTS)

This modification function is not available for use with the LP vocal tract model. It can be used only with the formant vocal tract model. Figures 7.38 and 7.39 show the windows for

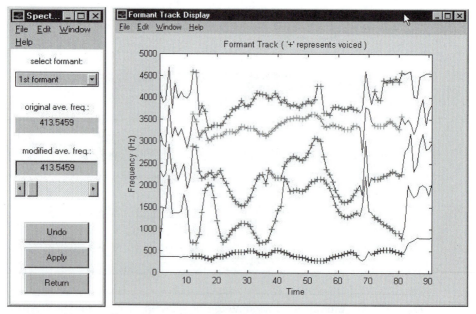

FIGURE 7.38–7.39 Formant modification window (left). Formant tracks (right).

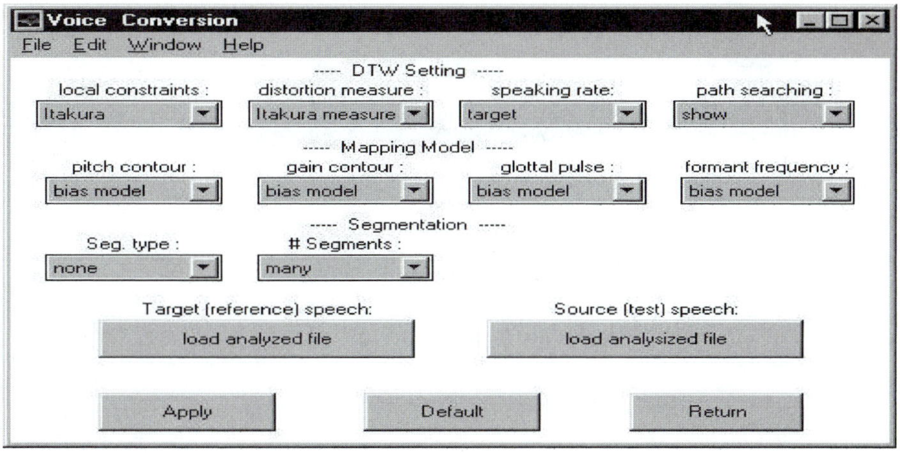

FIGURE 7.40 Voice conversion window.

modifying the spectrogram (formants) of the synthesized speech. Figure 7.38 provides the control buttons for shifting a formant track. Only the average value of the formant track can be modified. As a change is being made the new formant track is plotted along with the old track to facilitate the assessment of the change. The Undo button cancels any changes. After making the desired modifications, the user clicks the Apply button followed by a press of the Return button.

7.13 VOICE CONVERSION

Figure 7.40 shows the various options for voice conversion. The first row of four buttons is for dynamic time warping (DTW). These options let the user set the local constraints and the distortion measures, select the speaking rate as the source or the target, and display the search path. The second row of four buttons provide four models (bias, linear, copy, retain) for mapping the four acoustic features (pitch contour, gain contour, glottal pulse, formant frequency). The two buttons in the third row let the user set the type of segmentation for the speech file. We discuss the segmentation option later.

The fourth row of two buttons load the Target and Source speech files, which must be obtained via analysis and stored as mat files. The Default button resets the option choices to the default values. Once the various options are selected, and the necessary files loaded, press the Apply button to start the voice conversion process. The Return button closes the window and returns to the Modification window. As an example, suppose we select the default values (no segmentation) in the Voice Conversion window and load the target speech file, m11_poly_lp_anal.mat, and the source speech file, m21_poly_lp_anal.mat, and press Apply. As the voice conversion process calculations are underway, various windows are displayed related to the DTW calculations. These displays are only indicative of the results and serve only as indications that calculations are being made. Upon completion to the voice conversion process, a message appears in the MATLAB Command window informing the user that the synthesizer can be run. Close the Voice Conversion window and select Synthesis in the Main window. It is not necessary to load files for synthesis, since the files for voice conversion are in memory. Press the Synthesis button to synthesize the converted speech. Figure 7.41 shows the results.

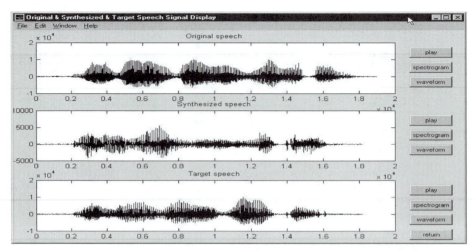

FIGURE 7.41   Synthesis window for voice converted speech waveforms.

Note that the target and source mat files must be for the same vocal tract model, that is, both must be obtained using either the LP or formant vocal tract model, but one file cannot be for the LP model and the other file for the formant model.

The results displayed in Figures 7.41 and 7.42 are for the original (m21.dat) and the target speech (m11.dat). The synthesized speech is the converted speech, that is, the data calculated using voice conversion. The synthesized results are obtained by converting

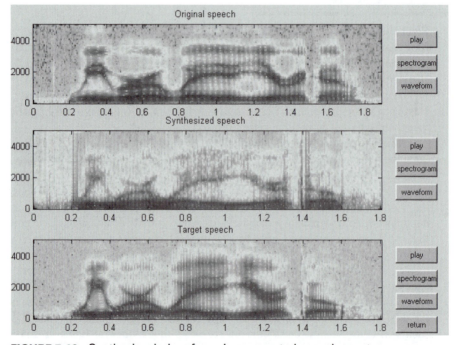

FIGURE 7.42   Synthesis window for voice converted speech spectrograms.

m21_poly_lp_anal.mat to m11_poly_lp_anal.mat. Figure 7.41 displays the speech data for the original and target, i.e., the m21.dat and m11.dat files, while Figure 7.42 shows the spectrograms for the same data. Another comparison is to show the synthesized files for all three, that is, original (source), converted, and target. We show how to do this below.

7.14 VOICE CONVERSION—SEGMENTATION

The purpose of the segmentation option is to improve the voice conversion process. To convert the parameters of one voice to that of another for a complete sentence can be a difficult task. However, doing voice conversion on a segment-by-segment basis can sometimes give improved results.

Segmentation is performed over the target speech using a simple normalized measure of spectral change given by:

$$\frac{\sum_{\omega}(|S_1(\omega)|)-|S_2(\omega)|)^2}{\sum_{\omega}|S_1(\omega)|^2} \geq \text{threshold} \qquad (7.14.1)$$

S_1 and S_2 are taken from two consecutive frames. If this value exceeds the threshold, then a new segment is specified. The segmentation buttons apply only if the speaking rate is set to the target. The Segmentation type choices include without segmentation (none), with segmentation (automatic), and with manual segmentation (manual). If the Segmentation option is selected, then the user specifies the number of segments via the # Segments pull down menu. The choices are many, medium, and few. The choice many corresponds to a threshold of 0.2, medium corresponds to a threshold of 0.4, and few to 1. A segmentation of one frame is not allowed. Furthermore, when the option many segments is selected, the analysis can calculate numerous one-frame segments. Thus, sometimes it is better to select a medium number of segments.

The segmentation option must be used as follows. The speaking rate must be set to target. The target and source files must be for the formant vocal tract model. The LP vocal tract model is not available for segmentation.

Once the user decides to segment the data, there are two options: automatic and manual. For each of these options, the threshold value can be set. This is done as described, by selecting the number of segments to be many, medium, or few. If the user selects, for example, automatic segmentation and medium number of segments, loads the desired data files, and presses the Apply button, then there are no other options. Upon completion of the calculations, a window shows the data with superimposed segments. The user can save the results and synthesize.

If the user selects, for example, manual segmentation and medium number of segments, loads the two files; m21_lf_for_anal.mat (target) and m11_lf_for_anal.mat (source), and presses the Apply button, then after the calculations are completed the window shown in Figure 7.43 appears. In this window the user can modify the segmentation results in a manner similar to that described for GCI and formant tract modification. The user can zoom in and out and add and delete a segment. To add a segment mark (or boundary), press the Add button, move the cross hair to the desired location and press the left mouse button. A new segment marker is placed at this location and the data between the new segment marker and the one to the left of this marker is played. Another option is to move the cross hair to the desired location and click the right mouse button. The data between the location of the

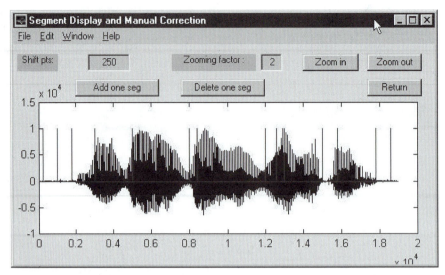

FIGURE 7.43 Manual segmentation window.

cross hair and the segment marker to the left of the cross hair is played. A new segment is not added until the left mouse button is clicked.

To delete a segment marker follow a similar procedure, clicking the left mouse button when the cross hair is placed on the segment marker to be deleted. When the user adds or deletes a segment marker the Apply button appears. Press this button to make the desired changes permanent. If the user presses the Return button without pressing the Apply button, then all changes are discarded. Pressing the Return button automatically initiates the continuation of the execution of the program.

The segmentation option does pitch conversion on a segment-by-segment basis by placing a restriction on the placement of the GCIs at segment boundaries to avoid discontinuities. The gain contour conversion process is smoothed with a median filter of length 7. The glottal source conversion is the same as that without segmentation, except the conversion is done on a segment-by-segment basis. The vocal tract conversion is performed only for the formant vocal tract model and is done in the same manner as that for no segmentation. In general, the formant bandwidths are not converted. The user can select either source model.

7.15 ALTERNATE LOAD, DISPLAY, AND PLAY OPTION

The m-file called speech_display.m in the display_speech folder within the VOCOS folder can be used to load, play, and compare four speech dat files. Change directory to the display_speech folder and type speech_display in the MATLAB Command window to bring up the window in Figure 7.44. The user can load speech or synthesized speech data (dat) files. These files can be played, and the waveforms or spectrograms can be compared. Figure 7.44 shows the files f11.dat, f21.dat, m11.dat, and m21.dat.

To use the speech_display.m file, first load the desired number of data files, that is, 1 to 4 files. A message appears in the MATLAB Command window reminding the user to strike any key on the key board to continue. Upon doing so, the window shown in Figure 7.45 appears. The user can select one of the various buttons, play or spectrogram, for each of the

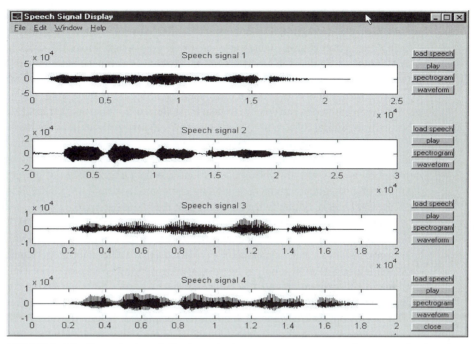

FIGURE 7.44 Speech signal display window to compare up to four speech signals.

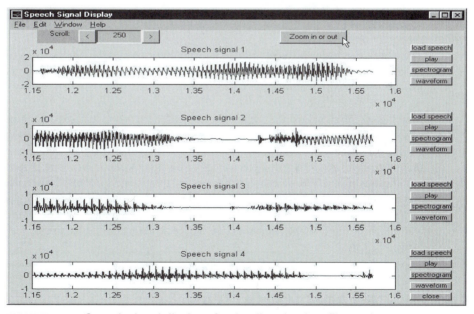

FIGURE 7.45 Speech signal display after loading the data files and pressing any key.

speech signals. The waveform button redisplays the speech signal if the spectrogram is plotted. Note that a Zoom-in or -out button is available to zoom the waveform data display. Pressing this button changes the mouse cursor to a cross hair. Move the cross hair to the desired initial x-axis data location to zoom in. Press the left mouse button once. Move the cross hair to the desired end of the zoom in x-axis segment. Press the left mouse button once again. This zooms in one level on all the loaded data files at the same time. The zoomed-in waveforms are plotted in each panel. Repeat these steps to zoom in another level. To stop the zoom-in process, press the right mouse button twice slowly while the cross hair is visible. The cross hair disappears and the standard mouse cursor reappears. To zoom out one level, press the Zoom-in or -out button. The cross hair reappears. Press the left mouse button, followed by a press of the right mouse button. The data are zoomed out one level. Each repetition of this sequence zooms out another level, until the original signal level is reached. Press the right mouse button twice slowly to exit the zoom out option.

The scroll option is to be used after zooming in on the data at least one level. The scroll option becomes available automatically after zooming in. The default scroll value is 250 data points. This value can be changed by typing in a new value in the scroll window panel. Press the left (right) mouse button to scroll the data to the left (right). You can continue scrolling in either direction until the end of the data record. The scroll option can be used at any zoom-in level, as long as the zoom-in level is at least one.

The play option is designed to play the data displayed. Thus, if the data are zoomed in, the data played are the data shown in the panel. Similar remarks apply to the spectrogram option.

The Close button closes out the Speech Signal Display window.

7.16 SUMMARY

There are three major features to the voice modification system: (1) the analysis and parameterization of acoustic features; (2) parameter visualization; and (3) windows for user interaction to modify parameters. There are four aspects of the software system: analysis, correction, modification, and synthesis. The analysis phase extracts the acoustic features from the speech signal using specific algorithms. The acoustic features are represented by sets of parameters, which are displayed in graphs. The correction phase allows the user to change parameter values. The modification phase is similar to the correction phase, but is more extensive, allowing parameter values to be altered (modified). One purpose for the modification phase is to alter the parameters to convert one speaker's voice to sound like that of another speaker's voice, that is, voice conversion. Finally, the synthesis phase allows the synthesis of speech using the measured parameters. Synthesis can be used for voice conversion or more simply to synthesize a speech file using the parameter file obtained using the analysis phase.

7.17 MISCELLANEOUS MATTERS

Sometimes the user will make an error in the use of the software, perhaps making an incorrect selection, or clicking the wrong option or the wrong mouse button. When this happens a software variable may be set incorrectly, leading to an incorrect software calculation or data plot. A beginning user may not notice such errors at first. However, as the user becomes more skilled such errors are apparent. The best way to reset the software is to quit the analysis program and restart, loading the data file and selecting the desired options again. This is necessary only occasionally.

Do not move and click the mouse button while the software is making calculations, since this can cause the data to be plotted in an incorrect window.

PROBLEMS

The following set of problems is to be completed using the default parameter settings in the analysis specification window. This means that the polynomial excitation model is used and the vocal tract filter is LP. Once Problems 7.1 through 7.6 (or a subset) have been completed, perform the same task using the polynomial model and the formant vocal tract model. Repeat using the LF and the LP models. Repeat using the LF and formant models

7.1 Use the software in this chapter to convert the sentence, "We were away a year ago," for a male voice in file m0125s.dat to sound like that of a female voice for the same sentence in file f0625s.dat.

7.2 Use the software in this chapter to convert the sentence, "We were away a year ago," for a female voice in file f0625s.dat to sound like that of a male voice for the same sentence in file m0125s.dat.

7.3 Repeat Problem 7.1 for the sentence, "Early one morning a man and woman ambled along a one mile lane," using files m0126s.dat and f0626s.dat.

7.4 Repeat Problem 7.2 for the sentence, "Early one morning a man and woman ambled along a one mile lane," using files m0126s.dat and f0626s.dat.

7.5 Repeat Problem 7.1 for the sentence, "Should we chase those cowboys?" using files m0127s.dat and f0627s.dat.

7.6 Repeat Problem 7.2 for the sentence, "Should we chase those cowboys?" using files m0127s.dat and f0627s.dat.

7.7 Analyze the sentence, "We were away a year ago," for the file m0125s.dat. Use the default settings for the polynomial and LP models. Experiment with altering the pitch contour and the jitter contour to create a breathy voice, a hoarse voice, a falsetto voice, and a vocal fry voice. It may be necessary to alter the gain contour as well.

7.8 Repeat Problem 7.7 for the sentence, "Should we chase those cowboys?" for file m0127.dat.

7.9 Analyze the sentence, "We were away a year ago," for the file m0125s.dat. Use the default settings for the polynomial and LP models. Experiment with altering the gain contour to determine the influence of this parameter on the quality of the voice.

7.10 Repeat Problem 7.9 for the sentence, "Should we chase those cowboys?" for file m0127.dat.

7.11 Analyze the sentence, "We were away a year ago," for the file m0125s.dat. Use the default settings for the polynomial and LP models. Experiment with altering the formant contours to determine the influence of this parameter on both the quality of the voice and the intelligibility of the synthesized speech.

7.12 Repeat Problem 7.11 for the sentence, "Should we chase those cowboys?" for file m0127.dat.

7.13 Return to Problem 7.1. Experiment with the various voice conversion parameters for dynamic time warping (DTW) to determine the effect these parameters have on voice quality and speech intelligibility.

7.14 Repeat Problem 7.13 except return to Problem 7.5.

7.15 Return to Problem 7.1. Experiment with the segmentation process to determine the effect this process has on voice quality and speech intelligibility.

7.16 Repeat Problem 7.15 except return to Problem 7.5.

7.17 Analyze the sentence, "We were away a year ago," for the file m0125s.dat. Use the default settings for the polynomial and LP models. Experiment with altering the formant contours to determine if you can create a voice with an accent or dialect, such as a Spanish accent (or one of your choice). It might be helpful to analyze the same sentence spoken by a person with such an accent.

7.18 Analyze the sentence, "We were away a year ago," for the file m0125s.dat. Use the default settings for the polynomial and LP models. Experiment with inserting pauses at various locations within the sentence to create a new effect in the synthesized speech.

CHAPTER **8**

TIME MODIFICATION OF SPEECH TOOLBOX

8.1 INTRODUCTION

The purpose of this chapter is to introduce a software-based, time modification system to independently and automatically modify the durations of the phonetic segments in a speech signal. The system can be used to create high-quality test tokens for use in speech perception studies, to examine the influence of various coding schemes on speech segments, to create speech that has a high "speaking rate" or is speeded-up, to create databases for speech recognition, and other applications.

The time modification system analyzes the speech signal, dividing the signal into phoneme-type segments, with each segment labeled as either vowel, semivowel, nasal, voice bar, voiced fricative, unvoiced fricative, unvoiced stop, or silent. The segmentation and labeling algorithms are based primarily on the short-term frequency distribution of the speech signal. The theory for the software toolbox is described in Appendix 9 and is based on White (1995).

The time modification system uses a graphical user interface that allows the user to specify, via slide-bar controls, both the desired time scale factor and minimum duration for each segment. The user can also specify a weighting function, or "map," for each segment. The map determines the portion of the segment that is modified. A linear predictive coding speech synthesizer creates the resulting time-modified speech. The mapping and synthesis algorithms are described in Appendix 9.

For certain speech applications, it is desirable to change the rate at which recorded or synthesized speech is presented to a listener. One example of this is a device that varies the playback rate of audio books for the blind. This allows the non-sighted listener to "read" at his or her own speed, independent of the rate at which the recording was originally made.

One of the driving forces behind research of time-modified speech is that it has long been known that a human being can comprehend speech at a rate greater than he or she can produce speech (de Haan, 1982; Foulke and Sticht, 1969; Goldman-Eisler, 1968; Goldstein, 1940). Therefore, a significant time savings results by increasing the rate of prerecorded speech in applications such as playback of academic lectures, conference papers, religious sermons, archived political speeches, and other such recordings. Because of this difference between the maximum speaking and perception rates, the large majority of published research has studied speech compression ("speeded up" speech) as opposed to speech expansion ("slowed down" speech). While there are applications for speech expansion, these are far fewer in number. The most common are expansion of speech for the hearing impaired or for foreign language learning.

Time-modified speech also has application as a research tool for the development of test data for use in perceptual studies of both normal and pathologic patients. The durations of different portions of the speech signal are modified in order to test theories of speech perception from either a psychological or phonological viewpoint.

In addition, time modification can be used to create databases for development of speech recognition systems. Many different variations (in terms of duration) of a test word or sentence can be systematically created from a single token. The different variations can then be used to further train the system, or to perform controlled tests of the system's ability to correctly detect data different from the training data.

8.2 OVERVIEW OF THE SPEECH TIME MODIFICATION TOOLBOX

The disadvantages of waveform editors and cut and paste methods, described in Appendix 9, to modify the duration of speech is that they are manually labor intensive. This provides the motivation for the software described in this chapter. The development of such a system could eliminate the need for computer-based waveform editors and the associated manual cutting and pasting processes. In addition, it would be desirable if the user could control the system with parameters that are closely related to the acoustic features of the speech signal. This could greatly decrease the user training time, and in general, would make the system easier for the speech researcher to operate. The system might also inspire new research into time-modified speech, due to the increased levels of efficiency and flexibility that were previously unavailable in a time modification system.

The time modification system allows the user to selectively modify certain portions of a speech signal based on the signal's time-varying acoustical composition. In order to aid the speech researcher, the segments are similar to the set of phoneme types (i.e., vowels, nasals, semivowels). To do this, a software tool is provided that first analyzes the speech signal to determine the identity of the different phonetic segments, and then independently modifies the durations of the phonetic segments according to global parameters specified by the user. The time modification is done automatically, without the use of waveform editors. In addition, the software is written in the MATLAB programming language and can be ported at relatively low cost to a wide variety of computing platforms. The theory for these algorithms is described in Appendix 9.

The time modification system incorporates a graphical user interface (GUI) that frees the user from having to remember complicated command-line syntax. All of the modification parameters are adjusted by using a mouse to move and select slide-bar and push-button controls that are displayed in various windows. After the modification parameters are specified, the resulting time-modified speech is synthesized and played with the click of the mouse button.

The three main stages of the system are (1) the speech analysis, segmentation, and labeling stage; (2) the manual correction stage (for optional correction of the segmentation and labeling results); and (3) the segment time modification and synthesis stage. Both the natural speech input signal and the synthesized speech output signal are sampled-data time–domain signals. The sampling frequency is fixed at $F_s = 10$ kHz.

The first stage works automatically with no input from the user, other than specifying the sampled-data input signal file. This stage divides the signal into pitch-synchronous frames (pitch-asynchronous for unvoiced and silent speech) and performs a linear predictive coding (LPC) analysis for each frame. The frames are then grouped into segments, and each segment is labeled with the most appropriate phonemic type label (i.e., vowel, semivowel, etc.). This entire process is accomplished by a series of software programs that extract the acoustic features from the signal and compare the relative contribution of each feature to the specific speech segment.

The second stage provides a means for the user to manually correct the automatic segmentation and labeling results. This is required only if the automatic segmentation and labeling stage makes mistakes. Determination of whether or not a mistake is made is left to the discretion of the user. In this stage, a set of software programs with a graphical user interface (GUI) allows the user to display and graphically edit the segment boundaries and labels. The user adjusts the results by moving sliders and pushing buttons (with a mouse) on the computer display.

The third stage performs the actual time modification process. It also performs the synthesis of the resulting, time-modified speech. This stage uses a set of software programs with a graphical user interface (GUI) that allows the user to graphically specify how the speech signal is to be modified. Each type of phoneme has its own modification parameters, and in addition, each segment can also have its own modification parameters independent of phoneme type, if desired. Once the parameters are all specified, the third stage synthesizes the time-modified speech using an LPC speech synthesizer.

While theoretically the toolbox can be used to time-modify sentences, it is used primarily to modify words. This is because the user can more easily identify the "phonemic" structure of a single word and verify that the analysis, segmentation, and labeling results are correct. As seen in previous chapters, the estimation of word boundaries in sentences can complicate this task.

8.3 TOOLBOX FOR THE TIME MODIFICATION OF SPEECH

The software is to be installed in a subdirectory (e.g., time) in MATLAB in a manner similar to that described for the toolboxes introduced in previous chapters.

The toolbox contains three software packages, which are outlined in Figure 8.1. The first package performs speech analysis, segmentation, and labeling, as described in Appendix 9. The second package allows the user to inspect the results of the first package

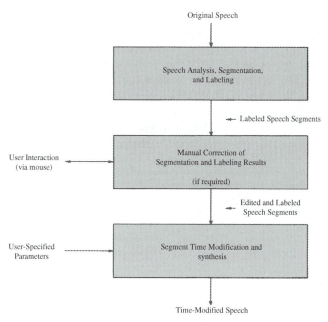

FIGURE 8.1 Block diagram of the speech time modification toolbox.

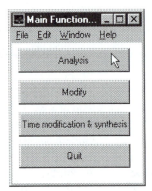

FIGURE 8.2 Main function window.

and make any desired corrections via manual interaction with the data. The third package synthesizes the time-modified speech.

8.3.1 Analysis Option

To start the software, change directory to the time directory and type main in the MATLAB command window. The Main Function window shown in Figure 8.2 appears. This window allows the user to select one of the three options: analysis, modify, or time modification and synthesis. The Quit button closes all previously opened windows and exits the user from the toolbox. The typical order of selection is analysis, modification, and finally time modification and synthesis. However, any of the options can be selected, provided the required data files are available. This will become more apparent as the details of the various options are described. Suppose we select the analysis option. This action opens a window (not shown) that allows the user to select a data file previously stored in the data folder. Suppose we select the b.dat file. The analysis starts automatically and opens a manual correction window, shown in Figure 8.3, which is similar to that for the voice conversion system. At the same time, the software prints messages in the MATLAB command window, as illustrated in Figure 8.4.

The appearance of Figure 8.3 and the messages in Figure 8.4 signal the completion of the first stage of the analysis phase, which is the identification and labeling of the glottal closure instants (GCIs). Recall that the MATLAB command window is not available with the stand-alone software. It is only seen with the regular version of the MATLAB software. This is the same as that used for the voice conversion system. Press the GCIs Inspection button. This opens two windows: the GCI display and manual correction window shown in Figure 8.5, and the pitch contour (reference) window shown in Figure 8.6.

FIGURE 8.3 Manual correction window for analysis.

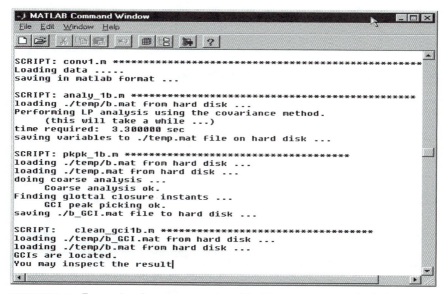

FIGURE 8.4 Example of messages printed in MATLAB command window.

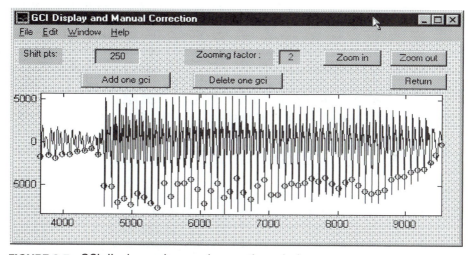

FIGURE 8.5 GCI display and manual correction window.

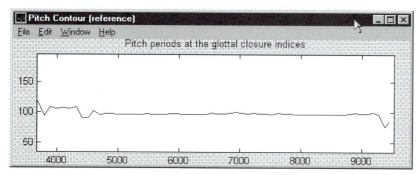

FIGURE 8.6 Pitch contour (reference).

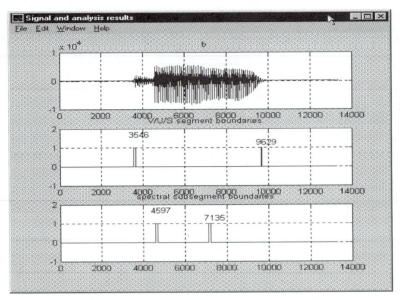

FIGURE 8.7   Signal and analysis results.

Figure 8.5 offers the same options as the window shown in Figure 7.11, that is, the user can shift the data, zoom in or out, add one GCI, delete one GCI, and return. Figure 8.6 displays the pitch contour, similar to that shown in Figure 7.12. The pitch contour is updated as the user makes manual corrections to the GCI display in Figure 8.5. The display in Figure 8.5 and 8.6 shows only the section of the data and excludes the preceding and following noise background. After zooming in or out, the scale changes to that of the complete data file, including the background noise. Once the desired manual corrections are made, press the Apply button (which appears if a GCI is added or deleted), then press the Return button in Figure 8.5. Next press the Continue button in Figure 8.3. The analysis phase continues, displaying a number of figures that are not saved. These figures, similar to the figures in Appendix 9, are provided as feedback to the user to illustrate that the analysis computation is being done. At the same time, messages are printed in the MATLAB command window.

The final result of the analysis phase is displayed in Figure 8.7, which is labeled as the signal and analysis results. Simultaneously, a summary of the results is printed in the MATLAB command window, as shown in Figure 8.8. Here the scores for the various types of phonations detected are presented along with a summary of the first and second choices for the type of phonations present in the original signal, which in this case is the data file b.dat. For a discussion of the algorithms that are used to obtain these results see Appendix 9.

In Figure 8.7, the name of the signal file (b) is displayed at the top middle of the figure. The next three panels show the data file (b.dat), the V/U/S segment boundaries, and the spectral subsegment boundaries. See Appendix 9 for a discussion of these results. The results shown in Figure 8.8 do not need to be recorded by the user. The results are stored along with the data vectors calculated by the software. This summary is presented to show the type of choices that the software determined for the data file.

8.3.2 Modify Option

At this point, the user can select to modify the results of the analysis phase. This is accomplished by selecting the Modify button in Figure 8.2 in the main function window. The

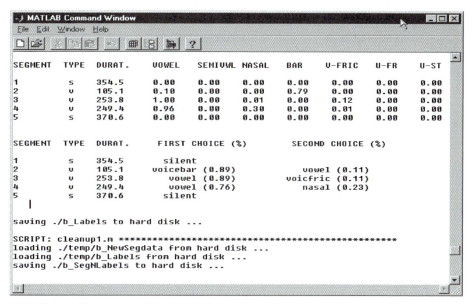

FIGURE 8.8 Summary of the analysis results printed in the MATLAB command window.

user can load the just analyzed data (in this case, b.dat) or a previously analyzed data file. This is accomplished by opening the data folder within the load window (not shown) and selecting the b.dat file. This selection tells the software the name of the original ASCII file that was analyzed. The software then selects the proper files that contain the analyzed data vectors. After the desired data file name is selected, Figure 8.9 appears, which for our case shows the original b.dat data, the original segment type and duration, and the modified segment type and duration. The modified segment type and duration is the same as the original until the user makes modifications to the data. The horizontal scale is the same as that for the original data file. Note that the segment labels are those for the first choice selection displayed in Figure 8.8. If we add the first four segment durations we obtain 9628, which is one less than the end of the voiced segment in Figure 8.7. The discrepancy of one sample is due to the manner by which the algorithms count segment intervals.

The user can now modify the analysis results using the tools shown in Figure 8.10, which is called main for main modify window. The signal name, b, is displayed. Next there is a row of four buttons: Quit (no save), Discard Changes, Save Changes, and Quit (after save). The display options are: top, which has a pull-down window that allows the original or the modified signal to be displayed; middle, which has a pull-down window that allows the display of the original time and duration, the modified time and duration, the original signal, the modified signal, the original segment boundaries, or the modified segment boundaries; and bottom, which has a pull-down window that allows the display of the original time and duration, modified time and duration, original segment boundaries, or modified segment boundaries. These selections correspond to the panels shown in Figure 8.9. The top, middle, and bottom panels in Figure 8.10 simply allow the user various options for displaying the original and modified data. The lower most row of options in Figure 8.10 is called parameter windows, with options to insert silence, change (move) segment boundaries, change (fix) segment labels, and merge like segments.

The selection of the insert silence option opens Figure 8.11, which allows the user to insert silence of various durations at various points. For example, the user can insert silence

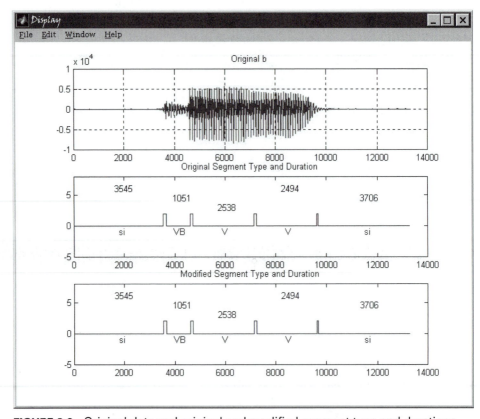

FIGURE 8.9 Original data and original and modified segment type and durations.

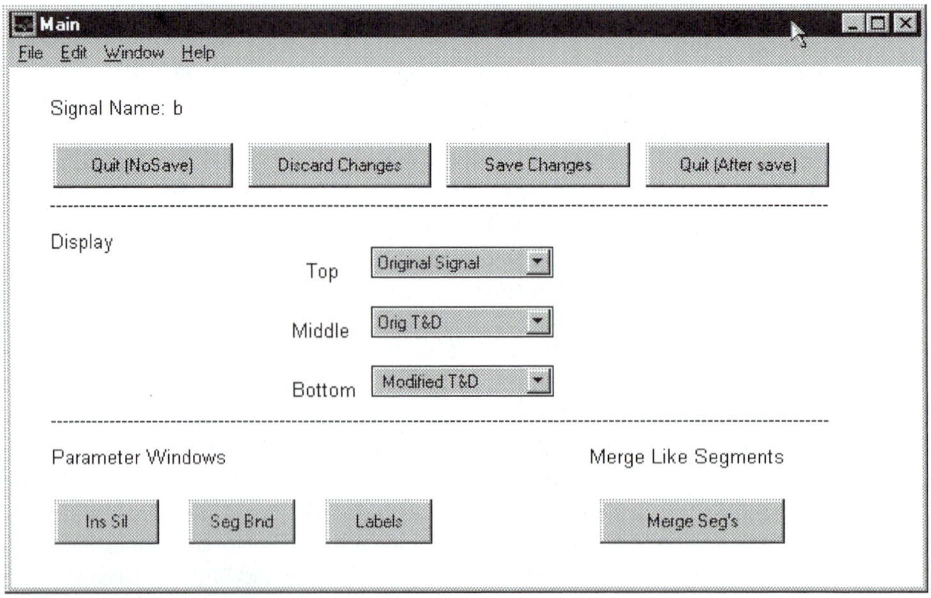

FIGURE 8.10 Main modify window.

FIGURE 8.11 Insert silence window.

of 100 msec duration starting at point 4597. This insertion will place a silent segment following the VB (voice bar), which will be displayed in the modification (lower) panel of Figure 8.9 as a modification. Selection of the OK button closes the insert silence window, Figure 8.11. Such changes will remain displayed in Figure 8.9 until the user selects one of the options: Quit (no save), Discard Changes, Save Changes, or Quit (after save).

Selection of the Segment Boundaries button in Figure 8.10 opens Figure 8.12, which displays the segment boundaries and allows the user to alter these boundaries. For example, the initial silent/semivowel boundary can be increased or decreased. However, the increase cannot exceed the duration of the voice bar segment, and so forth.

The user can change the labels assigned by the analysis (excluding silence segments) by selecting the labels button in Figure 8.10. This is shown in Figure 8.13. For example, the voice bar (bar) label can be changed to one of the eight possible segment types: vowel, nasal, semivowel, bar, voiced fricative, unvoiced stop, unvoiced fricative, or silent. Similar changes can be made for the two vowel segments.

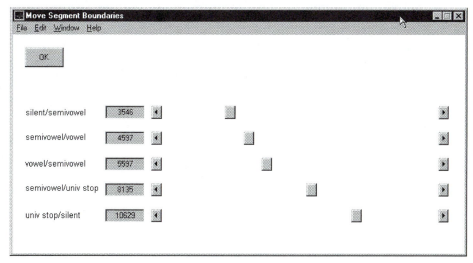

FIGURE 8.12 Change (move) segment boundaries window.

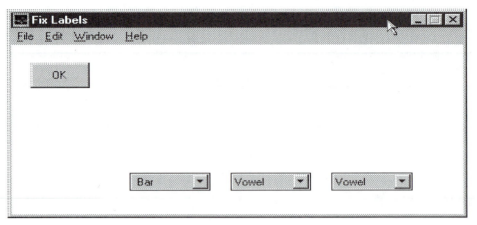

FIGURE 8.13   Change (fix) labels window.

Pressing the Merge Segments button in Figure 8.10, merges the two vowel segments shown in Figure 8.9. The result is shown in Figure 8.14. If there had been other adjacent like segments, they would have been merged as well. At the same time that Figure 8.14 appears, the Figure 8.12 is updated and displayed in Figure 8.15, allowing the user to adjust the segment boundaries for the newly modified data.

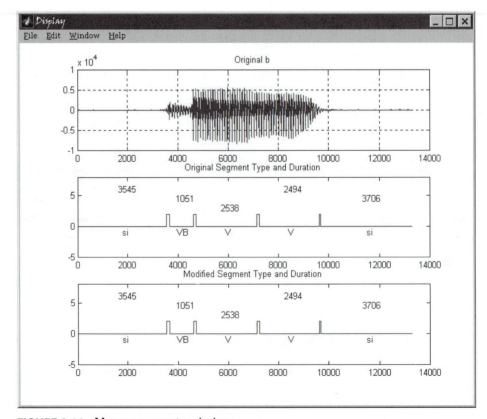

FIGURE 8.14   Merge segments window.

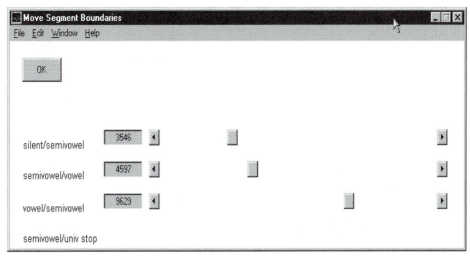

FIGURE 8.15 Updated change segment boundaries window after merging the two adjacent vowel segments.

At this point, the user can review the various displays by making various selections in the top, middle, and bottom panels in Figure 8.10. Next the user can quit (no save), discard changes, save changes, or quit (after save). The save changes option opens a save file window (not shown), asking the user for a file name. The file is saved as a *.m file, such as b.m. The quit (after save) closes the various windows opened during the modify session. The main window (analysis, modify, time modification and synthesis) remains open.

8.3.3 Time Modification and Synthesis

The primary option is time modification and synthesis, which allows the user to generate a speech signal with a desired time characteristic; that is, compressed or expanded. Pressing the time modification and synthesis button in Figure 8.2 opens a load input window (not shown) that allows the user to select the name of the original data file, b.dat. This process extracts the basename of the file, which in this case is b with the extension removed. The software then selects the required, previously analyzed and modified *.m files for time modification and synthesis. Thus, the name of the ASCII file (*.dat) must be the same as the file name assigned to the analyzed and modified data file. For example, if the original ASCII data file name was b.dat, but the user named the analyzed and modified file as bb.m, then be sure to make a copy of the original ASCII data file (b.dat) and name the copy bb.dat. Then be sure to select the file bb.dat, if this is the desired file. Note that the file basename.m must be in the temp folder. After loading the desired file, the main window opens, as shown in Figure 8.16. The signal name is b. The option to quit is obvious. The options available in the functions section include save, map, synthesize, and play. These options are usually selected after selecting one or more of the parameter windows options, which include preview, change scale factors (SF's), change minimum durations (MD's), manual scale factors (Man SF's), and select maps (Maps).

Selecting the preview option opens the window shown in Figure 8.17. This window displays the segment type and duration results as well as the segment boundaries. Recall that these results are obtained using the analysis and modification options. The horizontal scale is in samples. Figure 8.17 shows that there are three non-silent segments: a voicebar and two

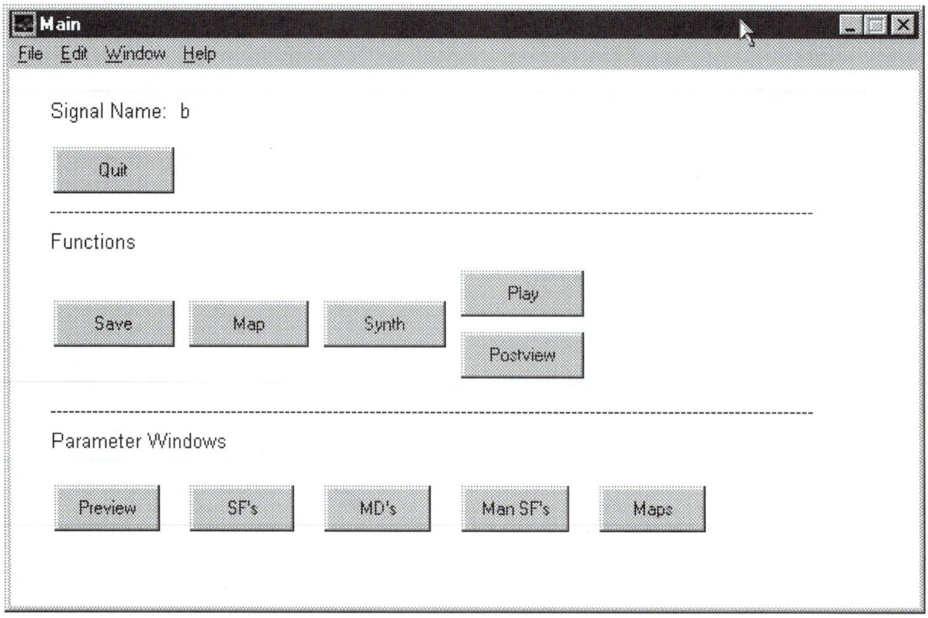

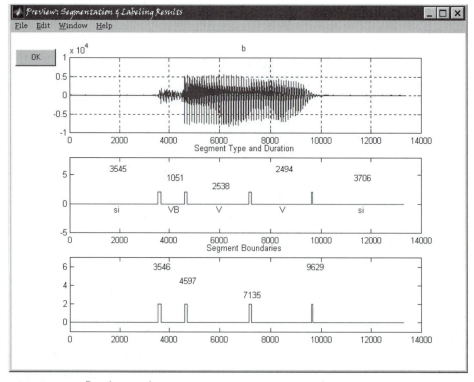

FIGURE 8.16 Main time modification and synthesis window.

FIGURE 8.17 Preview option.

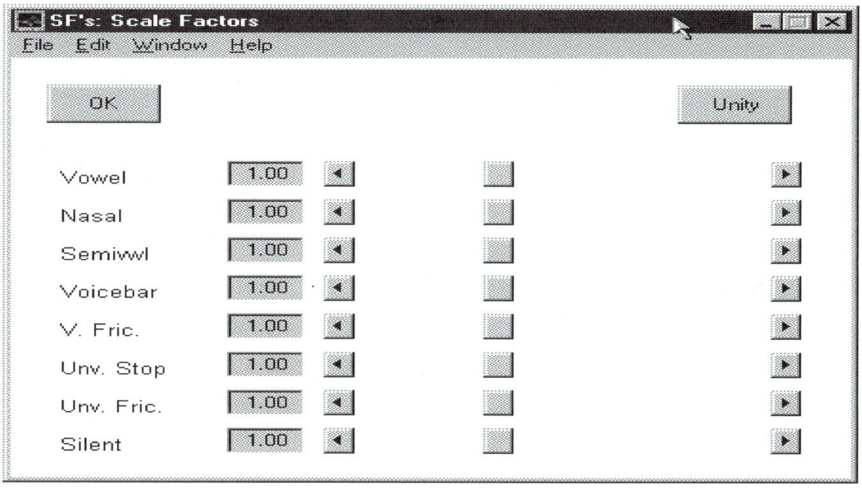

FIGURE 8.18 Scale factors option.

adjacent vowel segments that are not merged. Recall that the eight possible segment labels are si (silence), V (vowel), SV (semivowel), N (nasal), US (unvoiced stop), UF (unvoiced fricative), VB (voice bar), and VF (voiced fricative). The primary purpose for this option is to let the user verify that the proper data is selected. There are no options to change these results in this preview window. Pressing the OK button closes the window.

The scale factors option, shown in Figure 8.18, allows the user to alter the scale factors. The scale factors for the various possible phoneme-type segments can be reduced or increased, allowing the user to compress or expand a desired segment. There are eight sliders for the eight possible segment types. The desired scale factor is adjusted by moving the bar in the center of the slider with the mouse or by clicking the left or right arrow at either end of the slider. The slider position is automatically rounded to the nearest one-hundredth, and the rounded scale factor is displayed in the small box to the left of the slider. The range of values for the SF is limited to [0, 3.0]. Note, however, that modifying one line in the software can change this. If the user changes one or more of the SF parameters, then two new buttons appear (not shown). These buttons are labeled Update and Cancel. In order to accept and save the changes, the user must press the Update button. Pressing the Cancel button, resets the slider positions to their original values. The default for the scale factors is one (unity). The Unity button is also a reset button that resets all SF values to Unity. Pressing the OK button before pressing the Update button closes the window without saving the user-selected values.

Pressing the Minimum Durations (MD's) button opens the minimum durations window shown in Figure 8.19. There are several buttons at the top of the window and eight sliders for the eight possible segment types. The desired minimum duration (MD) is adjusted by moving the bar in the center of the slider with the mouse, or by clicking the left or right arrow at either end of the slider. The slider position is automatically rounded to the nearest millisecond, and the rounded minimum duration is displayed in the small box to the left of the slider. The range of values for MD is limited to [0, 250] msec. However, this can be changed in the software. If the user modifies one or more of the MD parameters, then Cancel and an Update buttons appear (not shown) next to the OK button. In order to accept and save changes, the user must press the Update button. If the user presses the Cancel button, the slider positions and MD values are reset to their original values. The Defaults

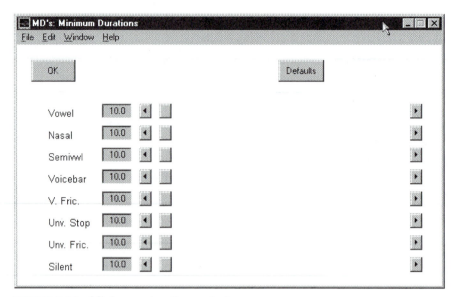

FIGURE 8.19 Minimum durations window.

button also resets all of the MD values to their predefined value of 10 msec. The default value can be changed in the software. Pressing the OK button before pressing the Update button closes the window without saving the user-selected values.

The Man SF's button opens the manual scale factors window shown in Figure 8.20. The window design is similar to that for Figures 8.18 and 8.19 except that in this window, the sliders are for the specific phoneme-type segments present in the data file, which in this case are voice bar, vowel, and vowel. The first and last silent segments are excluded. Recall that it is assumed that these silent segments are always present. For each row, both the index (segment number) and segment type are displayed within the Push Button box. There is also a diamond shaped box inside the Push button. If this box is filled, the push button is on, and the manual scale factor (MSF) parameter is active for the segment. If the box is not filled, then the push button is off, and the MSF is inactive. In addition, if the MSF parameter is active, then a slider and corresponding numerical display box are displayed to the right of the push button. These adjust and display the value of the MSF parameter

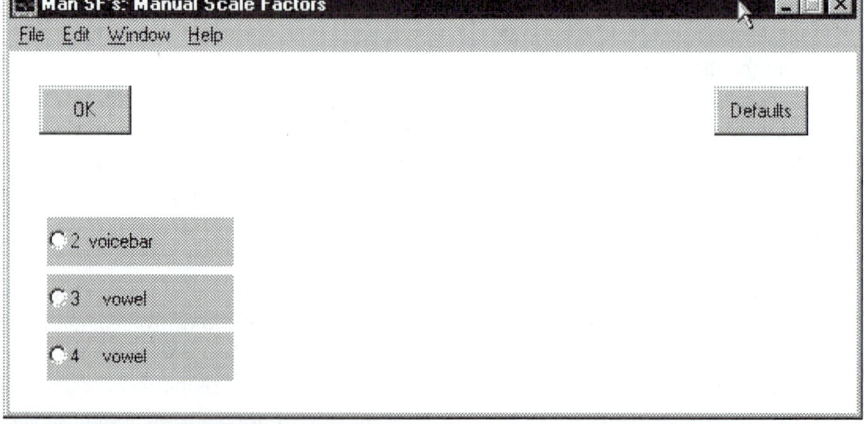

FIGURE 8.20 Manual scale factors window.

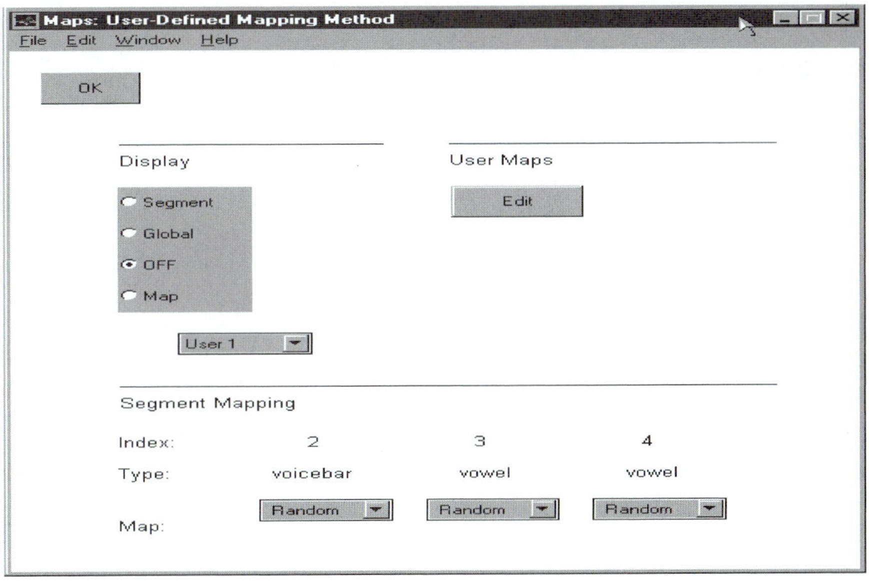

FIGURE 8.21 Maps window.

for the segment. The MSF value is adjusted by moving the bar in the center of the slider with the mouse, or by clicking the left or right arrow at either end of the slider. The slider position is automatically rounded to the nearest one-hundredth, and the rounded value is displayed in the small box to the left of the slider. The range of values for MSF is limited to [0, 2.50]. This can be changed in the software. If the user modifies one or more of the MSF parameters, then Update and Cancel buttons appear (not shown) next to the OK button. To accept and save changes, press the Update button. The Cancel button resets the slider positions and MSF values to their original values, which is unity. The Defaults button also resets the active MSF values to one. The inactive MSF values are not changed. Pressing the OK button before pressing the Update button closes the window without saving the user-selected values.

The Maps button opens the map window, shown in Figure 8.21, which is the top window in a hierarchy of sub-windows that control user-defined weighting functions, or maps. The map option provides the user with the ability to specify those frames of a segment that are to be removed or doubled in the time modification process. A weighting function, or map, is assigned by the user to each segment. For example, a map may specify that every other frame of a segment is to be removed to achieve compression; or a map may specify that every other frame of a segment is to be doubled to achieve expansion. The map option provides flexibility for modifying specific portions of a segment. Each frame is assigned a weight between zero and one. During synthesis, the frames with the lowest weights are eliminated (if SF < 1.00), and the frames with the highest weights are doubled (if SF > 1.00). If SF (or MSF) for the segment is 1.00, then the weight is ignored, since frames are neither eliminated nor doubled. See Appendix 9 for additional details on this option.

The maps window is divided into three sections. The display section controls another window that displays either (1) one or all of the speech segments and the associated interpolated map(s); or (2) one of the eight maps (with no interpolation). See Appendix 9 for an explanation of an interpolated map. The eight maps, discussed in Appendix 9, are: random, fixed_1, fixed_2, fixed_3, fixed_4, user_1, user_2, and user_3. The Edit button in the section designated as user maps opens a window that is used to edit and display one of the three user maps: user_1, user_2, or user_3. The segment mapping section offers

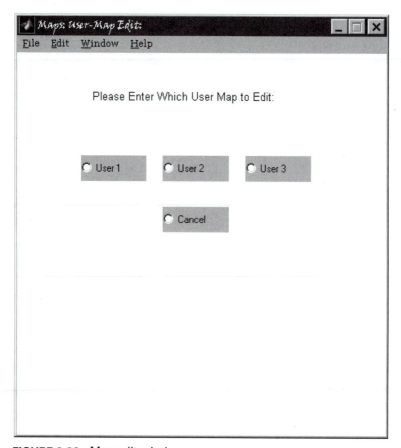

FIGURE 8.22 Map edit window.

a variable number of push buttons that correspond to the time sequence of segments in the speech data. There is one push button for each segment present. The push buttons are arranged from left to right, with the left most button being for the first non-silent segment, and the right most button being for the last non-silent segment. For example, for the speech file b.dat, we have voice bar, vowel, vowel. These push buttons offer a pop-up menu of the eight map choices available for each segment. The default map for each segment is random. The segment index and label (type) are displayed above each push button.

Pressing the Edit button opens the map edit window is shown in Figure 8.22. This option allows the user to edit the user maps to create customized weighting functions. There are four push buttons in this window. The user selects one of the three user maps to be edited, or selects cancel to close the window. If one of the three user push buttons is selected, the four buttons vanish, and a graph and a new combination of push buttons become available, as shown in Figure 8.23, where user_3 map is selected.

The graph of the map is displayed along with four push buttons. The # Targets push button controls a pop-up window that allows the user to select the number of targets displayed on the graph. In this case, two targets are shown as small circles at either end of the graph. The pop-up window (not shown) offers the choice of two, three, six, or eleven targets. If two targets are selected, they are located at 0 and 100 percent on the x-axis, as shown in the figure. If three targets are selected, they are located at 0, 50, and 100 percent on the x-axis. If six targets are selected, they are located at 0, 20, 40, 60, 80, and 100 percent on the x-axis. If eleven targets are selected, they are located at 0, 10, 20, 30, 40, 50, 60, 70,

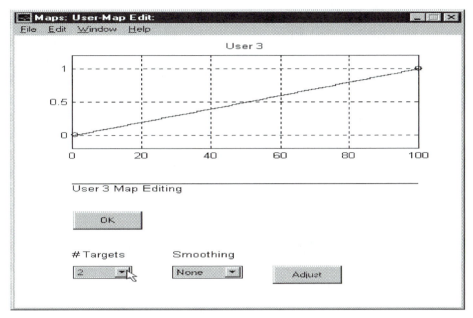

FIGURE 8.23 Map editing process: user_3.

80, 90, and 100 on the x-axis. The Adjust push button allows the user to move the targets vertically with the mouse. This is done by pressing the Adjust button, and then positioning the cursor over the graph at the desired location for a given target. When the cursor is at the desired position, the user presses the left mouse button, and the target is moved to the new location on the graph. The target positions are fixed along the x-axis, and quantized to the nearest one-tenth along the y-axis. After one target is moved, the adjust button must be pressed again to modify another (or the same) target.

In most cases, the weighting function is calculated as a best fit polynomial curve (in a least-squares sense) that is fitted to the targets. The process of curve fitting to the targets is called smoothing. The order of the polynomial is controlled by the Smoothing button, which provides a pop-up menu (to be illustrated in an example below) that displays the choices none, linear, poly_2, poly_3, poly_4, and poly_5. The default is none. None generates a straight line weighting function between each target. Linear provides a weighting function that is a best-fit straight line to the selected targets; that is, a first-order polynomial fit. If one of the polynomial choices is selected the weighting function is a best-fit polynomial curve, with the order being determined by the suffix of the polynomial choice. If the user modifies any of the default values, then Update and Cancel buttons appear. To accept the new settings, press Update. To return to the default values, press Cancel. Pressing the OK button before pressing the Update button closes the window without saving the user-selected values.

An example illustrating the selection of three targets and an order 2 polynomial fit to the user_3 map is shown in Figure 8.24.

Return to the maps window in Figure 8.21. The display section contains a group of four push button labeled segment, global, off, and map, as well as another button located below map. This latter button provides a pop-up menu selection for the eight possible maps: random, fixed_1, fixed_2, fixed_3, fixed_4, user_1, user_2, and user_3. The four buttons control the display in the map display window, and are known collectively as the display mode. When off is selected, the map display window is closed. When the segment display mode is selected, the map display window shows two graphs (not illustrated in the example below). The top graph displays the time–domain waveform of the segment, while

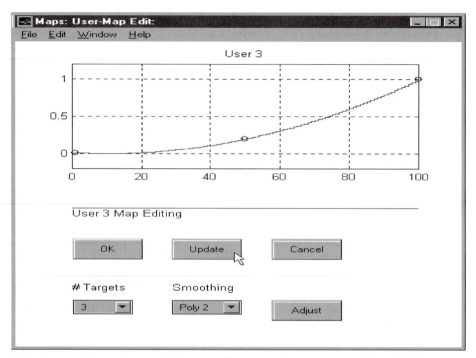

FIGURE 8.24 Example modifying user_3 map to three targets with poly_2 fit.

the bottom graph displays the user: selected interpolated weighting map associated with the segment. Both graphs are updated to display the next successive segment each time that the Segment button is pressed. This allows the user to scroll through the segments by repeatedly pressing the Segment button until the segment of interest is displayed. For example, for the b.dat file, the first segment display is segment 2 (voice bar) with the random map, since the first non-silent segment is the voice bar. The random map is shown because it is the map selected for that segment in the segment mapping section. Pressing the Segment button again changes the map display to segment 3 (vowel) with the random map, and so on. When the global display mode is selected, the map display window shows the same two graphs as for the segment mode, except that the graphs span the entire duration of the speech token. This is illustrated in the following example. When the map display mode is selected, only the non-interpolated map is displayed in the map display window. The specific map that is displayed is changed by the button directly under the Map button.

Upon completion of the desired changes in the various options, go to the functions section of Figure 8.16. Press the Save button, then the Map button, then the Synthesize button. Upon completion of the synthesis process the newly synthesized data file, syn_b.dat, is created and stored in the temp folder within the time directory. The file name for the synthesized data always contains the prefix syn_ for the data file being time modified. Thus, for the b.dat file the synthesized file name is syn_b.dat. This file can be played and/or viewed in the postview window. The manner by which this is done is illustrated in the following example.

For the b.dat file suppose we alter the scale factors (SF) as shown in Figure 8.25. The voicebar SF is changed from 1.00 to 0.18 (shortened), and the vowel SF is changed from 1.00 to 2.15 (lengthened). Press Update and then OK. Let the MDs and Man SFs remain as the default values.

Next, press the Maps button and change the voicebar, vowel, vowel maps to fixed_1, fixed_2, and fixed_3, as shown in Figure 8.26. Press the Global button. The result is shown

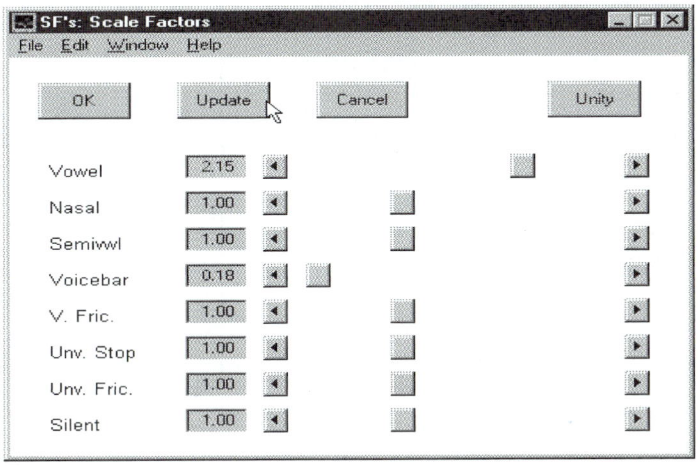

FIGURE 8.25 Example illustrating the changing of the SFs for b.dat.

in Figure 8.27, where the upper graph shows the time–domain waveform for the b.dat file, and the lower graph shows the global map, which is the fixed_1 map followed by the fixed_2 map followed by the fixed_3 map.

Next return to Figure 8.16 and press the Save button, then the Map button, then the Synthesize button. The newly synthesized data file can be viewed using the postview window shown in Figure 8.28. The upper graph shows the original time–domain waveform for the b.dat file. The second graph down shows the segment type and duration data obtained via the analysis software. The third graph down shows the cut/save/double frame (indicator of the number of times the input is used in the output synthesis file). This graph shows the number of times each frame of the unmodified speech token is used to synthesize the time-modified speech. In this case, it can be seen in Figure 8.28 that the latter portion of the voicebar is not used. It is removed. The first part of the first vowel segment is doubled, and

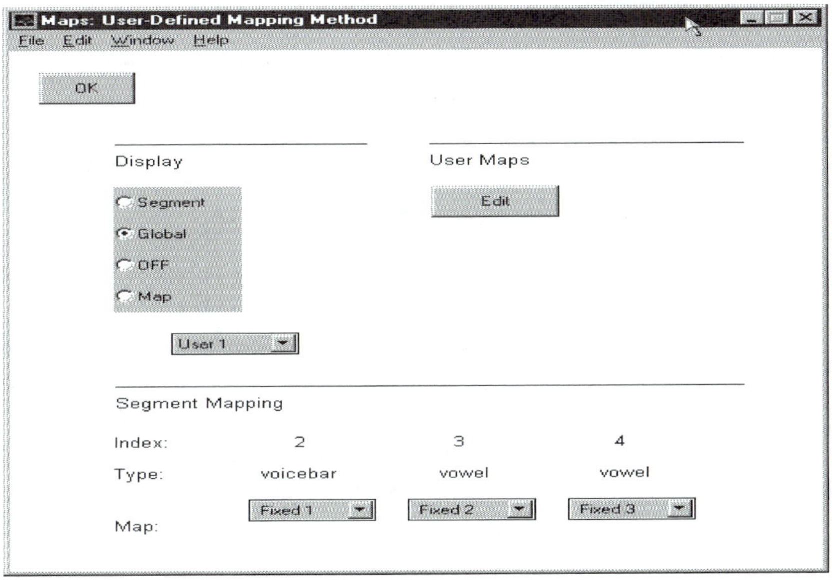

FIGURE 8.26 Continuation of example.

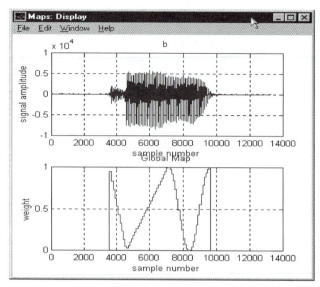

FIGURE 8.27   Continuation of example.

similarly for the second vowel segment. A few frames at the end of each vowel segment are tripled because an expansion of 2.15 was specified in Figure 8.25. Since 2.15 is greater than 2, then some frames are tripled to meet this specification. The last graph shows the synthesized data, with the voicebar shortened and the vowel segments lengthened. The user can select play or postview in any order and repeatedly. To play the synthesized data be sure to perform the following steps. Press the Play button. A load file window appears (not

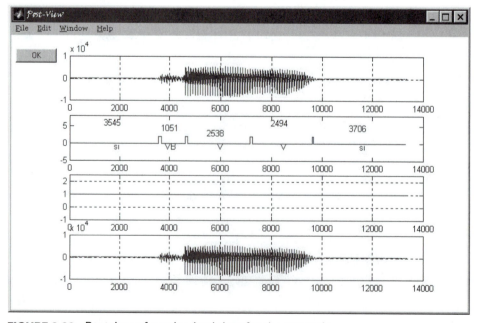

FIGURE 8.28   Postview of synthesized data for the example.

shown). Change directory to the temp folder and select syn_b.dat. For this example the syn_b.dat file sounds like b but with a longer duration because the vowel segment has been lengthened. The shortening of the plosive /b/ does not significantly alter the sound of /b/. In summary, the compression of the voicebar was performed using map fixed_1, and the expansion of the two vowel segment was performed using map fixed_2 and map fixed_3, respectively. For additional detail see Appendix 9.

The user can play the b.dat (or any) file in a similar manner. Press the Play button. Then load the desired data file. The selected data file is played.

8.4 SUMMARY

The time modification system is based on a LPC speech synthesizer. The first stage pitch-synchronously divides the speech signal into frames, and then performs the LPC analysis for each frame. The result for each frame is an ordered pair (A_i, R_i), where i is the frame index, A_i is a 1 by N vector of LPC coefficients, R_i is a 1 by M_i vector of the residue signal, and M_i is the length (in number of samples) of the frame. During synthesis, the ordered pairs are sent sequentially to the LPC speech synthesizer. To speed up or slow down the rate of speech, selected ordered pairs are either removed or duplicated, respectively, from the sequence. The synthesizer also incorporates a simple algorithm to prevent discontinuities (glitches) from being created in the synthesized output. See Appendix 9 for additional details.

The time modification algorithm modifies the durations of the segments that comprise the speech signal. Examples of the segment types include vowels, semivowels, unvoiced fricatives, and so forth. For each segment type the user can specify a manual scale factor (MSF). This allows a single occurrence of a specific type of segment to be modified independent of all other occurrences of the same type. For example, the word man is comprised of three segments: an initial nasal, a vowel, and a final nasal. The user has the option of specifying a separate MSF parameter for each of the three segments. Thus, the initial /m/ can be modified independently of the final /n/, even though both are the same segment type.

An important feature of the system is the ability to specify the frames of a segment that are to be removed (or doubled) in the time-modification process. This is accomplished by a weighting function, or map, that is assigned to each segment by the user. Several maps are provided and the user can edit these maps if desired. The time-modification process is controlled by a graphical user interface. This GUI consists of multiple windows to guide the user through the time-modification process.

PROBLEMS

8.1 Time modify the file b.dat to double the vowel segment. When you do the analysis and modification, merge the two vowel segments. Do the time modification using the three fixed maps: fixed_1, fixed_2, and fixed_3. Does the synthesized word still sound natural in all cases?

8.2 Time modify the b.dat file you analyzed and modified in Problem 8.1. However in this problem, modify the voicebar segment in two ways: (1) shorten the segment by 50%; and (2) lengthen the segment by 50%. Do each time modification using the three fixed maps: fixed_1, fixed_2, and fixed_3. Does the synthesized word still sound natural in all cases?

8.3 For the sentence, "Early one morning an man and a woman ambled along a one mile lane," isolate the word man by cutting it out of the sentence. Time modify the word by (1) simultaneously decreasing the two nasals by 50%; (2) simultaneously increasing the two nasals by 50%; (3) decrease only the /m/ by 50%; and (4) decrease only the /n/ by 50%. Discuss the results of your listening test.

8.4 Use the word man from Problem 8.3 again. Time modify the word by decreasing the first nasal, /m/, by (1) 50%; (2) 60%; (3) 80%; (4) 90%; (5) 95%; and (6) 100%. Use the fixed_3 map. Is the synthesized word recognized as man in all cases?

8.5 Repeat Problem 8.4, but time modify the word by decreasing last nasal, /n/, by (1) 50%; (2) 60%; (3) 80%; (4) 90%; (5) 95%; and (6) 100%. Use the fixed_3 map. Is the synthesized word recognized as man in all cases?

8.6 For the sentence, "Should we chase those cowboys?" isolate the word those by cutting it out of the sentence. Time modify the word by decreasing the /th/ by (1) 30%; (2) 50%; (3) 70%; (4) 90%; and (5) 100%. Use the random map. Describe the results of your listening test. Did the perception of the word change, i.e., did the word sound like doze, boze, loze, poze, or other words?

8.7 Repeat Problem 8.6 using the fixed_3 map.

8.8 For the sentence, "We saw the ten pink fish," in the Additional folder in the data set isolate the word saw by cutting it out of the sentence. Time modify the word by decreasing the /s/ by (1) 30%; (2) 50%; (3) 70%; (4) 90%; and (5) 100%. Use the random map. Describe the results of your listening test. Did the perception of the word change, i.e., did the ever sound baw, daw, aw, law, paw, thaw, or other words?

8.9 Repeat Problem 8.8 using the fixed_3 map.

8.10 Try to summarize some of your listening test results from the Problems 8.1 through 8.9. For example, does the duration of the time-modified consonant influence the perception of the identity of the consonant? As the duration of the consonant gets short (less than 50 msec) is the consonant perceived as either a weak fricative or a voiced stop? For longer durations was the initial consonant perceived as either an unvoiced stop or a liquid? When the consonant was nearly the same duration as the original duration, was there any perceptible change?

8.11 While you did not do many tests in Problems 8.1 through 8.9, can you determine if the position or portion of the initial consonant that was preserved affect the perception of the consonant?

8.12 If time permits, design a modification to the system that would control the loudness of the individual speech segments in the speech signal. You will have to first identify segments and then amplify or attenuate the segments. Since the system already identifies the segments, the only remaining task is to adjust the gain of the individual segments. You could assign a "gain" variable to each segment. Your design should include a means to avoid "glitches" at the boundary between segments. The gain could also be assigned on a global basis, like that for the scale factor (SF) and minimum duration (MD). Thus, the user could increase the gain of all occurrences of a particular segment type, for example, nasals, with one parameter.

8.13 In a manner similar to that for 8.12, design a means to modify the fundamental frequency contour. This may be difficult since the pitch period would have to be either lengthened or shortened by a varying amount for each voiced frame. Thus, the excitation would have to be lengthened or shortened. Shortening may be easier. This could be done by truncating the excitation. Why would lengthening be more difficult? (Hint: what would have to be done to the residue/excitation for each frame?)

8.14 Design a means to accomplish a pitch change. One way to do this is to modify the pitch contour without changing the number of frames in a given segment. This would result in a segment that is either longer or shorter than the original, depending on whether the average pitch period is either increased or decreased, respectively. This would require modification to the residue signal, but it could be done without having to interpolate the LPC filter coefficients. After doing the above, the existing system could be used to adjust the duration of the previously lengthened or shortened segment to be equal to the original length. This would involve discarding or repeating select frames from the segment. The frames could be selected at equally spaced intervals. This design probably would result in speech with no noticeable distortion.

CHAPTER 9

ANIMATED VOCAL FOLD MODEL TOOLBOX

9.1 INTRODUCTION

There are two types of glottal source models: (1) glottal models with acoustic parameters; and (2) vocal fold vibratory models. Two glottal models are described in Appendix 7: (1) the LF model; and (2) the polynomial model. These models generally describe the dynamic behavior of the glottal volume–velocity waveform, which is often used in speech synthesis. The vocal fold vibratory models depict the vibratory motion of the vocal folds and calculate estimates of data waveforms, such as the projected glottal area, the vocal fold contact area, and an estimate of the electroglottographic waveform. This chapter describes two models of the vocal folds using a graphic user interface and animation that is implemented in a software toolbox in MATLAB. The results are presented in the form of an animated movie of the vibratory motion of the vocal folds. The theory is provided in Appendix 10.

One model is referred to as the two-mass model (Ishizaka and Flanagan, 1972). This model depicts the vocal folds as two masses interconnected with springs and dash pots (dampers) (see Appendix 10). In our implementation of this model, the user can control the pre-phonatory shape (or configuration), the vocal fold tension, and the lung pressure of the vocal folds. Several other parameters can be adjusted by the user, such as the size of the two masses, their thickness, and their maximum excursion.

A second model is called the ribbon model (Titze, 1984). This model presents a more realistic 3-dimensional view of the vocal folds than the two-mass model. The model incorporates a pre-phonatory configuration of the vocal folds, maximum excursion profiles, the length and depth (thickness) of the vocal folds, and time lags in the movement of the vocal folds along their length and depth.

The vocal fold models depict the vibratory motion of the vocal folds in 3 dimensions. The ribbon model also calculates the projected glottal area, an estimate of the vocal fold contact area, and the electroglottographic waveform during vibratory motion. The toolbox creates movies of the vibratory motion that can be viewed by the user.

A digitized sequence of frames of one vibratory cycle of the actual vocal folds along with a movie of this sequence is contained in a folder on the accompanying CDs. See the README file.

9.2 TOOLBOX FOR THE VOCAL FOLD MODELS

The software is to be installed in a subdirectory (e.g., vocal_fold) in MATLAB in a manner similar to that described for the toolboxes introduced in previous chapters. The toolbox contains two software packages: (1) the two-mass model; and (2) the ribbon model. The

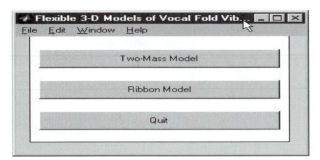

FIGURE 9.1   Main window for the vocal fold models.

two models are independent. To start the software, change directory to the vocal_fold directory and type main in the MATLAB command window. The Main function window shown in Figure 9.1 appears. This window allows the user to select one of the three options: (1) two-mass model; (2) ribbon model; or (3) quit. The Quit button closes all previously opened windows and exits the user from the toolbox.

9.3 TWO-MASS VOCAL FOLD MODEL

The theory for the two-mass vocal fold model is provided in Appendix 10. Here, we present the software toolbox and graphical user interface. To start the two-mass model, press the Two-Mass Model button in Figure 9.1. This opens the two-mass model control window shown in Figure 9.2. From top to bottom, the various parameters that the user can control

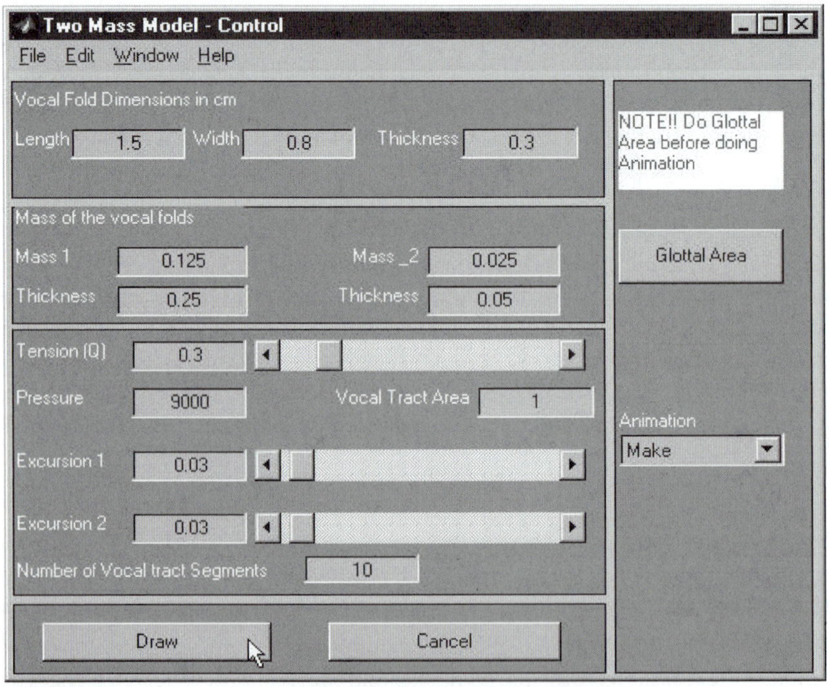

FIGURE 9.2   Two-mass vocal fold model control window.

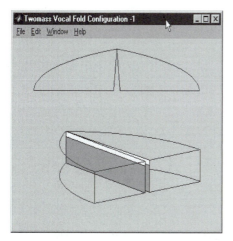

FIGURE 9.3 Static two-mass vocal fold model.

include the vocal fold length, width, and thickness. The length and thickness do not affect the model because of normalization of parameters within the model. However, the width does influence the vibratory motion of the vocal fold model. The size and thickness of the two masses of the model can be adjusted. The total thickness (the sum of the thickness 1 and thickness 2) can exceed 0.3 cm and can in fact, exceed 1 cm. There are no basic restrictions on the length, width, and thickness. However, unrealistic values can give unrealistic vibratory motion results. The tension parameter, Q, is the one suggested by Ishizaka and Flanagan (1972). This parameter is linearly related to the fundamental frequency of vibration of the vocal fold model. The pressure parameter is the lung pressure. The vocal tract area parameter controls the size of the vocal tract area, while the number of vocal tract segments sets the number of vocal tract area segments. The vocal tract area and the number of vocal tract segments present a load on the vocal fold model (see Appendix 10). However, these parameters do not greatly affect the vibratory motion of the vocal folds. The excursion parameters set the maximum excursion of the vocal folds in the pre-phonatory stage. Neither of these values should be zero. The parameter values shown in Figure 9.2 are the default values.

The Cancel button closes the two-mass model control window as well as any windows opened by the model. To start the model, first set the desired parameter values, then press the Draw button. This action opens the windows shown in Figures 9.3 and 9.4, which show the pre-phonatory vocal fold configuration for the two-mass model. Here we have used the default parameter values shown in Figure 9.2. Note that the vocal folds are plotted as though they are rigid ribbons or bands. The upper ribbon represents mass 2 with its corresponding thickness, while the lower ribbon represents mass 1 with its associated thickness. The plot of the vocal folds differs from the type of plot that is often used for the two-mass model. However, the equations for the two-mass model, given in Appendix 10, are used to make the model calculations. We chose this configuration because it is similar to the ribbon model, which is shown next.

Next press the Glottal Area button in Figure 9.2. This action opens the message window shown in Figure 9.5, which informs the user that the glottal area functions are being calculated. This window disappears once the calculations are completed. At the same time, the window shown in Figure 9.6 appears.

The lower graph shows the glottal area for mass 1 (the lower mass) of the two-mass model. The upper graph is the glottal area for mass 2 (the upper mass) of the two-mass

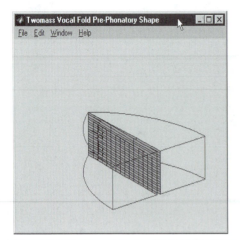

FIGURE 9.4 Pre-phonatory view of two-mass model.

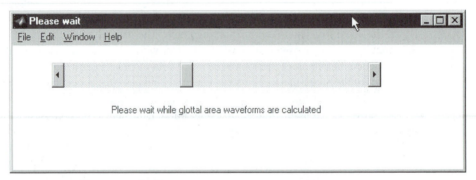

FIGURE 9.5 Message window.

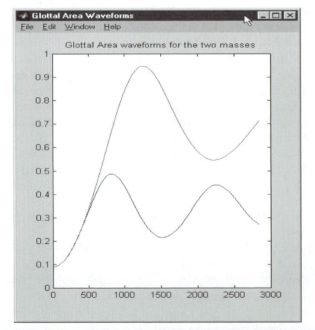

FIGURE 9.6 Glottal area waveforms calculated by the model.

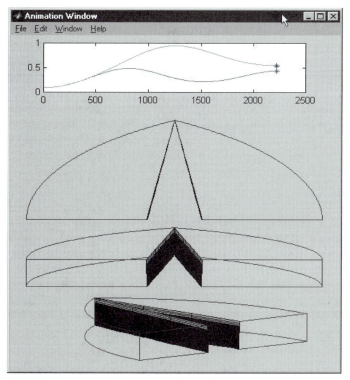

FIGURE 9.7 Animated movie of the vibratory motion of the two-mass vocal fold model.

model. The horizontal scale is arbitrary, with the value 3000 representing the end of a glottal vibratory cycle. Thus, the user can think of 3000 as the pitch period. The vertical scale is normalized to unity and depicts the amplitude of the area functions.

Once the glottal area functions are calculated, press the Make button. This action creates a 3-dimensional vibratory model of the motion of the vocal folds, as shown in Figure 9.7. This window is the animation window. As the vocal fold positions are being generated, a marker moves across the upper graphs of the two glottal area functions, while simultaneously the vibratory motion of the vocal folds is calculated. Three views of the vocal folds are provided. The top view is the same as that seen via a laryngeal mirror as shown in Chapter 3. The top of this view is anterior, while the bottom of this view is the posterior. The middle view of the vocal folds is looking at the vocal folds along a horizontal plane from posterior to anterior. The lower view is an isometric view at a 30-degree rotation from the other two views. The make action repeats the motion of the vocal folds twice upon completion of the necessary calculations. To view the animated movie, press the Play button, which is included in the pull down menu with the Make button. This action plays the animated movie of the vibratory motion of the vocal folds for 20 repetitions, that is, 20 pitch periods. The user may play the animated movie repeatedly. The colors shown for the two masses of the vocal fold model are generated by MATLAB, and are apparently not under user control. If the animated model does not appear in proper motion, that is, it appears as though the horizontal synchronization is not working properly, then reset the color resolution in the Windows 95 display setting to 256 colors or less. Note that the horizontal scale of the top graph (the glottal areas) in Figure 9.7 differs from that shown in Figure 9.6. This is because the values for mass 1, mass 2, the tension, excursion 1, and excursion 2 can affect the pitch period of the vibrating vocal folds. Thus, the final pitch period of the vibrating vocal folds is determined by the values of these parameters and is shown in the animation window. In this example the preliminary pitch period is 3000, while the final pitch period after the calculations by the model are completed is about 2500.

The user can experiment with various parameter settings to gain some experience with the manner by which the model parameters influence the vibratory motion of the vocal folds. One variation is to change the tension to $Q = 0.205$, excursion $1 = 0.05$, and excursion $2 = 0.01$, while all of the other parameter values remain as set in Figure 9.2. This will produce a very realistic vocal fold vibratory motion for a normal phonation.

9.4 THE RIBBON VOCAL FOLD MODEL

The ribbon vocal fold model has a control window similar to that of the two-mass vocal fold model, as shown in Figure 9.8. However, some of the model parameters are different. The vocal fold length, width, and thickness can affect the model. Changes in the width can cause strange vocal fold shapes to occur. It is recommended that the default settings of the length, width, and thickness be used. The vertical and horizontal phase difference parameters affect the motion of the vocal folds in the vertical and horizontal planes. The maximum excursion parameter controls the maximum excursion of the vocal folds (see Appendix 10). The fundamental frequency controls the relative pitch period of the vibratory motion of the vocal folds. For example, a setting of 300 causes the vocal fold model to vibrate twice as fast as a setting of 150. The abduction and shape quotients are defined in Appendix 10 and are calculated by the model. These values change with the values of

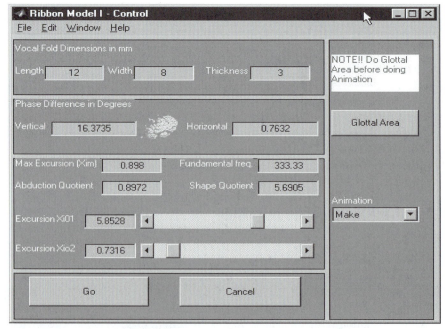

FIGURE 9.8 Ribbon vocal fold model control window.

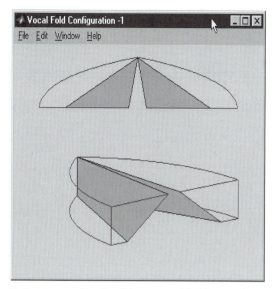

FIGURE 9.9 Static ribbon vocal fold model.

excursion xi01, excursion xi02, and maximum excursion xim. So the user does not set the values of the abduction and shape quotients. The excursion xi01 setting controls the pre-phonatory setting of the lower edge of the vocal fold model, while excursion xi02 controls the pre-phonatory setting of the upper edge of the vocal fold model.

After setting the desired parameter values, start the model by pressing the Go button. This action opens the windows shown in Figures 9.9 and 9.10, which are similar to those for the two-mass vocal fold model.

The parameter control settings used for Figures 9.9 and 9.10 are the default values shown in Figure 9.8. Next press the Glottal Area button in Figure 9.8. This opens a message window (not shown) similar to Figure 9.5. Upon completion of the model calculations, the message window closes and simultaneously opens the glottal area window shown in Figure 9.11. This figure shows three graphs: the projected glottal area, the vocal fold contact area, and the estimated electroglottographic waveform.

The next step is to press the make button in Figure 9.8, which creates a 3-dimensional vibratory ribbon model of the motion of the vocal folds that is similar to the two-mass vocal fold model. This model is shown in Figure 9.12, which is the animation window. As the vocal fold positions are being generated, a marker moves across the upper graph of the projected glottal area function, while simultaneously the vibratory motion of the vocal folds is calculated. Three views of the vocal folds are provided. The top view is the same as that seen via a laryngeal mirror as shown in Chapter 3. The top of this view is anterior, while the bottom of this view is the posterior. The middle view of the vocal folds is looking at the vocal folds along a horizontal plane from posterior to anterior. The lower view is an isometric view at a 30-degree rotation from the other two views. The make action repeats the motion of the vocal folds 20 times upon completion of the necessary calculations. To view the animated movie, press the Play button, which is included in the pull down menu with the Make button. This action plays the animated movie of the vibratory motion of the vocal folds for 20 repetitions; that is, 20 pitch periods. The user may play the animated movie repeatedly. While the figures in the text are black and white, the MATLAB software generates colors that will appear on a color computer monitor. Thus, the two masses of the

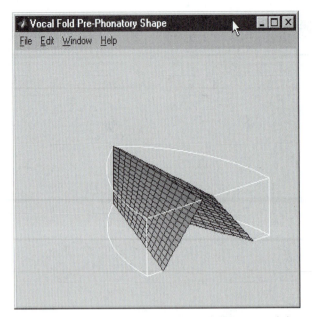

FIGURE 9.10 Pre-phonatory view of ribbon model.

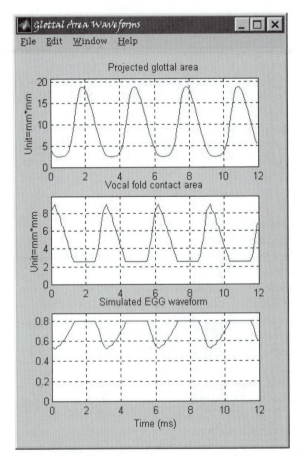

FIGURE 9.11 Glottal area waveforms calculated by the ribbon model.

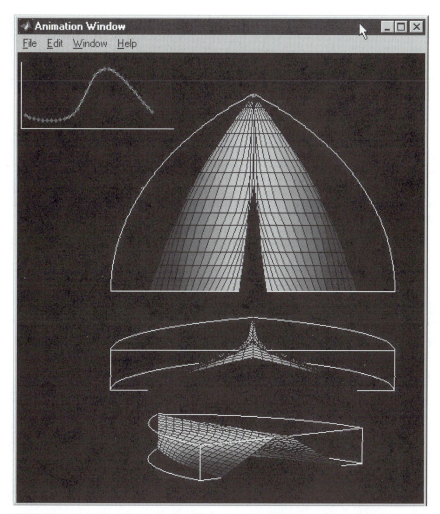

FIGURE 9.12 Animated movie of the vibratory motion of the ribbon vocal fold model.

vocal fold model will appear in color. If the animated model does not appear in proper motion, that is, it appears as though the horizontal synchronization is not working properly, then reset the color resolution in the Windows 95 display setting to 256 colors or less. Note that the horizontal scale of the top graph (the projected glottal area) in Figure 9.11 differs from that shown in Figure 9.12. However, the only difference is that the graph in Figure 9.12 is for one pitch period, otherwise the graphs are the same.

Several parameter values that result in a very realistic vocal fold vibratory motion are the following. Use the default settings except change vertical phase difference to 8. Another setting is maximum excursion xim = 1.3 (a setting of 2 is also good), excursion xi01 = 4.57 (or 3.09), and excursion xi02 = 0.73. The fundamental frequency can be left at 333 or changed to 150. The vertical phase difference can be left at 16 or changed to 8.

As the maximum excursion xim gets large, the user can see gaps occur between the vocal folds during vibration. As the excursion xi01 and excursion xi02 are made large, the user can see the vibratory motion of the vocal folds very clearly. No vocal fold contact takes place. The top view of the vocal folds is not good; however, the isometric view is very good.

A slight modification to the software allows the projected glottal area waveform to be saved and used as an excitation waveform for speech synthesis. Similarly, the movies of the vibratory motion of the vocal folds can be saved. However, this is not recommended since the movie files are very large (many MB) and can easily be regenerated at will.

Several other models of the vibratory motion of the vocal folds are briefly mentioned in Appendix 10. However, these models are not implemented here.

PROBLEMS

9.1 For the two-mass vocal fold model, change the thickness settings for mass 1 and mass 2 to 0.3 for each mass. Observe the pre-phonatory shape and the glottal area functions and the movie. Describe the differences between these results and those using the default settings.

9.2 For the two-mass vocal fold model, change the settings for mass 1 to 0.025 and that for mass 2 to 0.125. Observe the pre-phonatory shape and the glottal area functions and the movie. Describe the differences between these results and those using the default settings.

9.3 For the two-mass vocal fold model, experiment with various values for the pressure setting. Discuss the effect this parameter has on the vibratory motion of the vocal folds.

9.4 For the two-mass vocal fold model, experiment with various values of the vocal tract area and the number of vocal tract segments. Discuss the effect this parameter has on the vibratory motion of the vocal folds.

9.5 For the ribbon vocal fold model, change the vertical phase setting as follows: 2.0, 4.0, 8.0, 16.0, and 32.0. Let the other values remain at their default setting. Describe the effect this parameter has on the vibratory motion of the vocal folds.

9.6 For the ribbon vocal fold model, change the horizontal phase setting as follows: 0.1, 0.5, 1.0, 4.0, 8.0, 16.0, and 32.0. Let the other values remain at their default setting. Describe the effect this parameter has on the vibratory motion of the vocal folds.

9.7 For the ribbon vocal fold model, set the excursion xi02 to 2.5. Let the excursion xi01 be 1.0, 2.5, and 5.0. Let the other values remain at their default setting. Describe the effect this parameter has on the vibratory motion of the vocal folds.

CHAPTER 10

ARTICULATORY SPEECH SYNTHESIS TOOLBOX

10.1 INTRODUCTION

The aim of this chapter is to provide a flexible, articulatory speech synthesis toolbox, called ARTM, for articulatory speech synthesis model. One feature of this toolbox is to synthesize speech by matching the vocal tract parameters for a specified set of formant characteristics, called the target formants. A simulated annealing procedure is used to achieve the optimization. The derivation of the acoustic equations that include the subglottal system, the glottal impedance, the turbulence noise source, and the nasal tract with sinus cavities for the articulatory synthesizer is provided in Appendix 11. This toolbox can be useful for speech modeling, analysis, and synthesis.

The toolbox is designed with interfaces that provide for numerical specification of parameters and sliders that allow parameter adjustments. A transmission-line circuit model of the vocal system, which includes the vocal tract, the nasal tract with sinus cavities, the glottal impedance, the subglottal tract, the excitation source, and a turbulence noise source, are provided. A digital time–domain approach is used to simulate the dynamic properties of the vocal system as well as to improve the quality of the synthesized speech.

There are two components of the toolbox: the formant_track folder and the artm folder. These folders can be installed in the same or separate subdirectories of MATLAB. The use of these toolboxes is described in the sections that follow.

The analysis determines the articulatory parameters from the acoustic speech wave-form. The algorithm that is used is known as simulated annealing, which is constrained to avoid non-unique solutions and local minimal problems. The articulatory-to-acoustic trans-formation function and the boundary conditions for the articulatory parameters determine the constraints. The cost function is defined as a percentage of the weighted least-absolute-value error between the first four formant frequencies of the articulatory model (the model formants) and the first four formant frequencies determined from speech analysis (the target formants). A 1% error criterion is both practical and achievable.

In summary, this articulatory speech synthesis toolbox works as follows. The user analyzes a speech file, such as a word or a sentence to determine the target formants for a set of frames defined for the speech file. This preliminary task is performed with a special toolbox that is derived from the formant/pitch tracking option in the analysis toolbox described in Chapter 2. In this chapter, the target formants are derived using the formant_track folder, which is described later. Once the target formant tracks are obtained, ...iculatory speech synthesis toolbox and loads either a formant ...iculatory parameter vector

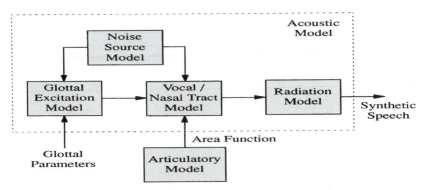

FIGURE 10.1 Overview of the articulatory speech synthesizer.

articulatory vector to a file, which can be loaded at a later time if desired. An example is given later. After the articulatory parameter vector has been determined, the user then specifies the excitation waveform using the LF parameters. The excitation can be voiced, unvoiced, or a mixture of voiced and unvoiced excitation. The voiced excitation can include jitter and shimmer, as well as aspiration and turbulence. After the excitation is specified, the speech is synthesized using the excitation to drive the vocal tract determined by the articulatory parameter vector. The speech synthesizer includes options to display an animated movie of the vocal tract used for speech synthesis, to display the vocal tract parameters, to play the synthesized speech, as well as other features. The toolbox provides the user with a method for observing the relationship between speech features (formants) and vocal tract shape.

While Appendix 11 provides the theoretical background for the articulatory speech synthesizer, we will from time-to-time give brief descriptions of important features of the system. Figure 10.1 shows an overview of the articulatory synthesizer, which consists of an articulatory model, an acoustic model, and an excitation source. The acoustic model is outline in Figure 10.2 and is described in more detail in Appendix 11. The major advantages of the articulatory approach to speech synthesis are that (1) the model is directly related to human speech production, therefore the model parameters vary slowly, and are easily interpolated; and (2) source-tract interaction is modeled in a natural manner. However, difficulties exist, for example (1) it is difficult to derive an accurate articulatory model, including an accurate representation of the losses; (2) the estimation of the model parameters by inverse filtering is difficult due to local minima and the solution can be non-unique; and (3) it is still difficult to model unvoiced sounds, although we have improved this feature (see Appendix 11). Nevertheless, our articulatory speech synthesizer is complete and user friendly.

There are four phases or steps in the use of the articulatory synthesizer: analysis, speech inverse filtering, excitation specification, and synthesis. The analysis phase extracts the target formants for the speech file by determining the formant tracts of the target speech signal. The user then marks the target formant trajectories at desired intervals (frames). This marked formant trajectory file is saved as the target formant file. Next, speech inverse filtering is performed to determine the articulatory model parameters. This is accomplished using a simulated annealing algorithm to minimize the ⟶⟶⟶⟶⟶⟶ the target formants and the model formants (see Appendix ⟶⟶⟶⟶⟶⟶⟶⟶ type of excitation

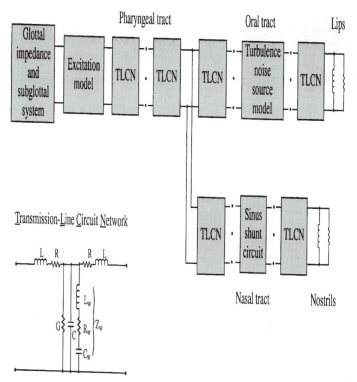

FIGURE 10.2   The acoustic model.

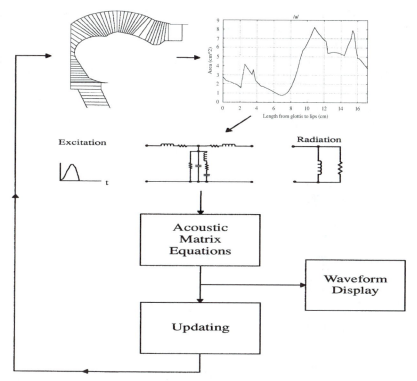

FIGURE 10.3   The steps in articulatory speech synthesis.

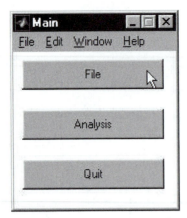

FIGURE 10.4   Main window for estimating the target formant tracks for a speech file.

10.2 DETERMINING THE TARGET FORMANTS

The target formants for a speech file are determined using the formant_track toolbox. As for previous toolboxes, start MATLAB, change directory to formant_track, and type main. The window shown in Figure 10.4 appears, which is similar to the main window for the analysis toolbox in Chapter 2.

Next, press the File button to open the file window shown in Figure 10.5. Pressing the Load button allows the user to load an ASCII speech file, such as b.dat or be_seg.dat, both of which are in the data folder in the formant_track toolbox. Figure 10.6 shows the b.dat file as the input signal file. Once the desired speech file is loaded, the user can play the file if desired or the option can be canceled.

After the desired speech file is loaded, the user returns to the main window and presses the Analysis button, which opens the analysis window shown in Figure 10.7. The options available are to calculate four formant contours (tracks), after which the user can save the formant contours by pressing the Save button. The file name is specified as *form.for, where the * is to be replaced by a name specified by the user, such as be_. The formant contour file name then becomes be_form.for. Such a file is already available in the data folder within

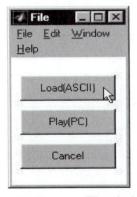

FIGURE 10.5   File window for loading a speech file.

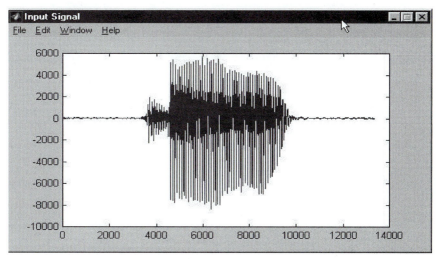

FIGURE 10.6 Input signal file.

the formant_track toolbox. Pressing the Cancel button performs the customary function. Pressing the Formant Contours button calculates the first four formant contours for the loaded speech file, which in this case is b.dat, shown in Figure 10.6. While the formant contours are being calculated several short message windows are displayed, as is done in the analysis toolbox in Chapter 2. Upon completion of the calculations, the formant contours are displayed as shown in Figure 10.8.

The saved formant tracks are used in the articulatory speech synthesizer toolbox, which is described next.

10.3 ARTICULATORY SPEECH SYNTHESIS TOOLBOX

After installing the articulatory speech synthesis toolbox in a subdirectory, start MATLAB, change directory to the subdirectory, and type main. Two windows appear. One is a menu window that contains the buttons shown in Figure 10.9. The other, shown in Figure 10.10,

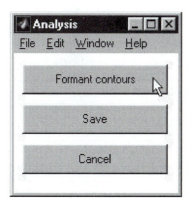

FIGURE 10.7 The analysis window.

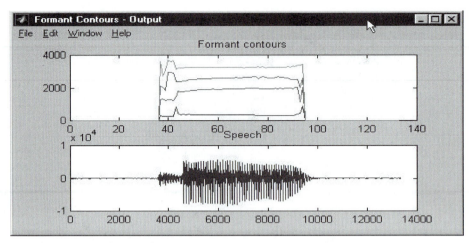

FIGURE 10.8 Formant contours and speech file.

is a canvas window that is used to display graphics for the various options. The user is not to close these two windows until he or she is ready to exit the toolbox.

There are six buttons (options) available in the main menu window: File is to load or to save a file, Shape is to set the articulatory model parameters, Optimize is to start the optimization process, Excitation is to set the desired excitation parameters for the excitation waveform, Synthesize is to start the synthesis process, and Quit is to quit and exit the toolbox. Generally these buttons are pressed in a left to right sequence.

10.3.1 The Load and Mark Option

The load feature allows the user to load one of three types of files: a formant tracks file (determined using the formant_track toolbox), an articulator parameter file, or a Fant area file (as shown in Appendix 5). The latter option is disabled at this time. See the comments that appear when this option is selected for additional information. The load articulatory parameter file is explained later. We start with the load formant tracks option. This option opens the data folder within the articulatory speech synthesis toolbox showing the files named *form.for, which includes be_seg_form.for. This latter file is the formant track file for the speech file be_seg.dat, which is a short segment of the b.dat file that starts just following the termination of the plosive /B/. The original speech file can be viewed using the formant_track toolbox by loading the be_seg.dat file. Load the be_seg_form.for file. The windows shown in Figures 10.11 and 10.12 appear. Figure 10.11 displays the four formant tracks for the be_seg_form.for file, while Figure 10.12 shows a menu window with five options. The primary options are contained in the pull-down menu, which are Mark and Add Mark. The term mark is used to designate the operation of marking temporal locations

FIGURE 10.9 The articulatory speech synthesizer menu window.

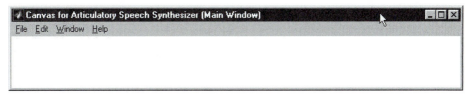

FIGURE 10.10   The canvas window.

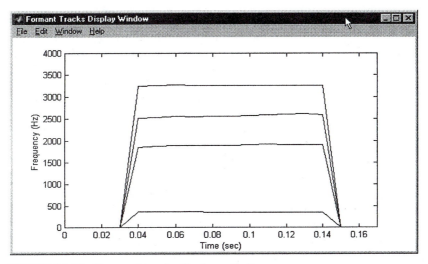

FIGURE 10.11   Formant tracks display window.

FIGURE 10.12   The formant track mark menu.

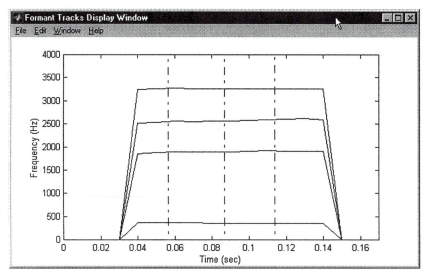

FIGURE 10.13 An example of a formant track file that has been marked using the add mark option.

for frame boundaries on the formant tracks. An example of this is shown in Figure 10.13, where the add mark option was used to add three frame boundaries on the formant tracks. This was accomplished by selecting the add mark menu option, then moving the cursor to the formant tracks display window, whereupon the cursor becomes a cross hair. The user places the cross hair at the desired location and clicks the left mouse button, a vertical frame boundary line is drawn at that location. This sequence of steps was repeated two more times to draw the three frame boundaries as shown in Figure 10.13. The purpose of the segmentation of the target formant tracks into boundaries is to facilitate the calculation of the shape of the vocal tract. This calculation minimizes the distance between the target formants and the formants of the articulatory model. The user can Unmark the file by pressing the unmark button. The operation of unmarking is complete; that is, all boundary marks are removed. The user must then initiate the add mark process again. Once the number of desired frame boundaries are marked at the desired locations, then save the file for future use by pressing the Save button. The save option opens the data folder and allows the user to name the marked file using the format *fmrk.mrk, where the * is to be replaced with a user selected name, such as be_seg_. After saving the marked frame boundary file, then press the Done button. The Done button is pressed only if a formant track file has been marked. The Done button can be pressed without saving the marked file. The marked file is retained in memory for the next phase of the toolbox. If a formant track file is not marked and the user wishes to exit this phase of the software, he or she can press the Cancel button. Finally, the purpose of the mark file option is to load a previously marked formant track file, such as be_seg_fmrk.mrk, which is the file shown in Figure 10.13 that was constructed using the add mark option and saved as described previously. Once such a previously marked file has been loaded, press the Done button to continue. The user can also unmark a loaded marked file. The original marked file is not destroyed, since it is retained in disk storage. However, all of the marked frame boundaries are removed from the software memory workspace and the user must add the desired frame boundaries in the manner described. In other words, it is as though the user had loaded the original formant track file, such as the be_seg_form.for file in the first place. So there is no advantage to unmarking a previously saved marked file.

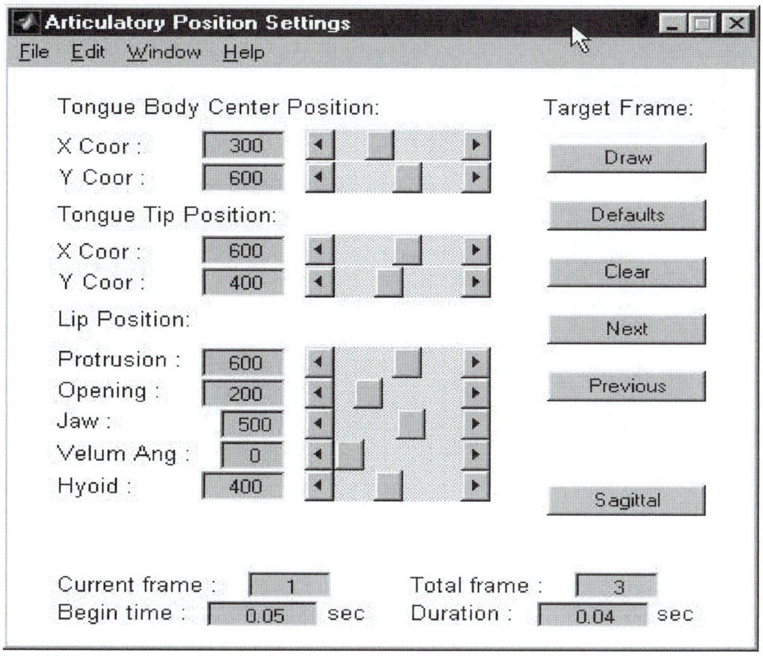

FIGURE 10.14 Articulatory positions settings window.

The number of frames to be marked on a formant track file is not critical. For example, the sentence, "We were away a year ago," is shown in Appendix A11-D for two different speakers marked at 26 and 22 locations for the entire sentence. The synthesis was quite satisfactory with so few frames. The word be can be synthesized with 3 frames.

10.3.2 The Shape Option

This option allows the user to experiment with altering the default articulatory vocal tract model shape settings to reduce the error between the target formants and the model formants. This option is manually operated and is not the optimization process, which is the next step (phase). There are two purposes for this step in the software: to allow user experimentation to obtain an idea of the effect the articulatory model parameters have on the error and to speed up the optimization process. Sometimes an experienced user can select parameter settings faster than the optimization process, thereby reducing the optimization search time. Press the Shape button. A Shape Settings button appears, which in turn is to be pressed. The articulatory positions settings window shown in Figure 10.14 appears. The default values are shown for the nine articulatory model positions. The default values can be altered by the user using the sliders. The bottom of the figure shows the current frame and the total number of frames. These values correspond to the first frame and the three frames, respectively, for the marked be_seg_fmrk.mrk file shown in Figure 10.13. The begin time at the lower left is the beginning of the formant tracks, while the duration is the duration of the frame measured from the beginning time to the frame boundary. The Draw push button at the top, right allows the user to draw the articulatory vocal tract model for the position settings. An example for the current frame using the default position values appears in Figure 10.15, where the vocal tract model is drawn on the canvas for the articulatory speech synthesizer.

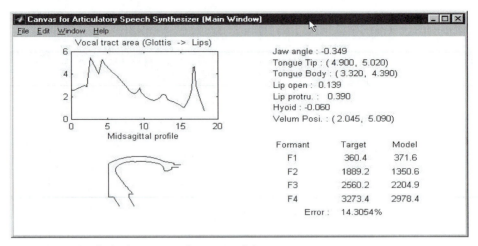

FIGURE 10.15   Articulatory vocal tract model.

If the user alters the default position values, then a new vocal tract model can be drawn by pressing the Draw button again. The Default button will restore the default values. The Clear button clears the canvas. It is not necessary to clear the canvas between successive draws, since the Draw command clears the canvas automatically. The Next button changes the frame to the next frame. The Previous button changes the frame back to the previous frame. The Sagittal button merely draws the number of sagittal sections on the vocal tract cross-sectional display. If the user alters the position values, then these values are not retained for the next frame, that is, the default values are restored automatically for the next frame. However, the user-set values are retained for the frame, unless altered again by the user.

Figure 10.15 shows the vocal tract area from the glottis to the lips, the nine position values for the articulatory model, the vocal tract cross-section (midsagittal profile), and the target and model formant values. The error for this configuration is shown as 14.3054%. This error can be easily reduced to less than 1% by adjusting the nine articulatory positions sliders.

The error is a weighted average of the absolute values of the difference between the target formants and the model formants. Define the first error term, e1, as the absolute value of the difference between the first target formant and the first model formant divided by the first target formant. Define similar error terms for the second, third, and fourth formants, e2, e3, e4. The error is (30)e1 + (30)e2 + (25)e3 + 15e4. Note that the first target formant cannot be zero.

10.3.3 The Optimization Option

The shape option must be selected and the window left open before the optimization option is selected. The default values for the shape option need not be altered and it is not required that the model be drawn. The software reads the values in the shape option window to start the optimization process. Pressing the Optimization button brings up two menu options: articulatory optimization setup and simulated annealing optimization. First, select the Setup Optimization button. This brings up the window shown in Figure 10.16.

The pull down menu offers four options for the dimension of the articulatory vector used in the simulated annealing optimization process (Figure 10.14): (1) eight position values for the tongue, lips, jaw, and hyoid, but with the velum closed; (2) nine position

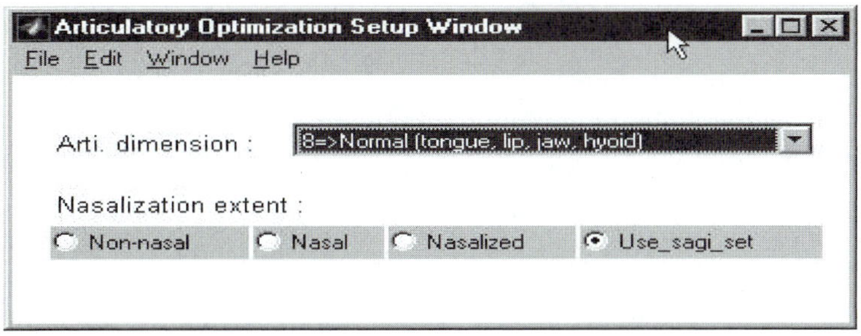

FIGURE 10.16   Articulatory optimization setup window.

values (the eight values above) and the velum, (3) eleven position values (the eight) and the pharynx (three values); and (4) twelve position values (the eight) and the velum and the pharynx (see Appendix 11 for more details). Briefly, as mentioned in Appendix 11-D, the eight position vector is best for front vowels; the nine position vector is best for nasalized front vowels; the eleven position vector is best for middle, back, and semivowels; and the twelve position vector is best for nasalized vowels. Usually, the velum position is set at a default position for nasal, non-nasal, or nasalized phonemes, but it can be optimized for some phonemes. The dimensions of the lower pharynx can also be optimized. These dimensions appear in Figure A11.4, and are given as the anterior–posterior movements of K and H (glk and wh in the software) and the height difference between K and H (hkl). When the pharynx is included, these three parameters, wh, glk, and hkl, are optimized by the software. There are default values for these parameters within the software.

The Nasalization Extent buttons allow the user to set the velum position for nasalization. The use_sag_set refers to the velum angle slider in Figure 10.14. This setting uses the value of the velum angle set by the user in that figure. The non-nasal setting closes the velum; that is, the slider setting is zero. The nasal setting is for a relatively large velum angle, namely, a slider setting of 492. The nasalized setting is for a moderate velum angle, namely, a slider setting of 324. The nasal tract design is included in the software, as discussed in Appendix 11. Appendix 11 shows some examples of the effect of no nasal tract, nasal tract coupling, and both nasal tract and sinus coupling; a sinus model is included.

After the user selects the desired articulatory vector dimension and nasalization, then select the Simulated Annealing Optimization button under optimization in the main window. This brings up the optimization window shown in Figure 10.17.

The annealing parameters that control the simulated annealing algorithm include the initial temperature T, the temperature reduction coefficient r_T, the number of steps to adjust the step length vector N_S, the number of step adjustments at each temperature N_T, the number of successive temperature reductions to test for termination N_ε, a small constant used for the termination criterion η, and the maximum number of function evaluations N_{tot}. The analogues between the annealing process and the articulatory problem can be identified as follows. First, the percentage of the weighted least-absolute-value (l_1-norm) error distance, equation (A11.3.3.1.3), corresponds to the energy of the material. The articulatory vector, equation (A11.3.3.1.1), corresponds to the configuration of particles. The change of articulatory parameters corresponds to the rearrangement of particles. Finding a near-optimal articulatory vector corresponds to finding a low-energy configuration. The temperature of the annealing process, T, becomes the control parameter for the speech inverse filtering process. Second, the Metropolis algorithm corresponds to the random fluctuations in energy. Third, the temperature reduction coefficient r_T corresponds to the cooling rate. Fourth, the finite number of moves at each downward control temperature value, $(N_S)(N_T)$, corresponds

Simulated Annealing Window			_ □ ✕
File Edit Window Help			

Initial Temp. :   10   ◀ ▮   ▶   *(0.01)

Temp. reduc. factor :   85   ◀   ▮ ▶   *(0.01)

Error Tol. :   50   ◀   ▮   ▶   *(0.001)

No. of Cycles :   20   ◀ ▮   ▶

No. of iterations :   5   ◀   ▮   ▶

Max. No. of Eval. :   5001   ◀ ▮   ▶

rand. seed 1 :   1   ◀▮   ▶

rand. seed 2 :   2   ◀ ▮   ▶

Default   Start

FIGURE 10.17 Simulated annealing window.

to the amount of time spent at each temperature. Reasonable values of the parameters (Table A11.4) are used as defaults for the optimization process, which are shown in Figure 10.17. However, a guideline of the optimization process is given in Appendix A11-D along with two examples. See Appendix 11 for more details. Figure 10.18 provides a summary of the effect of the temperature reduction factor on the simulated annealing algorithm and the Metropolis algorithm.

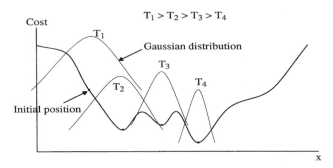

$$T_1 > T_2 > T_3 > T_4$$

Cost

T_1

Gaussian distribution

T_3

T_2 T_4

Initial position

x

T: Control parameter (artificial temperature)

x: Articulatory parameter, e.g., tongue tip x coordinate

Metropolis Algorithm:

$$p(\Delta E) = 1.0, \qquad \text{for } \Delta E \leq 0,$$
$$= e^{\left(\frac{-\Delta E}{T}\right)}, \qquad \text{for } \Delta E > 0.$$

$$\Delta E = E(\text{new state}) - E(\text{current state})$$

FIGURE 10.18 The temperature reduction factor and the simulated annealing algorithm.

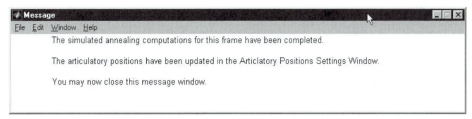

FIGURE 10.19 Message window stating that the frame calculations have been completed.

The parameter values shown in Figure 10.17 achieve a error on the order of 0.02%. But the optimization process takes on the order of 45 minutes with a 300 MHz machine. The parameter values can be altered by the user. For example, the optimization process is greatly speeded up by changing the number of cycles to 10 and the maximum number of iterations to 100. This results in an optimization on the order of a few minutes. However, such a change does not achieve as small an error as the default values because the search procedure is terminated prematurely. However, the error is still small, being on the order of 1% to 5% or so, compared to 0.02% or so.

To start the optimization, press Start in Figure 10.17. Figure 10.15 appears, where the error is shown as 14.3054%. The simulated annealing algorithm commences, and a succession of figures similar to Figure 10.15 are drawn on the canvas. Each successive figure has a smaller error. At the termination of the optimization process, a message window appears as shown in Figure 10.19.

At the same time Figure 10.20 appears, which is the final optimized articulatory configuration for the first target frame. The error is 0.0270%. The target and model formants are nearly identical in value. The vocal tract area is drawn along with the midsagittal profile and the values of the articulatory positions are given. The articulatory position settings in Figure 10.14 are also updated automatically and shown in Figure 10.21.

The user now presses the Next button in Figure 10.21 to advance the frame to frame 2 for optimization. The final articulatory position settings for frame 1, shown in Figure 10.21,

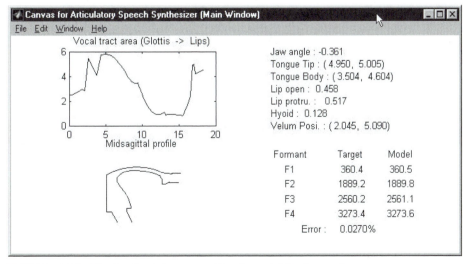

FIGURE 10.20 Final articulatory vocal tract model for frame 1.

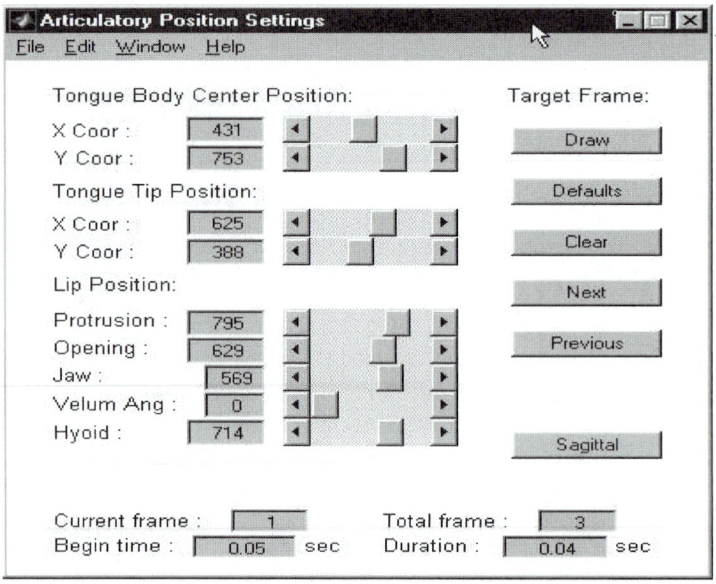

FIGURE 10.21 Final articulatory position settings for frame 1 after optimization.

are retained for frame 2. The reason for retaining the position settings from the last frame is that presumably, the position settings for the next frame are more similar to those of the last frame than the default values. This should speed-up the optimization process. The user can alter the simulated annealing algorithm values in Figure 10.17 or retain the default values. Press Start and the optimization process starts for frame 2. This time, the error begins at 1.3% and continues to reduce until the final error is 0.0385%. The same procedure is followed for frame 3. The final error is 0.0237%.

Upon completion of the optimization of the third and final frame, two message windows appear. One is the same as that shown in Figure 10.19. The other is shown in Figure 10.22.

The user now presses the File button in the main window and selects the Save Articulatory Parameters button to save the optimized articulatory parameter vector to a *.art file in the data folder. The art designation is for articulatory parameter vector. In this case, the saved file is called be_seg_opt.art. (The articulatory parameters vector is retained in working memory even if the file is not saved.) An *.art file can be loaded as an articulatory parameter vector, instead of a formant track file. In such a case, the Shape and

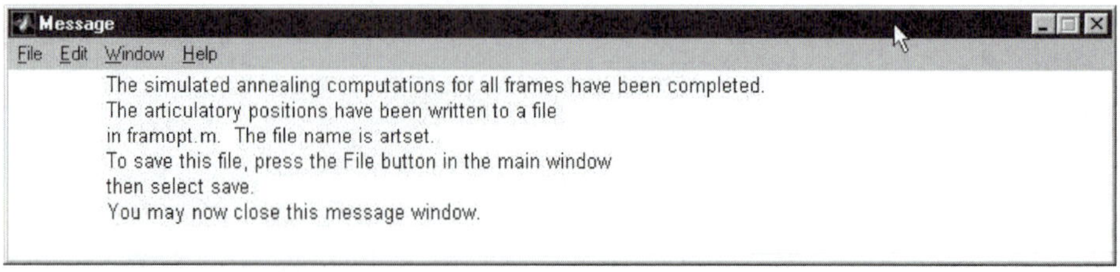

FIGURE 10.22 Message window stating that all simulated annealing computations are completed.

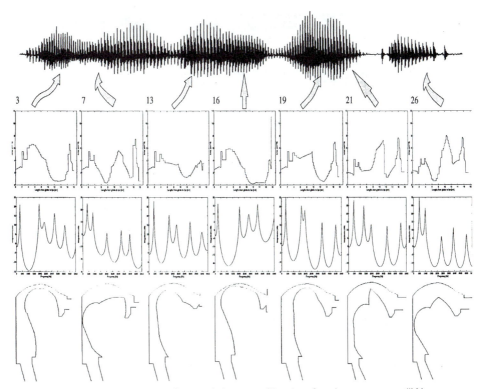

FIGURE 10.23 A summary of speech inverse filtering for the sentence, "We were away a year ago."

Optimization buttons are turned off (and therefore skipped) and the user progresses directly to the excitation option, which is described as follows.

Figure 10.23 is a summary of the speech inverse filtering results for the sentence, "We were away a year ago," using 26 frame boundaries. This example also appears in Appendix 11-D. The top panel in Figure 10.23 shows the time waveform for the sentence, the next panel down shows the vocal tract area functions, the next panel presents the formants, and the bottom panel shows the vocal tract cross-sectional (midsagittal) shapes for each frame. The results for only a few selected frames are presented due to space limitations. However, this figure is an illustration of how the toolbox is used to obtain the vocal tract configurations by minimizing the error between the target and model formants for each frame.

10.3.4 The Excitation Option

The excitation option can be started at any time. Neither a formant track file nor an articulatory parameter file needs to be loaded. However, the usual procedure is to design the excitation waveform after the articulatory parameter vector has been obtained via optimization. Prior to starting the excitation option, the user can close any open windows, except the main menu window and the canvas window. Next, press the Excitation button and the LF Model button. Figures 10.24, 10.25, and 10.26 appear.

In Figure 10.26, the user has pressed the Glottal Excitation Place button, which in turn activates the excitation mode options, namely, highlighting the Voiced, Unvoiced,

FIGURE 10.24   Excitation source menu window.

and Mixed Excitation buttons. If the Glottal Excitation button is activated, then the Vocal Tract (VT) button cannot be activated. This latter option is explained later. After the Glottal Excitation button is activated (pressed, i.e., the box is filled), the user can now select either voiced, unvoiced, or mixed excitation mode. These options are explained now.

10.3.4.1   *Voiced Excitation*

In the following, we illustrate the construction of a voiced excitation waveform. The user presses the Voiced button in Figure 10.26, which activates (highlights) the Jitter and Shimmer button, the Aspiration Noise button, and the Subglottal Model button. Imagine that the user has pressed these three buttons as well, as shown in Figure 10.27. Note that once any of the buttons is pressed (activated), they can be cleared (deactivated) by pressing them again. Also note that at this time, there is only one type of excitation model waveform, namely, the LF model.

Pressing the Voiced button brings up the menu window in Figure 10.28, which allows the user to adjust the LF waveform model parameters as well as the fundamental frequency of voicing, the gain, and another gain parameter, g0. If the user changes F0, then T0 changes automatically, and vice versa. The LF model is described in Appendix 7. The gain (av) determines the gain for the integrated LF model waveform. The parameter g0 provides additional flexibility to allow the user to scale the volume velocity waveform when mixed with aspiration noise. The default values are shown in Figure 10.28.

The jitter and shimmer menu window appears in Figure 10.29. The jitter and shimmer sliders specify the maximum jitter and shimmer values as a tenth percent of the fundamental frequency of voicing, F0, and the gain (av), respectively. The jitter and shimmer settings are only valid for sustained voicing. Jitter and shimmer are deactivated for the unvoiced option. The jitter and shimmer filter sliders specify the filter coefficients for the psuedo-random number generators, respectively. A negative coefficient means a high-pass filter, a zero coefficient represents no filtering, and a positive coefficient is a low-pass filter. The absolute value of the coefficient must be less than one.

Aspiration noise can be added to the excitation waveform through the use of the parameter window shown in Figure 10.30. The gain slider allows the user to set the gain for the noise, while the filter slider sets the filter coefficient for the aspiration filter in a similar manner to that described for the jitter and shimmer option.

Figure 10.31 summarizes the voiced excitation model.

The subglottal model is available only for voiced excitation, and is shown in Figure 10.32. There are two aspects to the subglottal model: the upper portion and the lower

FIGURE 10.25   Canvas for the excitation source window.

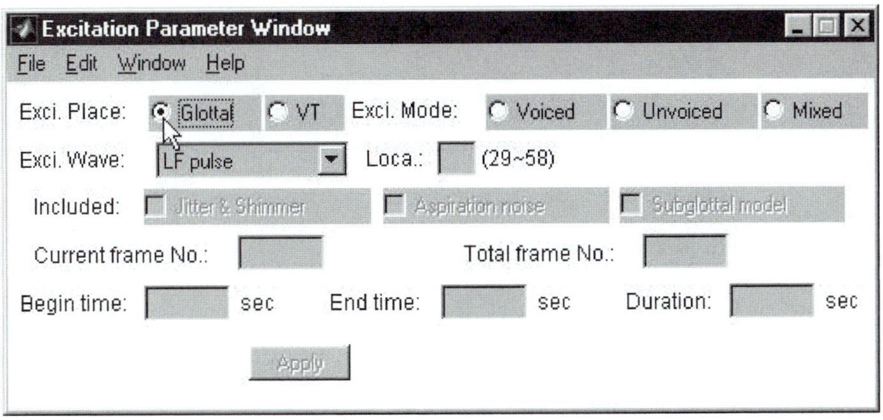

FIGURE 10.26 The excitation parameter window.

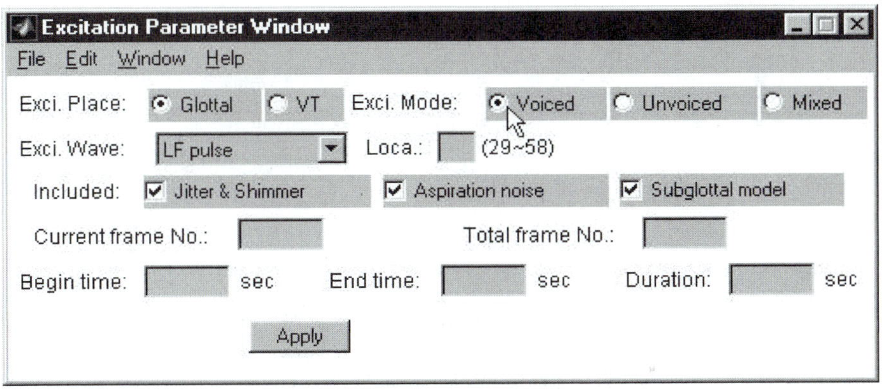

FIGURE 10.27 Excitation parameter window for voiced excitation with jitter and shimmer, aspiration noise, and subglottal model.

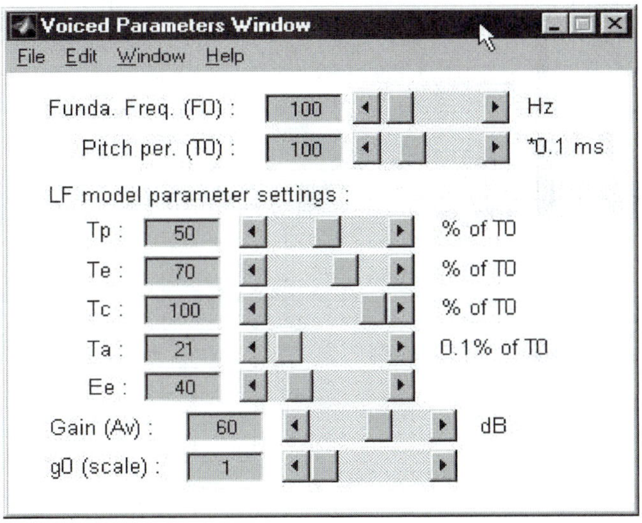

FIGURE 10.28 The voiced parameters window.

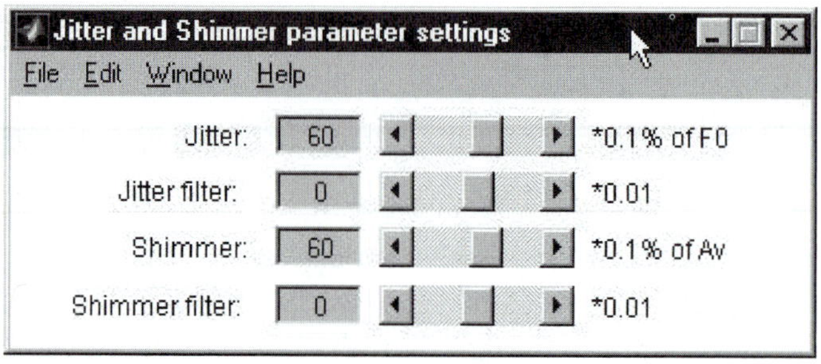

FIGURE 10.29 The jitter and shimmer parameters window.

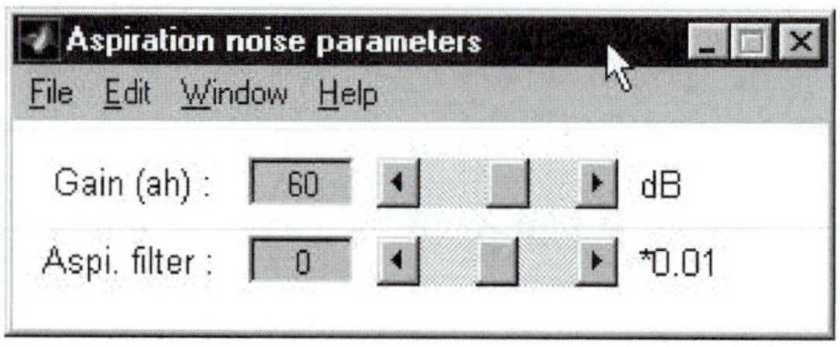

FIGURE 10.30 Aspiration noise parameters window.

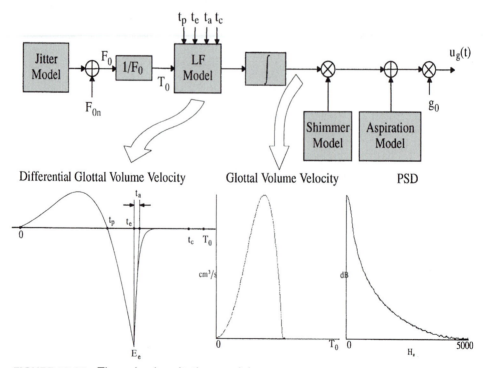

FIGURE 10.31 The voiced excitation model.

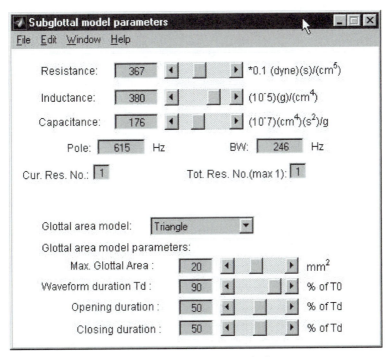

FIGURE 10.32   The subglottal parameters window.

portion. Since Foster RLC circuits are used to model the subglottal tract, the upper portion of Figure 10.32 enables the user to specify the RLC values and calculates and displays the resonant frequency (pole) and bandwidth. The current resonator is entered by the user and there is only one resonator allowed by the model at this time. For voiced and mixed excitation, the glottal area (lower) is a time-varying function. The user can select one of three glottal area functions using the pull-down window: triangular, sinusoidal, or raised cosine, as shown in Figure 10.33. For each glottal area model, the maximum area (in mm^2) and the waveform duration (in percent of T0) is to be specified by adjusting the appropriate sliders. For the triangular and raised cosine glottal area models the waveform can be unsymmetrical. This can be accomplished by adjusting the opening duration and closing duration sliders. Adjusting one slider automatically adjusts the other. For the sinusoidal model the waveform is always symmetric, so these sliders have no effect. For the unvoiced option, the glottal area is a constant. So only the maximum glottal area slider is activated to allow the user to specify the glottal opening area.

Once the user has set the desired parameter values, then type in the current frame number, the total number of frames, the beginning time, the duration, and the ending time (beginning time plus duration) as shown in Figure 10.34. The beginning time and duration are available from Figure 10.21 and subsequent such figures for each frame. Then press the Apply button. The excitation waveform can be viewed by pressing the Draw button in Figure 10.24, the main menu excitation source window. For the parameter settings used here, the excitation waveform appears as shown in Figure 10.35, which is the canvas for the excitation source waveform.

Figure 10.35 shows the LF model waveform, which is the differentiated glottal volume velocity waveform, the glottal waveform (glottal volume velocity), the glottal area model waveform, and the power spectral density (PSD) for the excitation waveform.

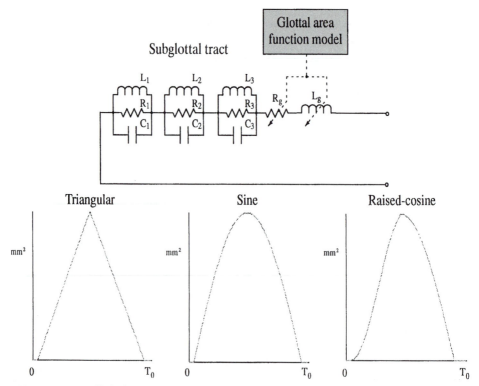

FIGURE 10.33 Subglottal system and glottal impedance model.

The user can design additional excitation waveforms for other frames, but one excitation waveform is sufficient for synthesis and is the recommended procedure. The jitter and shimmer are included in the synthesis process as well as the aspiration noise and the subglottal model. The excitation waveform can be saved by pressing the File button in Figure 10.24 and selecting the Save option. Excitation waveforms are saved as *.src files,

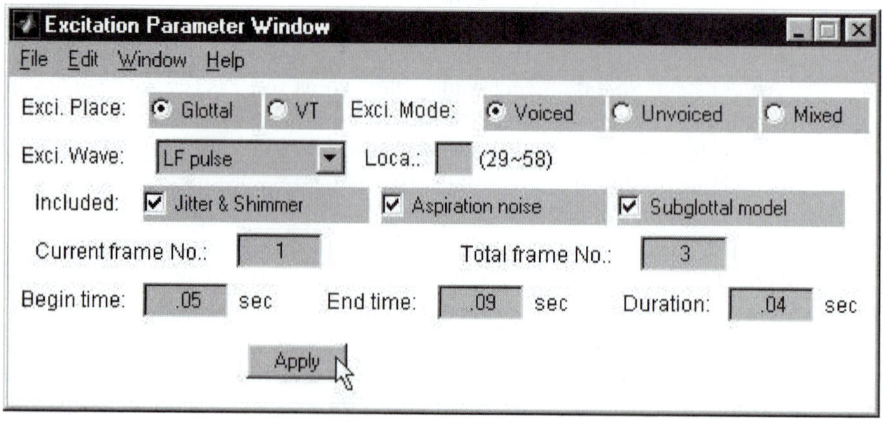

FIGURE 10.34 Excitation parameter window with the frame and time parameters set.

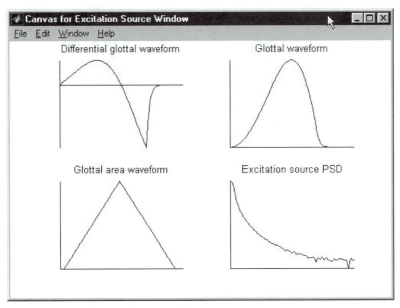

FIGURE 10.35 Excitation waveform.

where the * is replaced by a name selected by the user. A saved excitation waveform can be loaded by pressing the File button and selecting the Load option. Once a saved excitation file is loaded, the windows that were used to design the excitation open automatically. For example, if the excitation waveform was designed as voiced with jitter and shimmer and aspiration noise, then these windows also open, showing the design parameter values. Thus, the parameters of a loaded excitation waveform can be altered and saved as another file if desired. The Next and Previous buttons in Figure 10.24 are to allow the user to examine the parameter values and plot the excitation waveforms of a designed excitation or a loaded excitation waveform. These two options are typically not used since usually only one excitation waveform is used in the synthesis process. The Clear button erases the canvas window. An excitation waveform must be designed or loaded before the synthesis process is started. Similarly, an articulatory parameter vector must be designed or loaded before the synthesis process is initiated.

10.3.4.2 Unvoiced Excitation The options for unvoiced excitation are similar to that for voiced. However, certain options are not available, while others are. Pressing the Unvoiced option button in Figure 10.26 activates only the aspiration noise and subglottal model options. The jitter and shimmer option is not available for unvoiced excitation. The unvoiced option opens the turbulence noise parameters window shown in Figure 10.36. This window allows the user to design the turbulence excitation and set the place of the turbulence at 1) downstream from the maximal constriction point (the default); 2) the center of the maximal constriction; 3) upstream from the maximal constriction; or 4) distributed along the constriction region. See Appendix 11 for more details. The turbulence gain is adjustable, as is the critical Reynolds number (which sets the threshold at which the volume velocity becomes turbulent flow) and the magnitude of the glottal volume velocity, which sets the volume velocity at a constant value for the unvoiced excitation. If the aspiration and subglottal options are selected, then these are set as described for voiced excitation.

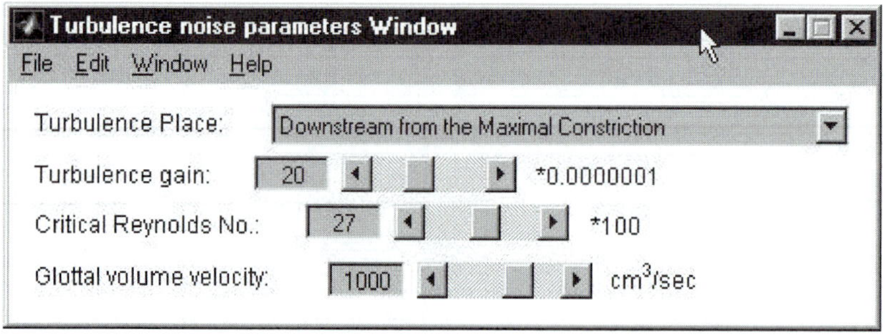

FIGURE 10.36 Turbulence noise parameters window for unvoiced excitation.

The unvoiced excitation can be saved or loaded as described previously. The waveform can be drawn by pressing the Draw button in Figure 10.24, the excitation source window.

10.3.4.3 Mixed Excitation This option provides for both voiced and turbulence excitation. Pressing the Mixed Excitation button brings up both the voiced parameters window and the turbulence noise parameters window. The user can also select jitter and shimmer, aspiration noise, and the subglottal model, as for voiced excitation. The designed excitation waveform can be saved, loaded, and drawn as described previously.

10.3.4.4 Turbulence Excitation Models Appendix 11 discusses the development of a model for unvoiced, or turbulence, excitation. The elements of this model are presented in Figures 10.37 and 10.38. The model for the turbulence noise source can be located at the center of the constriction, immediately upstream or downstream from the constriction, or spatially distributed within the constriction.

Examples of the vocal tract area and the transfer function for these unvoiced models are given in Appendix 11.

10.3.4.5 Excitation Within the Vocal Tract The final option available for the excitation waveform generation is the ability to use voiced excitation at a location within the vocal tract (see Appendix 11 for additional discussion of this option and its application to individuals who have no larynx to generate a voiced excitation). This is a special feature of this toolbox and will be useful to only a few researchers. The primary function of this option is to show that the excitation waveform can be designed (reshaped) to synthesize normal sounding speech even if the excitation is placed within the vocal tract, instead of at the glottis. To use this option select the VT button in Figure 10.26 instead of the Glottal Excitation button. This activates only the voiced option and the location option next to the excitation waveform on the second row in Figure 10.39. Press the Voiced option button. This opens the voiced parameters window and activates the jitter and shimmer and aspiration noise options as discussed previously for the voiced excitation option. The user may select either, both, or neither of these latter two options. The user must type in a vocal tract excitation location for this option. Figure 10.39 shows the location typed as 35. This designates that the voiced excitation is to occur at the 35th section of the vocal tract model, not at the glottis. The user designs the excitation waveform as described previously, enters the beginning time and other variables, and presses the Apply button. The designed waveform is plotted in Figure 10.40 by pressing the Draw button in Figure 10.24, the excitation source window. When the synthesis process is performed, the waveform excites the vocal tract at the 35th section and generates the appropriate synthetic speech with the excitation waveform at this vocal

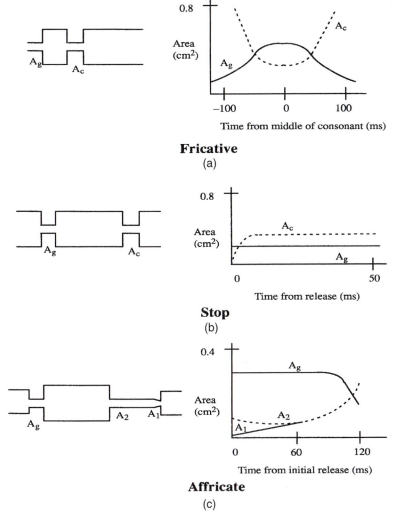

Fricative

(a)

Stop

(b)

Affricate

(c)

FIGURE 10.37 Unvoiced (turbulence) excitation models for (a) fricatives, (b) stops, and (c) affricates.

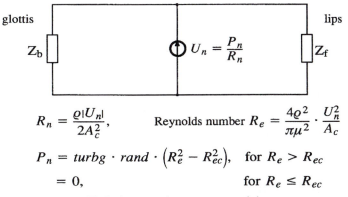

$$R_n = \frac{\varrho |U_n|}{2A_c^2},$$

Reynolds number $R_e = \dfrac{4\varrho^2}{\pi\mu^2} \cdot \dfrac{U_n^2}{A_c}$

$$P_n = turbg \cdot rand \cdot \left(R_e^2 - R_{ec}^2\right), \quad \text{for } R_e > R_{ec}$$

$$= 0, \qquad\qquad\qquad \text{for } R_e \leq R_{ec}$$

FIGURE 10.38 Turbulence noise source model.

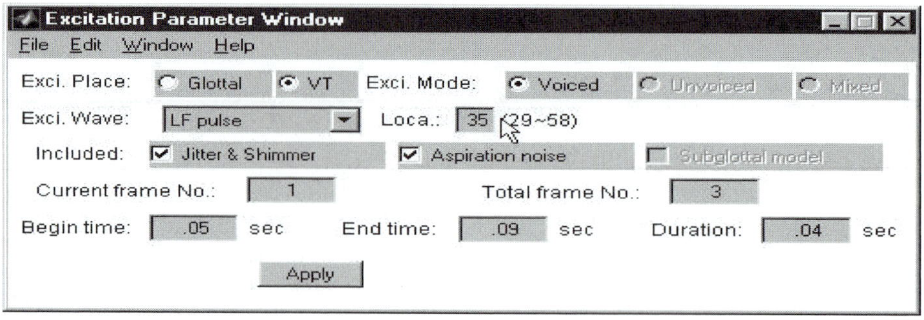

FIGURE 10.39 Excitation parameter window for vocal tract excitation at the 35th section.

tract location. Note that the designed excitation waveform provides the excitation; it is not prefiltered to compensate for the fact that the excitation is relocated.

While Appendix 11 discusses this feature of the articulatory speech synthesizer, we present several examples here. A model for calculating the transfer function with the excitation within the vocal tract is shown in Figure 10.41. The calculations for transfer function proceed from top to bottom in Figure 10.41. This model can be used to calculate a filter that can be used to prefilter the glottal excitation waveform to compensate for the fact that the excitation is relocated to a section within the vocal tract. The software does not provide the prefiltering; this must be done by the user. In the following examples if the excitation is relocated to a section within the vocal tract, it is not prefiltered unless otherwise stated.

Figure 10.42 shows the midsagittal profile for the vowel /AA/.

The optimized vocal tract cross-sectional area and the transfer function for the vowel /AA/ are shown in Figures 10.43 and 10.44.

Next, Figures 10.45 and 10.46 show the original and synthesized speech waveforms and spectrograms, respectively, with the excitation located at the 35th and 50th section of the vocal tract for the vowel /AA/.

Figure 10.47 shows the glottal pulse of the LF model immediately followed by the same glottal pulse prefiltered by the modified acoustic transfer function (top panel), the synthesized speech for the vowel /AA/ with the above excitation located at the 35th

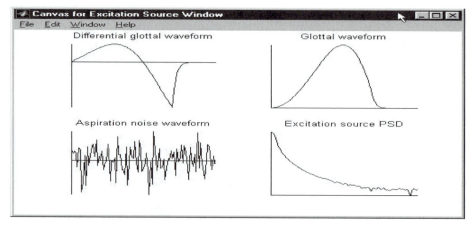

FIGURE 10.40 Excitation waveform for excitation at the 35th vocal tract section.

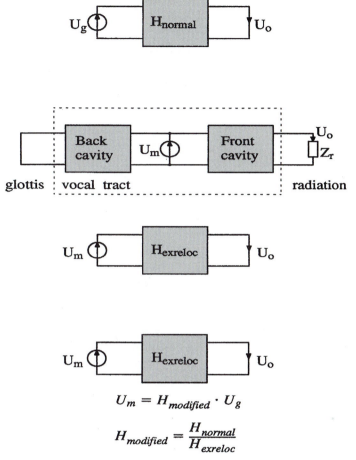

$$U_m = H_{modified} \cdot U_g$$

$$H_{modified} = \frac{H_{normal}}{H_{exreloc}}$$

FIGURE 10.41 Source model for excitation relocation in the vocal tract.

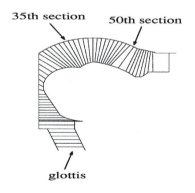

Midsagittal profile for vowel /a/.

FIGURE 10.42 Midsagittal profile for the vowel /AA/.

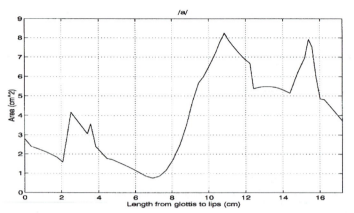

FIGURE 10.43 Optimized vocal tract cross-section area for vowel /AA/.

section of the vocal tract (middle panel), and the wideband spectrogram of the synthesized speech (lower panel). The purpose of this figure is to illustrate the effect of prefiltering the excitation waveform before exciting the vocal tract at the 35th section. In this case, the synthesized speech without prefiltering sounds very good, while the speech synthesized with the prefiltered excitation sounds like a tone. This is because the prefiltering introduces the fifth and sixth formants as artifacts. Consequently, the prefiltering needs to be improved.

The final two figures in this section, Figures 10.48 and 10.49, show the original and synthesized speech waveforms and spectrograms, respectively for the sentence, "We were away a year ago," with the excitation located at the 35th and 50th section of the vocal tract. The excitation is not prefiltered.

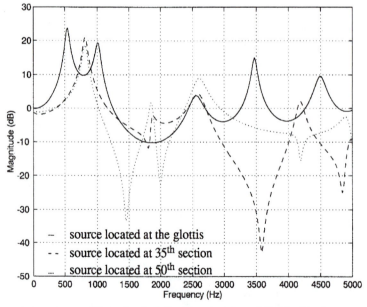

FIGURE 10.44 Transfer function for the vowel /AA/ with the source located at three locations.

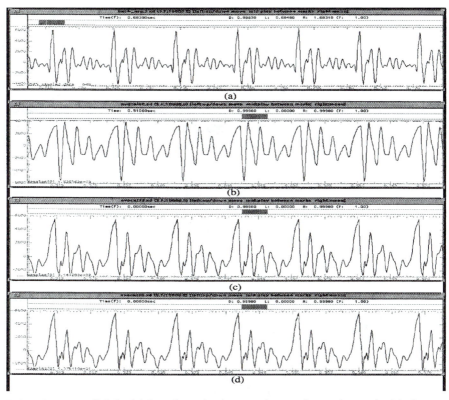

FIGURE 10.45 Original (*a*) and synthetic speech waveforms located with the source located at the,(*b*) glottis, (*c*) the 35th section, and (*d*) the 50th section of the vocal tract for the vowel /AA/.

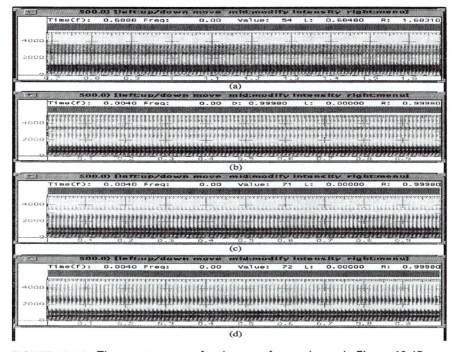

FIGURE 10.46 The spectrograms for the waveforms shown in Figure 10.45.

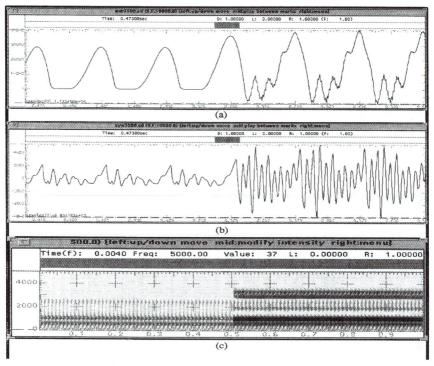

FIGURE 10.47 Illustration of prefiltering the excitation waveform before placing the excitation at the 35th section of the vocal tract for the vowel /AA/.

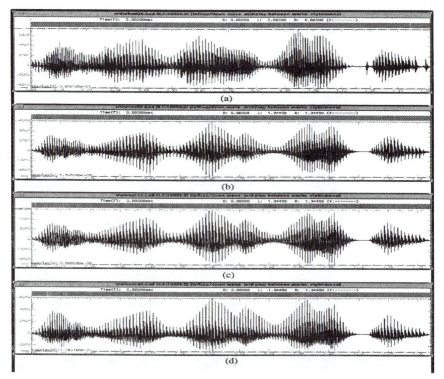

FIGURE 10.48 Original (*a*) and synthetic speech waveforms located with the source located at the, (*b*) glottis, (*c*) the 35th section, and (*d*) the 50th section of the vocal tract for the sentence, "We were away a year ago."

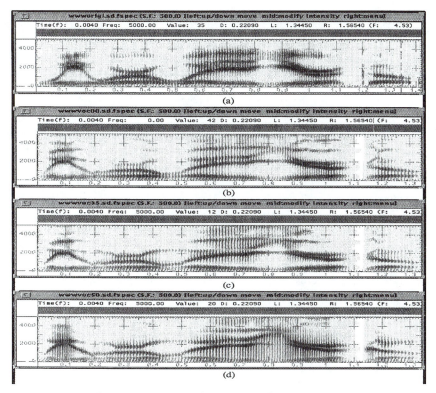

FIGURE 10.49 The spectrograms for the waveforms shown in Figure 10.48.

10.3.5 The Synthesis Option

10.3.5.1 Review of Steps Before Synthesis Before activating the synthesis option, several steps must be completed. The user must either load an articulatory parameter file obtained by following the optimization procedure, or load a target formant track file and mark the frame boundaries on the formant tracks. Next, the user opens the shape option, followed by the optimization option, thereby obtaining an articulatory parameter vector. Then the user either loads a previously designed excitation waveform or designs a new excitation. In summary, the working memory must contain 1) an articulatory parameter vector; and 2) an excitation waveform prior to the initiation of the synthesis process.

10.3.5.2 Starting the Synthesis Option Pressing the Synthesis button in the articulatory speech synthesizer (main window), closes all windows opened during the excitation option, and opens the three windows shown in Figures 10.50, 10.51, and 10.52, which are the articulatory synthesis menu window, the canvas for the articulatory synthesis, and the articulatory synthesis parameters window.

The menu selections in the articulatory synthesis window are start, animate, play, print, clear, and cancel. The Cancel button cancels the synthesis option and erases the

FIGURE 10.50 The articulatory synthesis menu window.

FIGURE 10.51   The canvas for the articulatory synthesis menu window.

three windows shown in Figures 10.50, 10.51, and 10.52. The Animate button opens the window shown in Figure 10.53 and plays a movie of the midsagittal profiles of the articulatory configuration for the articulatory parameter file in memory, which in this case, is the be_seg_opt.art file. For this case, the movie contains only three frames. The movie is repeated three times, ending with the last frame. The Play, Print, Clear, and Save buttons are explained later.

Prior to pressing the start button in Figure 10.50, the user selects one of three display options in Figure 10.52 and the synthesis sampling frequency. The default settings are shown as display choice 1 and a sampling frequency of 20 kHz. The three display options are:

1. The vocal tract cross-sectional area function, the midsagittal vocal tract outline of the current target frame, the target and model formants, the error, and the synthetic speech waveform.

2. The articulatory trajectories and the synthetic speech waveform.

3. The target and excitation frame data; the acoustic transfer function, the excitation waveform and its power spectrum; the pressure and volume velocity waveforms at the 20th, 30th, 40th, and 50th sections in the vocal tract; and the synthetic speech waveform.

After the Start button is pressed, it takes approximately 30 seconds before the display appears. The interpolation of the articulatory parameters is linear as explained in Appendix 11. The synthesis sampling rate is selected as 10, 20, 30, 40, 50, or 60 kHz. There are a total of 60 sections in the articulatory vocal tract model. This number is fixed and cannot be altered by the user. For display option 3, the pressure, volume velocity, and speech waveforms are updated (refreshed) every 400 samples. The acoustic transfer function, the excitation waveform and PSD, area function, and midsagittal outline are updated frame by frame. For display option 2, the articulatory trajectories are displayed for the entire duration of the

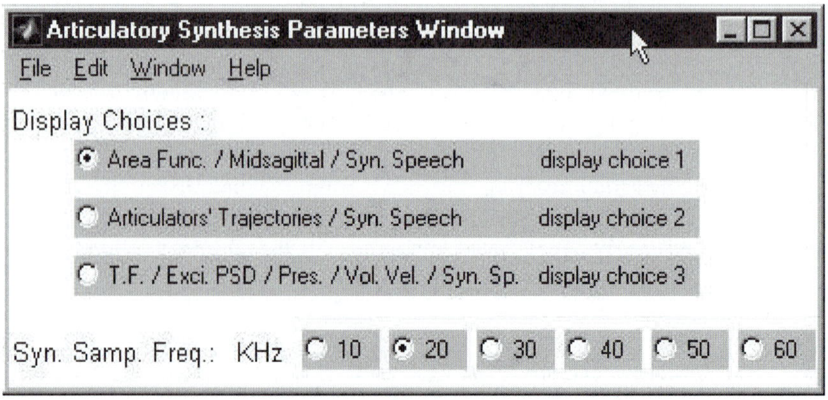

FIGURE 10.52   The articulatory synthesis parameters window.

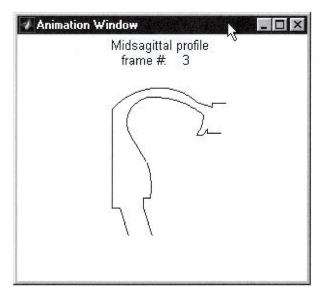

FIGURE 10.53 The animation window.

synthesis process. Display option 3 takes nearly twenty times as long to synthesize a speech sample as display options 1 or 2. Consequently, display options 1 or 2 are usually recommended. The be_total_form.for file was marked with 13 frames, then optimized, and the speech file be_total_syn_speech.dat was synthesized using display option 1. The synthesis took approximately 1.5 hours for the word be using 13 frames. Fewer frames takes much less time. Examples of the three display options appear in Figures 10.54, 10.55, and 10.56.

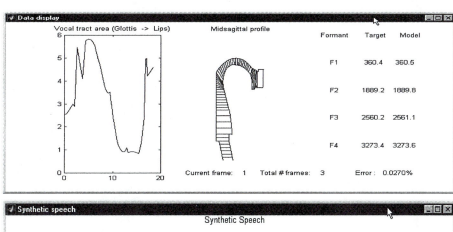

FIGURE 10.54 Display option 1.

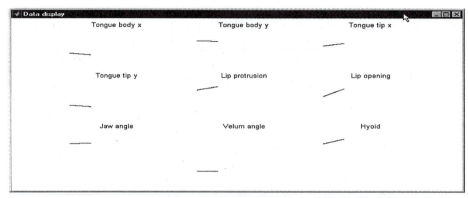

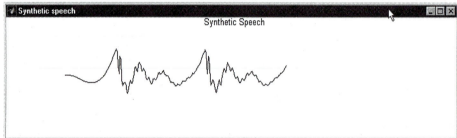

FIGURE 10.55 Display option 2.

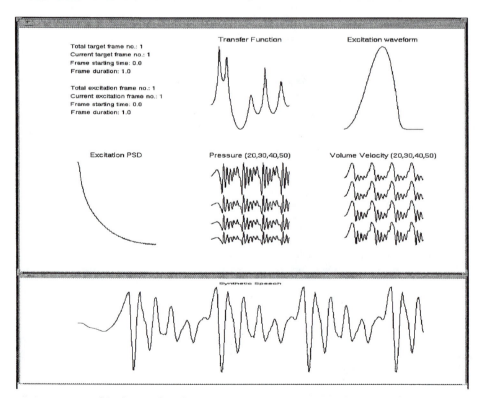

FIGURE 10.56 Display option 3.

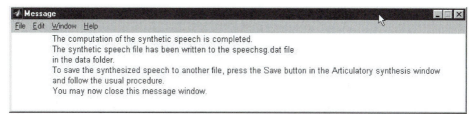

FIGURE 10.57   Message window stating that the synthesis is completed.

Upon completion of the synthesis process, a message window, shown in Figure 10.57, appears informing the user that the synthesis is completed and that the user can save the synthesized speech file. The user can print out a copy of the display (data and synthetic speech) using the Print button. The synthetic speech can be played by pressing the Play button. The speech file is saved automatically to a file called `speechsg.dat`. However, it is recommended that the user also save the speech file to another file by pressing the Save button, since the user may forget to rename the `speechsg.dat` file before performing another synthesis. A saved synthesized speech file cannot be loaded and played with this toolbox, rather the `analysis` toolbox or the `formant_track` toolbox can be used. The Clear button clears the canvas.

For each of the display options if the duration of the synthetic speech exceeds the length of the display window, then the window is cleared and the speech signal is drawn from the beginning of the synthesized speech window. For display option 3, if the duration of the synthetic speech exceeds the length of the display window, then the pressure and volume velocity waveforms overwrite the displayed waveforms.

The data display for display option 2 is not very interesting for the example shown, since the speech is a steady vowel. Consequently, the articulators do not move significantly. This display is more interesting for a sentence, such as, "Should we chase those cowboys?"

Figure 10.56 is for a vowel but is not the same data as shown in Figures 10.54 and 10.55.

10.4 SUMMARY

The operation of this toolbox entails the extraction of the first four formant tracks from a target speech file. The formant tracks are marked with frame boundaries. The shape of the articulatory vocal track is initially determined using a set of default articulatory position vectors. Inverse filtering is performed using a simulated annealing program that minimizes the error between the target formants and the model formants on a frame-by-frame basis. The inverse filtering step determines the optimized articulatory parameter vector for the target speech. Prior to synthesis, an excitation waveform is designed. The excitation can be voiced, unvoiced, or mixed excitation. The voiced and mixed excitation can include jitter and shimmer, aspiration, and a subglottal model. Mixed excitation can also include turbulence. The unvoiced excitation does not include voicing, but does include turbulence, aspiration, and the subglottal model. The synthesis phase constructs the synthesized speech waveform. The user can also view an animated movie of the midsagittal outlines for the articulatory vocal tract model. A summary of the toolbox is given in Figure 10.58.

10.5 THE DATA FOLDER

The data folder in the `artm` toolbox contains the following example data files, which are arranged by name: "aa" designates the vowel AA, "be" designates the word be, "m"

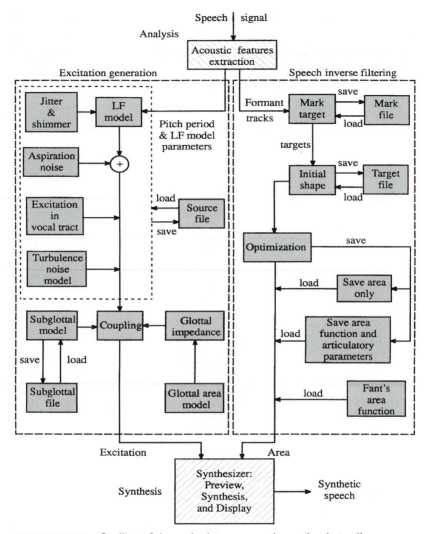

FIGURE 10.58 Outline of the articulatory speech synthesis toolbox.

designates the nasal M, "sh" designates the fricative SH, "we" designates the sentence, "We were away a year ago." The file extensions are:

*.fan	Fant area function. See Appendix 5, the Russian vowels.
*.art	Articulatory parameter file obtained by inverse filtering using simulated annealing.
*fmrk.mrk	A formant track file marked with frame boundaries.
*form.for	A formant track file obtained using the formant_track toolbox.
*.dat	A speech data file, either synthesized (syn) or natural.
aa.fan	The Fant area function file for the vowel AA.
aa_fmrk.art	The articulatory parameter file for the formants for AA.
aa_fmrk.mrk	The marked file for aa.
aa_form.for	The formant file for aa.
be_seg.art	An articulatory parameter file for a segment of the word be. Not optimized.

be_seg_fmrk.mrk	The marked file for be_seg.
be_seg_form.for	The formant file for be_seg.
be_seg_opt.art	The optimized articulatory parameter file for a segment of the word be.
be_seg_syn_speech.dat	The synthesized speech for the be_seg file using the be_seg.art file.
be_total.art	An articulatory parameter file for the total word be. Not optimized.
be_total_1_v_js_asp.src	The exicitation file used to synthesize be_seg_syn_speech.dat and be_total_syn_speech.dat. The file has one frame, voiced, jitter and shimmer, and aspiration noise.
be_total_fmrk.mrk	The marked file for be_total.
be_total_form.for	The formant file for be_total.
be_total_syn_speech.dat	The synthesized speech for be_total.
ex1_v_.src	An excitation file, one frame, voiced.
ex2_v_.src	An excitation file, two frames, voiced.
ex32_v_src	An excitation file, 32 frames, voiced.
m.fan	The Fant area function file for the nasal M.
sh.fan	The Fant area function file for fricative SH.
speechsg.dat	The speech file that is automatically saved after synthesis.
we_.art	The optimized articulatory parameter file for a segment of the sentence, "We were away a year ago."
we_fmrk.mrk	The marked file for we.
we_form.for	The formant file for we.

The following files are synthesized speech for various conditions.

avocal00.dat	AA with the excitation at the glottis (vocal tract section 00).
avocal35.dat	AA with the excitation at vocal tract section 35.
avocal50.dat	AA with the excitation at vocal tract section 50.
shcent54.dat	SH with the excitation noise source at the center of the constriction, section 54.
shdi5254.dat	SH with the excitation noise source distributed from section 52 to 54.
shdown59.dat	SH with the excitation noise source downstream from the constriction, section 59.
shup53.dat	SH with the excitation noise source upstream from the constriction, section 53.
swc123_1.dat	Sentence, "Should we chase those cowboys?"
syn3500.dat	Two successive vowels, both AA, both with excitation at section 35. The first is with the excitation designed by the software with no prefiltering. The second excitation is prefiltered to compensate for the relocation of the source to the 35th section. The prefiltered excitation produces a sound that is almost a tone because the fifth and sixth formants are introduced as artifacts by the prefiltering. Thus, the prefiltering needs to be improved.
wwwvoc00.dat	Sentence, "We were away a year ago," with excitation at the glottis (section 00).
wwwvoc35.dat	Sentence, "We were away a year ago," with excitation at section 35. The excitation has not been prefiltered to compensate for the relocation of the source.
wwwvoc50.dat	Sentence, "We were away a year ago," with excitation at section 50. The excitation has not been prefiltered to compensate for the relocation of the source.

To listen to and view the synthesized speech files use the analysis toolbox or the formant_track toolbox.

PROBLEMS

10.1   In Chapter 6 the synthesis tasks were primarily vowels. For this chapter, the tasks are to synthesize words and sentences. The purpose of this problem is to become familiar with the various options discussed in this chapter. To do so, you are asked to repeat some of the examples used in the text. Use the formant_track toolbox to obtain the formant tracks for the word be (file **b.dat**). Copy the target formant file to the data folder in the artm toolbox. Call this file be_test_form.for. Close the formant_track toolbox. Use the artm toolbox to load this file. Mark the formant tracks with two frame boundaries, approximately evenly spaced over the file, using the **add mark** option. Save the marked file as be_test_frmk.mrk. Start the **shape** option. Experiment with altering the sliders to reduce the error. Try to reduce the error manually to about 7% or less for each frame. Next, select and start the optimization option. Use the default settings in the **articulatory optimization setup** window. Next, select and start the **simulated annealing** window. Use the default values. Press Start in this window. Record the time. The optimization process may take nearly 45 minutes, depending on the type of machine you are using. Upon completion of the optimization, record the time required to do the optimization. You can save the optimized articulatory parameter file as be_test_.art if desired. Next, start the **excitation** option and load the **ex1.src** file. Alter the excitation parameters if desired, draw the waveform, and save the file to be_test_.src. Start the **synthesis** option. Select the **animation** option. Only two frames will be shown. Use the default values for the articulatory synthesis parameters. Press the Start button in the **articulatory synthesis window**. Record the time. The synthesis may take 1 to 1.5 hours. Upon completion of the synthesis, record the time required to do the synthesis. Save the synthesized speech file as be_test_syn_speech.dat. Play the file. Does it sound like the original? Print the file. Close out the artm toolbox. Open the formant_track toolbox. Load the file be_test_syn_speech.dat. Play the file and view its waveform and compare with the original.

10.2   Repeat Problem 10.1 but use display option 2 in the **articulatory synthesis parameters** window. Use different file names.

10.3   Repeat Problem 10.1 but use the 60 kHz sampling frequency and display option 1 in the **articulatory synthesis parameters** window. Use different file names. Compare the quality of the synthesized speech obtained from both problems.

10.4   The optimization process can be speeded up by altering the parameter settings in the **simulated annealing** window. For example, reduce the number of cycles to 10, the number of iterations to 5, and the maximum number of evaluations to 100. These changes will not give the optimum error, but will greatly reduce the computation time. Repeat Problem 10.1 with these changes and compare the two synthesized speech files. Be sure to use different file names. Note the errors and compare the location of the model formants with the target formants.

10.5   Repeat Problem 10.1 but change the excitation to use be_total_1_v_js_asp.src. Compare the two synthesized files. Be sure to use different file names.

10.6   Repeat Problem 10.1 but change the excitation to unvoiced. You will have to design an excitation file. Does the synthesized speech sound like whispered speech?

10.7   Use the analysis toolbox to obtain a segment "We were" from the sentence, "We were away a year ago." Repeat Problem 10.1 using the optimization process, and using the voiced excitation **ex1.src**. Compare the synthesized file with the original.

10.8   Repeat Problem 10.1 but synthesize the sentence, "We were away a year ago." This may take some time (see Appendix 11-D). Use 20 frames to mark the formant track file. Use display option 1 and then repeat using display option 2 so that you can observe the changes in the articulator movements.

10.9   Repeat Problem 10.8 but synthesize the sentence, "Should we chase those cowboys?"

APPENDIX *1*

GLOSSARY

Abduction: An opening of the vocal folds for voiceless sounds.

Accent: A characteristic manner of pronouncing sounds, the result of dialect or native language influence.

Adduction: A closing of the vocal folds necessary for voicing.

Affricate: A sound that combines a stop closure with a fricative release.

Allophone: A variation of a phoneme or a basic sound unit. May be a modified phoneme due to coarticulation.

Alveolar: An articulation involving the tip of the tongue and the alveolar ridge, as in "die."

Alveolar ridge: The part of the upper surface of the mouth immediately behind the front teeth.

Antiresonance: A filtering effect of the vocal tract that causes a loss of acoustic energy in a particular frequency region.

Aperiodic: Vibrations with irregular periods.

Articulation: Movements made with the articulators that change the shape of the mouth and the resonances of the vocal tract to control the production of sounds. The tongue, lips, teeth, jaw, palate, velum are the articulators. The type and effect of articulation are determined by the amount of constriction of the airflow through the vocal tract and the point at which the constriction occurs, that is, the place of constriction.

Aspirate: A sound with friction produced at the glottis; /HH/.

Assimilation: A change in the features of a speech sound toward the features of neighboring sounds.

Bernoulli effect: The pressure fall caused by increased velocity through a constricted passage.

Breathy voice: A whisper-like voice. It may have a slight vibratory excursion of the vocal folds with incomplete glottal closure.

Coarticulation: The influence of the production of one sound on another, usually an adjacent sound.

Consonant: One of a class of sounds, e.g., /B/, /D/, /G/, /F/, and so forth.

Continuant: A speech sound that can be sustained and retain its acoustic characteristics.

Creaky voice: *See Vocal fry.*

Demisyllables: A speech unit obtained by dividing syllables in half at the middle of the vowel.

Dental: A consonant sound produced with contact of the tongue and the teeth.

Dialect: A variety of spoken language caused by isolation of one group of speakers from another as a result of geographic, social, political, or economic barriers.

Diphones: A speech unit obtained by dividing the speech waveform into units of phonemes, but the division is at the middle of the phoneme.

Diphthongs: One of a class of sounds, for example, /OY/ as in boy.

Dyads: A speech segment during which the utterance changes from one phoneme to another. Within a dyad, the formants are in transition.

Electroglottograph: An instrument used to measure the impedance across the vocal folds.

Epiglottis: A flap of cartilage that closes the opening to the trachea to prevent food and liquids from entering the lungs.

Esophagus: The hollow muscular tube extending from the pharynx to the stomach.

Falsetto voice: A high-pitched voice. It has gradual glottal opening and closing phases with a short or no closed phase.

Formant: A major peak, or resonance, in the frequency spectrum of a sound.

Formant frequency: The frequency at which the formant resonance, or peak, occurs.

Fricative: One of a class of sounds, for example, sounds made by friction such as /F/.

Fundamental frequency: The frequency of voicing, that is, the frequency at which the vocal folds vibrate.

247

Glide: A phoneme produced by moving the tongue quickly from one relatively open position to another in the vocal tract.

Glottis: The opening between the vocal folds.

Glottograph: An instrument used to measure the relative amount of light transmitted through the glottis.

Hard palate: The bony partition between the mouth and the nose; the roof of the mouth.

Harsh voice: A voice with laryngeal strain and tension and low pitch. It has wide variations in jitter. Sometimes called a rough voice.

Hoarse voice: A voice with excessive breathiness and weak in intensity. It is sometimes defined as breathy plus harsh (rough). It is a combination of excessive air escapage and an aperiodicity of vocal fold vibration.

Idiosyncratic: A term used to refer to the sound changes that are unique to a particular individual but that may not be analyzable according to any standard phonological process or pattern.

Intensity: The power (amplitude) of a sound, which is perceived as its loudness by a listener.

Intonation: The pattern of pitch changes that occur during a phrase, which may be a complete sentence.

Labeling: Assigning a symbol to a parsed sound (*see Parse*).

Labiodental: An articulation involving the lower lip and the upper front teeth.

Language: The words, and rules for combining them.

Laryngograph: An instrument used to measure the impedance across the vocal folds.

Larynx: The cartilages and connective tissue that enclose the vocal folds.

Lexicon: A dictionary.

Mandible: The lower jaw.

Modal voice: A normal voice. During voicing, the glottis has a long opening phase and a rapid closing phase.

Morpheme: The minimally meaningful unit of language. Morphemes are either free or bound. Thus the word cats is comprised of two morphemes, cat (free morpheme) and plural "-s" (bound morpheme).

Motor theory: A theory put forth by A. M. Liberman that speech perception makes reference to speech production.

Myoelastic aerodynamic theory of phonation: Theory that vocal fold vibration is primarily due to air pressure forces acting on the elastic mass of the folds.

Nasals: One of a class of sounds that includes /M/, /N/, /NX/.

Nasal sounds: Sounds that are produced with an open velopharyngear port.

Neutral vowel: A vowel produced with the tongue in its rest position, typical of the position for the schwa.

Normal voice: Known as a modal voice.

Oral cavity: The space inside the mouth.

Orthography: The writing and spelling system of a language.

Palatal: An articulation involving the front of the tongue and the hard palate, as in you.

Palatalization: The modification of a sound by moving the tongue close to the palate or the roof of the mouth.

Parse: Analysis of speech or text into phonerues, syllables, phrases, clauses, and other segments, and labeling the parsed segments.

Pharynx: The section of the digestive tract from the vocal folds to the oral cavity.

Phon: A unit of equal loudness.

Phonation: The production of voice or voiced sounds by adduction of the vocal folds in relation to the airstream passing through the larynx from the lungs.

Phone: An articulated sound, may include whistles, grunts, and so forth.

Phoneme: An abstract linguistic unit representing the basic unit in a language. These are the smallest units that distinguish one utterance from another. For English, there are about 40 phonemes. These are grouped into classes such as consonants, fricatives, vowels, diphthongs, nasals, and so on.

Pitch: The subjective sensation of the frequency of a sound. The sensation includes the effects of the fundamental frequency of voicing and the intensity of the sound produced.

Place of articulation: The point of contact or near contact of the active and passive articulators in producing a speech sound.

Poles: An engineering term for resonances.

Prosody: The long-term timing or rhythmic variations of speech, controlled by the pitch, intensity, and timing of speech segments.

Rough voice: A grating voice, but very little, if any, breathiness, but with considerable jitter. Also called a harsh voice.

Semantics: The meaning of the speech message.

Semivowels: One of a class of speech sounds, for example, the liquids /L/ and /R/ and the glides /W/ and /Y/.

Sibilants: One of a class of speech sounds, for example, the hissing sounds, such as /S/.

Sound: The sensation produced by stimulation of the organs of hearing by vibrations transmitted through the air or another medium.

Sound pressure level: The pressure of a sound decibels derived from a pressure ratio; usually expressed as dynes/cm^2.

Source function: The origin of acoustic energy for speech; for vowels at the vocal folds, for voiceless consonants in the vocal tract, and for voiced consonants both at the folds and in the tract. The latter is sometimes called mixed excitation.

Spectrogram: A frequency spectrum of the speech signal that varies with time. The spectrogram is drawn in two dimensions with frequency along the ordinate and time along the abscissa. The intensity of the frequency signal is indicated by shades of gray with the darkest gray being the most intense or largest signal.

Stress: The emphasis placed on a particular segment of speech.

Suprasegmental: Extending over segments of speech such as stress and intonation.

Syllable: A vowel and its neighboring consonants.

Syntax: The grammer of forming phrases and sentences.

Tense: Referring to vowels. Phonetic property of vowels that are produced with slightly higher relative tongue position than lax vowels and with longer durations.

Trachea: The windpipe, a tube composed of horseshoe-shaped cartilages leading to the lungs.

Turbulence: Friction-like noise characteristic of consonants produced with incomplete closure of the vocal tract.

Unvoiced sounds: Sounds produced without vocal fold vibrations. The folds are held open.

Uvula: The cone-shaped projection hanging from the lower medial border of the soft palate.

Velopharyngeal closure: Closure of the nasal cavity from the oral cavity by contact between the soft palate and pharynx, thus directing the airflow through the mouth rather than the nose.

Velopharyngeal port: The passageway connecting oral and nasal cavities.

Velum: The soft palate. The soft tissue that controls the valving of air into the nasal passage.

Ventricular folds: The "false vocal folds"; the folds above the true vocal folds.

Vocal folds: The pair of folds, or cords, within the larynx that vibrate when air is passed from the lungs, thereby producing voiced or vocal sounds.

Vocal fry: A vocal mode in which the vocal folds vibrate at such low frequency that the individual vibrations can be heard. Also known as creaky voice. During voicing, the glottal area function has sharp, short pulses followed by a long closed glottal interval. The glottal opening phase may have one, two, or three opening/closing pulses.

Vocal tract: The oral cavity, usually including the pharynx to the lips.

Vocal tract model: A model of the vocal tract, which may be electronic or mechanical. This model may be used to synthetically generate speech.

Voice onset time (VOT): The time between the release of the closure for a stop consonant and the start of voicing of the following vowel.

Voiced sounds: Sounds that are produced with the vocal folds vibrating as in vowels or voiced fricatives, such as /Z/.

Voicing: The production of sound by the vibration of the vocal folds.

Vowels: One of a class of speech sounds produced usually with a fixed vocal tract configuration and with the vocal folds vibrating, such as /AA/.

Zeros: An engineering term for antiresonances.

REFERENCES

Abe, M., Nakamura, S., Shikano, K., and Kuwabara, H. (1988). Voice conversion through vector quantization. *Proc. ICASSP*, New York, 655–658.

Agrawal, A., and Lin, W. C. (1975). Effect of voiced speech parameters on the intelligibility of pb words. *J. Acoust. Soc. Am.*, **57**:217–222.

Alku, P., (1992). Glottal wave analysis with pitch synchronous iterative adaptive inverse filtering. *Speech Comm.*, **11**:109–118.

Allen, J., Hunnicutt, M. S., and Klatt, D. (1987). *From Text to Speech*. Cambridge University Press, Cambridge.

Allen, D. R., and Strong, W. J. (1985). A model for the synthesis of natural sounding vowels. *J. Acoust. Soc. Am.*, **78**(1):58–69.

The American Heritage Dictionary of the English Language, William Morris (Ed.), American Heritage Publishing Co. and Houghton Mifflin Co., Boston, 1969.

Ananthapadmanabha, T. V. (1982). Intelligibility Carried by Speech-Source Functions: Implication for Theory of Speech Perception, STL-QPSR, Royal Institute of Technology, Stockholm, Sweden, **4**:49–64.

Ananthapadmanabha, T. V., and Fant, G. (1982). Calculation of true glottal flow and its components. *Speech Comm.*, **1**:167–184.

Anderson, V. A. (1977). *Training the Speaking Voice*. Oxford University Press, Oxford.

Atal, B. S., Chang, J. J., Mathews, M. V., and Tukey, J. W. (1978). Inversion of articulatory-to-acoustic transformation in the vocal tract by a computer-sorting technique. *J. Acoust. Soc. Am.*, **63**:1535–1555.

Atal, B. S., Cuperman, V., and Gersho, A., Eds. (1991). *Advances in Speech Coding*. Kluwer Academic Publishing, Boston.

Atal, B. S., Cuperman, V., and Gersho, A., Eds. (1993). *Speech and Audio Coding for Wireless and Network Applications*. Kluwer Academic Publishing, Boston.

Atal, B. S., and David, N. (1979). On synthesizing natural sounding speech by linear prediction. *Proc.*

IEEE Int. Conf. Acoust., Speech, Signal Proc. 44–47.

Atal, B. S., and Hanauer, S. L. (1971). Speech analysis and synthesis by linear prediction of the speech wave. *J. Acoust. Soc. Am.*, **50**(2):637–655.

Atal, B. S., and Remde, J. R. (1982). A new model of LPC excitation for producing natural-sounding speech at low bit rates. *Proc. ICASSP*, 614–617.

Badin, P. (1989). Acoustics of voiceless fricatives: Production theory and data. *STL-QPSR*, Royal Institute of Technology, Stockholm, Sweden, **3**:33–55.

Badin, P. (1991). Fricative consonants: Acoustic and x-ray measurements, *J. Phonetics*, **19**:397–408.

Badin, P., and Fant, G. (1984). Notes on vocal tract computation. *STL-QPSR*, Royal Institute of Technology, Stockholm, Sweden, **2–3**:53–108.

Baer, T. (1981a). Investigation of the phonatory mechanism. *Proceedings of the Conference on the Assessment of Vocal Pathology*, ASHA Reports, **11**:38–47.

Baer, T. (1981b). Observations of vocal fold vibration: Measurement of excised larynges. In *Vocal Fold Physiology*, K. N. Stevens, and M. Hirano, Eds. University of Tokyo Press, Tokyo, Japan.

Baer, T., Gore, J. C., Gracco, L. C., and Nye, P. W. (1991). Analysis of vocal tract shape and dimensions using magnetic resonance imaging: Vowels. *J. Acoust. Soc. Am.*, **90**:799–828.

Bailly, G., and Benoit, C., Eds. (1992). *Talking Machines*. Elsevier Science Publishing, North-Holland, Amsterdam.

Baken, R. J. (1987). *Clinical Measurement of Speech and Voice*. College-Hill Press, San Diego, CA.

Barnwell, T. P. (1979). Objective measures for speech quality testing. *J. Acoust. Soc. Am.*, **66**:1658–1663.

Bavegard, M. (1996). Towards an articulatory speech synthesizer: Model development and simulations, *TMH-QPSR*, **1**:1–15.

Båvegård, M., and Högberg, J. (1992). Using artificial neural networks to relate articulation to acoustic features of the vocal tract. *STL-QPSR*,

Royal Institute of Technology, Stockholm, Sweden, **1**:33–48.

Båvegård, M., and Högberg, J. (1993). Using artificial neural nets to compare different vocal tract models. *STL-QPSR*, Royal Institute of Technology, Stockholm, Sweden, **2–3**:7–13.

Beker, H. J., and Piper, F. C. (1985). *Secure Speech Communications*. Academic Press, Orlando.

Benvenuto, N., Marchesi, M., and Uncini, A. (1992). Applications of simulated annealing to the design of special digital filters. *IEEE Trans. Signal Processing*, **40**(2):323–332.

Bocchieri, E. L. (1983). *An Articulatory Speech Synthesizer*. Ph.D. dissertation, University of Florida.

Bocchieri, E. L., and Childers, D. G. (1984). An animated interactive graphics editor for the study of speech articulation. *Speech Technology*, **2**:10–14.

Bogert, B. P., Heally, M. J., and Tukey, J. W. (1963). The quefrency alanysis of time series for echoes: Cepstrum, pseudo-autocovariance, cross-cepstrum, and saphe cracking. In *Time Series Analysis*, M. Rosenblatt, Ed. Wiley, New York, pp. 209–243.

Bohachevsky, I. O., Johnson, M. E., and Stein, M. L. (1986). Generalized simulated annealing for function optimization. *Technometrics*, **28**(3):209–217.

Bonder, L. J. (1983). The n-tube formula and some of its consequences. *Acustica*, **52**:216–226.

Bonomi, E., and Lutton, J.-L. (1984). The n-city travelling salesman problem: Statistical mechanics and the Metropolis algorithm. *SIAM Review*, **26**(4):551–568.

Boone, D. R. (1971). *The Voice and Voice Therapy*. Prentice-Hall Inc., Englewood Cliffs, NJ.

Borden, G. J., and Harris, K. S. (1980). *Speech Science Primer: Physiology, Acoustics, and Perception*. Williams & Wilkens, Baltimore.

Bristow, G. (Ed.) (1984). *Electronic Speech Synthesis*. McGraw-Hill Book Co., New York.

Broad, D. J. (1977a). Acoustics of speech production. In *Topics in Speech Science*, D. J. Broad, Ed., Speech Communications Research Laboratory, Inc., Los Angeles, CA, pp. 10–68.

Broad, D. J. (1977b). Theory of vocal fold vibration. In Topics in Speech Science, D. J. Broad, Ed., Speech Communications Research Laboratory, Inc., Los Angeles, CA, pp. 158–232.

Brownman, C. P., and Goldstein, L. (1989). Articulatory gestures as phonological units. *Phonology*, **6**:201–251.

Brownman, C. P., and Goldstein, L. (1990). Gestural specification using dynamically-defined articulatory structures. *J. Phonetics*, **18**:299–320.

Burrus, C. S., et al. (1994). *Computer-Based Exercises for Signal Processing Using MATLAB*. Prentice Hall, Englewood Cliffs.

Bush, M. A., Kopec, G. E., and Zue, V. W. (1983). Selecting acoustic features for stop consonant identification. *Proc. Int'l. Conf. on Acoust., Speech, and Signal Processing*, 742–745.

Carnevali, P., Coletti, L., and Patarnello, S. (1985). Image processing by simulated annealing. *IBM Journal of Research and Development*, **29**(6):569–579.

Carhart, R. T. (1940). Air-flow through the larynx. *Quart. J. Spch.*, **26**:606–614.

Carlson, R. (1993). Models of Speech Synthesis, STL-QPSR, 1, 1–14.

Carlson, R., Granstrom, B., and Klatt, D. H. (1979). Some notes on the perception of temporal patterns in speech. In *Frontiers of Speech Communication Research*, B. Lindblom and S. Ohman, Eds. Academic Press, New York, 233–243.

Cater, J. P. (1983). *Electronically Speaking: Computer Speech Generation*. Howard W. Sams, Indianapolis IN

Chan, K. (1989). *Modeling of Vocal Fold Vibration*. Ph.D. dissertation, University of Florida, Gainesville.

Charpentier, F. (1984). Determination of the vocal tract shape from the formants by analysis of the articulatory-to-acoustic nonlinearities. Speech Comm. 3(4):291–308.

Chiba, T., and Kajiyama, M. (1941). *The Vowel, Its Nature and Structure*. Tokyo-Kaiseikan, Tokyo, Japan.

Childers, D. G. (1977). Laryngeal pathology detection. *Critical Reviews in Bioengineering*, **2**:375–426.

Childers, D. G. (1991). Signal processing methods for the assessment of vocal disorders. Med. Life Sci. Engineering, **13**:117–130.

Childers, D. G. (1995). Glottal source modeling for voice conversion. *Speech Comm.*, **16**:127–138.

Childers, D. G. (1997). *Probability and Random Processes Using MALAB*. McGraw-Hill, Inc., New York.

Childers, D. G., and Ahn, C. T. (1995). Modeling the glottal volume-velocity waveform for three voice types. *J. Acoust. Soc. Am.*, **97**:505–519.

Childers, D. G., and Ding, C. (1991). Articulatory

synthesis: Nasal sounds and male and female voices. *J. Phonet.*, **19**:453–464.

Childers, D. G., and Hu, T. H. (1994). Speech synthesis by glottal excited linear prediction. *J. Acoust. Soc. Am.*, **96**(4):2026–2036.

Childers, D. G., and Krishnamurthy, A. K. (1985). A critical review of electroglottography. *CRC Critical Reviews in Bioengineering*, **12**:131–161.

Childers, D. G., and Larar, J. N. (1984). Electroglottography for laryngeal function assessment and speech analysis. *IEEE Trans. on Biomed. Eng.*, **31**(12):807–817.

Childers, D. G., and Lee, C. K. (1991). Vocal quality factors: Analysis, synthesis, and perception. *J. Acoust. Soc. Am.*, **90**(5):2394–2410.

Childers, D. G., and Wong, C. F. (1994). Measuring and modeling vocal source-tract interaction. *IEEE Trans. Biomed. Eng.*, **41**:663–671.

Childers, D. G., and Wu, K. (1990). Quality of speech produced by analysis–synthesis. *Speech Comm.*, **9**:97–117.

Childers, D. G., and Wu, K. (1991). Gender recognition from speech. Part II: Fine analysis. *J. Acoust. Soc. Am.*, **90**(4):1841–1856.

Childers, D. G., Hicks, D. M., Moore, G. P., and Alsaka, Y. A. (1986). A model for vocal fold vibratory motion, contact area, and the electroglottogram. *J. Acoust. Soc. Am.*, **80**:1309–1320.

Childers, D. G., Hicks, D. M., Moore, G. P., Eskenazi, L., and Lalwani, A. L. (1990). Electroglottography and vocal fold physiology. *J. Speech and Hearing Research*, **33**:245–254.

Childers, D. G., Skinner, D. P., and Kemerait, R. C. (1977). The cepstrum: A guide to processing. *Proc. IEEE*, **65**:1428–1443.

Childers, D. G., Smith, A. M., and Moore, G. P. (1984). Relationship between electroglottograph, speech, and vocal cord contact. *Folia Phoniatrica*, **36**:105–118.

Childers, D. G., Wu, K., Hicks, D. M., and Yegnanarayana, B. (1989). Voice conversion. *Speech Comm.*, **8**:147–158.

Coker, C. H. (1967). Synthesis by rule from articulatory parameters. *Proc. 1967 Conf. Speech Comm. Process.*, Boston, MA, pp. 52–53.

Coker, C. H. (1976). A model of articulatory dynamics and control. *Proc. IEEE*, **64**(4):452–460.

Coker, C. H., and Fujimura, O. (1966). Model for specification of the vocal-tract area function. *J. Acoust. Soc. Am.*, **40**:1271.

Cole, R. A., and Cooper, W. E. (1975). Perception of voicing in English affricates and fricatives. *J. Acoust. Soc. Amer.*, **58**(6):1280–1287.

Colton, R. H., and Estill, J. A. (1981). Elements of voice quality: Perceptual, acoustic and physiologic aspects. In *Speech and Language*, 5, N. J. Lass, Ed. Academic Press, NY, pp. 311–403.

Cook, P. R. (1991). *Identification of Control Parameters in an Articulatory Vocal Tract Model, with Applications to the Synthesis of Singing*. Ph.D. dissertation, Stanford University, CA.

Cook, P. R. (1993). SPASM, a real-time vocal tract physical model controller; and singer, the companion software synthesis system. *Computer Music Journal*, **17**(1):30–44.

Cooper, M. (1987). Human factor aspects of voice input/output. *Speech Technology*, **3**:82–86.

Corana, A. C., Marchesi, M., Martini, C., and Ridella, S. (1987). Minimizing multimodal functions of continuous variables with the simulated annealing algorithm. *ACM Transactions on Mathematical Software*, **13**(3):262–280.

Cranen, B., and Boves, L. (1987). On subglottal formant analysis. *J. Acoust. Soc. Am.*, **81**(3):734–746.

Davis, S. B., and Mermelstein, P. (1980). Comparison of parametric representations for monosyllabic word recognition in continuously spoken sentences. *IEEE Trans. on Acoust., Speech, and Signal Processing*, **28**(4):357–366.

Davis, L., and Ritter, F. (1987). Schedule optimization with probabilistic search. *Proceedings of the 3rd IEEE Conf. on Artificial Intelligence Applications*, pp. 231–236.

de Haan, H. J. (1982). The relationship of estimated comprehensibility to the rate of connected speech. *Perception and Psychophysics*, **32**(1):27–31.

Deller, J. R., Proakis, J. G., and Hansen, J. H. L. (1993). *Discrete-Time Processing of Speech Signals*. Macmillan, New York.

Denes, P. B. (1963). On the statistics of spoken English. *J. Acous. Soc. Am.*, **35**:892–896.

Denes, P. B., and Pinson, E. N. (1993). The Speech Chain, 2nd ed. W. H. Freeman, New York.

Dudley, H. (1936). Synthesis speech. *Bell Labs. Record*, **15**:98–102.

Dunn, H. K. (1950). The calculation of vowel resonances, and an electrical vocal tract. *J. Acoust. Soc. Am.*, **22**:740–753.

Edwards, H. T. (1992). *Applied Phonetics: The Sounds of American English.* Singular Publishing Co., San Diego.

Eggen, J. H. (1992). *On the Quality of Synthetic Speech: Evaluation and Improvements.* Ph.D. dissertation, University of Eindhoven, The Netherlands.

Erler, K., and Deng, L. (1993). Hidden Markov model representation of quantized articulatory features for speech recognition. *Computer Speech and Language,* **7**:265–282.

Eskenasi, L. (1988). *Acoustic Correlates of Voice Quality and Distortion Measure for Speech Processing.* Ph.D. dissertation, University of Florida.

Eskenasi, L., Childers, D. G., and Hicks, D. M. (1990). Acoustic correlates of vocal quality. *J. Speech and Hearing Res.,* **33**:298–306.

Espy-Wilson, C. Y. (1986). A phonetically based semivowel recognition system. *Proc. Int'l. Conf. on Acoust., Speech, and Signal Processing,* pp. 2775–2778.

Fairbanks, G. (1958). Test of phonemic differentiation: The rhyme test. *J. Acoust. Soc. Am.,* **33**:596–600.

Fairbanks, G., Everitt, W. L., and Jaeger, R. P. (1954). Method for time or frequency compression–expansion of speech. *Trans. of Inst. of Radio Engineers, Prof. Group on Audio,* **AU2**(1):7–12.

Fallside, F., and Woods, W. A., Eds. (1985). *Computer Speech Processing.* Prentice Hall, Englwood Cliffs.

Fant, G. (1960). *Acoustic Theory of Speech Production.* Moreton & Co, Gravenhage, Sweden

Fant, G. (1973). *Speech Sounds and Features.* M. I. T. Press, Cambridge.

Fant, G. (1979). Glottal source and excitation analysis. *STL-QPSR,* Royal Institute of Technology, Stockholm, Sweden, **4**:85–107.

Fant, G. (1985). The vocal tract in your pocket calculator. *STL-QPSR,* Royal Institute of Technology, Stockholm, Sweden, **2–3**:1–19.

Fant, G. (1986). Glottal flow: Models and interaction. *J. Phonetics,* **14**:393–399.

Fant, G. (1988). *Glottal Flow Parameters from the Frequency Domain,* The second symposium on advanced man-machine interface through spoken language, Hawaii.

Fant, G. (1993). Some problems in voice source analysis. *Speech Comm.,* **13**:7–22.

Fant, G. (1995). The LF model revisited: Transformations and frequency domain analysis. *STL-QPSR,* **4**:1–13.

Fant, G., and Lin, Q. G. (1988). Frequency domain interpretation and derivation of glottal flow parameters. *STL-QPSR,* Royal Institute of Technology, Stockholm, Sweden, **2–3**:1–21.

Fant, G., and Lin, Q. G. (1989). Comments on glottal flow modelling and analysis. *STL-QPSR,* Royal Institute of Technology, Stockholm, Sweden, **1**:1–8.

Fant, G., and Lin, Q. G. (1991). Comments on glottal flow modeling and analysis. In *Vocal Fold Physiology: Acoustic, Perceptual, and Physiological Aspects of Voice Mechanisms.* J. Gauffin and B. Hammarberg, Eds. Singular Publishing Co., San Diego, pp. 1–13.

Fant, G., Liljencrants, J., and Lin, Q. G. (1985). A four parameter model of glottal flow. *STL-QPSR,* **2–3**:119–156.

Fettweis, A., and Meerkötter, K. (1975). On adaptors for wave digital filters. *IEEE Transactions on Acoustics, Speech, and Signal Processing,* **23**(6): 516–525.

Flanagan, J. L. (1957). Note on the design of terminal analog speech synthesizers. *J. Acoust. Soc. Am.,* **29**:306–310.

Flanagan, J. L. (1972a). *Speech Analysis, Synthesis, and Perception.* Springer-Verlag, New York.

Flanagan, J. L. (1972b). Voices of men and machines. *J. Acoust. Soc. Am.,* **51**:1375–1387.

Flanagan, J. L. (1982). Talking with computers: Synthesis and recognition of speech by machines. *IEEE Trans. Biomed. Engr.,* **BME-29**:223–232.

Flanagan, J. L., and Cherry, L. (1968). Excitation of vocal-tract synthesizers. *J. Acoust. Soc. Am.,* **45**(3):764–769.

Flanagan, J. L., and Golden, R. M. (1966). Phase vocoder. *The Bell System Tech. J.* **45**:1493–1509.

Flanagan, J. L., and Ishizaka, K. L. (1976). Automatic generation of voiceless excitation to a vocal cord-vocal tract speech synthesizer. *IEEE Trans. Acoust., Speech, Signal Process.,* **24**(2):163–170.

Flanagan, J. L., and Ishizaka, K. (1978). Computer model to characterize the air volume displaced by the vibrating vocal cords. *J. Acoust. Soc. Am.,* **63**:1559–1565.

Flanagan, J. L., and Landgraf, I. L. (1968). Self-oscillating source for vocal tract synthesizers. *IEEE Trans. Audio and Electroacoustics*, **16**:57–64.

Flanagan, J. L., and Rabiner, L. R. (Eds) (1973). *Speech Synthesis*. Dowden, Hutchinson, and Ross, Stroudsburg, Pennsylvania.

Flanagan, J. L., Coker, C. H., and Bird, C. M. (1962). Computer simulation of a formant vocoder synthesizer. *J. Acoust. Soc. Am.*, **35**:2003(A).

Flanagan, J. L., Coker, C. H., Rabiner, L. R., Schafer, R. W., and Umeda, N. (1970). Synthetic voices for computers. *IEEE Spectrum*, **7**:22–45.

Flanagan, J. L., Ishizaka, K., and Shipley, K. L. (1975). Synthesis of speech from a dynamic model of the vocal cords and vocal tract. *Bell System Tech. J.*, **54**(3):485–505.

Flanagan, J. L., Ishizaka, K. L., and Shipley, K. L. (1980). Signal models for low bit-rate coding of speech. *J. Acoust. Soc. Am.*, **68**(3):780–791.

Fletcher, H. (1929). *Speech and Hearing*. Van Norstrand, Princeton, NJ, pp. 292–294.

Fletcher, H., and Steinberg, J. C. (1929). Articulation testing methods. *Bell System Tech, J.*, **8**:848–852.

Foulke, E., and Sticht, T. G. (1969). Review of research on the intelligibility and compression of accelerated speech. *Psychol. Bull.* **72**(1):50–62.

French, N. R., and Steinberg, J. C. (1947). Factors governing the intelligibility of speech sounds. *J. Acoust. Soc. Am.*, **19**:90–114.

Fry, D. (1979). How can you say that? *Human Nature*, February, pp. 38–43.

Fujisaki, H. (1983). Dynamic characteristics of voice fundamental frequency in speech and singing. In The Production of Speech, P. F. MacNeilage, Ed. Springer, New York, pp. 39–55.

Fujisaki, H., and Ljungqvist, M. (1986), Proposal and evaluation of models for the glottal source waveform. *Proc. IEEE Int. Conf., Acoust., Speech, Signal Process.*, 1605–1608.

Furui, S. (1989). *Digital Speech Processing, Synthesis, and Recognition*. Marcel Dekker, New York.

Furui, S., and Sondhi, M. M., Eds. (1992). *Advances in Speech Signal Processing*. Marcel Dekker, New York.

Garvey, W. D. (1949). Durational factors in speech intelligibility. In *Time-Compressed Speech: An Anthology and Bibliography*. Sam Duker, Ed. Scarecrow Press, 1974.

Garvey, W. D. (1951). An experimental investigation of the intelligibility of speeded speech. In *Time-Compressed Speech: An Anthology and Bibliography*. Sam Duker, Ed. Scarecrow Press, 1974.

Garvey, W. D. (1953a). The intelligibility of speeded speech. *J. Exp. Psychology*, **45**:102–108.

Garvey, W. D. (1953b). The intelligibility of abbreviated speech patterns. *Quarterly Journal of Speech*, **39**:296–306.

Geman, S., and Geman, D. (1984). Stochastic relaxation, Gibbs distributions, and the Bayesian restoration of images. *IEEE Transactions on Pattern Analysis and Machine Intelligence*, **6**(6):721–741.

Glass, J. R., and Zue, V. W. (1986). Detection and recognition of nasal consonants in American English. *Proc. Int'l. Conf. on Acoust., Speech, and Signal Processing*, pp. 2767–2770.

Glass, J. R., and Zue, V. W. (1988). Multi-level acoustic segmentation of continuous speech. *Proc. Int'l. Conf. on Acoust., Speech, and Signal Processing*, pp. 429–432.

Gobl, C. (1988). Voice source dynamics in connected speech. *STL-QPSR*, Royal Institute of Technology, Stockholm, Sweden, **1**:123–159.

Goffe, W. L., Ferrier, G. D., and Rogers, J. (1992). Simulated annealing: An initial application in econometrics. *Computer Science in Economics and Management*, **5**:133–146.

Goffe, W. L., Ferrier, G. D., and Rogers, J. (1994). Global optimization of statistical functions with simulated annealing. *J. Econometrics*, **60**:65–99.

Gold, B., and Rabiner, L. R. (1968). Analysis of digital and analog formant synthesizers. *Trans. Audio Electroacoustics*, **AU-16**:81–94.

Goldberg, D. E. (1989). *Genetic Algorithms in Search, Optimization, and Machine Learning*. Addison Wesley, Reading, MA.

Goldman-Eisler, F. (1968). *Psycholinguistics: Experiments in Spontaneous Speech*. Academic Press, New York, NY.

Goldstein, H. (1940). Reading and listening comprehension at various controlled rates. In *Time-Compressed Speech: An Anthology and Bibliography*. Sam Duker, Ed. Scarecrow Press, 1974.

Gopinath, B., and Sondhi, M. M. (1970). Determination of the shape of human vocal tract from acoustic measurements. *Bell System Tech. J.*, **49**:1195–1214.

Guo, Q., and Milenkovic, P. (1993). Optimal estimation of vocal tract area functions from speech signal constrained by x-ray microbeam data. *Proc. ICASSP*, **2**:538–541.

Gupta, S. K., and Schroeter, J. (1993). Pitch-synchronous frame-by-frame and segment-based articulatory analysis by synthesis. *J. Acoust. Soc. Am.*, **94**(5):2517–2530.

Gupta, V., Lennig, M., Mermelstein, P., Kenny, P., Seitz, P. F., and O'Shaughnessy, D. (1992). Use of minimum duration and energy contour for phonemes to improve large vocabulary isolated-word recognition. *Computer Speech and Language*, **6**:345–359.

Hamlet, S. L. (1981). Ultrasound assessment of phonatory function. *ASHA Reports*, **11**:128–140.

Hawley, M. E. (Ed.) (1977). *Speech Intelligibility and Speaker Recognition*. Dowden, Hutchinson, and Ross, Stroudsburg, PA.

Hegerl, G. C., and Höge, H. (1991). Numerical simulation of the glottal flow by a model based on the compressible Navier–Stokes equations. *Proc. ICASSP*, 477–480.

Heike, G. (1979). Articulatory measurement and synthesis—Methods and preliminary results. *Phonetica*, **36**:294–301.

Heinz, J. M., and Stevens, K. N. (1961). On the properties of voiceless fricatives. *J. Acoust. Soc. Am.*, **33**:589–596.

Heinz, J. M., and Stevens, K. N. (1964). On the derivation of area functions and acoustic spectra from cineradiographic films of speech, *J. Acoust. Soc. Am.*, **36**:1037.

Hermansky, H. (1990). Perceptual linear predictive (PLP) analysis of speech. *JASA*, **87**(4):1738–1752.

Hess, W. (1983). *Pitch Determination of Speech Signals: Algorithms and Devices*. Springer-Verlag, New York, NY.

Hildebrand, B. H. (1976). *Vibratory Patterns of the Human Vocal Cords During Variations of Frequency and Intensity*. Ph.D. dissertation, University of Florida, Gainesville.

Hilgers, F. J. M., and Schouwenburg, P. F. (1990). A new low-resistance, self-retaining prosthesis (ProvoxTM) for voice rehabilitation after total laryngectomy. *Laryngoscope*, **100**:1202–1207.

Hinton, G. E., and Sejnowski, T. J. (1983). Analyzing cooperative computation. *Proceedings of the Fifth Annual Conference of the Cognitive Science Society*, Rochester, NY.

Hirano, M. (1974). Morphological structure of the vocal cord as a vibrator and its variations. *Folia Phoniatrica*, **26**:89–94.

Hirano, M. (1975). Phonosurgery: Basic and clinical investigations. *Otologia* (Fukuoka), 21:239–440.

Hirano, M. (1977). Structure and vibratory behavior of the vocal folds. In *Dynamic Aspects of Speech Production*, M. Sawashima, and F. S. Cooper, Eds. University of Tokyo Press, Tokyo, Japan, pp. 13–27.

Hirano, M., and Kakita, Y. (1985). Cover-body theory of vocal cord vibration. In *Speech Science*. R. G. Daniloff, Ed. College-Hill Press, San Diego, pp. 1–46.

Hirano, M., Kurita, S., and Nakashima, T. (1981). The structure of vocal folds. In *Vocal Fold Physiology*, K. N. Stevens, and M. Hirano, Eds. University of Tokyo Press, Tokyo, Japan, pp. 33–41.

Hirano, M., and Sato, K. (1993). *Histological Color Atlas of the Human Larynx*. Singular Publishing, San Diego, CA.

Hirano, M., Matsuo, K., Kakita, Y., Kawasaki, H., and Kurita, S. (1983). Vibratory behavior versus the structure of the vocal fold. In *Vocal Fold Physiology*, I. R. Titze, Ed. The Denver Center for the Performing Arts, Iowa City, pp. 26–40.

Hogden, J., Lofqvist, A., Gracco, V., Zlokarnik, I., Rubin, P., and Saltzman, E. (1996). Accurate recovery of articulator positions from acoustics: New conclusions based on human data. *J. Acoust. Soc. Am.*, **100**:1819–1834.

Holmes, J. N. (1973). The influence of the glottal waveform on the naturalness of speech from a parallel formant synthesizer. IEEE Transactions on Audio & Electroacoustics, **21**:298–305.

Holmes, J. N. (1976). Formant excitation before and after glottal closure. *Proc. IEEE Int'l. Conf. Acoust., Speech, Signal Processing*, 39–42.

Holmes, J. N. (1983). Formant synthesizers: Cascade or parallel. *Speech Comm.*, **2**(4):251–274.

Holmes, J. N. (1987). Mechanisms and models of human speech production. In *Speech Synthesis and Recognition* Van Nostrand Reinhold, Wokingham, U.K., pp. 12–39.

Holmes, J. N. (1988). *Speech Synthesis and Recognition*. Van Nostrand Reinhold, Wokingham. U.K.

Horii, Y. (1979). Fundamental frequency perturbation observed in sustained phonation. *J. Speech and Hearing Research*, **22**:5–19.

Horii, Y. (1980). Vocal shimmer in sustained

phonation. *J. Speech and Hearing Research*, **23**:202–209.

Hoover, J., Reichle, J., van Tassell, D., and Cole, D. (1987). The intelligibility of synthesized speech: Echo II versus Votrax. *J. Speech and Hearing Research*, **30**:425–431.

House, A. S., Williams, C. E., Hecker, M. H. L., and Kryter, K. D. (1965). Articulation testing methods: Consonated differentiation with a closed-response set. *J. Acoust. Soc. Am.*, **40**:158–166.

Hsieh, Y. F. (1994). *A Flexible and High Quality Articulatory Speech Synthesizer*. Ph.D. dissertation, University of Florida.

Hsiao, Y. -S. (1996). *Speech Synthesis Algorithms for Voice Conversion*. Ph.D. dissertation, University of Florida.

Hsiao, Y. S., and Childers, D. G. (1996). *A New Approach to Formant Estimation and Modification Base on Pole Interaction*. 30th Asilomar Conf. On Signals, Systems and Computers.

Hu, H. T. (1993). *An Improved Source Model for a Linear Prediction Speech Synthesizer*. Ph.D. dissertation, University of Florida.

Huang, X. D., Ariki, Y., and Jack, M. A. (1990). *Hidden Markov Models for Speech Recognition*. Edinburgh University Press, Edinburgh.

Huggins, A. W. F., and Nickerson, R. S. (1985). Speech quality evaluation using phoneme-specific sentences. *J. Acoust. Soc. Am.*, **77**:1896–1906.

Iijima, H., Miki, N., and Nagai, N. (1992). Glottal impedance based on a finite element analysis of two-dimensional unsteady viscous flow in a static glottis. *IEEE Transactions on Signal Processing*, **40**(9):2125–2135.

Ince, A. N. (1992). *Digital Speech Processing: Speech Coding, Synthesis, and Recognition*. Kluwer Academic Publishing, Boston.

Ishizaka, K., and Flanagan, J. L. (1972). Synthesis of voiced sounds from a two-mass model of the vocal cords. *Bell System Tech. J.*, **51**(6):1233–1268.

Ishizaka, K., and Flanagan, J. L. (1977). Acoustic properties of longitudinal displacement in vocal cord vibration, *Bell System Tech. J.*, **56**(6): 889–918.

Ishizaka, K., French, J., and Flanagan, J. L. (1975). Direct determination of the vocal tract wall impedance. *IEEE Trans. on Acoust., Speech, and Signal Proc.*, **23**(4):370–373.

Ishizaka, K., Matsudaira, M., and Kaneko, T. (1976). Input acoustic impedance measurement of the subglottal system. *J. Acoust. Soc. Am.*, **60**:190–197.

Itoh, K., Kitawaki, N., and Kakehi, K. (1984). Objective quality measure for speech waveform coding systems. *Rev. Electr. Commun. Lab.*, 32, pt. **1**:220–228.

Iwahashi, N., and Sagisaka, Y. (1995). Speech spectrum conversion based on speaker interpolation and multi-functional representation with weighting by radial basis function networks. *Speech Comm.*, **16**:139–151.

Jayant, N. S., and Noll, P. (1984). *Digital Coding of Waveforms*. Prentice-Hall, Englewood Cliffs, NJ, Appendix F.

Johansson, C., Sundberg, J., Wilbrand, H., and Ytterbergh, C. (1983). From sagittal distance to area. *STL-QPSR*, Royal Institute of Technology, Stockholm, **4**:39–49.

Juang, B. H. (1984). On using the Itakura–Saito measures for speech coder performance evaluation. *AT&T Technical Journal*, **63**:1477–1498.

Juang, B. J., Wong, D. Y., and Gray Jr., A. H. (1982). Distortion performance of vector quantization for LPC voice coding. *IEEE Trans. Acoust., Speech, Signal Proc.*, **30**:294–304.

Kaburagi, T., and Honda, M. (1996). A model of articulatory trajectory formation based on the motor tasks of vocal-tract shapes. *J. Acoust. Soc. Am.*, **99**:3154–3170.

Kahn, M., and Garst, P. (1983). The effects of five voice characteristics on LPC quality. *Proc. IEEE Int. Conf. Acoust., Speech, Signal Proc.*, 531–534.

Kasuya, H., Ogawa, S., and Kikuchi, Y. (1986). An acoustic analysis of pathological voice and its application to evaluation of laryngeal pathology. *Speech Comm.*, **5**:171–181.

Kasuya, H., Ogawa, S., Mashima, K., and Ebihra, S. (1986). Normalized noise energy as an acoustic measure to evaluate pathological voice. *J. Acoust. Soc. Am.*, **80**(5):1329–1334.

Kay, S. M. (1988). *Modern Spectral Estimation*. Prentice-Hall, Englewood Cliffs, NJ.

Kelly, J. K., Jr., and Lochbaum, C. C. (1962). Speech synthesis. *Proc. Fourth Intern. Congr. Acoust.*, **G42**:1–4.

Kent, R. D., and Read, C. (1992). *The Acoustic Analysis of Speech*. Singular Publishing Co., San Diego.

Kirkpatrick, S., Gelatt, C. D., Jr., and Vecchi, M. P. (1983). Optimization by simulated annealing. *Science*, **220**(4598):671–680.

Kitawaki, N. (1992) Quality assessment of coded speech. In S. Furui and M. M. Sondhi, Eds. *Advances in Speech Signal Processing*. Marcel Dekker, Inc., New York, NY, pp. 357–385.

Kitzing, P. (1986). Glottography: The electrophysiological investigation of phonatory biomechanics. *Acta Oto-Rhino-Laryngologica*, **40**:863–878.

Klatt, D. H. (1977). Review of the ARPA Speech Understanding Project. *J. Acoust. Soc. Amer.*, **62**(6):1345–1366.

Klatt, D. H. (1979). Synthesis by rule of segmental durations in English sentences. In *Frontiers of Speech Communication Research*, B. Lindblom and S. Ohman, Eds. Academic Press, London.

Klatt, D. H. (1980). Software for a cascade/parallel formant synthesizer. *J. Acoust. Soc. Am.*, **67**(3):971–995.

Klatt, D. H. (1987). Review of text-to-speech conversion for English. *J. Acoust. Soc. Am.*, **82**(3):737–793.

Klatt, D. H., and Klatt, L. C. (1990). Analysis, synthesis, and perception of voice quality variations among female and male talkers. *J. Acoust. Soc. Am.*, **87**(2):820–857.

Klatt, D. H., and Stevens, K. N. (1973). On the automatic recognition of continuous speech: Implications from a spectrogram-reading experiment. *IEEE Trans. on Audio and Electroacoustics*, **AU-21**(3):210–217.

Koizumi, T., Taniguchi, S., and Hiromitsu, S. (1985). Glottal source-vocal tract interaction. *J. Acoust. Soc. Am.*, **78**:1541–1547.

Koizumi, T., Taniguchi, S., Hiromitsu, S. (1987). Two-mass models of the vocal cords for natural sounding voice synthesis, *J. Acoust. Soc. Am.*, **82**:1179–1192.

Koljonen, J., and Karjalainen, M. (1984). Use of computational psychoacoustical models in speech processing: Coding and objective performance evaluation. *Proc. IEEE Int. Conf. Acoust., Speech, Signal Proc.*, **1**:1.9.1–1.9.4.

Kryter, K. D. (1970). Masking and speech communication. In *Noise, The Effects of Noise on Man*, Academic Press, New York, NY, pp.

Kuwabara, H. (1984). A pitch-synchronous analysis/synthesis system to independently modify formant frequencies and bandwidths for voiced speech. *Speech Comm.*, **3**:211–220.

Kuwabara, H., and Sagisaka, Y. (1995). Acoustic characteristics of speaker individuality: Control and conversion. *Speech Comm.*, **16**:165–173.

Lalwani, A. L., and Childers, D. G. (1991a). A flexible formant synthesizer. *Proc. IEEE Int. Conf. Acoust., Speech, Signal Process.*, **2**:777–780.

Lalwani, A. L., and Childers, D. G. (1991b). Modeling vocal disorders via formant synthesis. *Proc. ICASSP*, **1**:505–508.

Larar, J. N., Schroeter, J., and Sondhi, M. M. (1988). Vector quantization of the articulatory space. *IEEE Trans. Acoust., Speech, and Signal Processing*, **36**:1812–1818.

Laver, J. (1980). *The Phonetic Description of Voice Quality*. Cambridge University Press, Cambridge.

Lawson, S., and Mirzai, A. R. (1990). *Wave Digital Filters*. E. Horwood, New York, NY.

Lee, C. K. (1988). *Voice Quality: Analysis and Synthesis*. Ph.D. dissertation, University of Florida.

Lee, F. F. (1972). Time compression and expansion of speech by the sampling method. *J. Audio Eng. Society*, **20**(9):738–742.

Lee, M. (1996). *Analysis and Synthesis of Speech Based on an Human Auditory Model*. Ph.D. dissertation, University of Florida.

Levinson, S. E., and Schmidt, C. E. (1983). Adaptive computation of articulatory parameters from the speech signal. *J. Acoust. Soc. Am.*, **74**(4):1145–1154.

Licklider, J. C. R. and Pollack, I. (1948). Effects of differentiation, integration, and infinite, peak clipping on the intelligibility of speech. *J. Acoust. Soc. Am.*, **20**:42–51.

Liljencrants, J. (1985). Dynamic line analogs for speech synthesis. *STL-QPSR*, **1**:1–14.

Lin, Q. (1992). Vocal-tract computation: How to make it more robust and faster. *STL-QPSR*, **4**:29–42.

Lin, Q. G. (1990). *Speech Production Theory and Articulatory Speech Synthesis*. Ph.D. dissertation, Royal Institute of Technology, Stockholm, Sweden.

Lindgvist-Gauffin, J. and Sundberg, J. (1976). Acoustic properties of the nasal tract, *Phonetica*, **33**:161–168.

Linggard, R. (1985). *Electronic Synthesis of Speech*. Cambridge University Press, Cambridge, New York.

Leung, H. C., Chigier, B., and Glass, J. R. (1993). A comparative study of signal representations and classification techniques for speech recognition. *Proc. Int'l. Conf. on Acoust., Speech, and Signal Processing*, 680–683.

Logan, J. S., Greene, B. G., and Pisoni, D. B. (1989).

Segmental intelligibility of' synthetic speech produced by rule. *J. Acoust. Soc. Am.*, **86**:566–581.

Lowry, L. D. (1981). Artificial larynges: A review and development of a prototype self-contained intraoral artificial larynx. *Laryngoscope*, **91**:1332–1355.

Luce, P. A., Feustel, T. C., and Pisoni, D. B. (1983). Capacity demands in short-term memory for synthetic and natural speech. *Human Factors*, **25**:17–32.

Mack, M. A., and Gold, B. (1985) The intelligibility of nonvocoded and vocoded semantically anomalous sentences. *MIT Tech. Rept.*, No. 703, July 28, ADA 160 401.

Mackenzie, J., Laver, J., and Hiller, S. (1983). Structural pathologies of the vocal folds and phonation. *Edinburgh University Dept. of Linguistics*, **16**:80–116.

Maeda, S. (1974). A characterization of fundamental frequency contours of speech. *QPR-RLE, MIT*, **114**:193–211.

Maeda, S. (1977). On a simulation method of dynamically varying vocal tract reconsideration of the Kelly–Lochbaum model. In *Articulatory Modeling and Phonetics*, R. Carre, R. Descout, and M. Wajskop, Eds. Groupe de Communication Parlee, Grenoble, France, pp. 281–288.

Maeda, S. (1982a). A digital simulation method of the vocal-tract system. *Speech Comm.*, **1**(3–4):199–229.

Maeda, S. (1982b). The role of the sinus cavities in the production of nasal vowels. *Proc. ICASSP*, 911–914.

Makhoul, J., Roucos, S., and Gish, H. (1985) Vector quantization in speech coding. *Proc. IEEE*, **73**:1551–1588.

Makhoul, J., Viswanathan, R., and Russell, W. (1976). A framework for the objective evaluation of vocoder speech quality. *Proc. IEEE Int. Conf. Acoust., Speech, Signal Proc.*, 103–106.

Malah, D. (1979). Time-domain algorithms for harmonic bandwidth reduction and time scaling of speech signals. *IEEE Trans. on Acoust., Speech, and Signal Processing*, **27**(2):121–133.

Markel, J. D., and Gray, A. H. (1974). A linear prediction vocoder simulation based upon the autocorrelation method. *IEEE Trans. Acoust., Speech, Signal Process.*, **22**:124–134.

Markel, J. D., and Gray, A. H. (1976). *Linear Prediction of Speech*. Springer-Verlag, New York, NY.

Mattingly, I. G. (1966). Synthesis by rule of prosodic features. *Lang. Speech*, **9**:1–13.

McAulay, R. J., and Quatieri, T. F. (1986). Speech analysis/synthesis based on a sinusoidal representation. *IEEE Trans. Acoust., Speech, Signal Proc.*, **ASSP-34**:744–754.

McCandless, S. S. (1974). An algorithm for automatic formant extraction using linear prediction spectra. *IEEE Trans. Acoust., Speech, Signal Process.*, **22**(2):135–141.

McGowan, R. S. (1994). Recovering articulatory movement from formant frequency trajectories using task dynamics and a genetic algorithm: Preliminary model tests. *Speech Comm.*, **14**:19–48.

McGowan, R. S., and Lee, M. (1996). Task dynamic and articulatory recovery of lip and velar approximations under model mismatch conditions. *J. Acoust. Soc. Am.*, **99**:595–608.

Meng, H. M., and Zue, V. W. (1991). Signal representation comparison for phonetic classification. *Proc. Int'l. Conf. on Acoust., Speech, and Signal Processing*, 285–288.

Mermelstein, P. (1967). Determination of the vocal tract shape from measured formant frequencies. *J. Acoust. Soc. Am.*, **41**(5):1283–1294.

Mermelstein, P. (1973). Articulatory model for the study of speech production. *J. Acoust. Soc. Am.*, **53**(4):1070–1082.

Mermelstein, P. (1977). On detecting nasals in continuous speech. *J. Acoust. Soc. Amer.*, **61**(2):581–587.

Metropolis, N., Rosenbluth, A. W., Rosenbluth, M. N., Teller, A. H., and Teller, E. (1953). Equation of state calculations by fast computing machines. *J. Chemical Physics*, **21**:1087–1092.

Meyer, P., Schroeter, J., and Sondhi, M. M. (1991). Design and evaluation of optimal cepstral lifters for accessing articulatory codebooks. *IEEE Trans. Acoust., Speech, and Signal Processing*, **39**(7):1493–1502.

Meyer, P., and Strube, H. W. (1984). Calculations on the time-varying vocal tract. *Speech Comm.*, **3**:109–122.

Meyer, P., Wilhelms, R., and Strube, H. W. (1989). A quasiarticulatory speech synthesizer for German language running in real time. *J. Acoust. Soc. Am.*, **86**(2):523–539.

Milenkovic, P. H. (1984). Vocal tract area functions from two-point acoustic measurements with formant frequency constraints. *IEEE Trans.*

Acoust., Speech, and Signal Processing, **32**(6):1122–1135.

Milenkovic, P. H. (1987). Acoustic tube reconstruction from noncausal excitation. *IEEE Trans. Acoust., Speech, and Signal Processing*, **35**(8):1089–1100.

Milenkovic, P. H. (1993). Voice source model for continuous control of pitch period. *J. Acoust. Soc. Am.*, **93**(2):1087–1096.

Miller, J. D. (1959). Nature of the vocal cord wave. *J. Acoust. Soc. Am.*, **31**:667–677.

Mizuno, H., and Abe, M. (1995). Voice conversion algorithm based on piecewise linear conversion rules for formant frequencies and spectrum tilt. *Speech Comm.*, **16**:153–164.

Moore, G. P. (1936). *A Stroboscopic Study of Vocal Fold Movement*. Ph.D. dissertation, Northwestern University, IL.

Moore, G. P. (1971). *Organic Voice Disorders*. Prentice-Hall Inc., Englewood Cliffs, NJ.

Moore, G. P. (1975). Ultra-high speed photography in laryngeal research. *Canadian J. Otolaryngol.*, **4**:793–799.

Moore, G. P., and Childers, D. G. (1983). Glottal area (real size) related to voice production. Transcripts of the Twelfth Symposium Care of the Professional Voice, The Voice Foundation, New York, pp. 92–96.

Moore, G. P., and von Leden, H. (1958). Dynamic variations of the vibratory pattern in the normal larynx. *Folia Phoniatrica*, **10**:205–238.

Morgan, N. (1984). *Talking Chips*, McGraw-Hill Book Co., New York, NY.

Morse, P. M. (1948). *Vibration and Sound*. McGraw-Hill, New York, NY.

Morse, P. M., and Ingard, K. U. (1968). *Theoretical Acoustics*. McGraw-Hill, New York, NY.

Morris, W., Ed. (1969). *The American Heritage Dictionary of the English Language*. American Heritage Publishing Co. and Houghton Mifflin Co., Boston.

Moulines, E., and Charpentier, F. (1990). Pitch-synchronous waveform processing techniques for text-to-speech synthesis using diphones. *Speech Comm.*, **9**:453–467.

Moulines, E., and Laroche, J. (1995). Non-parametric techniques for pitch-scale and time-scale modification of speech. *Speech Comm.*, **16**:175–205.

Murry, T., and Singh, S. (1980). Multidimensional analysis of male and female voices. *J. Acoust. Soc. Am.*, **66**:1294–1300.

Myrick, R., and Yantorno, R. (1993). Vocal tract modeling as related to the use of an artificial larynx. *Proceedings of the 19th Northeast Bioengineering Conference*, pp. 75–77.

Naik, J. K. (1984). *Synthesis and Evaluation of Natural-Sounding Speech Using the Linear Predictive Analysis-Synthesis Scheme*. Ph.D. dissertation, University of Florida.

Nakatsui, M., and Mermelstein, P. (1982). Subjective speech-to-noise ratio as a measure of speech quality for digital waveform coders. *J. Acoust. Soc. Am.*, **72**:1136–1144.

Narendranath, M., Murthy, H. A., Rajendran, S., and Yegnanarayana, B. (1995). Transformation of formants for voice conversion using artificial neural networks. *Speech Comm.*, **16**:207–216.

Neuburg, E. P. (1978). Simple pitch-dependent algorithm for high-quality speech rate changing. *J. Acoust. Soc. Amer.*, **63**(2):624–625.

Ning, T., and Whiting, S. (1990). Power spectrum estimation is a orthogonal transformation, *Proc. IEEE Conf. Acoust., Speech, Signal Process.*, 2523–2526.

Nocerino, N., Soong, F. K., Rabiner, L. R., and Klatt, D. H. (1985). Comparative study of several distortion measures for speech recognition. *Speech Comm.*, **4**:317–331.

Nooteboom, S. G. (1983). The temporal organization of speech and the process of spoken-word recognition. *IPO (Eindhoven) Progress Rept.*, No. 18, pp. 32–36.

Olive, J. (1971). Automatic formant tracking in a Newton–Raphson technique. *J. Acoust. Soc. Am.*, **50**:661–670.

Olive, J. P. (1992). Mixed spectral representation—Formant and linear predictive coding (LPC). *J. Acoust. Soc. Am.*, **92**(4):1837–1840.

Olive, J. P., Greenwood, A., and Coleman, J. (1993). *Acoustics of American English Speech*. Springer-Verlag, New York, NY.

Oppenheimer, A. V., and Schafer, R. W. (1975). *Digital Signal Processing*. Prentice-Hall, Inc., Englewood Cliffs, NJ.

O'Shaughnessy, D. (1987). *Speech Communication*. Addison-Wesley, Reading, MA.

Paget, R. (1930). *Human Speech*. Routledge and Kegan Paul, Ltd., London.

Paige, A., and Zue, V. W. (1970). Computation of

vocal tract area functions. *IEEE Trans. Audio Electroacoustics*, **18**(1):7–18.

Papamichalis, P. E. (1987). *Practical Approaches to Speech Coding*. Prentice-Hall, Englewood Cliffs, NJ.

Papcun, G., Hochberg, J., Thomas, T. R., Laroche, F., Zacks, J., and Levy, S. (1992). Inferring articulation and recognizing gestures from acoustics with a neural network trained on x-ray microbeam data. *J. Acoust. Soc. Am.*, **92**(2):688–700.

Parsons, T. (1986). *Voice and Speech Processing*. McGraw-Hill, New York, NY.

Parthasarathy, S., and Coker, C. H. (1990). Phoneme-level parameterization of speech using an articulatory model. *Proc. ICASSP*, 337–340.

Parthasarathy, S., and Coker, C. H. (1992). On automatic estimation of articulatory parameters in a text-to-speech system. *Computer Speech and Language*, **6**:37–75.

Peterson, G. E., and Barney, H. L. (1952). Control methods used in a study of the vowels, *J. Acoust. Soc. Am.*, **24**:175–184.

Pinto, N. B., Childers, D. G., and Lalwani, A. L. (1989). Formant speech synthesis: Improving production quality. *IEEE Trans. Acoust., Speech, Signal Process.*, **37**(12):1870–1887.

Pisoni, D. B. (1982). Perception of speech: The human listener as a cognitive interface. *Speech Technology*, **1**:10–23.

Pisoni, D. B., and Hunnicut, S. (1980). Perceptual evaluation of MITalk: The MIT unrestricted text-to-speech system. *Proc. IEEE Int. Conf. Acoust., Speech, Signal Process.*, 572–575.

Pisoni, D. B., Nusbaum, H. C., and Green, B. G. (1985). Perception of synthetic speech generated by rule. *Proc. IEEE*, **73**:1665–1676.

Pitas, I. (1994). Optimization and adaptation of discrete-valued digital filter parameters by simulated annealing. *IEEE Trans. on Signal Processing*, **42**(4):860–866.

Pohlmann, K. C. (1991). *Advanced Digital Audio*. SAMS, Carmel.

Pohlmann, K. C. (1995). *Principles of Digital Audio*, 3rd Ed. McGraw-Hill, Inc., New York, NY.

Portnoff, M. R. (1973). *A Quasi-One-Dimensional Digital Simulation for the Time-Varying Vocal Tract*. MS thesis, MIT, Cambridge, MA.

Portnoff, M. R. (1981). Time-scale modification of speech based on short-time fourier analysis. *IEEE*

Trans. on Acoust., Speech, and Signal Process*, **29**(3):374–390.

Potter, R. K., Kopp, G. A., and Kopp, H. G. (1966). *Visible Speech*. Dover, New York, NY.

Prado, P. P. L. (1991). *A Target-Based Articulatory Synthesizer*. Ph.D. dissertation, University of Florida.

Prado, P. P. L., Shiva, E. H., and Childers, D. G. (1992). Optimization of acoustic-to-articulatory mapping. *Proc. ICASSP*, **2**:33–36.

Quackenbush, S. R., Barnwell, T. P., and Clements, M. A. (1988). *Objective Measures of Speech Quality*. Prentice-Hall, Englewood Cliffs, NJ.

Quatieri, T. F., and McAulay, R. J. (1992). Shape invariant time-scale and pitch modification of speech. *IEEE Trans. on Signal Processing* **40**(3):497–510.

Rabiner, L. R. (1968). Digital-formant synthesizer for speech synthesis studies. *J. Acoust. Soc. Am.*, **43**:822–828.

Rabiner, L. R., and Juang, B. H. (1993). *Fundamentals of Speech Recognition*. Prentice-Hall, Englewood Cliffs, NJ.

Rabiner, L. R., and Shafer, R. W. (1978). *Digital Processing of Speech Signals*. Prentice-Hall, Englewood Cliffs, NJ.

Rahim, M. G., Goodyear, C. C., Kleijn, W. B., Schroeter, J., and Sondhi, M. M. (1993). On the use of neural networks in articulatory speech synthesis. *J. Acoust. Soc. Am.*, **93**(2):1109–1121.

Rayleigh, J. W. S. (1945). *The Theory of Sound*. Dover, New York, NY.

Riegelberger, E. L., and Krishnamurthy, A. K. (1993). Glottal Source Estimation: Methods of applying the LF-model to inverse filtering. *Proc. ICASSP*, **2**:542–545.

Rose, R. C., and Barnwell, T. P. III (1990). Design and performance of an analysis-by-synthesis class of predictive speech coders. *IEEE Trans. Acoust. Speech Signal Process.*, **38**(9):1489–1503.

Rose, K., Gurewitz, E., and Fox, G. C. (1992). Vector quantization by deterministic annealing. *IEEE Transactions on Information Theory*, **38**(4):1249–1257.

Rosenberg, A. E. (1971). Effect of glottal pulse shape on the quality of natural vowels. *J. Acoust. Soc. Am.*, **49**(2):583–590.

Rosenberg, A. E., Calson, R., Granstrom, B., and Gaufin, J. (1975). A three-parameter voice source for speech synthesis. *Speech Commun.*, **2**:235–243.

Rothauser, E. H., Urbanek, G. E., and Pachi, W. P.

(1968). The preference method for speech evaluation. *J. Acoust.Soc. Am.*, **44**:408–418.

Rothauser, E. H., et al. (1969). IEEE recommended practice for speech quality measurements, *IEEE Trans. Audio and Electroacoust.*, **17**:225–246.

Rothauser, E. H., et al. (1971). A comparison of preference measurement methods. *J. Acoust. Soc. Am.*, **49**, pt. 2:1297–1308.

Rothenberg, M. R. (1973). A New inverse filtering technique for deriving the glottal air flow waveform during voicing. *J. Acoust. Soc. Am.*, **53**:1632–1645.

Rothenberg, M. R. (1981). An interactive model for the voice source, *STL-QPSR*, Royal Institute of Technology, Stockholm, Sweden, **4**:1–17.

Rubin, P., Baer, T., and Mermelstein, P. (1981). An articulatory synthesizer for perceptual research. *J. Acoust. Soc. Am.*, **70**(2):321–328.

Rye, J. M., and Holmes, J. N. (1982). A versatile software parallel formant speech synthesizer. *JSRU Research Report*, no. 1016. Joint Speech Research Unit, Eastcote, England.

Saito, S. (1992). *Speech Science and Technology*. IOS Press, Amsterdam.

Saito, S., and Nakata, K. (1985). *Fundamentals of Speech Signal Processing*. Academic Press, Orlando.

Saito, S., Fukuda, H., Isogai, Y., and Ono, H. (1981). X-ray stroboscopy. In *Vocal Fold Physiology*, K. N. Stevens, and M. Hirano, Eds. University of Tokyo Press, Tokyo, Japan.

Saltzman, E. L. (1986). Task-dynamic coordination of the speech articulators: A preliminary research. *Experimental Brain Research*, **15**:129–144.

Saltzman, E. L., and Kelso, J. A. S. (1987). Skilled actions: A task-dynamic approach. *Psychological Review*, **94**(1):84–106.

Saltzman, E. L., and Munhall, K. G. (1989). A dynamic approach to gestural patterning in speech production. *Ecological Psychology*, **14**:333–382.

Savic, M., and Nam, L. H. (1991). Voice personality transformation. *Digital Signal Process.*, **1**:107–110.

Scherer, R. C. (1981). *Laryngeal Fluid Mechanics: Steady Flow Considerations Using Static Models*. Ph.D. dissertation, University of Iowa, Iowa City.

Schneiderman, C. R. (1984). *Basic Anatomy and Physiology in Speech and Hearing*. College-Hill Press, San Diego, CA.

Schroeder, M. R. (1967). Determination of the geometry of the human vocal tract by acoustic measurements. *J. Acoust. Soc. Am.*, **41**(4), p 2:1002–1010.

Schroeder, M. R. (1993). A brief history of synthetic speech. *Speech Comm.*, **13**:231–237.

Schroeder, M. R., and Atal, B. S. (1985). Code-excited linear prediction (CELP): High-quality speech at very low bit rates. *Proc. IEEE Int. Conf., Acoust., Speech, Signal Process.*, 937–940.

Schroeder, M. R., and Strube, H. W. (1979). Acoustic measurements of articulator motions. *Phonetica*, **36**:302–313.

Schroeter, J., and Sondhi, M. M. (1994). Techniques for estimating vocal-tract shapes from the speech signal. *IEEE Trans. Speech and Audio Processing*, **2**(1):133–150.

Schroeter, J., Larar, J. N., and Sondhi, M. M. (1987). Speech parameter estimation using a vocal tract/cord model. *Proc. IEEE Int. Conf., Acoust., Speech, Signal Process.*, 308–311.

Schroeter, J., Larar, J. N., and Sondhi, M. M. (1988). Multi-frame approach for parameter estimation of a physiological model of speech production. *Proc. IEEE Int. Conf., Acoust., Speech, Signal Process.*, **88**(S2.6):83–86.

Schroeter, J., Meyer, P., and Parthasarathy, S. (1990). Evaluation of improved articulatory codebooks and codebook access distance measures. *Proc. ICASSP*, 393–396.

Schwab, E. G., Nusbaum, H. C., and Pisoni, D. B. (1985). Some effects of training on the perception of synthetic speech. *Human Factors*, **27**:395–408.

Schwartz, R., and Makhoul, J. (1975). Where the phonemes are: Dealing with ambiguity in acoustic-phonetic recognition. *IEEE Trans. on Acoust., Speech, and Signal Processing* **23**(1):50–53.

Shadle, C. H. (1991). The effect of geometry on source mechanisms of fricative consonants. *J. Phonetics*, **19**:409–424.

Shearer, W. M. (1979). *Illustrated Speech Anatomy*. Charles C Thomas Publisher, Springfield, IL.

Shirai, K. (1993). Estimation and generation of articulatory motion using neural networks. *Speech Comm.*, **13**:45–51.

Shue, Y.-J. (1995). *A Formant-Based Linear Prediction Speech Synthesis/Analysis System*, Ph.D. dissertation, University of Florida.

Singh, S., and Murry, T. (1978). Multidimensional classification of normal voice qualities. *J. Acoust. Soc. Am.*, **64**:81–87.

Singhal, S., and Atal, B. S. (1989). Amplitude optimization and pitch prediction in multipulse coders. *IEEE Trans. on Acoust., Speech, Signal Process.*, **37**(3):317–327.

Smith, M. E., Robinson, K. E., and Strong, W. J. (1981). Intelligibility and quality of linear predictor and eigenparameter coded speech. *IEEE Trans. Acoust., Speech, Signal Proc.*, **29**:391–394.

Sondhi, M. M. (1974). Model for wave propagation in a lossy vocal tract. *J. Acoust. Soc. Am.*, **55**(5):1070–1075.

Sondhi, M. M. (1975). Measurement of the glottal waveform. *J. Acoust. Soc. Am.*, **57**(1):228–232.

Sondhi, M. M. (1979). Estimation of vocal-tract areas: The need for acoustical measurements. *IEEE Trans. Acoust., Speech, and Signal Processing*, **27**(3):268–273.

Sondhi, M. M. (1986). Resonances of a bent vocal tract. *J. Acoust. Soc. Am.*, **79**(4):1113–1116.

Sondhi, M. M., and Resnick, J. R. (1983). The Inverse problem for the vocal tract: Numerical methods, acoustical experiments, and speech synthesis. *J. Acoust. Soc. Am.*, **73**(3):985–1022.

Sondhi, M. M., and Schroeter, J. (1986). A nonlinear articulatory speech synthesizer using both time-and frequency-domain elements. *Proc. ICASSP*, 1999–2002.

Sondhi, M. M., and Schroeter, J. (1987). A hybrid time-frequency domain articulatory speech synthesizer. *IEEE Trans. Acoust., Speech, and Signal Processing*, **35**(7):955–967.

Sorokin, V. N. (1994). Inverse problem for fricatives. *Speech Comm.*, **14**:249–262.

Steeneken, H. J. M., and Houtgast, T. (1980). A physical method for measuring speech-transmission quality. *J. Acoust. Soc. Am.*, **67**:318–326.

Stevens, K. N. (1971). Airflow and turbulence noise for fricative and stop consonants: Static considerations. *J. Acoust. Soc. Am.*, **50**(4):1180–1192.

Stevens, K. N. (1993a). Modelling affricate consonants. *Speech Comm.*, **13**:33–43.

Stevens, K. N. (1993b). Models for the production and acoustics of stop consonants. *Speech Comm.*, **13**:367–375.

Stevens, K. N. (1998). *Acoustic Phonetics*, The MIT Press, Cambridge.

Stevens, K. N., and House, A. S. (1955). Development of a quantitative description of vowel articulation. *J. Acoust. Soc. Am.*, **27**(3):484–493.

Stevens, K. N., and Klatt, D. H. (1974). Current models of sound sources for speech. In *Ventilatory and Phonatory Control Systems*, B. Wyke, Ed. Oxford University Press, New York, pp. 279–292.

Stevens, K. N., Blumstein, S. E., Glicksman, L., Burton, M., and Kurowski, K. (1992). Acoustic and perceptual characteristics of voicing in fricatives and fricative clusters. *J. Acoust. Soc. Am.*, **91**:2979–3000.

Stevens, K. H., Kasowski, S., and Fant, G. (1953). An electrical analog of the vocal tract. *J. Acoust. Soc. Am.*, **25**(4):734–742.

Story, B. H., and Titze, I. R. (1995). Voice simulation with a body-cover model of the vocal folds. *J. Acoust. Soc. Am.*, **97**(2):1249–1260.

Story, B. H., and Titze, I. R. (1996). Vocal tract functions from magnetic resonance imaging. *J. Acoust. Soc. Am.*, **100**:537–554.

Strube, H. W. (1982). Time-varying wave digital filters and vocal-tract models. *Proc. ICASSP*, 923–926.

Sundberg, J., Johansson, C., Wilbrand, H., and Ytterbergh, C. (1987). From sagittal distance to area: A study of transverse, vocal tract cross-sectional area. *Phonetica*, **44**:76–90.

Tetschner, W. (1993). *Voice Processing* 2nd Ed. Artech House, Boston.

Therrien, C. W. (1992). *Discrete Random Signals and Statistical Signal Processing*. Prentice-Hall, Englewood Cliffs, NJ.

Thomas, I. B. (1968). The influence of first and second formants on the intelligibility of clipped speech. *J. Acoust. Eng. Soc.*, **16**:182–185.

Thomas, T. J. (1986). A finite element model of fluid flow in the vocal tract. *Computer Speech and Language*, **1**:131–151.

Tiffany, W. R., and Bennett, D. N. (1961). Intelligibility of slow-played speech. *J. Speech Hearing Research*, **4**:248–258.

Timcke, R., von Leden, H., and Moore, P. (1958). Laryngeal vibrations: Measurements of the glottic wave, Part I: The normal vibratory cycle. *Arch. Otolaryngo*, **68**:1–19.

Titze, I. R. (1973). The human vocal cords: A mathematical model, Part 1. *Phonetica*, **28**:129–170.

Titze, I. R. (1974). The human vocal cords: A mathematical model, Part 2. *Phonetica*, **29**:1–21.

Titze, I. G. (1976). On the mechanics of vocal-fold vibration. *J. Acoust. Soc.*, **60**(6):1366–1380.

Titze, I. R. (1982). Synthesis of sung vowels using a

time-domain approach. *Transactions of the Eleventh Symposium: Care of the Professional Voice*, 90–98.

Titze, I. R. (1984). Parameterization of the glottal area, glottal flow, and vocal fold contact area. *J. Acoust. Soc. Am.*, **75**(2):520–580.

Titze, I. R. (1988). The physics of small-amplitude oscillation of the vocal folds. *J. Acoust. Soc. Am.*, **83**(4):1536–1552.

Titze, I. R. (1989). A four-parameter model of the glottis and vocal fold contact area. *Speech Comm.*, **8**:191–201.

Titze, I. R. (1994). *Principles of Voice Production*. Prentice-Hall Inc., Englewood Cliffs, NJ.

Titze, I. R., and Strong, W. J. (1975). Normal modes in vocal cord tissues. *J. Acoust. Soc. Am.*, **57**(3):736–744.

Titze, I. R., and Talkin, D. T. (1979a). A theoretical study of the effects of the various laryngeal configurations on the acoustics of phonation. *J. Acoust. Soc. Am.*, **66**(1):60–74.

Titze, I. R., and Talkin, D. T. (1979b). Simulation and interpretation of glottographic waveforms. *ASHA Reports*, **11**:48–53.

Umeda, N. (1977). Consonant duration in American English. *J. Acoust. Soc. Amer.*, **61**(3):846–858.

Valbret, H., Moulines, E., and Tubach, J. P. (1992). Voice transformation using PSOLA technique, *Speech Commun.*, **11**:175–187.

van den Berg, J. W., Zantema, J. T., and Doorneball, P. (1957). On the air resistance and the Bernoulli effect of the human larynx. *J. Acoust. Soc. Am.*, **29**(5):626–631.

van den Berg, J., and Tan, T. S. (1959). Results of experiments with human larynxes. *Practica Oto-Rhino-Laryngologica*, **21**(6):425–450.

Vanderbilt, D., and Louie, S. G. (1984). A monte carlo simulated annealing approach to optimization over continuous variables. *J. Computational Physics*, **56**:259–271.

Van Santen, J., Sproat, R. W., Olive, J. P., and Hirschberg, J. (1997). *Progress in Speech Synthesis*, Springer, New York, NY.

van Riper, C., and Irwin, J. V. (1958). *Voice Articulation*. Prentice-Hall Inc., Englewood Cliffs, NJ.

Vecchi, M. P., and Kirkpatrick, S. (1983). Global Wiring by simulated annealing. *IEEE Trans. on Computer-Aided Design*, **2**:215–222.

Velhuis, R. (1998). A computationally efficient

alternative for the Liljencrants–Fant model and its perceptual evaluation. *J. Acoust. Soc. Am.*, **103**(1):566–571.

Viswanathan, V. R., Russell, W. H., and Huggins, A. W. F. (1984). *Objective Speech Quality Evaluation of Real-Time Speech Coders*. Bolt Beranek Newman, Cambridge, MA, 137, and AD-Al4l 194/1, 17.

Voiers, W. D. (1968). The present state of digital vocoding technique: A diagnostic evaluation. *IEEE Trans. Audio Electroacoust.*, **16**:275–279.

Voiers, W. D. (1977a). Diagnostic acceptability measure for speech communication systems. *Proc. IEEE Int. Conf. Acoust., Speech, Signal Process*, 204–207.

Voiers, W. D. (1977b). Diagnostic evaluation of speech intelligibility. In *Speech Intelligibility and Speaker Recognition*, M. E. Hawley, Ed. Dowden, Hutchinson, and Ross, Stroudsberg, PA, pp. 374–387.

Voiers, W. D. (1983). Evaluating processed speech using the diagnostic rhyme test. *Speech Technology*, Jan./Feb., 30–39.

von Leden, H., and Moore, P. (1961). The mechanics of the cricoarytenoid joint. *Arch. Otolaryngo.*, **73**:541–550.

von Kempelen—Not in additional references?

Waibel, A. (1988). *Prosody and Speech Recognition*. Pitman, London.

Wakita, H. (1973). Direct estimation of the vocal-tract shape by inverse filtering of acoustic speech waveforms. *IEEE Trans. Audio Electroacoustics*, **21**(5):417–427.

Wakita, H. (1979). Estimation of vocal-tract shapes from acoustical analysis of the speech wave: The state of the art. *IEEE Trans. Acoust., Speech, and Signal Proces.*, **27**(3):281–285.

Wakita, H., and Fant, G. (1978). Toward a better vocal tract model. *STL-QPSR*, Royal Institute of Technology, Stockholm, Sweden, **1**:9–29.

Wakita, H., and Gray, H. (1975). Numerical determination of the lip impedance and vocal-tract area functions. *IEEE Trans. Acoust., Speech, Signal Processing*, **23**(6):574–580.

Waterworth, J. A. (1983). Effect of intonation form and pause duration of automatic telephone number announcements on subjective and memory performance. *Applied Ergonomics*, **14**:39–42.

Weinstein, C. J., McCandless, S. S., Mondshein, L. F., and Zue, V. W. (1975). A system for acoustic-phonetic analysis of continuous speech. *IEEE Trans. on Acoust., Speech, and Signal Processing*, **23**(1):54–67.

White, J. M. (1995). *A System for Time Modification of Synthesized Speech*, Ph.D. dissertation, University of Florida.

Witten, I. H. (1982). *Principles of Computer Speech*. Academic Press, Orlando.

Wong, D. Y. (1980). On understanding the quality problems of LP speech. *Proc. IEEE Int. Conf. Acoust., Speech, Signal Proc.*, 725–728.

Wong, D. Y., and Markel, J. D. (1977). An intelligibility evaluation of several linear prediction vocoder modifications. *Proc. IEEE Int. Conf. Acoust., Speech, Signal Proc.*, 208–211.

Wong, D. Y., Markel, J. D., and Gray, A. H. (1979). Least squares glottal inverse filtering from the acoustic speech waveform. *IEEE Trans. on Acoust., Speech, and Signal Processing*, **27**(4):350–356.

Wright, G. T. H., and Owens, F. J. (1993). An optimized multirate sampling technique for the dynamic variation of vocal tract length in the Kelly–Lochbaum speech synthesis model. *IEEE Trans. on Speech and Audio Process.*, **1**(1):109–113.

Wu, C. J. (1996). *A Flexible 3-D Model of Vocal Fold Vibrations*. Ph.D. dissertation, University of Florida.

Wu, C. (1996). *Articulatory Speech Synthesizer*. Ph.D. dissertation, University of Florida.

Xue, Q., Hu, Y. H., and Milenkovic, P. (1990). Analyses of the hidden units of the multi-layer perceptron and its application in acoustic-to-articulatory mapping. *Proc. ICASSP*, 869–872.

Yu, Z. (1993). A method to determine the area function of speech based on perturbation theory. *STL-QPSR*, Royal Institute of Technology, Stockholm, Sweden, **4**:77–95.

Zemlin, W. R. (1964). *Speech and Hearing Science*. Stipes, Champaign, IL.

Zemlin, W. R. (1988). *Speech and Hearing Science: Anatomy and Physiology*. Prentice Hall Inc., Englewood Cliffs, NJ.

Zue, V. W., Glass, J., Phillips, M., Seneff, S. (1989). Acoustic segmentation and phonetic classification in the Summit system. *Proc. Int'l. Conf. on Acoust., Speech, and Signal Processing*, 389–392.

APPENDIX **3**

STANDARDS

3.1 STANDARDS ORGANIZATIONS

CCITT: International Consultative Committee for Telephone and Telegraph, now renamed International Telecommunications Union-Telecommunications Standardization Sector or ITU-TSS

GSM: Group Speciale Mobile

CTIA: Cellular Technology Industry Association

NSA: National Security Agency

EBU: European Broadcasters Union

ISO/MPEG: International Organization of Standardization, Motion Picture Expert Group, deals with audiovisual standards

CTIA: Cellular Technology Industry Association

CIF: Common Intermediate Format, digital television and images

CCIR: International Consultative Committee for Radio, digital television, and images

HDTV: High Definition Television Format

DAB: Digital Audio Broadcasting

3.2 STANDARDS

G.711 CCITT (1972) PCM at 64 kbits/s (8 bits/sample at 8 ksamples/s)

G.721 CCITT (1984) ADPCM 32 kbits/s uses pole-zero adaptive predictor

G.723 CCITT (1988) modification of G.721 to accommodate 24 and 40 kbits/s

G.722 CCITT (1988) 7 kHz audio at 64 kbits/s of integrated services digital network (ISDN) teleconferencing, based on 2 band sub-band/ADPCM coder

AT&T voice store-and-forward standard (1986), sub-band coding, 16 and 24 kbits/s, 5 band non-uniform tree-structured QMF bank in conjunction with APCM

AUSSAT Australian Satellite standard (1991), 6.4 kbits/s improved multiband excitation coder

INMARSAT-M International Maritime satellite standard (1991), uses same standard as AUSSAT, both of these can handle multirates (8, 4.8, 2.4 kbits/s)

LPC-10 DoD secure communications standard (1975), 2.4 kbits/s, tenth-order predictor

FS-1015 This is the LPC-10 standard

Skyphone British Telecom International (BTI) (1985), 9.6 kbits/s multipulse linear prediction (MPLP)

RPE-LTP GSM (1988) Pan-European digital mobile standard, 13 kbits/s coding using regular pulse excitation with long term prediction

FS-1016 Federal Standard CELP (1989), 4.8 kbits/s CELP, adopted by DoD for possible use in the third generation secure telephone unit (STU-III)

IS-54 North American Digital Cellular System standard (1989) VSELP vector sum excited linear prediction, 8 kbits/s

G.728 CCITT (1991) LD-CELP, 16 kbits/s, G. series of standards, low-delay CELP

JDC Japanese Digital Cellular standard, 6.7 kbits/s, based on VSELP

ISO standard for audiovisual (1991, 1993), calls for CD audio bandwidth of 20 kHz and a single channel bit rate of 96 or 128 kbits/s

AT&T perceptual audio coder (PAC) (1992), earlier version ASPEC, 128 k bits/s for CD quality

MUSICAM Philips coder (1992), 128 kbits/s CD quality

ATRAC Sony's Adaptive Transform Acoustic Coder (1992), MiniDisc

AC-2 Dolby's 128 kbits/s transform-based audio coder (1990)

3.3 REFERENCES

Jayant, N., Johnston, J., and Safranek, R. (1993) Signal compression based on models of human perception. *Proceedings of IEEE, 81*, Oct, 1385–1421.

Jayant, N. (1990) High-quality coding of telephone speech and wideband audio. *IEEE Communications Magazine*, Jan, 10–20.

Noll, P. (1993) Wideband speech and audio coding. *IEEE Communications Magazine*, Nov, 34–44.

Rabiner, L. R. (1994) Applications to voice processing to telecommunications. *Proceedings of IEEE, 82*, Feb, 199–228.

Spanias, A. (1994) Speech coding: A tutorial review. *Proceedings of IEEE, 82*, Oct, 1541–1582.

SPEECH, EGG, AND GLOTTAL AREA DATA

The data described herein are contained on two CD-ROMs. One CD-ROM contains the Normal folders, while the other CD-ROM contains the Disorder folder, the Mimic folder, the Other folder, the Additional folder, and the Area folder.

A4.1 Normal FOLDERS. PROCEDURES AND DATA

All data recordings were performed in a professional single-wall sound room. The speech and electroglottographic (EGG) signals were monitored synchronously and simultaneously. One of two microphones was used: an Electro-Voice (Buchanan, Michigan) RE-10 dynamic cardioid or a Bruel and Kjaer (A Division of Spectris Technologies, Inc., Norcross, GA) model 4113 condenser. The selected microphone was located 6 in. from the speaker's lips. The electroglottograph was a Synchrovoice, Inc. model. All data were directly digitized, thereby avoiding any low-frequency distortions that may have been introduced through the use of audio tape recordings. The speech and EGG signals were bandlimited to 5 kHz by antialiasing elliptic filters with a minimum stop-band attenuation of −55 dB and a passband ripple of ±0.2 dB. Both signals were amplified by a Digital Sound Corp. (Maynard, Massachusetts) DSC-240 audio control console. The two signals were sampled at 10 kHz per channel by a Digital Sound Corp. DSC-200 analog-to-digital system with 16-bit resolution. The data that were recorded using the Electro-Voice microphone were corrected for microphone distortions by deriving a microphone correction transfer function (Childers and Wong, 1994). The data that were recorded with the Bruel and Kjaer microphone did not require correction since its bandwidth characteristics were sufficiently broad enough that no frequency distortions were introduced into the data.

The total number of subjects was 52 (25 male, 27 female) with normal larynges. The subject's ages ranged from 20 to 80 years old. The complete speech protocol consisted of 28 tasks, including twelve sustained vowels /IY, IH, EY, EH, AE, UW, UH, OW, AO, AA, AH, ER/; five sustained unvoiced fricatives /HH, F, TH, S, SH/; and four sustained voiced fricatives /V, DH, Z, ZH/. This notation is the upper case version of the ARPAbet. The subjects were instructed to pronounce and sustain each vowel as it would be pronounced in the following words, respectively: beet, bit, bait, bet, bat, boot, book, boat, bought, Bach, but, Bert. For the fricatives, we used the following cue words: hat, fix, thick, sat, ship, van, this, zoo, and azure. The duration of each vowel and fricative approximated 2 sec. The additional tasks include counting from one to ten with a comfortable pitch and loudness, counting from one to five with a progressive increase in loudness, singing the musical scale using "la," and speaking three sentences (We were away a year ago; Early one morning a man and a woman ambled along a one mile lane; Should we chase those cowboys?). All of the above 27 tasks were recorded using the Electro-Voice RE-10 microphone. The 28th task used the Bruel and Kjaer model 4113 microphone, and repeated the sustained vowel, /AA/, as in Bach.

The data from subjects with a normal larynx are contained on the CD-ROM in the folders denoted as follows:

Normal_m folder (folder containing data folders for males) m01 to m25 folders

Normal_f folder (folder containing data folders for females) f01 to f27 folders

Each male (female) folder contains 56 data files: 28 speech and 28 EGG. The files are labeled m*xxyyz*.dat for male (with a similar notation for female), where *xx* denotes the subject number, *yy* denotes the speech task, and *z* denotes whether the file is speech or EGG data.

xx:	ranges from 01 to 25
yy:	ranges from 01 to 28
z:	is either s (speech) or e (EGG)
dat:	is the extension, denoting ASCII format

yy denotes the tasks, which were as follows.

01:	counting from 1 to 10 with a comfortable pitch and loudness
02:	counting from 1 to 5 with a progressive increase in loudness
03:	sustained vowel /IY/ as in beet
04:	sustained vowel /IH/ as in bit
05:	sustained vowel /EY/ as in bait
06:	sustained vowel /EH/ as in bet
07:	sustained vowel /AE/ as in bat
08:	sustained vowel /UW/ as in boot
09:	sustained vowel /UH/ as in book
10:	sustained vowel /OW/ as in boat
11:	sustained vowel /AO/ as in bought
12:	sustained vowel /AA/ as in Bach
13:	sustained vowel /AH/ as in but
14:	sustained vowel /ER/ as in Bert
15:	sustained fricative /HH/ as in hat
16:	sustained fricative /F/ as in fix
17:	sustained fricative /TH/ as in thick
18:	sustained fricative /S/ as in sat
19:	sustained fricative /SH/ as in ship
20:	sustained fricative /V/ as in van
21:	sustained fricative /DH/ as in this
22:	sustained fricative /Z/ as in zoo
23:	sustained fricative /ZH/ as in azure
24:	produce the musical scale singing "la," attempting to go up then down as one effort with a pause between the top two notes

Orally read the following sentences with a comfortable pitch and loudness.

25:	We were away a year ago.
26:	Early one morning a man and a woman ambled along a one mile lane.
27:	Should we chase those cowboys?
28:	Sustained vowel /AA/ as in Bach using a Bruel and Kjaer microphone, model 4113 condenser.

Thus, the example data file m0110s.dat would be the sustained vowel /OW/, as in boat, while m0110e.dat would be the EGG signal for the same sustained vowel.

The speech and EGG data were recorded simultaneously. However, the speech data lags the EGG data by the amount of time it takes for the sound to propagate from the vocal folds to the microphone (6 inches beyond the lips). This propagation delay varies with each subject due to variations in vocal tract length. A "typical" value in data samples would be 7 data samples at 10 kHz sampling frequency. This approximates the distance of 24 cm from the vocal folds to the microphone. Consequently, to align the speech and EGG data on a sample basis, shift the speech data to the left by 7 data samples while keeping the EGG data fixed (or shift the EGG data to the right by 7 data samples).

Missing Data

Several files are missing: f0801s.dat, f0802s.dat, f0803s.dat, f0828e.dat, f0828s.dat, f0928e.dat, f0928s.dat, f1028e.dat, and f1028s.dat.

A4.2 Disorder FOLDER PROCEDURES AND DATA

The data for the subjects with vocal disorders were collected in a manner identical to that described for the subjects with a normal voice. The file naming convention follows that specified for the normal data set. However, note that the subjects are not the same and the tasks were slightly different. I, however,

01:	Counting from 1 to 10 with a comfortable pitch and loudness
02:	Counting from 1 to 5 with a progressive increase in loudness
03:	Sustained vowel /IY/ as in beet. (Some subjects phonated three vowels in succession: /IY/, /AA/, /UW/. This can be determined by examining the speech file.)
04:	Produce the musical scale singing "la," attempting to go up then down as one effort with a pause between the top two notes

Orally read the following sentences with a comfortable pitch and loudness.

05:	We were away a year ago.
06:	Early one morning a man and a woman ambled along a one mile lane.
07:	Should we chase those cowboys?

The vocal disorder database consists of 16 patients with vocal disorders. Two patients were remeasured twice and 3 patients were remeasured once after treatment. A description of the vocal disorders can be found in Table A4.1 below. The range of voices varied from mildly deviant to very deviant as determined by a clinician. The patients were asked to phonate the vowel /IY/ for about 2 sec.

The file names follow the convention adopted for the Normal data set, namely mxxyyz.dat for male speakers (with a similar notation for female speakers).

xx:	ranges from 01 to 08 for male and from 01 to 15 for female
yy:	ranges from 01 to 07
z:	is either s(speech) or e (EGG)
dat:	is the extension denoting ASCII data

A4.3 Mimic FOLDER

In addition to the above disorder data, this folder contains a set of data that mimics vocal disorders.

m01xxms.dat:	Male subject 01, task xx, m (mimic), s (speech). The tasks range from 01 to 07. The data in this set is a normal modal voice.
m02xxms.dat:	This is the same subject as above but the subject mimicked a breathy voice for all tasks.
m03xxms.dat:	Male subject 03, task xx, m (mimic), s (speech). The subject mimicked a vocal fry for all tasks.
m04xxms.dat:	This is the same subject as m03. The subject mimicked a breathy voice for all tasks.
m05xxms.dat:	Male subject 05, task xx, m (mimic), s (speech). The subject mimicked a hoarse voice for all tasks.
m06xxms.dat:	Same subject as m05. The subject mimicked "computer voice" for all tasks.

TABLE A4.1. List of Vocal Disorder Speakers

Subject number	Symptoms
m01	True vocal cords (TVC) contact ulcer
m02	Hoarse
m03	Vocal fry
m04	Breathy; hoarse
m05[a]	Hoarse unilateral TVC carcinoma
m06	m04 1 month post-injection
m07	m02 5 months post-injection
m08[a]	Bilateral paralysis of TVC
f01[a]	Posterior cyst
f02	Bilateral nodule
f03	Mild hoarseness
f04[a]	Unilateral paralysis
f05	f04 1 month post-injection
f06[a]	Hyper functional
f07	Breathy, weak
f08[a]	Vocal fry
f09[a]	f04 4 months post-injection
f10	f08 1 month post-injection
f11	f07 1 month post-injection
f12[a]	Enlarged vocalus muscle
f13[a]	Right TVC unilateral paralysis
f14	f08 3 months post-injection
f15	Nodules

[a]Very deviant voices.
  Missing files: f0901e and f0901s, m0707e and m0707s.

m07xxms.dat:	Same subject as m05. The subject mimicked a rough voice for all tasks.
m08xxms.dat:	Same subject as m01. The subject mimicked a rough voice for all tasks.

Note: subjects m01, m02, and m08 are the same male. Subjects m03, m04, m05, m06, and m07 are the same male, but differs from m01. These two mimics are professional voice clinicians with many years experience and are recognized as expert mimics of various vocal disorders. Missing files include m0602me, m0602ms, m0604me, and m0604ms.
  Note: the EGG data are denoted with an e in place of the s.

A4.4 Other FOLDER

The data contained in this folder contains only speech data for six adult speakers (three male and three female) each speaking the following sentence.

That zany van is azure. (voiced fricatives)

The data were collected as described for the Normal data set. Note that this data set does not include the EGG signal. The subjects had no vocal disorders.
  The data files are denoted as follows.

m01zs.dat:	Male subject 01, sentence: That zany van is azure, (s) speech.
m01ze.dat:	Male subject 01, sentence: That zany van is azure, (e) EGG.
f01zs.dat	Female subject 01, sentence: That zany van is azure, (s) speech.

These files continue to be denoted but are not shown here. The subjects for this data correspond to the subjects listed for the Normal data as follows.

Other folder	Normal folder
m01	m01
m02	m03
m03	m05
f01	f06
f02	not in Normal set
f03	f01

A4.5 Additional FOLDER

Nine sentences were recorded. The same male speaker spoke each sentence twice. The male speaker is not in the Normal set. The file naming convention is such that the first number denotes the sentence number, which is followed by an underscore and a second number that denotes the repetition. Thus, there are eighteen speech and corresponding EGG data files in this corpus. The file name starts with "spe" if it is a speech file; it starts with "egg" if it is an EGG file.

An example data file name for sentence 1 is the following.

egg1_1.dat for the egg file for sentence 1, first repetition of two

egg1_2.dat for the egg file for sentence 1, second repetition of two

spe1_1.dat for the speech file for sentence 1, first repetition of two

spe1_2.dat for the speech file for sentence 1, second repetition of two

1. The rain in Spain falls mainly on the plain.
2. A bird in the hand is worth two in the bush.
3. The supplies were stored for everyone.
4. Early one morning a man and a woman ambled along a one mile lane.
5. We were away a year ago.
6. I owe you a yo-yo.
7. Should we chase those cowboys?
8. The zany van is azure.
9. We saw the ten pink fish.

Sampling frequency: 10,000 Hz
Data format: 16-bit ASCII.

A4.6 Area FOLDER

The data in this folder contains the glottal area measured from ultra-high speed laryngeal films, the EGG signal, and the speech signal. These data were measured simualtaneously. The procedures are described in Childers and Krishnamurthy (1985). Also see the references in this paper for additional detail.

A4.7 PROCEDURES AND AREA DATA/Normal

The subjects were four adult males, coded as m01, m02, m03, and m04, who had no vocal disorders. The experimental tasks for each subject consisted of phonation of the vowel /IY/ at three different intensities at each of three different fundamental frequencies. The recorded phonation was sustained for about three seconds. The three target fundamental frequencies used were 125 Hz, 170 Hz, and

340 Hz. To control the fundamental frequencies during the experiments, the subjects were asked to match a pure tone of the appropriate frequency that they heard over a pair of headphones. The three different intensities at each fundamental frequency represent a "comfortable" intensity, an intensity approximately 4 dB above the comfortable level, and another intensity about 4 dB below the comfortable level. The actual intensities were monitored with a sound meter. Thus, there were nine tasks for each subject, giving a total of 36 tasks for the four subjects.

The procedures are described in Childers and Krishnamurthy (1985) and references. Briefly, each subject was filmed using ultra-high speed photography. The EGG waveform was filmed simultaneously as a trace from an oscilloscope. This trace appeared on the film along with a timing signal. The EGG signal and the speech signal were also tape recorded along with the same timing signal. The microphone was attached to the handle of the laryngeal mirror at the point where the mirror frame joins the handle. The distance of the glottis from this point varies from subject to subject, but was approximately 11 cm. The audio bandwidth of the microphone was about 6 kHz.

For these data, the glottal area and traced EGG have been aligned as a pair of signals. Similarly, the tape recorded EGG data and speech data have been aligned as a pair of signals. However, these two data set pairs have not been aligned, but can be if desired. Perhaps the simplest procedure to align the two pairs of signals is to shift one pair relative to the other until the EGG waveforms are aligned.

The glottal area waveform is sampled at 5 kHz, so it must be interpolated to twice its data record length before it can be compared (plotted) with the other three data records. After interpolation, the glottal area waveform will be approximately 250 samples, which is approximately the same length as the traced EGG waveform. These two waveforms are much shorter in duration than the taped EGG and speech data, which are several seconds in duration.

All four waveforms can be plotted as a superimposed figure or as four separate figures for comparison purposes.

The convention for the file names for subject m01 is as follows.

am01nn.dat

tm01nn.dat

em01nn.dat

sm01nn.dat

The first letter denotes the following.

a:	glottal area data sampled at 5 kHz
t:	traced EGG data sampled at 10 kHz
e:	EGG data sampled from tape at 10 kHz
s:	speech data sampled from tape at 10 kHz

The letters nn denote the tasks which totaled nine.

The file names less the leading preface (a, t, e, s) are:

Filename	Measured Intensity (dB)	Target Frequency (Hz)	Measured Frequency (Hz)
m0101.dat	64	125	186.6
m0102.dat	68	125	198.0
m0103.dat	72	125	189.0
m0104.dat	68	170	156.3
m0105.dat	70	170	214.0
m0106.dat	80	170	
m0107.dat	70	340	201.0
m0108.dat	74	340	202.0
m0109.dat	79	340	330.9
m0201.dat	68	125	(The e file is missing.)
m0202.dat	74	125	(The e file is missing.)

(Continued)

Filename	Measured Intensity (dB)	Target Frequency (Hz)	Measured Frequency (Hz)
m0203.dat	77	125	114.5
m0204.dat	64	170	162.6
m0205.dat	70	170	161.3
m0206.dat	74	170	162.6
m0207.dat	63	340	333.3
m0208.dat	68	340	356.0
m0209.dat	74	340	336.1
m0301.dat	52	125	
m0302.dat	56	125	161.3
m0303.dat	60	125	
m0304.dat	64	170	161.3
m0305.dat	70	170	175.4
m0306.dat	76	170	180.2
m0307.dat		All files missing for this task.	
m0308.dat		All files missing for this task.	
m0309.dat	73	340	321.1
m0401.dat	56	125	122.9
m0402.dat	60	125	
m0403.dat	64	125	
m0404.dat	68	170	
m0405.dat	72	170	165.3
m0406.dat	75	170	
m0407.dat	54	340	277.7
m0408.dat	58	340	330.6
m0409.dat	62	340	346.0

A4.8   PROCEDURES AND AREA DATA/Disorder

Also contained in this folder are data for three patients with vocal disorders. The procedures are the same as that for the normal data. The subjects were three adults, one female and two males. There was only one task for these subjects, namely to phonate the vowel /IY/ at a comfortable intensity and fundamental frequency. The same data were collected as for the normal subjects. The file names are similar to those used for the normal subjects: a prefix of a, t, e, or s; the subject's number (f01, f05, f06,f06); task number (01 for all subjects); and the extension (.dat). Thus, there are four files for each subject, for a total of 12 files. The vocal disorders for these patients are as follows.

f01:   Unilateral nodule, which has an observable affect on the EGG waveform.

m05:   Unilateral polyp with companion bulge on the other fold, which results in some loss of voice.

m06:   Voice production problem, loss of voice.

A4.9   REFERENCES

Childers, D. G., and Krishnamurthy, A. K. (1985) A critical review of electroglottography. *CRC Critical Reviews in Bioengineering*, **12**, 131–161.

Childers, D. G., and Wong, C. F. (1994) Measuring and modeling vocal source-tract interaction. *IEEE Transactions on Biomedical Engineering*, **41**, 663–671.

WAVEFORMS, SPECTRA, SPECTROGRAMS, AND VOCAL TRACT AREA FUNCTIONS

This appendix contains a file of waveforms, spectra, and spectrograms for most vowels and some fricatives for real speech samples. The vocal tract area functions and their corresponding waveforms and spectra are for synthesized vowels using the articulatory speech synthesizer. There are also the waveforms and spectrograms for four sentences that have been phonetically parsed. The labels used for this parsing are not the conventional phoneme symbols, rather we have chosen to use conventional spelling of the words in each sentence, but space the letters according to the phoneme boundaries. Finally, the spectra and the vocal tract area functions for five Russian vowels are given. The vocal tract area functions for these vowels were measured from x-ray data taken in the 1950s. These data have been used by numerous investigators, including Fant (1960), who describes the data and how it was collected.

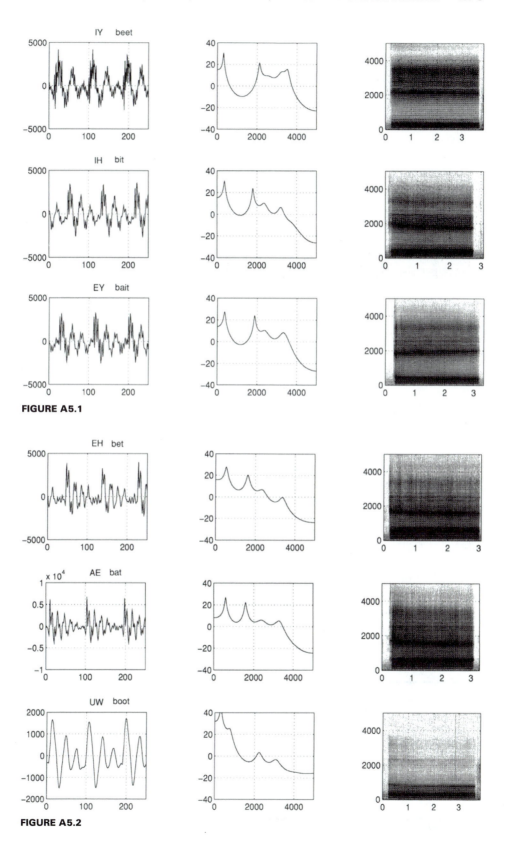

FIGURE A5.1

FIGURE A5.2

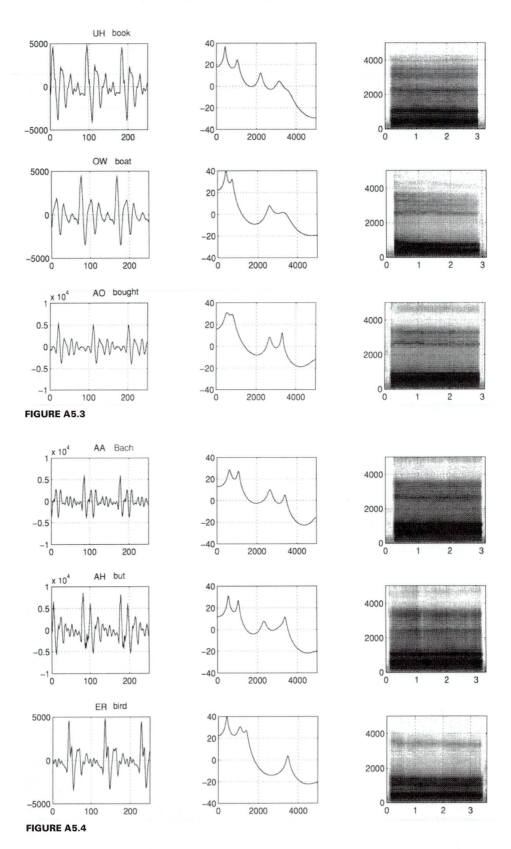

FIGURE A5.3

FIGURE A5.4

IY beet

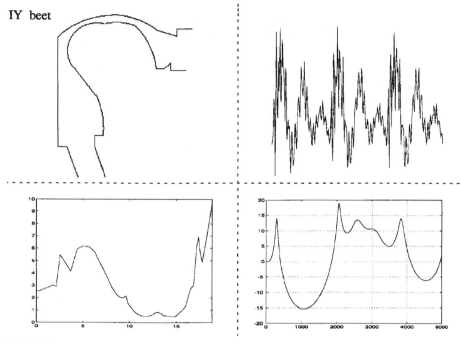

FIGURE A5.5

IH bit

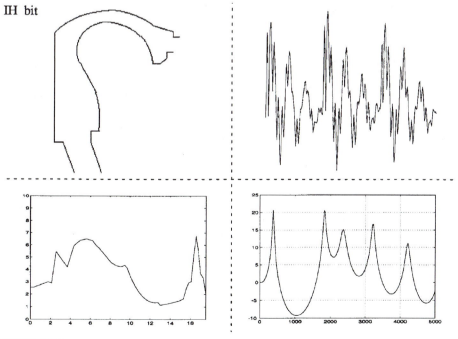

FIGURE A5.6

EY bait

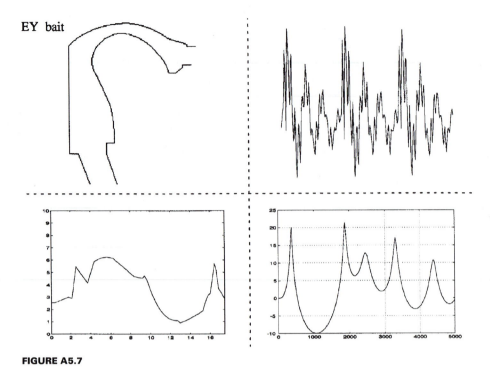

FIGURE A5.7

EH bet

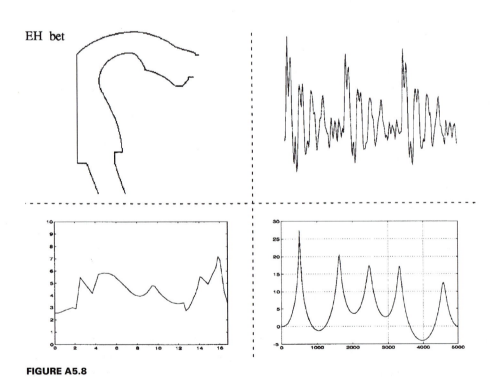

FIGURE A5.8

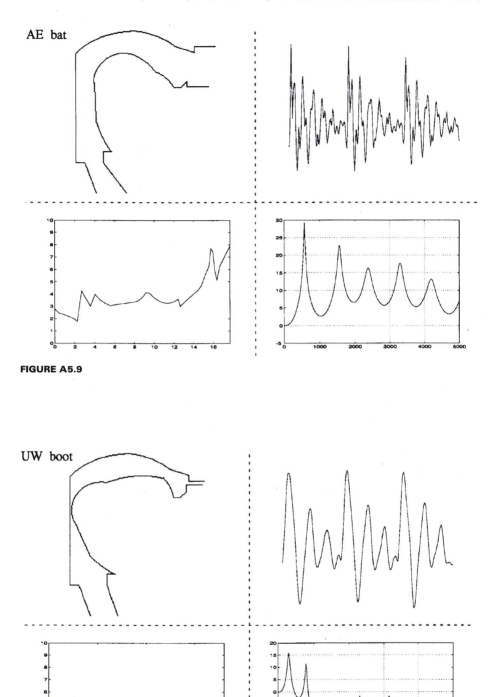

FIGURE A5.9

FIGURE A5.10

UH book

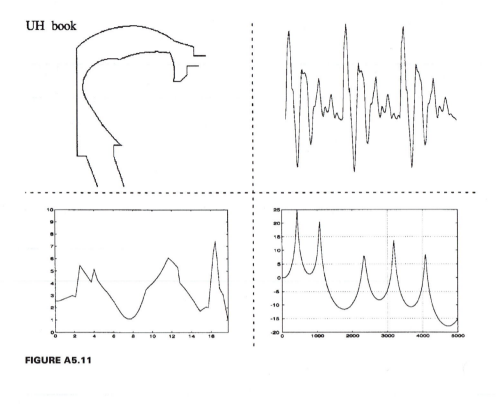

FIGURE A5.11

OW boat

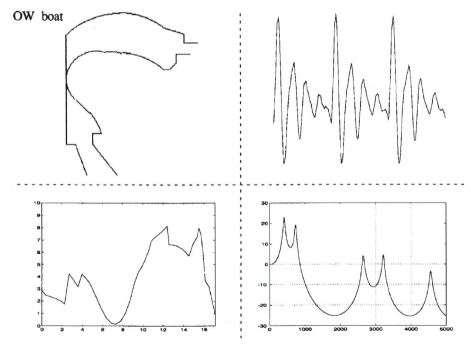

FIGURE A5.12

AO bought

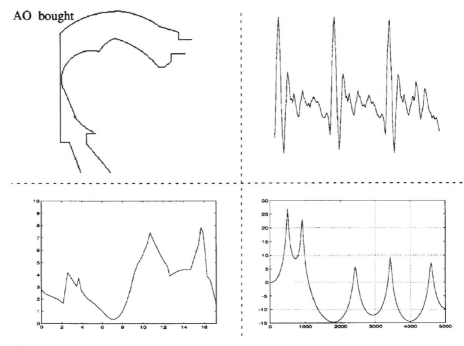

FIGURE A5.13

AA Bach

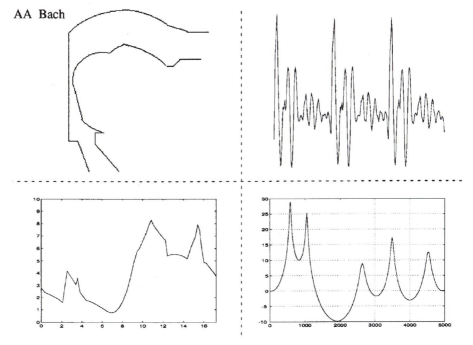

FIGURE A5.14

AH but

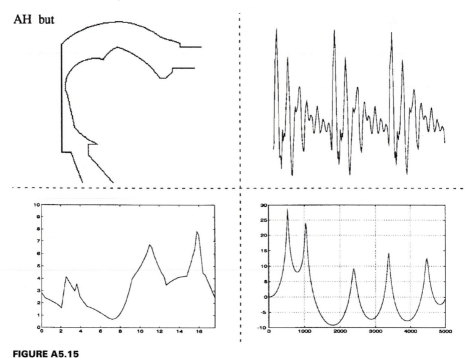

FIGURE A5.15

ER bird

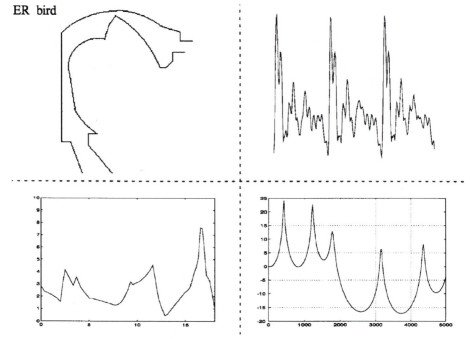

FIGURE A5.16

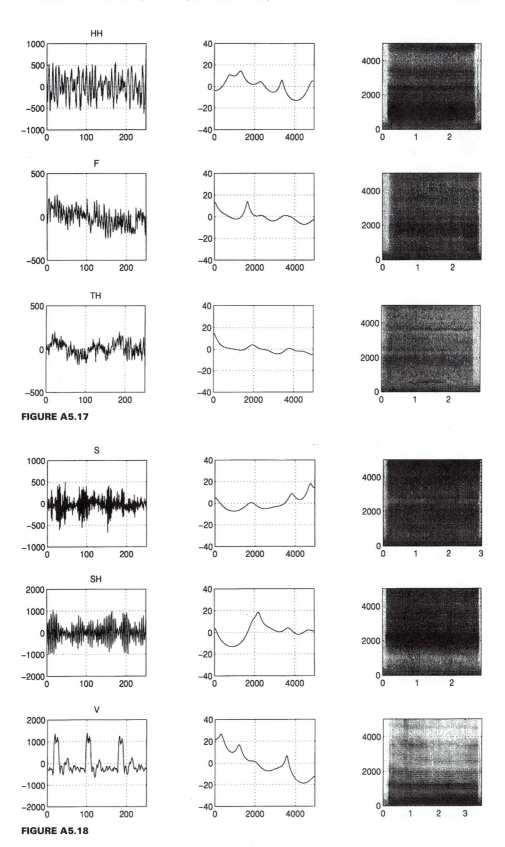

FIGURE A5.17

FIGURE A5.18

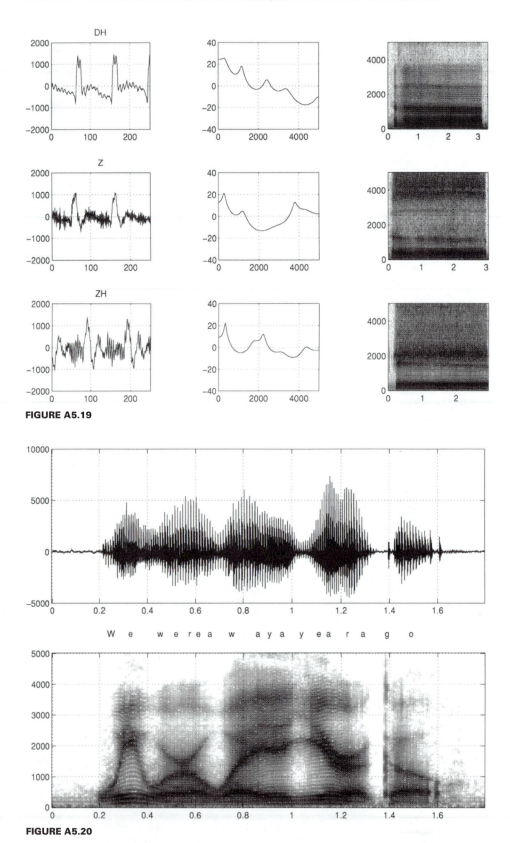

FIGURE A5.19

W e w e r e a w a y a y e a r a g o

FIGURE A5.20

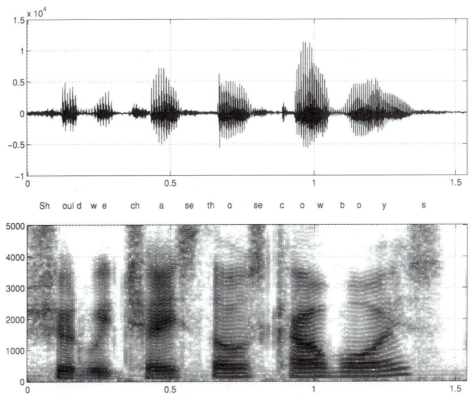

Sh oul d w e   ch   a   se th o   se   c   o   w   b   o   y   s

FIGURE A5.21

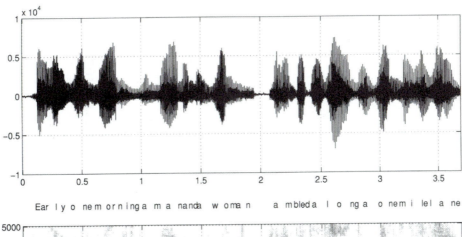

Ear l y o ne m orn i ng a m a nan da w oma n   a mble d a l o ng a o ne m i le l a ne

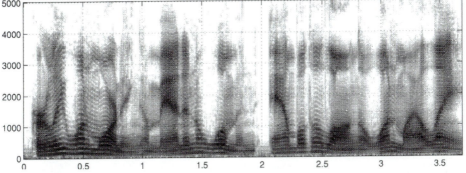

FIGURE A5.22

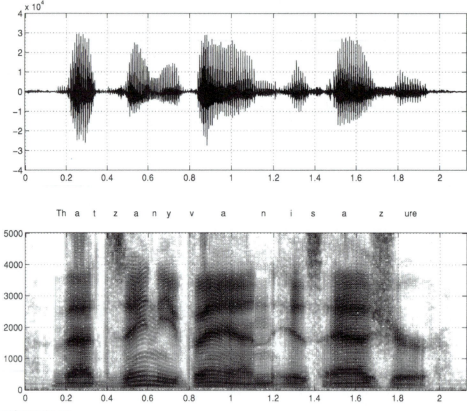

Th a t z a n y v a n i s a z ure

FIGURE A5.23

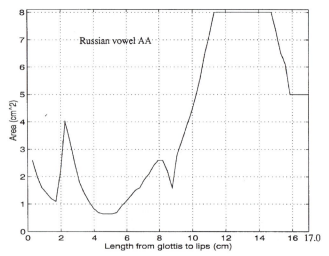

Russian vowel AA

FIGURE A5.24a

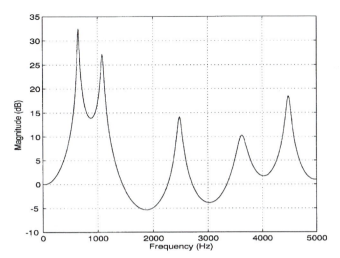

FIGURE A5.24b   (*Continued.*)

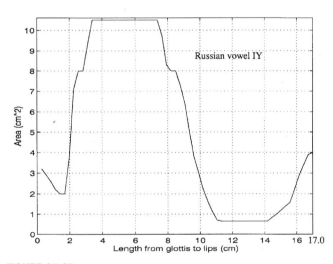

FIGURE A5.25a

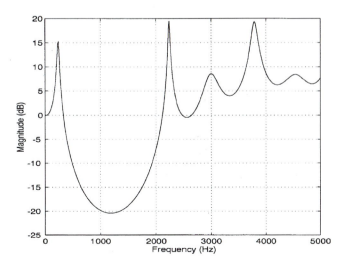

FIGURE A5.25b

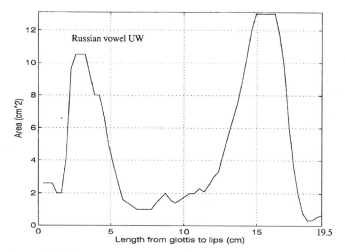

FIGURE A5.26a

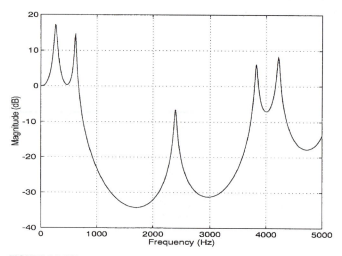

FIGURE A5.26b

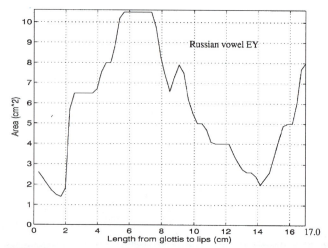

FIGURE A5.27a

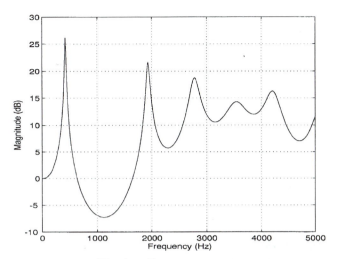

FIGURE A5.27b   (*Continued.*)

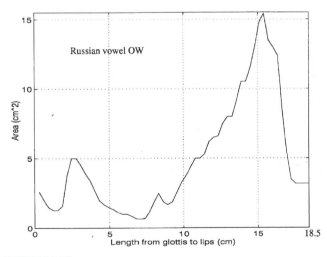

FIGURE A5.28a

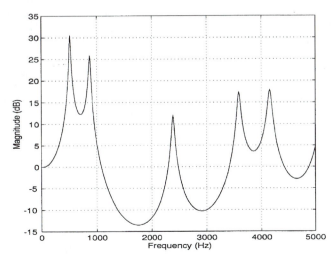

FIGURE A5.28b

ANALYSIS ALGORITHMS

The speech analysis software provided with this book is implemented in MATLAB using a graphic user interface. The software provides basic display tools as well as editing tools and also has features for glottal inverse filtering.

A6.1 TIME DOMAIN ANALYSIS

For time domain analysis, the energy and zero-crossing rate can be calculated as well as the biased autocorrelation function. For our use, all data are windowed. A frame is a windowed segment of data. In speech research, we use short-time analysis of data, that is, the data are windowed. Thus, we use an analysis data frame. Data frames can be overlapped or not. The result of the frame analysis of data is usually a number, or perhaps a small set of numbers, that is less than the number of data samples within the data window. This set of numbers varies with time, that is, it forms a waveform. So we analyze data to reduce the number of data samples in a given data waveform to a small set of values or features. These features can be used to determine characteristics of speech, for example, if voicing or non-voicing is present in the speech segment being analyzed.

The window option under time (domain) analysis illustrates how frames of data are generated. The reader should experiment with this option.

One of the most common short-time analysis methods is the short-time average energy of a signal. If w(n) is the data window for $0 \le n \le N - 1$ and is zero otherwise, and x(n) is the data record, then the short-time average energy is

$$E_n = \sum_m [x(m)w(n - m)]^2 = \sum_m x^2(m)h(n - m) \qquad (A6.1.1)$$

where $h(n) = w^2(n)$. Thus, h(n) is a window (filter) that affects the short-time energy function. If this window is very short, then its bandwidth is large and there is little filtering of the data by the window. If the window is very long, then the window bandwidth is narrow, and there is a great deal of filtering of the data. We select the window parameters (or type of window) to provide "good" data smoothing. The type of window selected can depend on the application or task but for speech, the window used is usually the hamming window.

An illustration of windowing the data using the window option under time analysis in the software provided with this book is shown in Figure A6.1. In this figure, the energy at point n1 of the data is calculated and labeled as E_{n1}. The point n1 is at the end of the window shown directly above the windowed data. Then the window is shifted to point n2 and the energy is calculated and labeled as E_{n2}. This procedure is repeated for point n3. A plot of these values is called the energy contour. Note in Figure A6.1 that the x-axis scales are not perfectly aligned with one another. The scales for the data and the windowed data align. However, the scale of the window is slightly off. This is a problem in MATLAB. This explains why the "notch" in the windowed data does not occur where two successive windows overlap. One can see that the energy contour is a considerable data reduction over that of the actual data. This is quite useful for certain applications.

Since the squaring of the data in the calculation of the energy can make small data values even smaller, the average magnitude function is often used instead of the average energy function. The

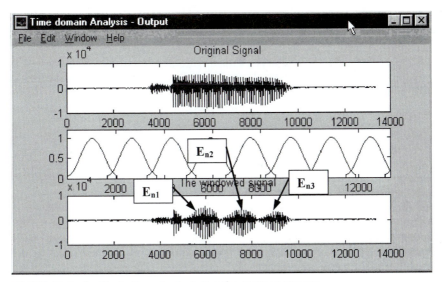

FIGURE A6.1   An illustration of calculating the average energy.

average magnitude function is

$$M_n = \sum_m |x(m)|w(n-m) \qquad (A6.1.2)$$

The advantage of these short-time functions is that they have a smaller bandwidth than the signal, so they can be sampled less frequently. For example, a speech signal sampled at 10,000 samples/sec can be reduced to approximately 100 samples/sec using the short-time energy function for some applications.

The short-time average zero-crossing rate is used to estimate the frequency of voicing. Suppose we are given a sampled sinewave of frequency F_0 that is sampled at rate F_s. The number of samples in a period is $N = F_s/F_0$. For a periodic signal, we have two zero-crossings per period, which is $2F_0/T$, where $T = 1/F_s$. To estimate F_0, count the number of zero-crossings in the sampled data over some time interval, T_1. Then T_1 is the number of samples in T_1 divided by F_s, that is, $T_1 = N_1/F_s$ or $N_1 = F_s T_1$. The average number of zero-crossings per sample is $2F_0/F_s$, thus $F_0 = $ (average number of zero-crossings)$/(2T_1)$. So the zero-crossing rate algorithm is the following.

- Count the number of samples in the data between sign changes. This interval (number of samples) represents $\frac{1}{2}$ the period or $\frac{1}{2F_0}$.
- The average number of samples between sign changes is N_1.
- $N_1/F_s = \frac{1}{2F_0}$
- $F_0 = \frac{F_s}{2N_1}$

The short-time energy and zero-crossing functions are useful for estimating word and phoneme boundaries.

The short-time autocorrelation function is

$$R_n(k) = \sum_m x(m)w(n-m)x(m+k)w(n-k-m) \qquad (A6.1.3)$$

which is the autocorrelation function for the windowed data. This function is useful for estimating the fundamental frequency of voicing for voiced speech segments.

A6.2 FREQUENCY DOMAIN ANALYSIS

Linear prediction (LP) analysis is based on an all pole model of the data, as discussed in Chapter 5. This analysis technique is also known as autoregression (AR) analysis, which is one of the most popular time series modeling methods because accurate estimates of the model parameters can be calculated by solving a set of linear equations using well known algorithms. The LP spectral estimator provides less bias and has less variability than Fourier-based spectral estimators.

The autocorrelation method (Yule–Walker) is used to estimate the LP parameters by minimizing an estimate of the prediction error power.

The covariance method estimates the LP parameters by minimizing an estimate of the prediction error power as well. However, the summation over the observed data samples differs from that for the autocorrelation method.

The modified covariance method estimates the LP parameters by minimizing the average of the estimated forward and backward prediction error powers.

There are three methods for calculating an estimate of the order of the LP model, namely, final prediction error (FPE) due to Akaike, the Akaike information criterion (AIC), and the criterion autoregressive transfer function (CAT) due to Parzen. However, these, and other, methods for estimating the LP model order are not particularly useful in practical situations. One rule to follow is that the LP model order, p, should be such that $N/3 < p < N/2$, where N is the data record length. For further discussion of these issues see Kay (1988).

In contrast to the above methods, which estimate the LP parameters directly, the Burg method estimates the reflection coefficients first, and then uses the Levinson recursion to obtain the LP parameter estimates. The reflection coefficient estimates are obtained by minimizing estimates of the prediction error power for different order predictors in a recursive manner. For more information, see Kay (1988), where a discussion of the recursive maximum likelihood estimation procedure can also be found.

The perceptual linear prediction (PLP) analysis technique (Hermansky, 1990), is based on well established psychophysical concepts of hearing. The speech signal is filtered by a critical band filter bank followed by an equal loudness pre-emphasis, and an intensity to loudness adjustment using the intensity-loudness power law. The auditory spectrum is then modeled by an LP model. The PLP analysis yields an auditory spectrum with relatively low-frequency resolution. Furthermore, the frequency resolution of the auditory spectrum is nonuniform. In the higher frequency range it has less resolution, which agrees with the characteristics of the human auditory system. The PLP method is more consistent with human hearing than the conventional LP method. The PLP method is computationally efficient and yields a low dimensional representation of speech. The block diagram of PLP analysis is shown in Figure A6.2.

A moving average (MA) process has a power spectral density with wide peaks and/or sharp nulls. Therefore, the MA spectral estimator is not a high resolution spectral estimator for processes with narrowband spectral features. When the underlying process is an MA process, the MA estimator is more accurate than conventional Fourier-based estimators. Durbin's algorithm (maximum likelihood estimation) for the estimation of the MA parameters of an MA(q) process uses the data $\{x[O], x[l], \ldots, x[N-1]\}$ to fit a large order AR model to the data by the autocorrelation method (see Kay, 1988).

ARMA spectral estimation is a more general model than AR. Since nearly all data are corrupted by some amount of observation noise, the ARMA model is nearly always the appropriate one. However, optimal estimators based on maximum likelihood for an ARMA model require the solution of nonlinear equations. As a result, suboptimal but easily implementable algorithms have been emphasized. In this software, the Akaike approximate MLE method, modified Yule–Walker equations, least squares modified Yule–Walker equations, and the Mayne–Firoozan method are included. For more detail on these algorithms, refer to (Kay, 1988).

The spectrogram is calculated using the MATLAB function specgram.

The menu window for the spectrum option includes FFT analysis, periodogram estimation, Blackman–Tukey analysis, the MUSIC spectrum estimation (uses a MATLAB function), and the ESPRIT spectral estimation method. The periodogram is an inconsistent estimator in that even though the average value converges to the true value as the data record length becomes large, the variance is

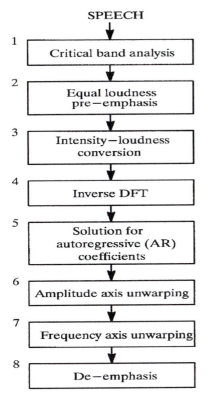

SPEECH

1 Critical band analysis

2 Equal loudness pre—emphasis

3 Intensity—loudness conversion

4 Inverse DFT

5 Solution for autoregressive (AR) coefficients

6 Amplitude axis unwarping

7 Frequency axis unwarping

8 De—emphasis

FIGURE A6.2   Block diagram of PLP analysis.

a constant. To circumvent this problem, we can reduce the variance by segmenting the data by non-overlapping blocks and averaging the periodograms of each block. However, this method increases the bias. The Blackman–Tukey spectral estimator also has a bias to variance trade off. The trade off is affected by the choice of the lag window. For more information on the MUSIC estimator and ESPRIT estimator, refer to Kay (1988) and Therrien (1992).

A6.3 GLOTTAL INVERSE FILTERING

The manual inverse filtering option is implemented to assist the analysis of the speech signal with an automatic glottal inverse filtering algorithm. The filter coefficients of the inverse filter can be adjusted by hand using keyboard entry or with a mouse selection so that the glottal volume velocity has a maximally flat spectrum. To construct the inverse filter manually, the vocal tract filter is modeled with a cascade model. First, by controlling the five formant frequencies and bandwidths, the spectrum of the cascade vocal tract model is matched to the spectrum of the pre-emphasized speech. Then the inverse filter transfer function is the reciprocal of the transfer function of the cascade model. The software program provides initial estimates of the formant frequencies and bandwidths. The filter parameters can be calculated pitch synchronously as well as pitch asynchronously.

The pole-zero plot and spectrum of the speech signal can be displayed along with the result obtained by glottal inverse filtering. Using these tools, accurate values of formant frequencies and bandwidths along with the glottal excitation signal can be obtained. The estimates can be saved for further research, such as speech synthesis and coding, and other uses.

Alku (1992) suggested a new glottal wave analysis algorithm PSIAIF (pitch synchronous iterative adaptive inverse filtering) based on IAIF (iterative adaptive inverse filtering). In the PSIAIF method, the glottal pulse is computed by applying the IAIF twice to the same speech signal. The first analysis gives a glottal excitation that spans several pitch periods, which is used to determine the positions and lengths of frames for pitch synchronous analysis. The final result for the glottal waveform is obtained by analyzing the original speech pitch period by pitch period.

The IAIF method can be summarized as follows.

Step 1: The effect of the glottal pulse on the speech is estimated by a first-order LPC analysis.

Step 2: Remove the effect of the glottal pulse by inverse filtering.

Step 3: Estimate the vocal tract transfer function by applying LPC analysis to the output of Step 2.

Step 4: Inverse filter the speech to obtain the differentiated glottal waveform.

Step 5: Get a first glottal waveform estimate by canceling the lip radiation effect.

Step 6: Estimate a more accurate glottal pulse by applying LPC analysis to the output of Step 5.

Step 7: Inverse filer to remove the effect of the glottal pulse contribution to the speech.

Step 8: Do LPC analysis on the output of the previous step to get the final vocal tract model.

Step 9: Inverse filter the original speech with the output of Step 8.

Step 10: Integrate the output to cancel the lip radiation effect.

The PSIAIF method can be summarized as follows.

Step 1: Apply IAIF to get the glottal pulse and the pitch period for pitch synchronous analysis.

Step 2: Apply IAIF pitch synchronously.

One advantage of pitch synchronous analysis is that it is free of the harmonic structure of the source and, therefore, more accurate modeling is possible, especially when a vowel changes rapidly to another vowel. The iterative method seems to overcome the weak points of conventional linear predictive analysis. When the fundamental frequency is high and the first formant is low, the tract is excited by a source pulse in a way that affects the previous pulse. In other words, the effects of one excitation pulse will not be entirely attenuated before the next one occurs.

A6.4 PITCH/JITTER/FORMANT TRACKING

First, the glottal closure instants are obtained using the glottal excitation waveform, that is, the linear prediction residual signal and a peak picking algorithm. Using this information, the pitch contour can be constructed. The pitch tracking algorithm is similar to the SIFT algorithm (see Markel and Gray, 1976 and Rabiner and Schafer, 1978). The jitter contour is calculated using the pitch perturbation, order 1, method. The perturbation method is expressed as follows. Let a_i be any cyclic parameter, such as amplitude, pitch period, and so on, in the ith cycle of the waveform. The arithmetic mean over N cycles is

$$\bar{a} = \frac{1}{N} \sum_{i=1}^{N} a_i \tag{A6.4.1}$$

The zeroth-order perturbation function is the arithmetic difference, given as

$$p_i^0 = a_i - \bar{a}, \quad i = 1, \ldots, N \tag{A6.4.2}$$

where the subscript i denotes the order of the perturbation function. Higher order perturbation functions are obtained by taking backward and forward differences of lower order functions, for example

$$\begin{aligned}
p_i^k &= p_i^{k-1} - p_{i-1}^{k-1}, \quad k \text{ odd} \\
&= p_{i+1}^{k-1} - p_i^{k-1}, \quad k \text{ even}
\end{aligned} \tag{A6.4.3}$$

where

$$\begin{aligned}
&k_1 + 1 \leq i \leq N - k_2 \\
&k_1 = (k+1)/2, \quad k_2 = (k-1)/2, \quad k \text{ odd} \\
&k_1 = k/2, \quad k_2 = k/2, \quad k \text{ even}
\end{aligned} \tag{A6.4.4}$$

For example, the first-order perturbation function is

$$p_i^1 = p_i^0 - p_{i-1}^0 = a_i - a_{i-1}, \quad i = 2, \ldots, N \tag{A6.4.5}$$

and the second-order perturbation function is

$$p_i^2 = p_{i+1}^1 - p_i^1 = a_{i+1} - 2a_i + a_{i-1}, \quad i = 2, \ldots, N \tag{A6.4.6}$$

The pitch and jitter contours are smoothed with a 5-point median filter.

The formant frequency is calculated using the iterative glottal inverse filtering algorithm. From the inverse filtering algorithm, accurate estimates of the vocal tract transfer function are obtained. The formant frequency and bandwidth are calculated by root solving for the AR model coefficients.

A6.5 CEPSTRUM

Historically the cepstrum is concerned with the general problem of the deconvolution of two or more signals. Bogert, Heally, and Tukey (1963) described the power cepstrum for finding echo arrival times in a composite radar signal. They showed that if you have a signal composed of a waveform and an echo of the waveform, then the effect in the log spectrum turns out to be a "ripple." The "frequency" of this ripple is easily determined by calculating the spectrum of the log spectrum wherein this "frequency" will appear as a peak. However, the units of "frequency" of this ripple in the log spectrum are in units of time; thus, the independent variable (abscissa) in the spectrum of the log spectrum is time. Other parameters were also observed to undergo similar transformations of units. To avoid confusion these authors introduced the following paraphrased terms according to a syllabic interchange rule.

frequency....quefrency

spectrum.....cepstrum

phase...........saphe

amplitude....gamnitude

filtering.......liftering

harmonic.....rahmonic

period..........repoid

among others. Today the two most common terms that remain are cepstrum and quefrency. Filtering in the cepstral domain is often not called liftering, but is simply called filtering. The remainder of this discussion is based on a paper by Childers et al. (1977).

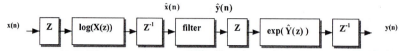

FIGURE A6.3 Calculating the complex cepstrum.

We define the complex cepstrum of a data sequence as the inverse z-transform of the complex logarithm of the z-transform of the data sequence

$$\hat{x}(n) = \frac{1}{2\pi j} \oint_c \log(X(z)) z^{n-1} \, dz \qquad (A6.5.1)$$

where $\hat{x}(0) = \log[x(0)]$ and $X(z)$ is the z-transform of the data sequence $x(n)$. Frequently, $\hat{X}(z)$ is used to denote the $\log[X(z)]$. Then $\hat{x}(n)$, the complex cepstrum, is the inverse z-transform of $\hat{X}(z)$. The contour integration lies within an annular region in which $\hat{X}(z)$ has been defined as singular valued and analytic. If we have the convolution of two sequences, then

$$x(n) = f(n) * g(n) \qquad (A6.5.2)$$

or

$$X(z) = F(z)G(z) \qquad (A6.5.3)$$

and

$$\hat{X}(z) = \log X(z) = \log F(z) + \log G(z) \qquad (A6.5.4)$$

or

$$\hat{x}(n) = \hat{f}(n) + \hat{g}(n) \qquad (A6.5.5)$$

Further, if $\hat{f}$ and $\hat{g}$ occupy different quefrency ranges, then the complex cepstrum can be liftered (filtered) to remove one or the other of the convolved sequences. We will show how this applies to speech later. Since the phase information is retained, the complex cespstrum in invertible. Thus, if $\hat{g}$ is rejected from $\hat{x}$ by liftering, then $\hat{x} = \hat{f}$ and we can then z-transform, exponentiate, and inverse z-transform to obtain the sequence f, that is, f and g have been deconvolved. Figure A6.3 illustrates an overall wavelet recovery or deconvolution (filtering) system. MATLAB has functions to calculate both the real (rceps) and complex (cceps) cepstrum, which we use in our software. The real cepstrum is similar to the complex cepstrum but take the absolute magnitude of $X(z)$ first.

Examples of long-pass, short-pass, and notch lifters (filters) are shown in Figure A6.4. These lifters are analogous to high-pass, low-pass, and notch (comb) filters, respectively, in the frequency domain.

The computation of the complex cepstrum is complicated by the fact that the complex logarithm is multivalued. If the imaginary part of the logarithm is computed modulo 2π; that is, evaluated as its principal value, then discontinuities can appear in the phase curve. This is not allowed since the $\log(X(z))$ is the z-transform of $\hat{x}$ and thus must be analytic in some annular region of the z-plane. This problem may be rectified by making the following observations.

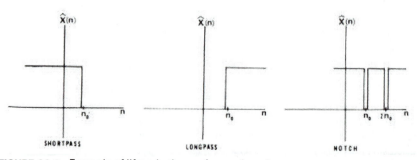

FIGURE A6.4 Example of lifters in the quefrency domain.

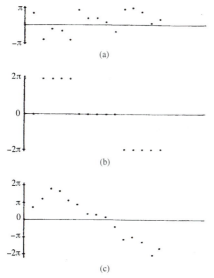

FIGURE A6.5 Phase unwrapping. (*a*) Phase modulo 2π. (*b*) C(k), the correction sequence. (*c*) Unwrapped phase.

- The imaginary part of $\log(X(z))$ must be a continuous and periodic (evaluated on the unit circle) function of ω with period $(2\pi/T)$ since it is the z-transform of $\hat{x}$.

- Since it is required that the complex cepstrum of a real function be real, it follows that the imaginary part of $\log(X(z))$ must be an odd function of ω.

Subject to these conditions, we may compute the unwrapped phase curve as follows (provided the phase is sampled at a rate sufficiently great to assure that it never changes by more than π between samples). A correction sequence $C(k)$ is added to the modulo 2π phase sequence $P(k)$ where $C(k)$ is

$$
\begin{aligned}
C(0) &= 0 \\
C(k) &= C(k-1) - 2\pi, && \text{if } P(k) - P(k-1) > \pi \\
&= C(k-1) + 2\pi, && \text{if } P(k-1) - P(k) > \pi \\
&= C(k-1), && \text{otherwise}
\end{aligned}
\tag{A6.5.6}
$$

This is illustrated in Figure A6.5.

Alternately, the phase can be unwrapped by computing the relative phase between adjacent samples of the spectrum. These phases may be added to achieve a cumulative (unwrapped) phase for each point. Both methods have the drawback that the computation must be done sequentially. It is also noted that if the phase never changes by more than $\frac{\pi}{2}$ between samples, the phase modulo π could be computed and unwrapped with algorithms similar to the above. This is interesting since it is slightly easier to calculate the phase modulo π than the phase modulo 2π (the arctangent algorithm is simpler) and many signals have this property (though noise generally does not). There are other phase unwrapping procedures.

Phase unwrapping is unnecessary for the class of minimum phase signals, that is, a sequence with a z-transform that has no poles or zeros outside the unit circle, which implies that $\hat{x}(n) = 0$ for $n < 0$. The complex cepstrum of such a sequence is zero at negative quefrencies. Analogously, a maximum phase sequence can be defined (the z-transform has no poles or zeros inside the unit circle). The complex cepstrum of such a sequence is zero for positive quefrencies. The signals of general interest are of mixed phase. It is difficult to properly analyze or process such signals in the presence of noise.

We now show that the impulses that appear in the complex cepstrum can be caused by the presence of a single additive echo. These impulses are non-zero on only one side of the origin and are therefore referred to as minimum or maximum phase impulse trains.

For $x(n) = f(n) * g(n)$, let

$$g(n) = \delta(n) + a\delta(n - n_0) \tag{A6.5.7}$$

then

$$x(n) = f(n) + af(n - n_0) \tag{A6.5.8}$$

or

$$X(z) = F(z)(1 + az^{-n_0}) \tag{A6.5.9}$$

Taking the log of both sides, we have

$$\hat{X}(z) = \log(F(z)) + \log(1 + az^{-n_0}) \tag{A6.5.10}$$

If $a < 1$ (corresponding to a minimum phase sequence), then we may expand the right most term above in a power series, giving

$$\hat{X}(z) = \log(F(z)) + az^{-n_0} - \frac{a^2}{2}z^{-2n_0} + \frac{a^3}{3}z^{-3n_0} - \cdots \tag{A6.5.11}$$

Inverse z-transforming, we have the complex cepstrum

$$\hat{x}(n) = \hat{f}(n) + a\delta(n - n_0) - \frac{a^2}{2}\delta(n - 2n_0) + \frac{a^3}{3}\delta(n - 3n_0) - \cdots \tag{A6.5.12}$$

Thus the complex cepstrum of the composite signal consists of the complex cepstrum of the basic wavelet, $\hat{f}$, plus a train of δ functions located at positive quefrencies at the echo delay (and its multiples) with amplitudes that are directly related to the echo amplitude. Notch liftering and interpolation (smoothing) can be performed to remove the δ functions. The basic wavelet can then be recovered by inverting the operations used to compute the complex cepstrum (see Figure A6.3). If the complex cepstra of the basic wavelet and the impulse train are sufficiently separated in quefrency, then short-pass liftering can be used to recover the basic wavelet. Analogously, the impulse train, g, can be recovered by using long-pass liftering.

If the echo amplitude is greater than or equal to unity, $a \geq 1$ (corresponding to a maximum phase sequence), then we have

$$\hat{X}(z) = \log(az^{-n_0}F(z)) + \log\left(1 + \frac{1}{a}z^{n_0}\right) \tag{A6.5.13}$$

which can be expanded to give

$$\hat{X}(z) = \log(az^{-n_0}F(z)) + \frac{1}{a}z^{n_0} - \frac{1}{2a^2}z^{2n_0} \ldots \tag{A6.5.14}$$

which can be rewritten as

$$\hat{X}(z) = \log(F(z)) - \log z^{n_0} + \log a + \frac{1}{a}z^{n_0} - \frac{1}{2a^2}z^{2n_0} \ldots \tag{A6.5.15}$$

If we remove the linear phase term $(-\log z^{n_0})$, the complex cepstrum is

$$\hat{x}(n) = \hat{f}(n) + \log(a)\delta(n) + \frac{1}{a}\delta(n + n_0) - \frac{1}{2a^2}\delta(n + 2n_0) \ldots \tag{A6.5.16}$$

Thus, the complex cepstrum again has peaks at the echo delay (and its multiples), however, these peaks now occur at negative rather than positive quefrencies and their gamnitudes (amplitudes) are related to $\frac{1}{a}$ rather than a. If these peaks are removed by liftering and the wavelet recovery procedure is followed including the reinsertion of the linear phase term, then the echo is recovered rather than the basic wavelet. The effect of liftering on the complex cepstrum is schematized in Figure A6.6 for $a > 1$ and $a < 1$.

It will be noticed that the peaks in the complex cepstrum due to the impulse train can never have an amplitude greater than unity regardless of the value of a. Furthermore, note that multiplying the original composite signal by a scale factor only changes the coefficient of the $\delta(n)$ term in the

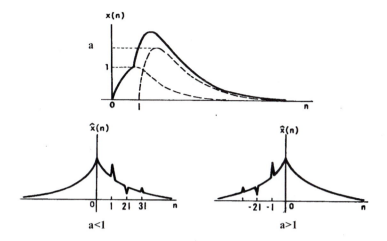

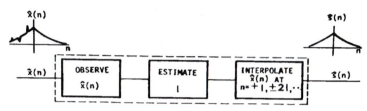

FIGURE A6.6 The superposition of two wavelets to form x(n); the complex cepstra for a < 1 and a > 1; the liftering of the complex cepstrum by notch filtering.

complex cepstrum, since the scale factor appears as a shift in the mean of the log spectrum. Therefore, the complex cepstrum does not depend on the composite signal scale factor, but does depend on the signal-to-noise ratio (SNR).

Another interesting, and perhaps, more representative example is the one with an infinite series of decaying echoes. Here we have

$$g(n) = \delta(n) + \alpha_1^{n_0}\delta(n - n_0) + \alpha_1^{2n_0}\delta(n - 2n_0) + \cdots \tag{A6.5.17}$$

where $0 \leq \alpha_1 \leq 1$. Then with the simplification $\alpha = \alpha_1^{n_0}$ we have

$$g(n) = \delta(n) + \alpha\delta(n - n_0) + \alpha^2\delta(n - 2n_0) + \cdots$$

$$= \sum_{m=0}^{\infty} \alpha^m \delta(n - mn_0) \tag{A6.5.18}$$

which when convolved with f(n) will give us a minimum phase sequence. Then

$$G(z) = 1 + \alpha z^{-n_0} + \alpha^2 z^{-2n_0} + \cdots$$

$$= \frac{1}{1 - \alpha z^{-n_0}} \tag{A6.5.19}$$

Thus the complex cepstrum is

$$\hat{x}(n) = \hat{f}(n) + \alpha\delta(n - n_0) + \frac{\alpha^2}{2}\delta(n - 2n_0) + \cdots \tag{A6.5.20}$$

This complex cepstrum is minimum phase and is nearly identical to that calculated previously, except that the signs of the train of pulse functions are all positive rather than alternating in sign. The remarks made previously apply here as well. When α is near unity, this example might be considered more representative of speech data for the situation of a sustained vowel phonation such as /IY/.

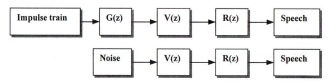

FIGURE A6.7   Two models of speech production.

In the general multiple echo case, the delays become "mixed" via the series expansion of the logarithm. This greatly complicates the proper estimation of the true echo delay times. The estimation is further complicated if aliasing is severe.

For speech production we have two basic models shown in Figure A6.7. The upper model is for voiced speech, while the lower model is for unvoiced speech. In either case, we have the convolution of a signal with transfer functions to produce speech. For voiced speech, if we are given the speech signal and asked to estimate the glottal excitation function, then we must deconvolve the radiation and vocal tract system functions from the glottal transfer function. The complex cepstrum can be applied to this task. A similar task is present for unvoiced speech.

The complex cepstrum will be affected by the vocal tract, the glottal pulse, and the radiation. We will generally have a non-minimum phase signal. This means that the complex cepstrum will be non-zero for both negative and positive quefrencies. The complex cepstrum will decay as $\frac{1}{n}$ with quefrency. The periodic excitation will cause pulses in the complex cepstrum at integer multiples of the spacing between impulses, that is, the complex cepstrum will have pulses at multiples of the fundamental period of voicing.

The impulses due to the fundamental frequency of voicing in speech tend to be separated from the other components (glottal, vocal, and radiation filters) of the complex cepstrum. Thus one can filter the complex cepstrum with a lifter defined as $h(n) = 1$ for $-n_0 < n < n_0$ (and zero elsewhere), where n_0 is less than the pitch period n_p. This lifter tends to select the glottal, vocal tract, and radiation effects rather than the low quefrency material. Therefore it is useful for formant estimation. A lifter defined such that $h(n) = 1$ for $n \geq |n_0|$ and zero elsewhere will retain the impulse excitation train. Thus, this lifter is useful for pitch period estimation. For pitch detection for voiced speech, one measures the interval between the origin and the first peak in the cepstrum. No pulses appear in the cepstrum for unvoiced speech. Thus, we can use the cepstrum to detect voiced/unvoiced segments and to estimate the pitch period.

A6.6   WRLS-VFF

The WRLS_VFF algorithm is discussed in the following paper titled "Adaptive WRLS_VFF for Speech Analysis" previously published in Childers, D. G., Principe, J. C., and Ting, Y. T. (1995). "Adaptive WRLS-VFF for Speech Analysis," *IEEE Transactions on Speech and Audio Processing*, 209–213.

A6.7   SILENT AND VOICED/UNVOICED/ MIXED EXCITATION (FOUR-WAY) CLASSIFICATION OF SPEECH

One algorithm for the four-way classification of speech using the speech and electroglottographic (EGG) signals is described in the following paper titled "Silent and Voiced/ Unvoiced/Mixed Excitation (Four-Way) Classification of Speech," previously published in Childers, D. G., Holm, M., and Larar, J. N. (1989). "Silent and Voiced/Unvoiced/Mixed Excitation (Four Way) Classification of Speech," 37, 1771–1774. The algorithm can be modified to eliminate the use of the EGG signal and rely only on the speech signal.

IEEE TRANSACTIONS ON SPEECH AND AUDIO PROCESSING, VOL. 3, NO. 3, MAY 1995

209

Adaptive WRLS-VFF for Speech Analysis

D. G. Childers, J. C. Principe, and Y. T. Ting

Abstract— The purpose of this correspondence is to show that an adaptive weighted recursive least squares algorithm with a variable forgetting factor (WRLS-VFF) will adjust the size of the data segment to be analyzed according to its time-varying characteristics, as during the transitions between vowels and consonants. The algorithm can accurately estimate the vocal tract formants, anti-formants, and their bandwidths, be used for glottal inverse filtering, perform voiced (V)/unvoiced (U)/silent (S) classification of speech segments, estimate the input excitation (either white noise or periodic pulse trains), and estimate the instant of glottal closure.

I. ALGORITHM DESCRIPTION

We assume that the speech signal is generated by an autoregressive, moving average (ARMA) model

$$y_k = -\sum_{i=1}^{p} a_i(k) y_{k-i} + \sum_{j=1}^{q} b_j(k) u_{k-j} + u_k \qquad (1)$$

where y_k denotes the kth sample of the speech signal, u_k is the input excitation, (p, q) are the order of the poles and zeros, respectively, and $a_i(k)$ and $b_j(k)$ are the time-varying AR and MA parameters, respectively. Measurement noise is ignored in this model but could be included [10], [11]. We assume that the values of p and q are predetermined. Note that the measured speech signal, y_k, depends on the input, u_k. The excitation, u_k, is usually considered to be white Gaussian noise. In this paper we allow u_k to be either a zero-mean, white, Gaussian noise process, u_k^w, with variance σ_c^2, or a train of periodic pulses, u_k^p. We must estimate the input excitation, either u_k^w or u_k^p, so that the ARMA parameters can be estimated accurately from y_k.

Let us define a parameter vector, θ_k, and a data vector, ϕ_k, by the following equations

$$\theta_k^t = [a_1(k), \cdots, a_p(k), b_1(k), \cdots, b_q(k)] \qquad (2)$$

$$\phi_k^t = [-y_{k-1}, \cdots, -y_{k-p}, u_{k-1}, \cdots, u_{k-q}] \qquad (3)$$

where the superscript t denotes transpose. The corresponding estimated quantities will be denoted by $\hat{\theta}_k$. Then

$$y_k = \phi_k^t \theta_k + u_k \qquad (4)$$

$$\hat{y}_k = \hat{\phi}_k^t \hat{\theta}_k + \hat{u}_k. \qquad (5)$$

Let r_k be the residual error of the ARMA process, namely,

$$r_k = y_k - \hat{y}_k = y_k - \hat{\phi}_k^t \hat{\theta}_k - \hat{u}_k. \qquad (6)$$

The predicted signal, $\hat{y}_{k/k-1}$, determined from the estimated ARMA parameters at time $(k - 1)$, is

$$\hat{y}_{k/k-1} = \hat{\phi}_k^t \hat{\theta}_{k-1}. \qquad (7)$$

Manuscript received August 13, 1993; revised November 15, 1994. This work was supported by NIH Grant NIDCD R01 DC00577, NSF Grant IRI-9215331, the University of Florida Center of Excellence Program in Information Transfer and Processing, and the Mind Machine Interaction Research Center. The associate editor coordinating the review of this paper and approving it for publication was Dr. Amro El-Jaroudi.

D. G. Childers and J. C. Principe are with the Department of Electrical Engineering, University of Florida, Gainesville, FL 32611 USA.

Y. T. Ting is with the Chung San Institute of Science and Technology, Taiwan, Republic of China.

IEEE Log Number 9410229.

Consequently, the prediction error is

$$e_k = y_k - \hat{y}_{k/k-1} - \hat{u}_k. \qquad (8)$$

Note that $\hat{u}_k$ is usually assumed to be unavailable at $(k - 1)$ and is set to zero [10], [11]. We will address this issue again later and modify the algorithm accordingly. An approach to deal with time varying regression coefficients is to minimize a weighted estimation (or residual) error [10], [14]

$$E_k = w(1, k)[\hat{\theta}_k^t P_1^{-1} \hat{\theta}_k] + \sum_{i=1}^{k} w(i, k) r_i^2 \qquad (9)$$

where P_1 is an arbitrary real symmetric positive definite matrix. The weighting coefficient ($w(i, k)$) is called a forgetting factor [10], [14]

$$w(i, k) = \prod_{j=i+1}^{k} \lambda_j, \qquad i = 1, 2, \cdots, k-1$$
$$= 1, \qquad i = k. \cdots \qquad (10)$$

The coefficient λ_j decreases the weight of past estimation errors provided $0 < \lambda_j < 1$. Note for fixed $\lambda_j = \lambda$ that $w(i, k)$ becomes an exponentially weighted coefficient, e.g., $\lambda^{k-1}, \lambda^{k-2}, \cdots, \lambda, 1$. Consequently, the estimation error, E_k, becomes the exponentially weighted sum of squares of the estimation errors [3], [5], [14], i.e.

$$E_k = \sum_{i=1}^{k} \lambda^{k-1}(y_i - \hat{y}_i)^2 + \lambda^{k-1}[\hat{\theta}_k^t P_1^{-1} \hat{\theta}_k]. \qquad (11)$$

Minimizing the least square weighted estimation error, E_k, with respect to the ARMA parameter vector, $\hat{\theta}_k$, assuming that $\hat{u}_k$ is available, gives the following algorithm [10], [11], [14].
Residual error

$$r_k = y_k - \hat{\phi}_k^t \hat{\theta}_k - \hat{u}_k. \qquad (12)$$

Prediction error

$$e_k = y_k - \hat{\phi}_k^t \hat{\theta}_{k-1} - \hat{u}_k. \qquad (13)$$

Gain update

$$K_k = P_{k-1} \hat{\phi}_k [\lambda_{k-1} + \hat{\phi}_k^t P_{k-1} \hat{\phi}_k]^{-1}. \qquad (14)$$

Parameter update

$$\hat{\theta}_k = \hat{\theta}_{k-1} + K_k e_k. \qquad (15)$$

Covariance matrix

$$P_k = \lambda_k^{-1}[P_{k-1} - K_k \hat{\phi}_k^t P_{k-1}]. \qquad (16)$$

The above algorithm updates the ARMA parameters at each instant k and has been shown to be stable and to provide a unique solution [3], [5], [10], [11], [14].

E_k can be calculated recursively, thereby allowing λ to be calculated recursively. Since $w(k, k) = 1$ and $w(k-1, k) = \lambda_k$, then (9) leads to the following recursive expression

$$E_k = \lambda_k E_{k-1} + (y_k - \hat{y}_k)^2 + w(1, k)$$
$$\times [\hat{\theta}_k^t P_1^{-1} \hat{\theta}_k - \hat{\theta}_{k-1}^t P_1^{-1} \hat{\theta}_{k-1}]. \qquad (17)$$

Then rearranging, we have

$$\lambda_k = \frac{E_k}{E_{k-1}} - \frac{(y_k - \hat{y}_k)^2}{E_{k-1}} - \frac{w(1, k)}{E_{k-1}}$$
$$\times [\hat{\theta}_k^t P_1^{-1} \hat{\theta}_k - \hat{\theta}_{k-1}^t P_1^{-1} \hat{\theta}_{k-1}]. \qquad (18)$$

210

IEEE TRANSACTIONS ON SPEECH AND AUDIO PROCESSING, VOL. 3, NO. 3, MAY 1995

Using the previous expressions leads to the following approximation to compute and update λ_k

$$\lambda_k = \frac{E_k}{E_{k-1}} - \frac{e_k^2}{E_{k-1}}[1 - \hat{\phi}_k^t K_k]^2 \qquad (19)$$

where we have assumed that the third term in (18) becomes negligible with increasing k since $w(1, k) = \lambda_2 \lambda_3 \cdots \lambda_k \ll 1$ for $0 \le \lambda_i < 1$.

In (19), λ_k depends upon the ratio of the weighted estimation errors at step k and $k - 1$ (E_k and E_{k-1}), and on the prediction error at step k. A simplifying strategy to compute λ_k can be defined if we require that $E_k = E_{k-1} = \cdots = E_1$ [4]. This means that the forgetting factor compensates at each step k for the new error information in the latest measurement. This has the added benefit of normalizing the forgetting factor with respect to the same error information, yielding

$$\lambda_k = 1 - \frac{e_k^2}{E_1}[1 - \hat{\phi}_k^t K_k]^2. \qquad (20)$$

The WRLS-VFF algorithm is specified by a set of equations similar to those for the WRLS algorithm. However, the weighting factor, λ_k, is estimated by (20). We recommend that the estimation of E_1 be determined using an initial block of data. The minimum length of the window should be related to the size of the ARMA model, therefore, we further limit the smallest value of λ by

$$\lambda_{\min} = 1 - \frac{1}{N_a}, \text{ if } \lambda_k < \lambda_{\min}, \text{ then } \lambda_k = \lambda_{\min}. \qquad (21)$$

We have determined empirically that the value of N_a should be approximately twice the model order to obtain good results.

We now estimate the residual and prediction errors as well as the gain, ARMA parameters, the covariance matrix, and the excitation, u_k. Furthermore, we must determine whether $\hat{u}_k$ is $\hat{u}_k^p$ or $\hat{u}_k^w$. Since the previous results assumed that we had an estimate for u_k, we now modify our definitions of the residual and prediction errors to account for the fact that an estimate for the excitation, $\hat{u}_k$, is not available at k. Thus, we define the following two errors

$$r_k = y_k - \hat{y}_k = y_k - \hat{\phi}_k^t \hat{\theta}_k \qquad (22)$$

$$\xi_k = y_k - \hat{y}_{k/k-1} = y_k - \hat{\phi}_k^t \hat{\theta}_{k-1}. \qquad (23)$$

The equation for the forgetting factor (eq. (20)) remains as before with e_k replaced by ξ_k. Once we make a decision regarding the input excitation, we define a new error as

$$e_k = \begin{cases} \xi_k & \text{for } \hat{u}_k = u_k^w \\ \xi_k - u_k^p & \text{for } \hat{u}_k = u_k^p \end{cases} \qquad (24)$$

We update the parameter estimates and the covariance matrix estimate. The data/excitation vector $\hat{\phi}_k^t$ is updated using the new speech sample y_k and $\hat{u}_k$.

From (20), we see that an increase in the prediction error, e_k, results in a decrease in λ_k. A small value of λ_k indicates that the input has undergone an abrupt change, typically indicating that a glottal pulse excitation has occurred. Hence, we can determine the time of occurrence of a pulse by determining the instant at which λ_k falls below a minimum threshold λ_0. When this occurs we set $\hat{u}_k = \hat{u}_k^p$ and $\hat{u}_k^w = 0$. The magnitude of the pulse excitation is determined from the prediction error ξ_k, at the estimated time of the input pulse by assuming that $\xi_k = \hat{u}_1^r$ [10]. Thus,

$$\hat{u}_k^p = y_k - \hat{\phi}_k^t \hat{\theta}_{k-1}. \qquad (25)$$

For white noise input, λ_k is close to unity upon convergence [14]. Under this condition the residual error, r_k, of the adaptive process

TABLE I
ADAPTIVE WRLS-VFF ALGORITHM WITH INPUT ESTIMATION

Prediction error:	$\xi_k = y_k - \hat{\phi}_k^t \hat{\theta}_{k-1}$
Gain update:	$K_k = P_{k-1} \hat{\phi}_k [\lambda_{k-1} + \hat{\phi}_k^t P_{k-1} \hat{\phi}_k]^{-1}$
Forgetting Factor:	$\lambda_k = 1 - \xi_k^2 (1 - \hat{\phi}_k^t K_k)^2 / E_1$
Input estimate:	

a) Pulse input
If $\lambda_k < \lambda_0$ then
$$\hat{u}_k^w = 0$$
$$\hat{u}_k = \hat{u}_k^p$$
$$= y_k - \hat{\phi}_k^t \hat{\theta}_{k-1}$$
b) White noise input
If $\lambda_k > \lambda_0$ then
$$\hat{u}_k^p = 0$$
$$\hat{u}_k = \hat{u}_k^w$$
$$= \xi_k (1 - \hat{\phi}_k^t K_k)$$

Parameter update:	$\hat{\theta}_k = \hat{\theta}_{k-1} + K_k (y_k - \hat{\phi}_k^t \hat{\theta}_{k-1} - \hat{u}_k^p)$
Covariance matrix:	$P_k = \lambda_k^{-1} [P_{k-1} - K_k \hat{\phi}_k^t P_{k-1}]$

can be used as the estimate of the white noise input, $\hat{u}_k^w$, as indicated in Morikawa's method [12], [13], i.e., from (22)

$$r_k = y_k - \hat{y}_k = \xi_k(1 - \hat{\phi}_k^t K_k) = \hat{u}_1^w \qquad (26)$$

and $\hat{u}_k^p = 0$. This is similar to estimates used previously [5], [12], while a different approach was used in [10]. However, our approach uses only one adaptive algorithm instead of two as in [10].

In order to select the threshold value, λ_0, that determines whether the speech was voiced or unvoiced, we adopted the following strategy. We compute a running average of the last M values of the forgetting factor as follows, where M is typically the number of samples in a frame

$$L_k = \frac{1}{M} \sum_{i=0}^{M-1} \lambda_{k-i}. \qquad (27)$$

If $L_k < 0.9$, then $\lambda_0 = 0.99^* L_k$; if $L_k > 0.9$, then $\lambda_0 = 0.9^* L_k$; should $\lambda_0 < \lambda_{\min}$, then $\lambda_0 = \lambda_{\min}$. Thus, the threshold may be made adaptive, whereby, it is adjusted on a frame-by-frame basis.

We summarize the WRLS-VFF algorithm including input estimation in Table I. The algorithm in Table I differs from previous algorithms in that we: 1) update the variable forgetting factor at each step, 2) let the prediction error, e_k, be the estimate for the pulse magnitude, $\hat{u}_k^p$, and 3) let the residual error, r_k, be the estimate for the noise excitation, $\hat{u}_k^w$. The algorithm may be shown to be stable and to provide a unique solution following the method given in Appendix III of [10]. Several factors affect the convergence of the WRLS-VFF algorithm: 1) model order, 2) stationarity of the signal, and 3) size of the data analysis interval. We have assumed that the model order may be determined a priori and that the data analysis interval is sufficiently large for the algorithm to work. One can show from (22) and (23) that

$$c_k^2 = r_k^2 (1 + \lambda_{k-1}^{-1}[\hat{\phi}_k^t P_{k-1} \hat{\phi}_k])^2. \qquad (28)$$

If 1) the covariance matrix P_{k-1} is positive definite and 2) $[\hat{\phi}_k^t P_{k-1} \hat{\phi}_k]$ converges to zero as k goes to infinity, then the variance of the prediction error, e_k, converges to the variance of the residual error, r_k. These two conditions can be shown to be satisfied for a

IEEE TRANSACTIONS ON SPEECH AND AUDIO PROCESSING, VOL. 3, NO. 3, MAY 1995

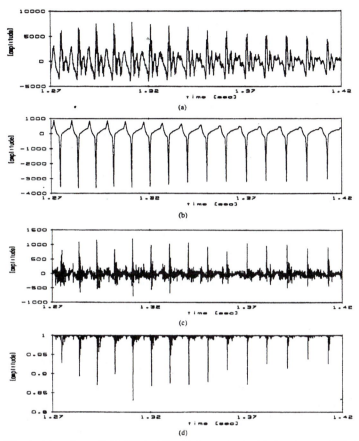

Fig. 1. Speech (a), differentiated electroglottographic (DEGG) signal (b), residual error (c), and variable forgetting factor (VFF) (d).

stationary ARMA process with white noise, zero mean excitation [9], [10], [16] under the assumption that λ_k approaches unity.

The parameter estimation error is usually large when there is a large glottal open phase during speech production. Small values of λ_k occur at the instants of glottal closure where the prediction error is maximum. Fig. 1 shows that the minima of λ_k occur at nearly the same instant as the negative peaks of the differentiated electroglottographic (DEGG) signal. Since the large negative peaks in the DEGG are known to occur at glottal closure (or the minimum glottal aperture) [1], [7], then comparisons such as those in Fig. 1 serve as a partial validation of the WRLS-VFF algorithm. With such validation we have concluded that the minima of λ_k can be used to predict the instant of glottal closure, and, therefore, the presence of voiced excitation. When λ_k does not fall below the threshold value, λ_0, then we may decide that the excitation is unvoiced.

The WRLS-VFF algorithm requires on the order of $(5(p+q)^2 + 6(p+q))$ floating point multiplications and additions (flops) per data point for an ARMA model [16]. By using the idea of shift low rank

the WRLS-VFF algorithm can be implemented with $O(N(p+q))$ flops instead of $O(N(p+q)^2)$ [8]. Consequently, the WRLS-VFF algorithm can be made of the same complexity as other recursive algorithms.

II. CONCLUSION

We have implemented a closed phase adaptive WRLS-VFF algorithm, which is able to track formants and anti-formants for various speech sounds, e.g., vowels, diphthongs, nasals and some consonants for either isolated words or sentences. In [16] we have shown that this algorithm accomplishes these tasks with greater accuracy than other methods, such as LPC [9], Iterative Inverse Filtering (ITIF) [15], the two-stage least squares modified Yule–Walker equations (MYWE) [6], and the recursive algorithms: sequential estimation ARMA (SEARMA) [13], weighted recursive least squares (WRLS) [5], weighted least squares lattice (WLSL) [8], and modified WRLS

IEEE TRANSACTIONS ON SPEECH AND AUDIO PROCESSING, VOL. 3, NO. 3, MAY 1995

(MRLS) [4]. The WRLS-VFF algorithm has also been shown to estimate accurately the values of the formants and their bandwidths for vowels and fricatives [2]. The algorithm is able to perform glottal inverse filtering automatically as well as or better than other procedures that require two channels of data (e.g., speech and EGG) or that require two-passes of the data. The WRLS-VFF algorithm uses the VFF and the estimation error to determine the time at which an excitation pulse occurs and excludes that interval from parameter updating. This provides an improved estimation of the vocal tract transfer function and, consequently, an improved estimation of the glottal volume-velocity waveform.

REFERENCES

[1] D. G. Childers, D. M. Hicks, G. P. Moore, L. Eskenazi, and A. L. Lalwani, "Electroglottography and vocal fold physiology," *J. Speech and Hearing Res.*, vol. 33, pp. 245–254, June 1990.
[2] D. G. Childers and K. Wu, "Gender recognition from speech: Part II. Fine analysis," *J. Acoust. Soc. Am.*, vol. 90, no. 4, pp. 1841–1856, 1991.
[3] C. F. N. Cowan and P. M. Grant, *Adaptive Signal Processing.* Englewood Cliffs, NJ: Prentice-Hall, 1985, ch. 3.
[4] T. R. Fortescue, L. S. Kershenbaum, and B. E. Ydstie, "Implementation of self-regulators with variable forgetting factors," *Automatica*, vol. 17, pp. 831–835, 1981.
[5] B. Friedlander, "A recursive maximum likelihood algorithm for ARMA spectral estimation," *IEEE Trans. Inform. Theory*, vol. 28, no. 4, pp. 639–646, 1982.
[6] S. Kay, *Modern Spectrum Estimation.* Englewood Cliffs, NJ: Prentice-Hall, 1987.
[7] A. K. Krishanamurthy and D. G. Childers, "Two channel speech analysis," *IEEE Trans. Acoust., Speech, Signal Processing*, vol. 34, no. 4, pp. 730–743, 1986.
[8] D. T. L. Lee, M. Morf, and B. Friedlander, "Recursive least squares ladder estimation algorithms," *IEEE Trans. Acoust., Speech, Signal Processing*, vol. 29, no. 3, pp. 627–641, 1981.
[9] J. Makhoul, "Linear prediction: A tutorial review," *Proc. IEEE*, vol. 64, pp. 99–118, 1976.
[10] Y. Miyanaga, N. Miki, N. Nagai, and K. Hatori, "A speech analysis algorithm which eliminates the influence of pitch using the model reference adaptive system," *IEEE Trans. Acoust., Speech, Signal Processing*, vol. 30, no. 1, pp. 88–95, 1982.
[11] Y. Miyanaga, N. Miki, and N. Nagai, "Adaptive identification of a time-varying ARMA speech model," *IEEE Trans. Acoust., Speech, Signal Processing*, vol. 34, pp. 423–433, 1986.
[12] H. Morikawa and H. Fujisaki, "Adaptive analysis of speech based on a pole-zero representation," *IEEE Trans. Acoust., Speech, Signal Processing*, vol. 30, no. 1, pp. 77–87, 1982.
[13] ——, "System identification of the speech production process based on a state-space representation," *IEEE Trans. Acoust., Speech, Signal Processing*, vol. 32, no. 2, pp. 252–262, 1984.
[14] T. Soderstrom and P. Stoica, *System Identification.* Englewood Cliffs, NJ: Prentice-Hall, 1989.
[15] K. Steiglitz, "On the simultaneous estimation of poles and zeros in speech analysis," *IEEE Trans. Acoust., Speech, Signal Processing*, vol. 25, pp. 194–202, 1977.
[16] Y. T. Ting, "Adaptive estimation of time-varying signal parameters with applications to speech," Ph.D. dissertation, University of Florida, 1989.

IEEE TRANSACTIONS ON ACOUSTICS, SPEECH, AND SIGNAL PROCESSING, VOL. 37, NO. 11, NOVEMBER 1989

Silent and Voiced/Unvoiced/Mixed Excitation (Four-Way) Classification of Speech

D. G. CHILDERS, M. HAHN, AND J. N. LARAR

Abstract—We present an algorithm for automatically classifying speech into four categories: silent and speech produced by three types of excitation, namely, voiced, unvoiced, and mixed (a combination of voiced and unvoiced). The algorithm uses two-channel (speech and electroglottogram) signal analysis and has been tested on data from six speakers (three male and three female), each speaking five sentences. An overall correct classification accuracy of approximately 98.2 percent was achieved when compared to skilled manual classification. This is superior to previously reported automatic classification schemes. If word boundary errors, including the beginning and ending of sentences, are excluded, then the algorithm's performance improves to 99.5 percent.

INTRODUCTION

In previous work [1], we described a two-channel, two-way (V/U–S) algorithm for automatically classifying speech. This algorithm used the speech and electroglottogram (EGG) signals. One of our objectives has been to demonstrate that two-channel-based algorithms can lead to computational and performance improvements over algorithms based on acoustic-signal-only analysis methods. We recognize that in many situations, the EGG signal is either unavailable or cannot be used. However, both the speech and EGG signals can be used in the laboratory to help benchmark the performance of numerous speech systems. We advocate this approach.

Previous research has focused on three-way speech classification, i.e., either V/U/S or V/U/M [2]–[13]. The speech classification problem is important because its solution affects other speech analysis, synthesis, and recognition problems. For example, speech classification can help reduce the number of lexical candidates in speech (word) recognition, improve speech synthesis by selecting the proper excitation, and improve the performance of phoneme boundary detection in speech analysis. Consider the large vocabulary isolated word recognition problem. By using only four-way (V/U/M/S) classification and stress analysis, one can define an equivalence class of words having the same representation or "coding" [9], [10], [14]. For example, the words speed, steep, scout, and stop all belong to the same equivalence class of U/S/ stressed-U/V/U. In an isolated word recognition system, the search to identify a test word among all possible candidates can be reduced by using such a simple coding technique. Following such a reduction of the lexical candidates, one may perform other, more detailed analyses to match the test word with one of the remaining words.

Some of the problems with classifying speech as V/U using acoustic-signal-based algorithms are caused by the use of a large analysis frame, a low level of voicing, or even the strength of the first formant. Classification of U/S segments is even more difficult for such algorithms. Typically, researchers have adopted sophisticated approaches to overcome these problems, using additional features, a statistical approach, or an optimized set of parameters [2], [3], [7], [12], [13].

Manuscript received March 8, 1988; revised February 13, 1989. This work was supported in part by NSF Grant ECE-8413583, NIH Grants NINCDS RO1 NS17078 and NS 27022, the University of Florida Center of Excellence Program in Information Transfer and Processing, the Florida High Technology and Industry Council under Grant 4505208-12, and Mind-Machine Interaction Research Center.

D. G. Childers and M. Hahn are with the Department of Electrical Engineering, University of Florida, Gainesville, FL 32611.

J. N. Larar was with AT&T Bell Laboratories, Murray Hill, NJ. He is now with the Department of Medicine, University of Miami, Miami, FL 33101.

IEEE Log Number 8930546.

IEEE TRANSACTIONS ON ACOUSTICS, SPEECH, AND SIGNAL PROCESSING, VOL. 37, NO. 11, NOVEMBER 1989   1771

A Speech-EGG-Based Algorithm for V/U/M/S Classification

A. Algorithm Overview

The properties and some applications of the EGG to speech analysis appear in [1], [8], [15]-[18]. The EGG offers advantages not readily available from a microphone, even a throat contact microphone. The EGG is not susceptible to environment noise, providing instead a direct measure of vocal fold contact [19], while the throat contact microphone provides an acoustic signal similar in form to that provided by other microphones.

The EGG amplitude varies both within and across speakers. Baseline variations in the EGG may be removed by differentiating the EGG. For voiced segments, the EGG usually has only two zero crossings per fundamental (pitch) period of voicing. One exception is vocal fry. For unvoiced segments, the electroglottograph output is a very low-level high-frequency noise-like signal generated by the internal electronics of the device that is easily distinguished from the excitation for voiced speech. Thus, V/U-S classification is achieved using a combination of EGG amplitude and level-crossing rate [1], [8]. Mixed excitation detection is accomplished by noting that the EGG signal appears similar to that for voiced sounds, but the speech signal is small in amplitude and has a high level-crossing rate (see Fig. 2 and other examples in [1]). Silent intervals are detected by observing that the EGG waveform appears as it does for unvoiced speech and that the speech signal is below a predetermined energy threshold. The two-channel, four-way speech classification algorithm appears in Fig. 1. Note that the algorithm does not use endpoint classification, but this could be added if desired [4], [11].

B. Algorithm Details

Fig. 2 depicts both illustrative results and some difficulties encountered in attempting to evaluate the level-crossing rate (LCR) and energy of the EGG signal. Some fluctuations in the EGG data may be removed by simple differentiation, a procedure which also yields a waveform with enhanced positive and negative peaks. These peaks occur approximately at the instants of glottal opening and closure, respectively [1], [8], [18]. The differentiation is implemented as a backward difference equation. The differentiated EGG is then normalized by dividing by its maximum positive value. The resulting waveform is denoted as the DEGG and is shown at the bottom of Fig. 2.

The two-way, V-M and U-S classification uses the LCR and energy information from the DEGG as follows. The DEGG is segmented into 10 ms frames of 100 samples each (10 kHz sampling rate). The energy of each frame n of the DEGG signal is given by

$$E_D(n) = \sum_{i=1}^{100} \left(D((n-1)100 + i) \right)^2 \qquad (1)$$

where $D(j)$ denotes the sample value of the DEGG signal. The LCR of the DEGG is calculated at the -0.5 level, also on a frame basis. The energy and LCR contours are smoothed with a three-point zero-phase filter with coefficients (0.19, 0.62, 0.19), such that the filter output is $y_n = 0.19x_{n-1} + 0.62x_n + 0.19x_{n+1}$. The energy and LCR for the speech signal are calculated in a similar manner and the contours are also smoothed.

Other calculations include the following:

1) average energy levels for both the speech and DEGG signals for voiced segments,

2) average of the rectified voiced speech,

3) average LCR for voiced segments that exceed the 10 percent level of signal calculated in step 2) above.

4) smooth the preliminary four-way classification to remove obvious errors in a string. For example, if the preliminary classification is . . . VVVUVV . . . , then "smooth" this string to give . . . VVVVVV . . . .

Various threshold values are determined empirically, but are fixed once selected. The thresholds we selected are shown in Fig. 1 enclosed in the parentheses. Samples of silent intervals are required to establish several threshold values. The other numbers in Fig. 1 were determined experimentally to establish the decision levels for steps (b), (d), and (f).

Results and Discussion

The algorithm has been tested with data from six speakers (three male and three female), each speaking five sentences. The sentences were as follows.

1) We were away a year ago. (Voiced.)

2) Early one morning a man and a woman ambled along a one mile lane. (Voiced and nasals.)

3) Should we chase those cowboys? (Fricatives and plosives.)

4) That zany van is azure. (Voiced fricatives, i.e., mixed.)

5) We saw the ten pink fish. (Unvoiced plosives and fricatives.)

This data set is more extensive than that used in [2], [3], [7], [13].

The threshold values were determined with data from two speakers (one male and one female), each speaking the first three sentences only. An example comparing the algorithm classification to manual classification is shown in Fig. 2. Note that the classification task is aided by the EGG and the DEGG.

The detailed classification results appear in Table I. The designation for the test data sets is as follows.

Complete: Refers to all the data from all six speakers.

Threshold: The subset of Complete used to establish the threshold values. one male and one female speaker, for the first three sentences only.

Nonthreshold: The subset of Complete not included in the Threshold set.

Male: The male speaker subset of Complete.

Female: The female speaker subset of Complete.

The overall correct recognition rate is 98.2 percent when compared to manual classification of the data. The recognition rate is an improvement over the overall 95 percent rate reported in [3], [7] and 88 percent reported in [2], [13]. Nearly 83 percent correct classification of the mixed excitation frames was achieved in [7], which we have increased to 89 percent.

Table II provides a breakdown of the types of errors. The most troublesome classifications for the algorithm were unvoiced and

TABLE I
Classification Results

TEST DATA SETS	TOTAL NUMBER OF FRAMES	NUMBER OF FRAMES IN ERROR	ERROR RATE (%)	CORRECT RATE (%)
COMPLETE	7785	146	1.88	98.12
THRESHOLD	1622	20	1.23	98.77
NON-THRESHOLD	6163	126	2.04	97.96
MALE	3887	47	1.21	98.79
FEMALE	3898	99	2.54	97.46

TABLE II
Error Analyses in Number of Frames

MANUAL CLASSIFICATION	CLASSIFICATION OUTPUT V	U	M	S	CORRECT RATE (%)
V	5298	24	39	6	98.71
U	20	710	5	6	95.82
M	5	5	81	0	89.01
S	27	9	0	1550	97.73

1772   IEEE TRANSACTIONS ON ACOUSTICS, SPEECH, AND SIGNAL PROCESSING, VOL. 37, NO. 11, NOVEMBER 1989

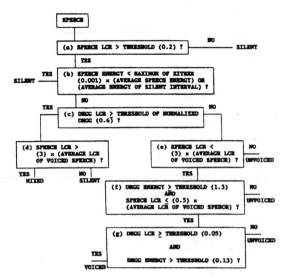

Fig. 1. Speech-EGG algorithm for V/U/M/S (four-way) classification of speech.

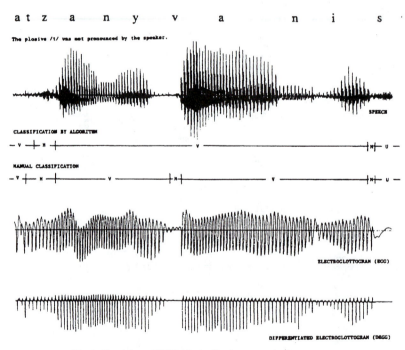

Fig. 2. Comparison of V/U/M/S classification by the algorithm and manual procedures. The speech is a segment of the sentence "That zany van is azure" spoken by a male subject.

IEEE TRANSACTIONS ON ACOUSTICS, SPEECH, AND SIGNAL PROCESSING, VOL. 37, NO. 11, NOVEMBER 1989   1773

mixed excitation frames. A large number of errors (38.7 percent) occurred at the beginning and ending of the sentences. If these errors are ignored, then the overall performance of the algorithm becomes 99.23 percent. If we further ignore the errors that occurred at the boundaries between words, then the overall performance increases to 99.5 percent. The major cause of errors at word boundaries (approximately 60 percent), including the beginning and ending of the sentences, was due to U to V and S to V misclassifications. These errors were caused by a failure to properly recognize voice-onset and voice-offset intervals. An algorithm for recognizing voice onset and offset using the EGG is described in [11] and is an extension of the one in [3], [4]. Note that there is a slight tendency for the algorithm to perform better using male speech than female speech.

CONCLUSIONS

We advocate the laboratory use of this algorithm to benchmark speech system performance. The benchmarking can be done automatically and the results compared to acoustic-signal-only-based algorithms. A useful improvement to the algorithm would be a diagnostic capability. For example, perhaps the algorithm could identify and label the frames that were particularly difficult to classify. Such information could conceivably be used to improve system designs. We believe a spectral distance metric can be used to improve the V/U (U/V) and V/S (S/V) classifications.

REFERENCES

[1] A. K. Krishnamurthy and D. G. Childers, "Two-channel speech analysis," *IEEE Trans. Acoust., Speech, Signal Processing*, vol. ASSP-34, pp. 730–743, Aug. 1986.

[2] B. S. Atal and L. R. Rabiner, "A pattern recognition approach to voiced-unvoiced-silence classification with applications to speech recognition," *IEEE Trans. Acoust., Speech, Signal Processing*, vol. ASSP-24, pp. 201–212, June 1976.

[3] L. R. Rabiner and M. Sambur, "Application of an LPC distance measure to the voiced-unvoiced-silence detection problems," *IEEE Trans. Acoust., Speech, Signal Processing*, vol. ASSP-25, pp. 338–343, Aug. 1977.

[4] L. R. Rabiner and R. W. Schafer, *Digital Processing of Speech Signals*. Englewood Cliffs, NJ: Prentice-Hall, 1978.

[5] J. D. Markel and A. H. Gray, Jr., *Linear Prediction of Speech*. New York: Springer-Verlag, 1976.

[6] R. W. Schafer and J. D. Markel, Eds., *Speech Analysis*. New York: IEEE Press, 1979.

[7] L. J. Siegel and A. C. Bessey, "Voiced/unvoiced/mixed excitation classification of speech," *IEEE Trans. Acoust., Speech, Signal Processing*, vol. ASSP-30, pp. 451–460, June 1982.

[8] D. G. Childers and J. N. Larar, "Electroglottography for laryngeal function assessment and speech analysis," *IEEE Trans. Biomed. Eng.*, vol. BME-31, pp. 807–817, Dec. 1984.

[9] J. N. Larar, "Towards speaker independent isolated word recognition for large lexicons: A two channel, two-pass approach," Ph.D. dissertation, Univ. Florida, Gainesville, 1985.

[10] ——, "Lexical access using broad acoustic-phonetic classifications," *Comput. Speech Language*, vol. 1, pp. 47–59, 1986.

[11] N. B. Pinto, D. G. Childers, and A. L. Lalwani, "Formant speech synthesis: Improving production quality," *IEEE Trans. Acoust., Speech, Signal Processing*, to appear, Dec. 1989.

[12] F. Daaboul and J. P. Adoul, "Parametric segmentation of speech into voiced-unvoiced-silence intervals," in *Proc. IEEE Conf. Acoust., Speech, Signal Processing*, Hartford, CT, May 1977, pp. 327–331.

[13] L. R. Rabiner, C. E. Schmidt, and B. S. Atal, "Evaluation of a statistical approach to voiced-unvoiced-silence analysis for telephone quality speech," *Bell Syst. Tech. J.*, vol. 56, pp. 455–482, Mar. 1977.

[14] D. W. Shipman and V. W. Zue, "Properties of large lexicons: Implications for advanced word recognition systems," in *Proc. IEEE Conf. Acoust., Speech, Signal Processing*, Paris, France, May 1982, pp. 546–549.

[15] W. Hess, *Pitch Determination of Speech Signals*. New York: Springer-Verlag, 1983.

[16] W. Hess and H. Indefrey, "Accurate pitch determination of speech signals by means of a laryngograph," in *Proc. Int. Conf. Acoust., Speech, Signal Processing*, vol. 1, 1984, pp. 1813.1.1–1813.1.4,

[17] ——, "Accurate time domain pitch determination of speech signals by means of a laryngograph," *Speech Commun.*, vol. 6, pp. 55–58, Mar. 1987.

[18] D. G. Childers and A. K. Krishnamurthy, "A critical review of electroglottography," *CRC Crit. Rev. Bioeng.*, vol. 12, no. 2, pp. 131–164, 1985.

[19] D. G. Childers, D. M. Hicks, G. P. Moore, and Y. A. Alsaka, "A model for vocal fold vibratory motion, contact area, and the electroglottogram," *J. Acoust. Soc. Amer.*, vol. 80, pp. 1309–1320, Nov. 1986.

7

GLOTTAL EXCITATION MODELS

A7.1 INTRODUCTION

Chapters 3 through 7 often refer to the need for models of the excitation waveform, especially for speech synthesis. In this appendix, we introduce two glottal models that are used in the software that accompanies this book: the Liljencrants-Fant (LF) model (Fant et al., 1985) and the polynomial model (Childers and Hu, 1994; Milenkovic, 1993). Also covered is a noise model for generating aspiration and fricative noise for unvoiced sounds. This noise model is also used for enhancing the naturalness of voiced sounds for certain voice types. This appendix presents some typical values for these models that have been found useful for generating several voice types, such as modal, breathy, vocal fry, whisper, harsh, and falsetto.

A typical glottal excitation volume–velocity waveform (glottal flow) obtained by inverse filtering is shown in Figure A7.1 along with its spectrum. As discussed in Chapters 3 through 6, some time domain acoustic features of the glottal source are the pitch period, pitch perturbation (jitter), amplitude perturbation (shimmer), the glottal flow pulse width, the glottal flow skewness, the abruptness of closure of the glottal flow, and aspiration noise. While some important frequency domain features are the spectral tilt (slope), harmonic richness factor, and harmonic-to-noise ratio. The models in this appendix generally can account for these features via the manipulation of the model parameters.

The relationship between the volume–velocity (glottal flow), the differentiated glottal flow, and the speech waveforms is illustrated in Figure A7.2. Note that the main excitation occurs at the initiation of the glottal closed phase. The speech waveform decays beyond this point until the next excitation occurs.

A7.2 THE LF MODEL

Instead of modeling the glottal flow, Liljencrants and Fant (Fant et al., 1985) chose to model the differentiated glottal flow. This model has become known as the LF model and is illustrated in Figure A7.3. The glottal waveform parameters are given as

$$
\begin{aligned}
g(t) &= E_0 e^{\alpha t} \sin(\omega_g t) \quad \text{for } 0 \leq t \leq t_e \\
&= -\frac{E_e}{\varepsilon t_a}\left[e^{-\varepsilon(t-t_e)} - e^{\varepsilon(t_c-t_e)}\right] \quad \text{for } t_e \leq t \leq t_c \leq T_0
\end{aligned}
\tag{A7.2.1}
$$

where T_0 is the pitch period, which is typically larger than t_c. The following conditions hold.

$$
\int_0^T g(t)\,dt = 0, \quad \omega_g = \frac{\pi}{t_p}, \quad \varepsilon t_a = 1 - e^{-\varepsilon(t_c-t_e)}, \quad \text{and} \quad E_0 = -\frac{E_e}{e^{\alpha t_e}\sin(\omega_g t_e)} \tag{A7.2.2}
$$

The modeled waveforms can be specified by either the direct synthesis parameters (E_0, α, ω_g, and ε) or the timing parameters (t_p, t_e, t_a, and t_c). The parameter t_p denotes the instant of the maximum glottal flow model waveform. The parameter t_e is the instant of the maximum negative differentiated glottal flow model. The parameter t_a is the time constant of the exponential curve of the second segment of the LF model. The parameter t_c is the instant at which complete glottal closure is reached for the model. One reason this model is important and useful is that it is related to actual glottal waveforms via the

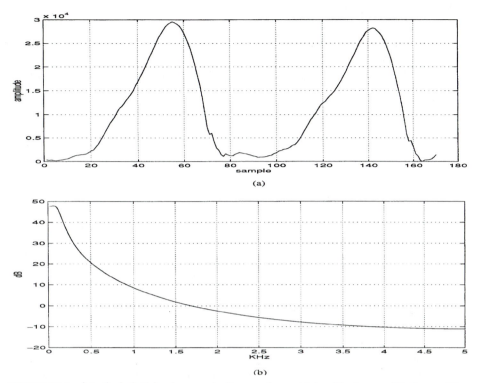

(a)

(b)

FIGURE A7.1 A typical glottal volume–velocity waveform obtained by inverse filtering and its spectrum.

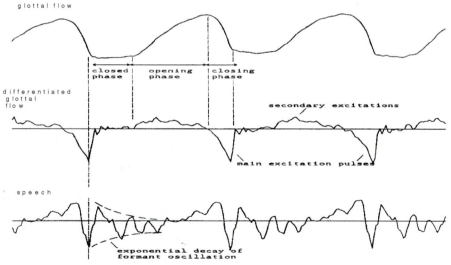

FIGURE A7.2 Illustration of the relationship between the glottal flow, differentiated glottal flow, and speech waveforms.

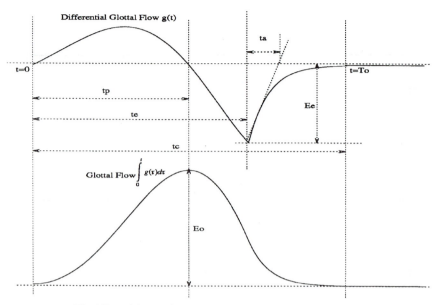

FIGURE A7.3 The LF model waveforms.

timing parameters, which have been shown to be important in psychoacoustic studies (Childers and Ahn, 1995). Lalwani and Childers (1991a,b) added modulated aspiration noise, jitter, and shimmer to the LF model to form a more complete glottal model. The LF model parameters are such that $0 \leq t_p \leq t_e \leq t_c$, and $t_a \leq 0$.

The first segment of the LF model characterizes the differentiated glottal flow over the interval from the glottal opening to the maximum negative excursion of the waveform. The second segment represents a residual glottal flow that comes after the maximum negative excursion. It can be shown that the spectrum of the first segment is dominated by the exponential component, $e^{\alpha t}$, of which the "negative bandwidth" equals $\frac{\alpha}{\pi}$. Likewise, the frequency response of the second segment can be approximated by a first-order low-pass filter with a cutoff frequency $F_a = 1/(2\pi t_a)$ (Fant and Lin, 1988). As a result, the bandwidths of the first and second segments are, respectively

$$B_1 \cong \frac{\alpha}{\pi} \quad \text{and} \quad B_2 \cong \frac{1}{2\pi t_a}$$

Thus, the combination of the two segments of the differentiated glottal flow LF model can be approximated with a two pole filter. The center frequency of the poles and ω_g are nearly the same. Thus, the bandwidths control the filter. The bandwidth of inverse filtered data is in general close to B_1, making it difficult to estimate the waveshape of the first segment. However, B_2 is much greater than the bandwidth of the inverse filtered data. Thus, the second segment retains its waveshape after inverse filtering. However, the phase may be different. A typical example is given in Figure A7.4, which shows the spectra of the first and second segments of the LF model as well as the spectrum of the two pole model. The residue derived from the LF model has a flat spectral envelope and the waveform exhibits a sharp pulse at the conjunction between the two segments, where glottal closure occurs in the LF model. A knowledge of the relationship between the LF model and the residue serves as an aid for retrieving one from the other. Although the residue does not appear to be informative, its integral tends to favor the shape of the LF model waveform, as seen in Figure A7.5, where a known excitation waveform is used to synthesize the vowel /IY/. The synthesized speech is inverse filtered to obtain the residue, which in turn is integrated. This latter waveform is quite similar to the original excitation, as it should be.

Three differentiated glottal waveforms obtained by inverse filtering are compared in Figure A7.6 with their respective LF model waveforms for three voice types: modal, vocal fry, and breathy. The

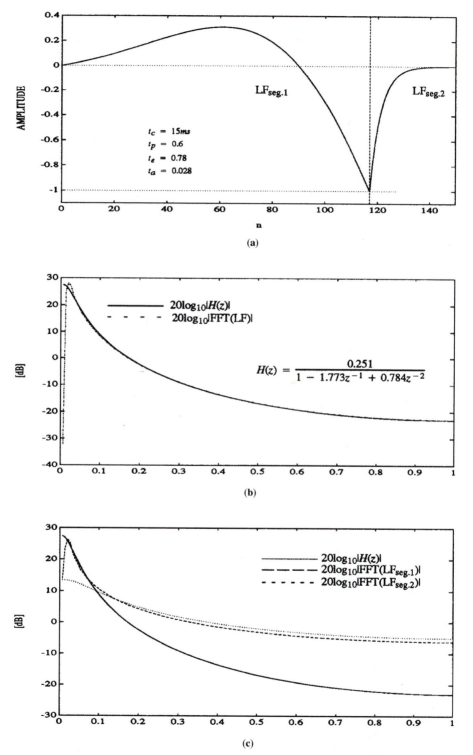

FIGURE A7.4 (a) The LF model waveform. (b) The FFT spectra of the LF model waveform and two pole model, H(z). (c) The FFT spectra of the segments of the LF model.

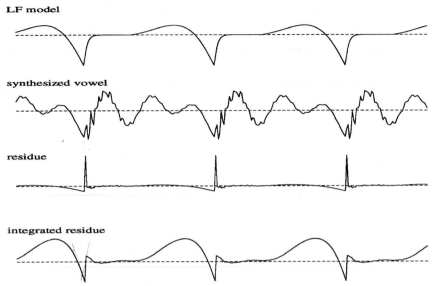

FIGURE A7.5 Illustration of the similarity between the LF model waveform and the integrated residue.

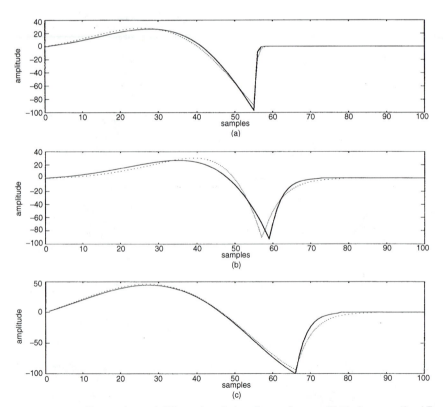

FIGURE A7.6 Comparison of differentiated glottal waveforms with their respective LF model waveforms for three voice types. The solid line represents the data and the dotted line is the model.

LF model parameters are: (*a*) $t_p = 41.3$, $t_e = 55.4$, $t_a = 0.4$, $t_c = 58.2$ (modal); (*b*) $t_p = 48.1$, $t_e = 59.6$, $t_a = 2.7$, $t_c = 72.0$ (vocal fry); and (*c*) $t_p = 46.2$, $t_e = 66.0$, $t_a = 2.7$, $t_c = 77.1$ (breathy).

From experiments by Childers and Lee (1991), Childers and Hu (1994), and Childers and Ahn (1995), some general characteristics of the time domain and frequency domain properties of the glottal waveform have been determined. Before these characteristics are tabulated, we define some terms that often occur in the literature. The **open quotient** of the glottal vibratory cycle is

$$OQ = \frac{\text{open glottal interval}}{\text{pitch period}} \tag{A7.2.3}$$

The open quotient is primarily determined by the **glottal pulse width**. The OQ can be expressed as follows using the LF model parameters.

$$OQ_{LF} = \frac{(t_e + kt_a)}{T_0} \tag{A7.2.4}$$

where the value of k is a function of the parameter t_a. It has been found that k has values in the range 2.0 to 3.0 when $0\% < t_a < 10\%$, where t_a is represented by a percentage of the pitch period. Note that $k = 0$ when $t_a = 0$. In using actual data to make calculations, a definition of various events is needed. For example, for the LF model calculations, the instant of glottal closure is defined as the instant at which the glottal flow waveform falls below 1% of the peak value of the waveform.

The **speed quotient** is

$$SQ = \frac{\text{opening interval}}{\text{closing interval}} = SQ_{LF} = \frac{t_p}{t_e + kt_a - t_p} \tag{A7.2.5}$$

The SQ is a measure of **glottal pulse skewness**. Another definition often used for the speed quotient is

$$SQ = SQ_{LF} = \frac{t_p}{t_c - t_p} \tag{A7.2.6}$$

However, this definition is such that often $t_e + kt_a$ is nearly equal to t_c, so the two definitions reduce to the same.

The **abruptness of closure** of the glottal pulse is measured by the value of t_a. If t_a is small, then the abruptness of closure is fast.

The **aspiration noise** is determined by the **signal-to-noise ratio** (SNR).

The **spectral tilt** is defined as the slope of the spectrum in dB of the glottal waveform. This is typically −12 dB/octave, but can vary from −6 dB to −18 dB. Low spectral tilt is defined as −6 dB, medium as −12 dB, and high as −18 dB.

The **normalized noise energy** (NNE) is 10 log [NE/SE], where SE is the total signal (speech plus noise) energy. The NNE is often measured by calculating the energy in the noise spectrum and the energy in the total signal spectrum. The **harmonics-to-noise ratio** (HNR) is 10 log [HE/NE], where NE is the noise energy and HE is the harmonic energy. The HE is calculated by subtracting NE from SE. The **harmonics richness factor** (HRF) is 10 log of the ratio of the energy of the sum of all the harmonics in the speech signal (less the fundamental) to the energy of the fundamental frequency. The NNE, HNR, and HRF are all measured in dB.

Table A7.1 summarizes the features of the time domain glottal factors, while Table A7.2 summarizes the frequency domain factors for several voice types. Some typical numerical values

TABLE A7.1. Time Domain Glottal Factors

Voice	Pitch period	Pulse width	Pulse skewness	Abruptness of closure	Aspiration noise	Jitter	Shimmer
Modal	Medium	Medium	Medium	Medium	High	Low	Low
Vocal fry	Long	Short	High	Fast	Medium	High	High
Breathy	Medium	Long	Low	Slow	Low	Low	Medium
Rough	Medium	Medium	Medium	Medium	Medium	High	High
Hoarse	Medium	Medium	Medium	Medium	Low	High	High

TABLE A7.2.   Frequency Domain Glottal Factors

Voice	Fundamental frequency	Spectral tilt	Harmonic richness factor	Harmonic to noise ratio
Modal	Medium	Medium	Medium	High
Vocal fry	Low	Low	High	Medium
Breathy	Medium	High	Low	Medium
Rough	Medium	Medium	Medium	Low
Hoarse	Medium	Medium	Medium	Low

for some the LF and noise model parameters are summarized later after we introduce the noise model.

Another form of the LF model is called the transformed LF model (Fant, 1995), which is a simplified version of the LF model described previously. However, the features of the model are the same.

An alternative to the LF model was recently introduced by Velhuis (1998), which is computationally more efficient than the LF model.

In summary, in some research applications it is desired to fit a LF model to actual data. In this case, the LF model parameters are determined by a least square error fit of the data to the model. For the speech synthesizers in Chapters 6 and 7, the user is provided with a means to adjust the model parameters to obtain a desired LF excitation model waveform.

A7.3 NOISE MODEL

Both the LF model and the polynomial glottal waveform models tend to account for the low-frequency component of the glottal waveform. By subtracting the low-frequency component (as a model) from the differentiated glottal waveform (actual data), a noise-like waveform is obtained. This noise-like signal is attributed to turbulent noise and is important for the naturalness of both real and synthetic speech (Holmes, 1976; Kasuya et al., 1986; Klatt, 1980; Lalwani and Childers, 1991a,b). The parameters that are important for this noise are intensity, spectral shape, and timing (Childers and Lee, 1991; Klatt and Klatt, 1990; Lalwani and Childers, 1991a,b). The turbulent noise model used here is shown in Figure A7.7.

The noise model uses Gaussian white noise along with six parameters, defined as follows.

- snr: The power ratio of the low-frequency component to the aspiration noise.
- amp1: The amplitude modulation index 1 such that $0 \le amp1 \le 1.0$.

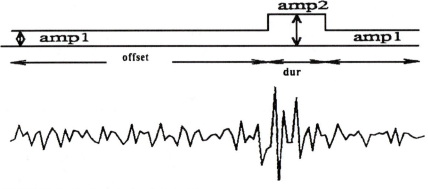

FIGURE A7.7   The turbulent noise model.

TABLE A7.3. Typical LF and Noise Model Parameter Values for Several Voice Types

Voice	t_p (%)	t_e (%)	t_a (%)	t_c (%)	Jitter (%)	snr (dB)	amp1 (%)	amp2 (%)	offset (%)	dur (%)
Modal	41.0	55.0	0.4	58.0	2.0	40.0	0.0	100.0	50.0	50.0
Vocal fry	48.0	59.0	2.7	72.0	10.0	20.0	0.0	100.0	20.0	20.0
Breathy	46.0	66.0	2.7	77.0	5.0	20.0	100.0	100.0	50.0	50.0
Whisper	50.0	80.0	8.0	100.0	2.0	−20.0	100.0	100.0	50.0	50.0
Falsetto	50.0	80.0	8.0	100.0	2.0	50.0	0.0	0.0	50.0	50.0
Harsh	25.0	30.0	1.0	50.0	10.0	10.0	100.0	100.0	50.0	50.0

- amp2: The amplitude modulation index 2 such that $0 \leq amp2 \leq 1.0$.
- offset: The duration of the noise with amp1. This starts from the instant of the opening of the glottis.
- dur: The duration of amp2. The dur starts from the end of offset. The sum of the offset and dur must be less than or equal to the pitch period.

The term offset is used for the duration of amp1 for historical reasons in the author's research and has no other special significance.

The values used for the noise parameters are speaker dependent and speech dependent. Generally, for modal voices the snr is small, being about 0.25%. Some examples for these parameters are summarized in Table A7.3. This table is a summary of values obtained from the analysis of data (Childers and Ahn, 1995) as well as used in speech synthesis experiments in voice conversion (Childers et al., 1989; Childers and Hu, 1994; Childers and Lee, 1991; Childers and Wu, 1990; Shue, 1995).

The values for modal, vocal fry, and breathy voice were determined experimentally by analyzing data (Childers and Ahn, 1995), while the values for the three other voice types were determined via listening tests with synthesized speech. The 100% value for amp1 and amp2 means that the sliders on the noise settings in the source specification option are to be set to the maximum value of 1.0, or 100%. Jitter is a parameter that is not available in the formant synthesizer in Chapter 6.

A7.4 SOME EXAMPLES

Figure A7.8 shows some example excitation LF model waveforms with noise added for six voice types. These waveforms were created using the formant synthesizer. Figure A7.9 shows the synthesized vowel /IY/ waveforms using the excitation waveforms in Figure A7.8 as the input to the formant synthesizer. The model parameters are given in Table A7.4, which is similar to Table A7.3.

TABLE A7.4. LF and Noise Model Parameters for the Synthesized Data Shown in Figures A7.8 and A7.9

Voice	t_p (%)	t_e (%)	t_a (%)	t_c (%)	Jitter (%)	snr (dB)	amp1 (%)	amp2 (%)	offset (%)	dur (%)
Modal	45.0	60.0	0.5	65.0	2.0	40.0	0.0	100.0	50.0	50.0
Vocal fry	20.0	25.0	0.2	35.0	10.0	20.0	0.0	100.0	20.0	20.0
Breathy	50.0	80.0	8.0	100.0	5.0	20.0	100.0	100.0	50.0	50.0
Whisper	50.0	80.0	8.0	100.0	2.0	−20.0	100.0	100.0	50.0	50.0
Falsetto	50.0	80.0	8.0	100.0	2.0	50.0	0.0	0.0	50.0	50.0
Harsh	25.0	30.0	1.0	50.0	10.0	10.0	100.0	100.0	50.0	50.0

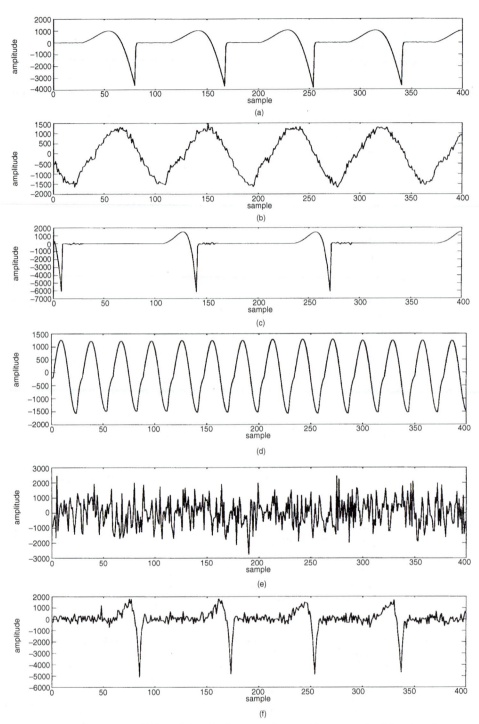

FIGURE A7.8 Examples of LF model excitation waveforms with noise added for six voice types: (*a*) modal, (*b*) breathy, (*c*) vocal fry, (*d*) falsetto, (*e*) whisper, and (*f*) harsh.

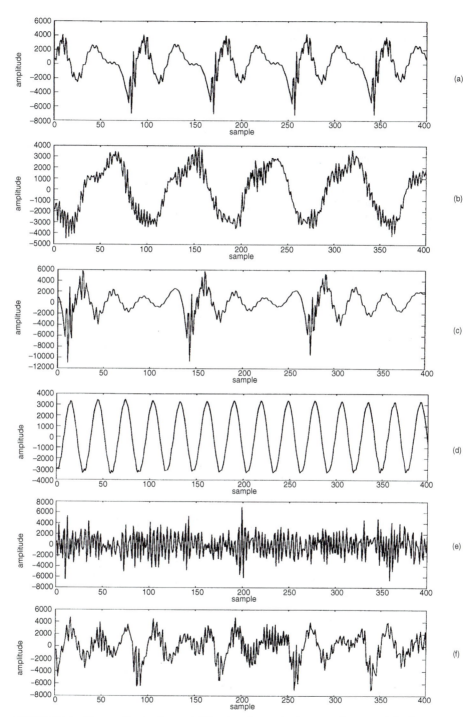

FIGURE A7.9 The synthesized vowel /IY/ using the excitation waveforms and the formant synthesizer: (*a*) modal, (*b*) breathy, (*c*) vocal fry, (*d*) falsetto, (*e*) whisper, and (*f*) harsh.

A7.5 THE POLYNOMIAL MODEL

The polynomial model described here appears in Childers and Hu (1994) and is similar to the Milenkovic (1993) polynomial model. The model is

$$p(t) = c_0 + c_1\tau + c_2\tau^2 + c_3\tau^3 + c_4\tau^4 + c_5\tau^5 + c_6\tau^6 \qquad (A7.5.1.)$$

where $\tau = \frac{t}{T}$, t is the independent variable and T is the period of the pulse waveform (or the pitch period). If the model is being used to fit data, then the coefficients are determined by fitting the polynomial function, p(t), to the estimated differentiated glottal waveform in a least squares sense, which is similar to that described in Milenkovic (1993). However, for speech synthesis using the software in Chapters 6 and 7, the user can adjust the coefficients to obtain a desired excitation waveform. An example of a waveform obtained by this model is shown in Figure A7.10 for both the glottal flow (lower) and its derivative (upper). While the coefficients of the polynomial have no obvious physical or physiologic interpretation, they have been shown to be suitable for the synthesis of high-quality speech (Milenkovic, 1993).

An overview of a system that uses the features of the LF, polynomial, and noise models for a glottal excited LP speech synthesizer is given in Figure A7.11 (Childers and Hu, 1994). In such a system, the excitation and noise models are designed as codebooks. The features of this system include an all pole synthesis filter, voiced and unvoiced excitations and gains, a glottal pulse codebook for voiced sounds, a stochastic codebook for unvoiced sounds, and an analysis to determine the glottal closure instants (GCIs).

For additional details on the formant and LP synthesizers consult the following PhD dissertations: Hu (1993), Shue (1995), and Hsiao (1996).

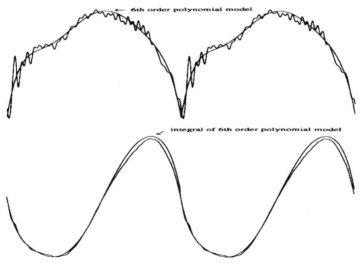

FIGURE A7.10 An example of the polynomial model waveforms: (upper) derivative of the glottal flow, (lower) the glottal flow.

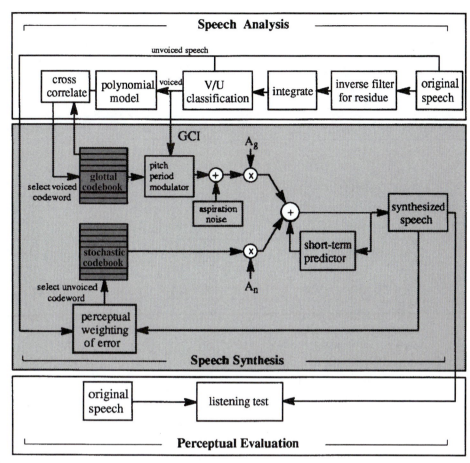

FIGURE A7.11   A glottal excited, linear prediction (GELP) speech synthesis system.

VOICE MODIFICATION AND SYNTHESIS

A8.1 INTRODUCTION

A speech synthesis procedure is outlined in Chapter 6 and discussed in Appendix 7. Chapter 7 describes the software that implements this system. This appendix provides some additional background material on the algorithms used in the software for the voice conversion and synthesis system.

A8.1.1 Pitch Detection and Glottal Closure Instants

A simple classification algorithm for voiced/unvoiced decision is the following. The energy of the prediction error and the first reflection coefficient are used to classify a segment as voiced. The first reflection coefficient is

$$r_1 = \frac{R_{SS}(1)}{R_{SS}(0)} \qquad (A8.1.1.1)$$

where

$$R_{SS}(0) = \frac{1}{N} \sum_{n=1}^{N} s(n)s(n)$$

$$R_{SS}(1) = \frac{1}{N} \sum_{n=1}^{N-1} s(n)s(n+1)$$

where N is the number of samples in the analysis frame and $s(n)$ is the speech sample.

The decision rules are as follows.

- If the first reflection coefficient is greater than 0.2 and the prediction error energy is greater than twice the threshold, e.g., 10^7, then the current frame is classified as voiced.
- If the first reflection coefficient is greater than 0.3 and the prediction error energy is greater than the threshold used in rule 1 and the previous frame is also voiced, then the current frame is classified as voiced.
- If the above conditions are not valid, then the current frame is classified as unvoiced.

The above algorithm generates a sequence of 1s and 0s. Patterns of 101 and 010 seldom occur in real speech and are corrected to strings of 111 and 000, repectively, to reduce the classification error rate.

The two-pass glottal closure instant detection algorithm in Childers and Hu (1994) is outlined below. This method uses the results from both a voiced/unvoiced classification procedure and the prediction error waveform, $e(n)$, to detect the pitch period and the glottal closure instants (GCIs). The detection algorithm consists of two stages: pitch period estimation and peak picking.

Pitch Period Estimation

- Low-pass filter one segment of the predicition error waveform, $e(n)$. The filtered waveform is denoted as $e_{Lp}(n)$.

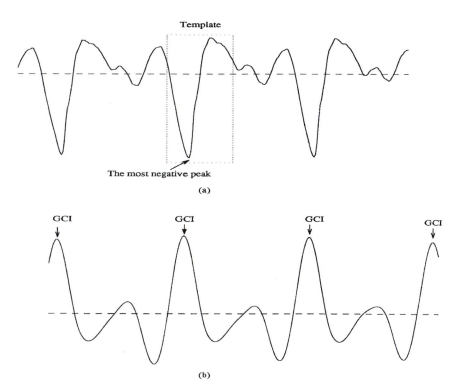

FIGURE A8.1 Illustration of the pitch period and GCI detection algorithm. (*a*) The filtered prediction error sequence, $e_{Lp}(n)$. (*b*) The correlation output sequence, $C_{te}(n)$.

- Calculate the cepstrum-like sequence, $C_e(n)$.

$$C_e(n) = \text{IFFT}(|\text{FFT } e_{LP}(n)|) \quad 1 \leq n \leq N \qquad (A8.1.1.2)$$

where N is the frame size, FFT is the fast Fourier transform, and IFFT is the inverse FFT.

- Search for the index m, where $C_e(m)$ is the maximum amplitude in the subset $\{C_e(i) \mid 25 \leq i \leq N\}$.
- Search for the index k, where $C_e(k)$ is the maximum amplitude in the subset $\{C_e(i) \mid 25 \leq i \leq m - 25\}$.
- If $C_e(k) > 0.7C_e(m)$, k is the estimated pitch period, otherwise m is the estimated pitch period.
- If an abrupt change in the pitch period is observed, compared to previous pitch periods, then low-pass filter (or median filter) to smooth the abrupt change.

Peak Picking

- In each analysis frame (256 samples), search for the most negative peak of the $e_{Lp}(n)$ waveform.
- Build a template as illustrated in Figure A8.1(*a*). This template is formed by the waveform around the most negative peak of $e_{Lp}(n)$. The length of the template is 46 samples, including 15 samples before the peak, 30 samples after the peak, and the peak itself.
- Correlate the template with the $e_{Lp}(n)$ waveform to generate a new sequence, $C_{te}(n)$, as shown in Figure A8.1(*b*).
- The positive peaks of the $C_{te}(n)$ sequence provide initial estimates for the GCIs. The estimated pitch period from the pitch period estimation (stage 1) can assist in correcting the erroneous peak detection.

- Adjust the position of each GCI under the criterion that no two GCIs are located within 25 samples of on another.

A8.2 PITCH CONTOUR MODIFICATION

The glottal closure instants (GCIs) are measured in a manner as described previously. Then the GCIs are sorted into a vector, which is a GCI sequence. The distance between the GCIs is the length of the glottal pulse or the pitch period. For pitch synchronous analysis and synthesis, these instants determine the timing for the generation of the glottal pulses as well as the times at which the vocal tract parameters are to be updated. The method for modifying the pitch contour is one that alters or modifies the GCI vector. This modification is one factor, if not the major factor, for creating or mimicking a voice type. For example, changing the length of the GCI sequence alters the fundamental frequency of voicing of the synthesized speech. Furthermore, if we alter segments of the GCI sequence, then we can alter the intonation pattern. For example, by sequentially decreasing the distance between successive GCIs in the sequence, one can synthesize speech that has a rising intonation. This is the method adopted here; that is, the pitch contour is altered by altering the GCI sequence.

A8.2.1 Pitch Contour Model

A plot of the GCI sequence is the pitch period contour. This plot is constructed with the horizontal axis being the GCI points, while the vertical axis is the pitch period; that is, the interval between successive GCIs. An example is shown in Figure A8.2. The top two panels show a segment of the sentence, "We were away a year ago," and the corresponding pitch contour for that segment. The third panel from the top shows the waveform for the sentence, while the fourth panel shows the pitch contour for the entire sentence. One can model the pitch contour with three waveforms. One waveform models the average value of the pitch contour, which is a constant over the sentence and is the fundamental pitch period. Another model waveform is the steady state or long time variation of the pitch contour, which is called the pitch wave. The pitch wave is related to the intonation. A third model waveform is the short time perturbation of the pitch contour, which is the jitter. The three waveform models are illustrated in Figure A8.3. One advantage of this model is that each waveform is related to certain perceptual features and each waveform can be independently controlled or modified. For example, the fundamental pitch period can be altered without affecting the pitch wave or the jitter.

The three pitch waveform models can be measured as follows. The fundamental pitch period is the mean value of the pitch contour. The pitch wave is estimated by a 5th-order median filter after subtracting the pitch contour from the fundamental pitch period. The jitter is the standard deviation of the difference between the pitch contour and the sum of the pitch wave and the fundamental pitch period.

Several approaches have been taken to model the pitch contour (Fujisaki, 1983; Maeda, 1974; Mattingly, 1966). One method models the contour as consisting of three parts or patterns: a falling waveform, a rising waveform, and a baseline trend. Each of these patterns can be fit with a second-order polynomial using a least squares approach to find the polynomial coefficients. Another approach is to fit a third-order spline function to the pitch contour. Both approaches are available in the software.

In summary, the pitch contour model is estimated as follows.

- Transform the GCI sequence to the pitch contour. Calculate the mean value, which is the fundamental pitch period.
- Smooth the pitch contour with a 5th-order median filter and subtract the result from the fundamental pitch period. This is the pitch wave.
- Subtract the original pitch contour from the sum of the pitch wave and the fundamental pitch period. Calculate the standard deviation of the resultant sequence. This is the jitter.
- Segment the pitch wave into three pitch patterns: rising, falling, and baseline trend. Approximate each pattern with a second-order polynomial or a third-order spline function.

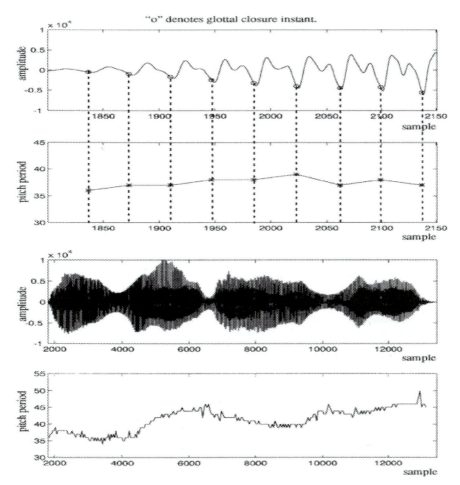

FIGURE A8.2 Speech signal and the corresponding pitch contour.

A8.2.2 Pitch Contour Modification

The pitch contour model is modified as follows.

- Fundamental pitch period. Scale this value upward or downward.
- Pitch wave. Segment the pitch wave manually into several pitch patterns, with each pattern modeled by a polynomial or a spline. Modify each pattern by altering the coefficients of the model.
- Jitter. Scale this factor upward or downward.

A8.2.3 Synthesis

The procedure for synthesis is as follows.

- Form the fundamental pitch period as a contour for a selected (voiced) segment or for the entire sentence.
- Add the jitter to the contour.
- Add the pitch wave to the contour. This forms the modeled pitch contour.

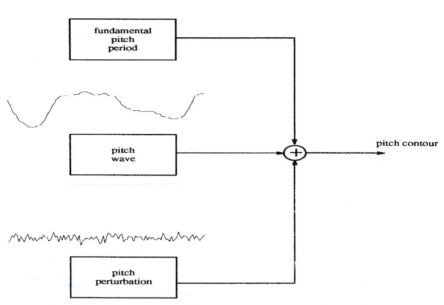

FIGURE A8.3 The pitch contour model.

- Construct a GCI sequence from the pitch contour model. Initialize the first GCI with its pitch period. This is done by sampling the pitch contour model at the first point (or a desired first point). Call this value $GCI(1) = T_1$. The next GCI is located at $(1 + T_1)$ with a value $GCI(1 + T_1) = T_2$, which is sampled from the pitch contour model. The next GCI is located at $(1 + T_1 + T_2)$ with a value $GCI(1 + T_1 + T_2) = T_3$, and so on. Continue this process until the new GCI sequence is constructed for the selected voiced speech segment. This new GCI sequence (vector) is used to synthesize a new voice or to convert one voice to another.

These algorithms are implemented in the software provided.

A8.3 GAIN CONTOUR MODIFICATION

The gain parameter is the average value of the speech energy for each pitch period. Its function is to control the energy transitions through out an utterance. This parameter is related to the intensity (or loudness) of the speech signal. As the loudness of the speech increases, the gain increases. The software provides a mechanism for controlling the gain to create or alter various voice types.

A8.3.1 Analysis

The gain parameter is pitch synchronous and has a contour, which is defined as the value of the gain versus its GCI location along the time axis. The gain contour is calculated using pre-emphasized speech. The gain contour can be divided into two factors or models, analogous to the pitch contour model. One model is the gain envelope, which is the smoothed envelope of the voiced speech segments. The second model is the gain perturbation, or the pitch period-to-pitch period variability of the gain. This model is related to shimmer. These two models, in combination, form the gain contour, and are estimated as follows.

- Construct the gain contour using the gain parameter.

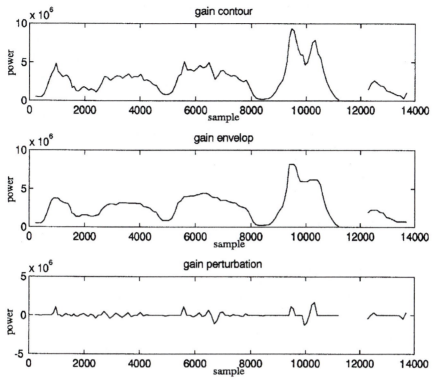

FIGURE A8.4 An illustration of the gain contour and its models (factors).

- Smooth the gain contour with a 5th-order median filter. The result is the gain envelope.
- Subtract the original gain contour from the gain envelope. The standard deviation of the resultant difference is the gain perturbation.

Figure A8.4 shows the gain contour for a sentence and its corresponding gain envelope and gain perturbation models.

A8.3.2 Modification of Gain Models

The gain models are modified independently as follows.

- The gain envelope. Segment the gain envelope manually into one to three gain patterns in an analogous manner as to that used for the pitch patterns. Model the patterns with a second-order polynomial or a third-order spline.
- The gain perturbation. Scale this model upward or downward.

A8.3.3 Synthesis

The gain contour is synthesized in a manner analogous to that used for the pitch contour. Interpolation of the gain contour may be required for the new GCI sequence. If the pitch contour is modified, then the gain contour must be modified to avoid discontinuities in the synthesized speech.

A8.4 VOCAL TRACT MODIFICATION

The formants and their bandwidths are found by computing the roots of the filter polynomial. This is done frame-by-frame to obtain the formant tracks. The formant tracks can be modified by a scale factor, or by using the mouse to draw a new track, or by loading a file that contains a desired track. The vocal tract filter is calculated from this information. However, this calculation can result in a filter design with improper pole positions. For example, the second formant may incorrectly merge with the first formant. This is called the pole interaction problem. This is solved using an algorithm presented in Hsiao and Childers (1996). This algorithm is included in the software.

A8.5 GLOTTAL PULSE MODIFICATION

There are two models for the glottal pulse: the polynomial model and the LF model. It is difficult to do glottal pulse waveform design using the coefficients of the polynomial model because there is little physical relationship between the volume–velocity waveform and the polynomial model. For voice conversion, the procedure recommended is to model the volume–velocity waveform of both the target and source speakers with separate polynomial models. Then, when converting the source speech to the target speech, use a linear mapping of the polynomial parameters of the source speaker to that of the target speaker using linear regression. In this way, the glottal parameters of the source speaker are mapped (modified) to match those of the target speaker. For the LF model, the source speaker model parameters can be modified either by using a linear regression approach like that described above or by using the graphic user interface provided in the software.

A8.6 VOICE CONVERSION

Voice conversion is the process of transforming the speech of one speaker to sound like that of another speaker. The objective is to develop methods for creating new synthetic voices, study factors responsible for synthetic voice quality, and to determine methods for speaker adaptation.

A8.6.1 Speaker Translation Models

The approach taken here is to model speaker characteristics with parametric models. To alter or convert the speech of one speaker to that of another, the parameters of one speaker (the source) are mapped to match the parameters of another speaker (the target). This is accomplished as follows.

A phoneme can be represented by an n-dimensional acoustic vector. The mean of this phoneme vector over various speakers is denoted as μ. The phoneme for speaker i is denoted as vector s^i and is given as

$$s^i = \mu + \delta \tag{A8.6.1.1}$$

where the vector δ is a "bias" considered to be a characteristic of the speaker, and thus the name for this model.

Assuming that the acoustic features are independent and time invariant, then one can convert the acoustic parameters of one speaker to those of another provided the offset value between the two speakers is known, that is

$$X = Y + B \tag{A8.6.1.2}$$

where X and Y are the n-dimensional acoustic parameter vectors for the target and source speakers, respectively. B is the offset vector and n is the number of measured acoustic features. This is called the translation model.

The task is to estimate B, which is the difference between the two speakers, including any channel effect such as a telephone line characteristic. Equation (A8.6.1.2) can be written as a set of equations

$$x_1 = y_1 + b_1$$
$$x_2 = y_2 + b_2$$
$$\vdots \qquad \vdots$$
$$x_n = y_n + b_n$$

(A8.6.1.3)

where x_i and y_i are the ith acoustic features for the target and the source speakers, respectively, and b_i is the offset scalar. If m samples are available for the two speakers, the estimate of the value of b_i is

$$\hat{b}_i = \frac{1}{m} \sum_{k=1}^{m} (x_{ik} - y_{ik}) \quad i = 1, \ldots, n$$

(A8.6.1.4)

where x_{ik} is the kth sample of the acoustic feature x_i, and similarly for y_{ik}.

A8.6.2 Affine Model

Although the translation model discussed in the previous section is simple, the assumption of the model is that a speaker's speech can be modeled as a single invariant transformation. This assumption may not be valid for some situations. A more detailed model is presented here.

As described previously, the acoustic features of one speaker may be modeled as a linear combination of another speaker's features, that is

$$X = AY + B$$

(A8.6.2.5)

where X and Y are n-dimensional acoustic feature vectors for the target and source speakers, respectively. A is an n by n matrix and B is an n-dimensional vector. This is known in modern algebra as an affine transformation between the vectors X and Y. If we hypothesize that the acoustic features are linearly independent and the transformation is time invariant, then A is a diagonal matrix, and of course, Equation (A8.6.1.2) is a special case of Equation (A8.6.2.1).

The process for determining the mapping function between X and Y is known as the training process. Due to variations in the speaking rate from one speaker to another, we use dynamic time warping (DTW) to adjust the parameters of the source to be in accord with those of the target along the time axis. A diagram illustrating the training process is shown in Figure A8.5. The DTW algorithm is discussed in Rabiner and Juang (1993).

The training process consists of the following procedures.

- The target and source speakers pronounce the same set of sentences.

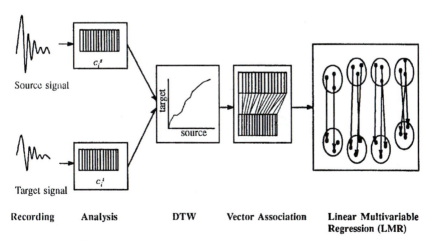

FIGURE A8.5 The training process.

- The acoustic features are measured, providing a set of framed-based vectors.
- The source vectors are time aligned with the corresponding target vectors by DTW.
- Linear multiple regression (LMR) is used to estimate the coefficients of the mapping function.

A8.7 VOICE CONVERSION ALGORITHMS

The algorithms described here focus on modification of the segmental parameters of the speech signal. We modify five measured acoustic features so that the voice conversion process is simulated with a parameter mapping process. This is illustrated in Figure A8.6.

We offer four types of mapping methods: the translation transformation, the affine transformation, the copy method, and the retain method. These mapping methods may be intermixed, that is, one method may be used for the pitch contour, another for gain contour, and so on. The copy method uses the acoustic features of the target speaker to synthesize the converted speech. This method can serve as a basis for examining the effectiveness of the translation and affine methods. The retain method retains the source parameters so that the synthesized speech contains certain features of the source speech. However, the source parameters are warped in accord with the target's speaking rate.

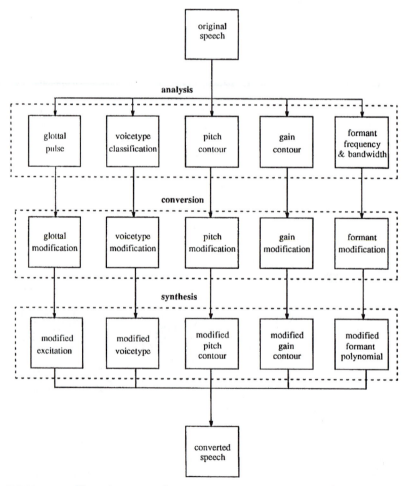

FIGURE A8.6 The voice conversion process.

A8.7.1 Overview of Conversion Algorithms for Each Acoustic Feature

- Voice type conversion. For this process, the voiced parameters (pitch, gain, glottal pulse, and formants) are transformed, while the unvoiced parameters (gain, stochastic codeword, and LP coefficients) are copied from the target to eliminate unwanted noise in the unvoiced segments.

- Pitch contour conversion. For transformation, the average value of the pitch period of each frame is used for the source and target. After training, the source vector is converted to a new value using a specified mapping function. Since the synthesizer is pitch synchronous, the pitch vector is transformed to the timing instants that correspond to the glottal closure instants (GCIs). To eliminate the unvoiced-to-voiced transition noise, the first GCI is fixed at the beginning of the voiced segment. For the voiced-to-unvoiced transitions, the last GCI is extended to the next unvoiced segment. In the voiced/unvoiced transition region, the voiced speech is overlapped with and added to the unvoiced speech.

- Gain contour conversion. The gain parameter controls the excitation energy for each pitch period and is pitch synchronous. Thus, the frame-based gain data must be interpolated for each GCI after transformation. Furthermore, once the pitch contour is modified, the gain contour must be modified, although the gain contour itself is not changed. To eliminate any discontinuities in the voiced-to-unvoiced transition, the gain value for the first voiced pulse is linearly interpolated between the last unvoiced gain and the second voiced gain, and vice versa.

- Glottal pulse conversion. The LF timing parameters (t_p, t_e, t_c, t_a) have constraints as described in Appendix 7. If the user tries to override these constraints, the synthesizer software uses either the previous LF timing parameters or the average values calculated for the entire speech record.

- Formant frequency transformation. Only the formant frequencies are converted by the linear mapping function. The bandwidths of the formants are determined by the algorithm we developed to counter the pole interaction problem (Hsiao and Childers, 1996). Unlike other transformation methods, our algorithm is independent of the pitch contour and the speaking rate (Childers et al., 1989).

A8.8 SUMMARY COMMENTS

From experimental results using the voice conversion system, it is concluded that the quality of the synthesis using the LP synthesizer is superior to that using the formant synthesizer. The implication is that the analysis algorithms for extracting the formants and their bandwidths is not as accurate as needed. The affine transformation method is superior to that using the simpler translation method. This is especially so for male-to-female and female-to-male voice conversion tasks. Generally, it is easier to convert words than it is to convert entire sentences. This is probably due to the fact that there are more dynamic changes over a sentence, than over a word. The gain contour conversion is nearly the same for both the translation and affine transformation methods. The glottal pulse transformation is a critical factor in achieving high-quality voice conversion. Generally, the quality of female synthesized speech is inferior to that for male synthesized speech. Possibly, this is due to the fact that female speech has a higher fundamental frequency than male speech, and therefore there may be more interaction of the fundamental frequency with the first formant in female speech than in male speech.

Consult Hsiao (1996) for additional details on the algorithms and experiments and Mizume and Abe (1995) for other voice conversion algorithms.

APPENDIX **9**

TIME MODIFICATION OF SPEECH. THEORY: SPEECH ANALYSIS, SEGMENTATION, AND LABELING

A9.1 INTRODUCTION

The time-modification system varies the duration of selected segments of the speech signal. The segment duration can be lengthened or shortened using the software toolbox described in Chapter 8. Possible segments include vowels, nasals, unvoiced fricatives, and so forth. Each segment duration is modified according to parameters specified by the user. These parameters apply to either all occurrences of a specific type of segment, or to a single occurrence of a segment. One example of a parameter that applies to a vowel is the vowel scale factor, SF_{vowel}. The vowel scale factor specifies the desired ratio of the duration of the vowel segment(s) in the time-modified word to the duration of the corresponding vowel segment(s) in the original, unmodified word.

Since the speech segments are the basis of the time-modification system, it is important that the segments are accurately detected and identified. (For brevity, the detection and identification processes will hereafter be called detection.) To accomplish this goal, the system designer is faced with one of two choices: manual detection (by hand) or automatic detection (software). While manual detection provides good results, it is extremely tedious and time consuming. This limits the usefulness of the overall time-modification system. Automatic detection is quick and relatively "painless," but is more prone to mistakes than the manual method and requires significant initial development time.

As a compromise, we use automatic detection of the speech segments with subsequent manual editing to correct errors that may occur in the detection process. Automatic detection consists of three main steps: (1) speech analysis; (2) segmentation of the word or sentence into segments of unknown type; and (3) appropriate labeling of these segments. The manual editing process allows the user to display and edit the automatic segmentation and labeling results using a set of software programs with a convenient, easy-to-use, graphical user interface (GUI). The GUI allows the user to insert silent segments, change segment boundaries, change segment labels, and merge adjacent segments that have been labeled as the same type. This is accomplished with the click of a workstation mouse. The software toolbox is described in Chapter 8.

This appendix describes both the automatic and manual algorithms used to analyze, segment, and label the speech segments. It begins with a discussion of the selection of speech segment categories. Next, a brief overview of the automatic segment detection process is presented with an example of automatic detection of the voiced/unvoiced/silent (V/U/S) parameter for the word sue. The example illustrates the methods associated with automatic detection of a single parameter, or "feature," of speech. The relevance and application of these methods to the general set of feature detection programs is described. Each of the automatic feature detection algorithms is then described in detail. The segmentation and labeling processes are also presented in detail. The mistakes produced by the automatic algorithms are discussed. The editing system and corresponding GUI are described in Chapter 8.

A9.1.1 Time Modification Methods of the Past

The methods used to accomplish time modification have evolved through several stages over the last 40 years. Although there are variations in the specific implementations, almost all of the methods used to date can be assigned to one of four main categories.

A9.1.2 Variable-Playback-Rate Method

The variable-playback-rate method is relatively simple, and accomplishes a rate change by playing back previously recorded speech at a rate different from the original recording rate. An example of this is playing an LP phonograph record at 45 RPM, instead of its intended rate, 33 1/3 RPM. A modern, digital signal processing (DSP) analogy of this is digital-to-analog conversion (with appropriate filtering) at a sampling rate different from the signal's original sampling rate, without interpolation or decimation. Note that for this technique, the rate change is always accompanied by a linear shift in the frequency content of the signal. For speech that has been slowed down or speeded up by a factor of about two or more, a frequency shift leads to undesirable perceptual effects that mask the identity of the speaker, and, in general, causes a decrease in intelligibility (Garvey, 1953a, 1953b; Tiffany and Bennett, 1961).

The variable-playback-rate method was never popular among researchers, mainly because of the detrimental perceptual effects attributed to the accompanying pitch shift. However, it was the only method available until about 1950, when the sampling method was introduced.

A9.1.3 Sampling Method

The general class of rate-change techniques known as the "sampling method" involves the periodic removal or duplication of small segments of recorded speech. The remaining segments are then spliced together to form the rate-altered speech. The main advantage of this method is that the frequency content of the resulting speech is not affected, and as a result, many of the speaker-dependent characteristics are preserved. It has also been shown that for a variety of different rates, the speech produced by this method is significantly more intelligible than speech produced by the variable-playback-rate method (Fletcher, 1929; Garvey, 1951; Lee, 1972).

One way of implementing the sampling method is by manually cutting and splicing magnetic recording tape (Garvey, 1949). A disadvantage of this method is the time required to perform the task. Another disadvantage is that it is impossible to guarantee waveform continuity across the splice boundary. This results in audible "pops" and "clicks," although these clicks can be reduced by cutting the tape at a 45-degree angle relative to the edge of the tape. This method does, however, have the advantage of allowing the user to (manually) select the locations and durations of the discarded segments. To mark the tape, the operator manually passes the tape back and forth across the playback head of a tape recorder and listens for the starting and stopping points of a syllable. Once these points are found, they are marked and labeled on the back of the tape with a grease pencil. This method is flexible, but extremely time consuming. As a result, it is often impractical for extensive use and is typically used only for "proof of concept."

Today, this "cut and splice" method is typically implemented on a digital computer using a video display. Although the physical inconvenience of the magnetic tape is no longer present, the problems of identifying the desired segment and maintaining waveform continuity across the splice boundary remain.

An automatic time-modification method exists that is based on a modified magnetic tape recorder (Fairbanks et al., 1954). It involves rotating the playback head assembly that contains four playback heads. As the head assembly rotates, each one of the four playback heads individually and sequentially contacts the moving magnetic tape. The outputs from the four heads are wired in parallel. Time compression results from the fact that there are gaps between each of the four heads. Therefore, as the head assembly rotates, every segment of the tape that these gaps contact (instead of a playback head) are not reproduced by the playback head. The spacings in the head assembly are calculated so that as the head assembly turns, one playback head is always beginning to make contact with the tape just as the previous playback head is loosing contact with the tape. The output of the rotating head assembly is rerecorded onto a second, conventional, tape recorder. The second tape then contains the compressed speech.

The initial automatic time-modification device introduced by Fairbanks in 1954 has some major limitations. One problem is that the duration of the segments that are discarded or duplicated is fixed. This was changed in later adaptations and copies of the original machine (Lee, 1972; Neuburg, 1978). However, other problems ultimately limit the fidelity and usefulness of the machine. The first of these problems is the lack of repeatability of the process. In order to repeat the process of compressing a tape segment, the user has to know exactly the starting phase of the rotating head assembly with respect to the beginning of the tape. The second problem is the noise created by the rotating head assembly's slip rings and distortion due to the heads' misalignment with the moving tape. Despite the shortcomings of the method, the large majority of published research on the intelligibility and comprehensibility of compressed speech was done using the sampling method.

A9.1.4 Vocoder Methods

A third class of rate change techniques is accomplished by the use of vocoders (VOice CODERS). Vocoders were originally designed to reduce the bandwidth requirements for transmission of a normal voice signal. Their ability to modify the rate of speech is thought of as a secondary benefit. Of all of the vocoders, the phase vocoder is the best suited for rate modification (Flanagan and Golden, 1966; Rabiner and Schafer, 1978).

Vocoders implement an analysis–synthesis speech transmission scheme. In the analysis stage, natural speech is analyzed, typically by a bank of bandpass filters. The output of each bandpass filter in the bank is coded by one of a variety of different methods, and this coded information is transmitted across a channel. At the synthesis stage at the receiving end of the channel, the coded information is decoded, and is used to control a bank of tuned oscillators. The outputs of the oscillators are then summed to produce synthesized speech (Rabiner and Schafer, 1978).

Typically, the synthesis oscillators are tuned to the same frequencies as the bandpass filters in the analysis stage. However, this one-to-one match in tuning is not strictly required, and if the oscillator frequencies are tuned to multiples of the analysis stage's bandpass filters, it is possible to implement a modification of the synthesized speech. For example, the phase vocoder can be used to implement a rate change in the following two-stage manner. In the first stage, speech is analyzed by a bank of equally spaced bandpass filters with center frequencies at ω_i for $i = 1, \ldots, N$. The outputs of the bank of bandpass filters are then used to control a bank of oscillators tuned to center frequencies $\omega_i/2$. At this point, the rate of the synthetic speech is identical to that of the original speech, but the frequency spectrum of the synthetic speech is shifted down to one-half that of the original speech. The second stage of the process is to double the playback speed of the speech synthesized by the first stage. The resulting speech is twice the rate of the original speech, and has the same spectrum as the original speech.

While vocoders are able to modify the rate of speech, they suffer from the fact that their analysis–synthesis schemes create unwanted artifacts in the speech signal (Portnoff, 1981). The speech produced by vocoders is often described as sounding artificial or "buzzy." Another problem previously associated with the use of vocoders for research applications is that in the 1960s and 1970s, vocoders were relatively expensive and not a cost-effective option for many researchers.

A9.1.5 Recent Methods

Over the last 10 to 15 years, other methods for modifying the rate of speech have evolved. While the speech produced by these methods is seldom tested in formal intelligibility or comprehensibility tests, the methods are being studied due to their low cost, low computational requirement, and relative ease of implementation. Note that some of the newer methods are hybrids of older vocoder technology and recent waveform coding technology.

The simplest new method consists of "a pitch detector followed by an algorithm that discards (or repeats) pieces of speech equal in length to a pitch period" (Neuburg, 1978). This is a minor variation of the sampling method. The method does not operate pitch-synchronously, which means that the beginning of the segment that is either duplicated or discarded does not occur at the instant of glottal closure. The method relies upon the fact that for the majority of time, the speech signal does not vary greatly across a single pitch period (about 10 msec for a male speaker). Therefore, as long as the duration of the discarded (or repeated) segment is exactly equal to the pitch period, the ear can not discern any significant distortion from the process. No formal listening tests have been conducted for this method.

Another speech rate modification method, by Malah (Malah, 1979), is similar in principle to the two-step process implemented by the phase vocoder. For both of these methods, if the speech rate is modified by an integer scale factor n, the first step shifts the frequency spectrum by a factor of n, and the second step plays the frequency-shifted speech at a speed of n times the original rate. Since the second step of this process is essentially trivial, the success of this method relies on the ability to shift the frequency spectrum of the speech. Malah developed numerically efficient algorithms that shift the frequency spectrum of speech (without changing the rate). These algorithms are known as time-domain harmonic scaling (TDHS) algorithms. In most cases, the algorithms require only one multiplication and two additions per output sample of speech. The primary problem in this implementation is that rate modification is only implemented in integer multiples (i.e., 2:1, 3:1). Thus, only a small, finite set of compression and expansion ratios can be implemented. Malah claims that for the two-step rate modification scheme, rates of greater than 2:1 are impractical, "due to perceptual limitations." Because of this limitation to integer multiples, the algorithm is not that applicable to speech research. Although no formal tests were conducted, the author states that "Simulation results with a scaling factor of two for different speakers and texts have been informally judged to be very good. . . ."

A recent and popular approach for time modification is based upon the short-term Fourier transform (STFT). The method is composed of three parts. The first part models the speech signal with the STFT. The second part modifies the STFT parameters to implement the rate change. The third part synthesizes the modified speech signal from the modified STFT parameters (Portnoff, 1981). While no formal listening tests were performed, the author claims that the system "is capable of producing high quality rate-changed speech . . . for compression ratios as high as 3:1 and expansion ratios as high as 4:1." For ratios outside this range, the method introduces reverberation for expanded speech, and exhibits a "rough" quality for compressed speech.

Another recent approach models the speech signal by a set of time-varying sine waves (Quatieri and McAulay, 1992). In terms of the general procedure, this method is very similar to the STFT approach. The speech is modeled by a set of time-varying parameters, namely sine wave amplitudes and phases. The algorithm adjusts the speech parameters, and then resynthesizes the modified speech by controlling a set of sine wave generators. Again, no formal listening tests were conducted, and the authors state that for time-modified speech, "the synthesized speech was generally natural sounding and free of artifacts." Perhaps the most interesting point in the Quatieri and McAulay study is that they experiment with what they call "speech-adaptive time-scale modification." In essence, they implement rate change by modifying only the voiced portions of speech. In addition, they measure the "degree" of voicing, and concentrate their time-base modification on the frames that exhibit the highest degree of voicing. Note that their measurement of the degree of voicing is based upon how little the harmonic structure varies across multiple frames, that is, the less the harmonics vary, the higher the "degree of voicing." However, no formal listening tests were conducted.

A9.1.6   Phonological and Psychological Testing

Time modification of speech is a research methodology that is used to study certain aspects of speech. In general, psychological testing is concerned with such issues as (1) determination of the highest rate at which speech can be presented to a listener and still be understood; (2) an explanation of why our perceptual process fails at higher speaking rates; and (3) the role of short-term and long-term memory in speech perception. In contrast, phonological testing is often conducted by speech pathologists. The goal is to model the human perception of phonemes (or similar segments of speech). The results are discussed in terms of measurable acoustic features and how their presence (or absence) affects perception. The literature in these areas is reviewed in White (1995). Many of the studies in the literature use a computerized cut and paste method. However, such methods rely on the user to identify various speech segments within a word. In effect, previous systems were manually operated waveform editors.

A9.2 SELECTION OF SPEECH SEGMENT CATEGORIES

The complexity of the algorithms for segmentation and labeling in any speech analysis task depends upon the degree of recognition that must be achieved (Davis and Mermelstein, 1980). For a given speech sound, algorithms that determine only the phoneme category will be less complicated than algorithms that determine not only the category, but the identity of the phoneme as well. Likewise, for a given speech sound, algorithms that determine the allophonic variation of a particular phoneme can be expected to be complicated, due to the number of variations of the pronunciation of a single phoneme that can occur during conversational speech (Klatt and Stevens, 1973). Thus, the complexity of the segmentation and labeling task is dependent upon both the number and choice of categories used to subdivide the speech.

There are several possibilities for selecting speech segment categories. In a top-down paradigm, the simplest choice is to classify speech as voiced, unvoiced, or silent (V/U/S). The next, more complex choice is to classify each speech sound as a member of one of the basic phoneme types. Although the exact description and number of different phoneme categories vary slightly depending upon the school of thought, these categories usually include vowels, nasals, semivowels, voiced fricatives, voiced stops, unvoiced fricatives, unvoiced stops, and silence. An even more complex categorization requires identification of the exact phoneme. This requires matching the segment under consideration with one of the 41 or so phonemes in English.

For the system described here, speech is divided into eight segment categories: vowels, semivowels, nasals, voiced fricatives, voice bars, unvoiced stops, unvoiced fricatives, and silence. Overall, this choice is a compromise between the complexity of the segment recognition algorithms and the resolution of the resulting speech segments. We assume that recognition of individual phonemes is not required. This greatly reduces the complexity of the segment recognition algorithms, since the choices in the matching process are reduced from 41 to eight. In addition, it is easier to recognize the typically large differences between phoneme categories than it is to recognize smaller differences between various phonemes of the same category (Schwartz and Makhoul, 1975).

A9.3 OVERVIEW OF AUTOMATIC SEGMENT DETECTION

Automatic detection of the speech segments is accomplished by a series of software programs that sequentially analyze the speech signal. These programs are grouped according to the task they perform in the detection process. The three main tasks are shown in Figure A9.1 as: (1) speech analysis; (2) segmentation of the word or sentence into segments of unknown type; and (3) labeling of these segments with the most appropriate segment label. A brief overview of each of these tasks follows, with a more detailed discussion provided later.

Speech analysis is the most complicated of the three tasks, and is divided into several steps. A block diagram of the speech analysis task is shown in Figure A9.2. The initial analysis and decomposition of the speech waveform is derived from a two-pass method developed by Hu (1993) (see also Childers and Hu, 1994) and is essentially the same method as that used for the voice conversion toolbox. In the first pass, the sampled waveform is divided asynchronously into 5 msec frames. A 13th-order, linear predictive coding (LPC) analysis is performed for each frame, and the residue is processed to determine the glottal closure points (Hu, 1993). Only the glottal closure indices (GCI) are retained for use in the second pass of the algorithm. In the second pass, the sampled waveform is again divided into frames. The frames are chosen pitch asynchronously for unvoiced speech and silence, and pitch synchronously for voiced speech, using the glottal closure indices as a reference. A 13th-order, LPC analysis is performed for each frame, and the residue, LPC coefficients, and power are saved for later modification and synthesis. Next, a set of feature detection algorithms analyzes each frame individually. Each algorithm in the set detects a different acoustic feature. For example, one of the algorithms detects the presence or absence of nasals, while another algorithm detects the presence or absence of semivowels. Each feature detection algorithm uses a combination of fixed thresholds, median filtering, and empirical rules to calculate the final result, or "feature score."

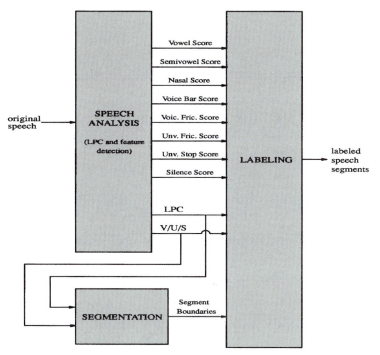

FIGURE A9.1 Block diagram of automatic speech detection.

    The second automatic detection task shown in Figure A9.1 is determination of the time-domain boundaries that separate the segments of the speech signal. This process is known as segmentation. The boundaries are chosen such that each segment has relatively stable acoustic properties for the duration of the segment. Segmentation is accomplished by combining the results of two different algorithms. The first algorithm determines the changes in the "trend" of the short-term frequency spectra, and the second uses the results of the voiced/unvoiced/silent (V/U/S) feature detector.

    The third task in Figure A9.1 is labeling of the segments using the results obtained from the feature detection algorithms. Examples of labels are vowel, semivowel, and unvoiced fricative. Labeling is done in two steps. First, the average feature detection scores are calculated. Next, empirical rules are applied to the average scores to determine the most appropriate label for each segment.

A9.4 FEATURE DETECTION ALGORITHMS—GENERAL DEVELOPMENT

Acoustic feature detection is the search for different (acoustic) features. Examples of acoustic features include voicing, nasality, and sonorance. While acoustic features are used to help differentiate between the various segment categories, it is important to realize that individual acoustic features may not be unique to one particular segment category. For example, nasality may indicate the presence of a nasal, or it may indicate the presence of a nasalized vowel. Thus, in this example, one acoustic feature is common to two different segment categories. This lack of one-to-one correspondence between acoustic features and segment categories requires that multiple acoustic features be evaluated and weighed when attempting to match an unknown speech segment with the most appropriate segment label.

    Although it is logical to define the term "segment detector" as an algorithm that detects one of the eight segment types listed in Section A9.2, this term is misleading. It can be confused with the previously defined definition of segmentation, which is the task of dividing the speech signal into segments of unknown type. Therefore, the term "feature detector" is used in a broad sense, and implies both an algorithm that detects a single acoustic feature, as well as an algorithm that detects multiple acoustic features in order to detect one of the eight segment types listed in Section A9.2.

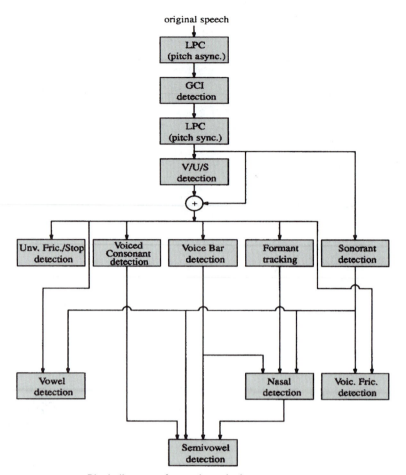

FIGURE A9.2   Block diagram of speech analysis.

Feature detection is achieved by algorithms that examine the short-term frequency spectra of the speech signal. The spectra are calculated from the LPC coefficients that are, in turn, calculated for each frame of the signal during the initial analysis stage. It has been shown that the short-term frequency spectra method is a reliable technique that is used in a wide variety of recognition systems (Bush et al., 1983; Glass and Zue, 1986; Glass and Zue, 1988; Klatt, 1977; Leung et al., 1993; McCandless, 1974; Meng and Zue, 1991; Mermelstein, 1977; Weinstein et al., 1975; Zue et al., 1989).

Each feature detection algorithm utilizes a sequence of processing stages to calculate the resulting feature score. In many instances, the structure of each of the feature detection algorithms is similar, although the exact numerical values may differ.

The detection of acoustic features from the speech signal is the most complicated portion of the analysis, segmentation, and labeling process. Because of the complexity of the feature detection algorithms, the explanation of the algorithms is broken down into two sections. In this section, a simple example is given to explain one feature detection algorithm and its development and implementation. Although the example is for a single feature detector, it illustrates the general structure of the majority of the feature detectors. The example also discusses the problems and considerations associated with the set of feature detection algorithms as a whole. In the next section, the algorithms are detailed individually, examining the specific equations of the algorithms.

The feature detection algorithms use a combination of methods to produce the final results. These methods include bandpass filters, fixed thresholds, median filter smoothing, and empirical pattern recognition rules. These methods are used in a similar manner in each of the algorithms. The example that follows illustrates how these methods work in a feature detection algorithm that detects the voiced/unvoiced/silence (V/U/S) feature of speech.

A9.4.1 Input Data and V/U/S Pre-Processing

All of the feature detection algorithms require the LPC results as input data. Most of the feature detectors also require the results from other feature detectors (specifically the V/U/S results), as shown in Figure A9.2.

V/U/S classification is different from the other feature detection algorithms in that a portion of the algorithm is accomplished during the initial LPC analysis algorithm. During the first pass of the LPC algorithm, the first reflection coefficient is calculated for each pitch-asynchronous frame. The frame is classified as voiced (V) if the reflection coefficient is greater than 0.2, and is classified as unvoiced (U) if the reflection coefficient is less than or equal to 0.2. This threshold was determined empirically by Hu (1993). In addition, Hu makes no distinction between unvoiced and silent frames. Therefore, all silent frames are classified as unvoiced. During the second pass of the LPC algorithm, certain frames are labeled as transitional frames (T). The first voiced frame in an unvoiced-voiced sequence, and the last voiced frame in a voiced-unvoiced sequence are changed to transitional frames. Hu's explanation for this is that mistakes may be made in the simple V/U decision process, so the frames at the transition regions are marked, since this is typically where the mistakes are made. Since it has been observed that the transition frames are always voiced, all transition frames are converted to voiced frames in this system.

A9.4.2 Volume Function

A volume function, V(i), similar to one presented by Weinstein et al. (1975) is calculated for each frame to determine a quantity analogous to the loudness, or acoustic volume, of the signal at the output of a hypothetical bandpass filter. This is the first processing step in the majority of feature detectors. The volume function is normalized by the number of samples in the frame, and is given as

$$V(i) = \frac{1}{N_i} \sqrt{\sum_{m=A}^{B} \left| H_i \left(e^{j\pi \frac{m}{256}} \right) \right|^2} \qquad (A9.4.2.1)$$

where i is the current frame index, N_i is the number of samples in frame i, A is the index of the low cutoff frequency of the bandpass filter, B is the index of the high cutoff frequency of the bandpass filter, and $H_i(e^{j\pi \frac{m}{256}})$ is the complex, single-sided, frequency response of the IIR filter, $H_i(z)$, produced by the LPC coefficients and evaluated at the points $\exp(j\pi m/256)$, for $0 \leq m \leq 255$. $H_i(z)$ is given as

$$H_i(z) = \frac{G(i)}{a_0 + a_1 z^{-1} + a_2 z^{-2} + \cdots + a_N z^{-N}} \qquad (A9.4.2.2)$$

where $N = 13$, $a_0 = 1$, and G(i) is given as

$$G(i) = \sqrt{\sum_{n=s}^{t} r^2(n)} \qquad (A9.4.2.3)$$

where r(n) is the value of the LPC residue at sample n, i is the current frame index, s is the beginning sample number of the current frame, and t is the ending sample number of the current frame.

The volume function of Equation (A9.4.2.1) is used extensively in the software, although the frequency range of the bandpass filter varies depending upon the specific detector. In addition, many of the feature detection algorithms calculate the ratio of two volume functions, each with its own frequency range. This compares the energy in one frequency band to the energy in a second frequency band.

In the majority of feature detectors, median filtering is done to smooth any large, short-term fluctuations in the volume function. The fluctuations are caused by a variety of sources including incorrect GCI determination, incorrect V/U/S classification, and recording artifacts such as background noise. Although the majority of the feature detectors use a 5th-order median filter for smoothing, the exact order is given in the detailed description of each detector. The filter order is determined empirically in each case.

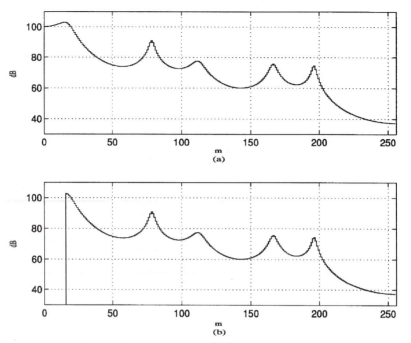

FIGURE A9.3 $H_i(z)$ used to calculate the V/U/S volume function for one pitch period of the vowel portion of the word sue. (*a*) $H_i(z)$ before filtering. (*b*) $H_i(z)$ after filtering.

The V/U/S detector uses a single volume function of Equation (A9.4.2.1) with the values A = 17 and B = 255. The lower limit of A = 17 serves to highpass (HP) filter the frequency response with a cutoff frequency of 312 Hz (the upper limit B = 255 corresponds to one-half the sampling rate, thus a highpass instead of a bandpass filter). Weinstein claims that the HP filter is needed to reduce the sensitivity to voiced stops, but experiments show that its primary effect is to reduce low-frequency artifacts such as wind noise and other pop-like sounds caused by non-optimum microphone placement during the recording process. In general, the volume function is used in the V/U/S detector as a relatively wide-band integrator that calculates the approximate energy in each frame. The role of this integrator is discussed in the next section.

A graph of the frequency response of $H_i(z)$ before and after the hypothetical highpass filter for one pitch period of the vowel portion of the word "sue" spoken by a male speaker is shown in Figure A9.3. As described in Equation (A9.4.2.1), the single-sided frequency response of $H_i(z)$ is evaluated at 256 equally spaced points around the upper half of the unit circle in the z-plane. Although it is not shown on the graph, the value of $V(i)$ for the pitch period analyzed in Figure A9.3(*b*) is 70.62 dB. It is seen from the graph that this is approximately the average level of the frequency response.

Note also that unlike the other feature detectors, median filtering is not performed on the V/U/S volume function. This is to ensure that any short-term energy fluctuations, such as those produced by stops, are not inadvertently smoothed.

A9.4.3 Fixed Thresholds and Feature Scores

Each feature detection algorithm calculates a feature score to indicate the presence of the corresponding acoustic feature in a given frame of speech. The feature score is typically continuous over the range [0, 1], although there are exceptions (several of the feature scores are discrete, either binary or ternary). In general, the feature score is calculated by comparing the value of the volume function (or the ratio of two volume functions) with one or more fixed thresholds. The values of the thresholds are determined empirically by trial and error during the analysis of approximately 100 words of the Diagnostic Rhyme Test (DRT) spoken by two male and one female speakers (Voiers, 1983). In some cases, initial estimates for the thresholds are taken from literature (sources are listed in discussion of the specific algorithms), and the thresholds are then "fine tuned" using the DRT speech data. The empirical determination of these thresholds constitutes a type of "learning" phase in the algorithm development. This contrasts with one of the initial goals of the automatic segmentation and labeling process, which is to not require "training" of the algorithms. However, given the nature and variability of the speech signal, in hind sight it now appears that to create a set of reliable segmentation and labeling algorithms based upon frequency distributions (i.e., volume functions) that some type of training, or parameter "tuning," is required.

The advantages and disadvantages of training are obvious. If the training data does not accurately represent the set of intended users, the algorithms will not function as expected in practice. If the training data completely represents the set of intended users, the algorithms will work efficiently and accurately. Since the topic of training of speech recognition algorithms is beyond the scope of this software toolbox, it will be accepted that training is mandatory, regardless of the particular algorithm.

In general, if two thresholds are used, the feature score for each frame is determined by

$$
\text{Feature\_Score}(i) = \begin{cases} 1, & \text{if Vol\_Fcn}(i) \geq T_{upper} \\ 0, & \text{if Vol\_Fcn}(i) < T_{lower} \\ \dfrac{\text{Vol\_Fcn}(i) - T_{lower}}{T_{upper} - T_{lower}}, & \text{if } T_{lower} \leq \text{Vol\_Fcn}(i) < T_{upper} \end{cases} \tag{A9.4.3.1}
$$

for a feature score that increases as the volume function increases. If the feature score decreases as the volume function increases, the feature score is given by

$$
\text{Feature\_Score}(i) = \begin{cases} 0, & \text{if Vol\_Fcn}(i) \geq T_{upper} \\ 1, & \text{if Vol\_Fcn}(i) < T_{lower} \\ \dfrac{T_{upper} - \text{Vol\_Fcn}(i)}{T_{upper} - T_{lower}}, & \text{if } T_{lower} \leq \text{Vol\_Fcn}(i) < T_{upper} \end{cases} \tag{A9.4.3.2}
$$

For both equations, i is the current frame index, T_{lower} is the fixed lower threshold, T_{upper} is the fixed upper threshold, and Vol_Fcn(i) is the volume function (or the ratio of two volume functions) for the current frame. Both Equations (A9.4.3.1) and (A9.4.3.2) are used in practice. If only one threshold is used to calculate a binary feature score (zero or one), then either Equation (A9.4.3.1) or (A9.4.3.2) is used with $T_{lower} = T_{upper}$.

The original LPC analysis described in Section A9.4.1 distinguishes only between voiced and non-voiced frames. The non-voiced frames denoted by Hu (1993) as "unvoiced" are either unvoiced or silent. The procedure used in the V/U/S detector to classify non-voiced frames as either unvoiced or silent is based upon a single volume function, the background noise power in the speech signal, and Equation (A9.4.3.1). First, the mean and the standard deviation of the background noise power, BNP, are calculated as

$$
\text{BNP}_{mean} = \frac{1}{20} \sum_{n=1}^{20} p(n) \tag{A9.4.3.3}
$$

$$
\text{BNP}_{std\,dev} = \sqrt{\frac{1}{20} \sum_{n=1}^{20} (p(n) - \text{BNP}_{mean})^2} \tag{A9.4.3.4}
$$

where p(n) is the frame power in decibels (dB). It is assumed that the first 100 msec (20 frames) of the speech signal are silence.

The V/U/S volume function for each non-voiced frame is then compared to a constant threshold, $T_{U/S}$, using Equation (A9.4.3.1) with $T_{lower} = T_{upper}$ and $T_{upper} = T_{U/S}$. The V/U/S feature score for each non-voiced frame is given as

$$
\text{VUS\_Score}(i) = \begin{cases} 1, & \text{if } 20 \log_{10}(V(i)) \geq T_{U/S} \\ 0, & \text{if } 20 \log_{10}(V(i)) < T_{U/S} \end{cases} \tag{A9.4.3.5}
$$

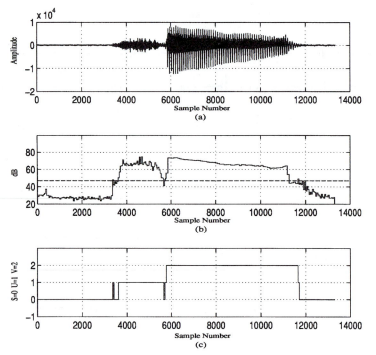

FIGURE A9.4   Voiced/unvoiced/silent classification using the volume function and a single fixed threshold for the non-voiced portion of the word sue. (*a*) Time-domain waveform. (*b*) Volume function and $T_{U/S}$ threshold (dashed line). (*c*) V/U/S score.

where V(i) is calculated from Equation (A9.4.2.1) with A = 17 and B = 255, and i is the index of the current frame. If VUS_Score(i) = 1, the frame is classified as unvoiced, and if VUS_Score(i) = 0, the frame is classified as silent. Note that this method only separates unvoiced from silent frames. The value of VUS_Score(i) is arbitrarily set equal to two for all voiced frames. As a result, the V/U/S feature score is different from many of the other feature scores in two ways: First, it spans the range [0, 2] instead of [0, 1]. Second, it can have only one of three discrete values, while most of the other feature scores are continuous.

It is found (empirically) that the best results are obtained when

$$T_{U/S} = BNP_{mean} + (k)(BNP_{std\,dev}) \tag{A9.4.3.6}$$

where k = 2.0. Obviously, the value used for k is dependent upon the statistical properties of the background noise in the speech signal. However, the absolute level of the background noise is compensated for automatically since the value of BNP_{mean} is calculated before analysis of each word.

Figure A9.4 shows the V/U/S classification for the word "sue" spoken by a male speaker. Part (*a*) shows the time-domain speech waveform, part (*b*) shows the volume function (in dB) from (A9.4.2.1) and the fixed threshold, $T_{U/S}$, and part (*c*) shows the resulting V/U/S score.

A9.4.4   Automatic Correction Rules

The feature scores produced by Equation (A9.4.3.1) [or Equation (A9.4.3.2)] sometimes require additional processing to help eliminate false detection of features. The processing is accomplished by the application of pattern recognition rules. These rules are used sparingly, and counteract specific, regularly occurring, incorrect classifications of feature types. They are developed empirically on an algorithm-by-algorithm basis.

In the case of the V/U/S detector, the unprocessed, or "raw," V/U/S score produced by Equation (A9.4.3.5) has the undesirable effect of sometimes oscillating between states during word onsets

TABLE A9.1   Rules to Modify Initial Voiced/Unvoiced/Silent (V/U/S) Results[a]

Rule number	Initial pattern	Requirements for modification	Final pattern
1	VUS	length V > 100.0 msec, length U < 25.1 msec	VVS
2	xSy	length S < 10.1 msec	xxy
3	SUV	length U < 7.5 msec	SSV
4	xUy (except SUV)	length U < 10.0 msec	xyy (if x = S), else xxy

[a] The symbols x and y denote an arbitrary segment type.

and offsets, as well as during transitions between unvoiced and voiced regions. Since this does not accurately model the human speech process, rules are applied to smooth the V/U/S score, or "track." Table A9.1 lists the rules.

The first rule eliminates an incorrect unvoiced classification at the end of a long voiced segment. Since the energy often decreases quickly during the last few glottal cycles of a voiced-silent transition, the classification algorithm sometimes (incorrectly) labels these pitch periods as unvoiced. The second rule smoothes out momentary "drop outs" that occur when the signal level drops below the $T_{U/S}$ threshold for a brief period of time. The third rule is similar to the first rule, except that it smoothes out the beginning of the segment, instead of the end. It reclassifies the very short, low energy frame at the beginning of a silence-unvoiced-voiced transition from unvoiced to silent. The fourth rule smoothes out momentary unvoiced segments of very short duration. This rule does not eliminate the short noise bursts exhibited by plosives, since it only acts if the unvoiced segment is less than 10 msec, which is far shorter than the average duration of the plosive burst (Klatt, 1979; Umeda, 1977).

Figure A9.5 shows the V/U/S score for the word "sue" spoken by a male speaker before and after the application of the pattern recognition rules. The rules reduce the number of non-silent segments from five to two.

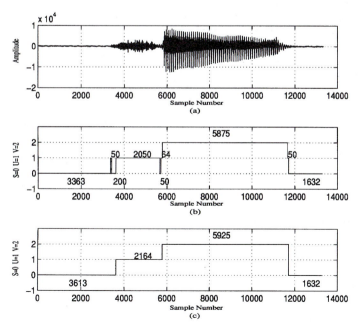

FIGURE A9.5   Voiced/unvoiced/silent (V/U/S) classification for the word sue. Segment durations are indicated in number of samples. (*a*) Time-domain waveform. (*b*) Before application of pattern recognition rules. (*c*) After application of pattern recognition rules.

A9.4.5 Summary

Each feature detection algorithm consists of a similar sequence of processing stages. In general, the first stage calculates one or more volume functions. The volume function (or ratio of volume functions) is smoothed by a median filter to remove any short-term fluctuations. The second stage calculates a feature score, typically over the range [0, 1], by comparing the volume function with one or two fixed thresholds. The third stage applies pattern recognition rules to correct for any known deficiencies in the algorithms.

The following section details the individual feature detection algorithms. Each of the detectors in Figure A9.2 is described, and any differences from the V/U/S detector are discussed.

A9.5 FEATURE DETECTION ALGORITHMS—DETAILED DESCRIPTIONS

This section gives the details of the feature detection algorithms. The algorithms are, for the most part, similar in form to the V/U/S feature detection algorithms described in Section A9.4.

A9.5.1 Sonorant Detection

To be classified as sonorant, a frame must be voiced and must also have a high ratio of low-frequency to high-frequency energy. The group of sonorants typically include vowels, voice bars, nasals, and semivowels. The non-sonorants include unvoiced fricatives, unvoiced stops, and strong, voiced fricatives. Weak voiced fricatives are classified as sonorant if they have a relatively large proportion of low-frequency energy (Weinstein et al., 1975).

The volume function from Equation (A9.4.2.1) is calculated for each frame with $G = 1$, $A = 5$, and $B = 46$. This is termed the low-frequency volume function, or LFV. The LFV is equivalent to a bandpass filter from 98 Hz to 898 Hz. A second volume function from Equation (A9.4.2.1) is calculated for each frame with $G = 1$, $A = 189$, and $B = 255$. This is termed the high-frequency volume function, or HFV. The HFV is equivalent to a bandpass filter from 3691 Hz to 5000 Hz. The sonorant ratio, $R(i)$, is calculated for each frame as

$$R(i) = \frac{LFV(i)}{HFV(i)} \qquad (A9.5.1.1)$$

where i is the index of the current frame.

The sonorant ratio is then smoothed by a fifth-order median filter. The smoothed sonorant ratio is compared to a threshold, T_{son}, and a binary (zero or one) sonorant score, $SS(i)$, is calculated for each frame as

$$SS(i) = \begin{cases} 0, & \text{if } R(i) < T_{son} \\ 1, & \text{if } R(i) \geq T_{son} \end{cases} \qquad (A9.5.1.2)$$

where i is the frame index and $T_{son} = 10$. The threshold T_{son} is determined empirically.

Figure A9.6 shows the sonorant detection results for the word "sue" spoken by a male speaker. The threshold $T_{son} = 10$ is shown as a dashed line in Figure A9.6(*b*). Note that the sonorant ratio is nearly zero for the entire duration of the /s/.

A9.5.2 Vowel Detection

Vowel detection is accomplished in a manner similar to that for sonorant detection. A LFV function from Equation (A9.4.2.1) is calculated with $G = 1$, $A = 1$, and $B = 51$. The LFV is equivalent to a bandpass filter from 20 Hz to 996 Hz. A HFV function from Equation (A9.4.2.1) is calculated for $G = 1$, $A = 52$, and $B = 255$. The HFV is equivalent to a bandpass filter from 1016 Hz to 5000 Hz.

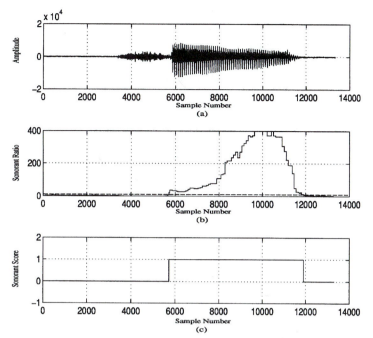

FIGURE A9.6 Sonorant detection for the word sue. (*a*) Time-domain waveform. (*b*) Sonorant ratio and threshold. (*c*) Sonorant score.

A vowel ratio, VWL(i), is calculated for each frame by

$$VWL(i) = \frac{LFV(i)}{HFV(i)} \tag{A9.5.2.1}$$

where i is the frame index. The vowel ratio is then smoothed with a fifth-order median filter. A vowel score, VWLS(i), within the continuous range [0, 1] is calculated for each frame by comparing the smoothed vowel ratio with two thresholds. The score is given by

$$VWLS(i) = \begin{cases} 0, & \text{if } VWL(i) \geq T_{upper} \\ 1, & \text{if } VWL(i) < T_{lower} \\ \dfrac{T_{upper} - VWL(i)}{T_{upper} - T_{lower}}, & \text{if } T_{lower} \leq VWL(i) < T_{upper} \end{cases} \tag{A9.5.2.2}$$

where $T_{upper} = 18$ and $T_{lower} = 8$. The two thresholds are determined empirically.

In a final processing stage, the vowel score is automatically set to zero for all frames in any vowel segment that is 150 samples (15 msec) or less in length. This helps to reduce false vowel detection.

Figure A9.7 shows the vowel detection results for the word "said" spoken by a male speaker. The two thresholds are shown as dashed lines in Figure A9.7(*b*). Note that the voiced /d/ exhibits a high vowel score for a short duration. Since there is no voiced stop segment category in this system, voiced stops are typically classified as either a vowel-unvoiced stop sequence, or a vowel-unvoiced fricative sequence.

A9.5.3 Voiced Consonant Detection

Voiced consonant detection is accomplished in a manner almost identical to vowel detection. A LFV function from Equation (A9.4.2.1) is calculated with G = 1, A = 1, and B = 51. The LFV is equivalent to a bandpass filter from 20 Hz to 996 Hz. A HFV function from Equation (A9.4.2.1) is

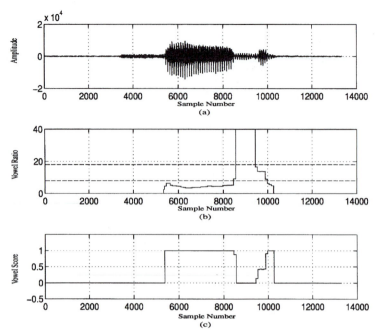

FIGURE A9.7 Vowel detection for the word said. (*a*) Time-domain waveform. (*b*) Vowel ratio and thresholds. (*c*) Vowel score.

calculated for G = 1, A = 52, and B = 255. The HFV is equivalent to a bandpass filter from 1016 Hz to 5000 Hz. These filter values are the same as those used for vowel detection. A voiced consonant ratio, VC(i), is calculated for each frame as

$$VC(i) = \frac{LVF(i)}{HVF(i)} \tag{A9.5.3.1}$$

where i is the frame index. The voiced consonant ratio is then smoothed with a fifth-order median filter. A voiced consonant score, VCS(i), within the continuous range [0, 1] is calculated for each frame by comparing the smoothed voiced consonant ratio with two thresholds. The score is given by

$$VCS(i) = \begin{cases} 1, & \text{if } VC(i) \geq T_{upper} \\ 0, & \text{if } VC(i) < T_{lower} \\ \dfrac{VC(i) - T_{lower}}{T_{upper} - T_{lower}}, & \text{if } T_{lower} \leq VC(i) < T_{upper} \end{cases} \tag{A9.5.3.2}$$

where $T_{upper} = 18$ and $T_{lower} = 8$. The thresholds are determined empirically.

Note that VCS can be calculated during the VWLS calculation for each frame, since VCS = 1 − VWLS, provided that the value of VWLS is used before the short segment (<15 msec) vowel detection and elimination of Section A9.5.2 is done. This is because the same filters and thresholds are used for both vowel and voiced consonant detection.

However, calculating VCS directly from VWLS eliminates the possibility of future experiments with the filter characteristics and thresholds for voiced consonant detection independent of the vowel detection algorithm.

Figure A9.8 shows the voiced consonant detection results for the word "said" spoken by a male speaker. The two thresholds are shown as dashed lines in Figure A9.8(*b*). The voice bar that occurs before the release of the /d/ is clearly classified as a voiced consonant. However, the algorithm assigns a score of slightly greater than 0.5 to the initial portion of the release of the /d/. This shows the difficulty of detecting voiced stops using a single acoustic feature.

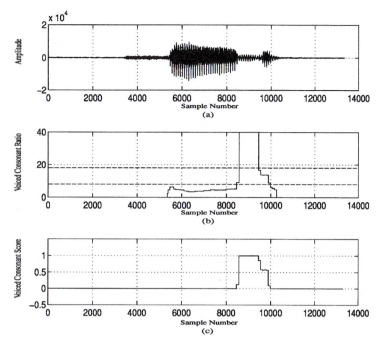

FIGURE A9.8 Voiced consonant detection for the word said. (*a*) Time-domain waveform. (*b*) Voiced consonant ratio and thresholds. (*c*) Voiced consonant score.

A9.5.4 Voice Bar Detection

Voice bar detection is accomplished in a manner similar to both vowel and voiced consonant detection. A LFV function from Equation (A9.4.2.1) is calculated with G = 1, A = 1, and B = 33. The LFV is equivalent to a bandpass filter from 20 Hz to 645 Hz. A HFV function from Equation (A9.4.2.1) is calculated for G = 1, A = 34, and B = 255. The HFV is equivalent to a bandpass filter from 664 Hz to 5000 Hz. A voice bar ratio, VB(i), is calculated for each frame as

$$VB(i) = \frac{LFV(i)}{HFV(i)} \qquad (A9.5.4.1)$$

where i is the frame index. The voice bar ratio is then smoothed with a fifth-order median filter. A voice bar score, VBS(i), within the continuous range [0, 1] is calculated for each frame by comparing the smoothed voice bar ratio with two thresholds. The score is given by

$$VBS(i) = \left\{ \begin{array}{ll} 1, & \text{if } VB(i) \geq T_{upper} \\ 0, & \text{if } VB(i) < T_{lower} \\ \dfrac{VB(i) - T_{lower}}{T_{upper} - T_{lower}}, & \text{if } T_{lower} \leq VB(i) < T_{upper} \end{array} \right\} \qquad (A9.5.4.2)$$

where $T_{upper} = 30$ and $T_{lower} = 10$. The thresholds are determined empirically.

In a final processing stage, the voice bar score is automatically set to zero for all frames in any voice bar segment that is 300 samples (30 msec) or less in length. This helps to reduce false voice bar detection.

Figures A9.9 and A9.10 show voice bar detection for the words "said" and "bond," respectively, spoken by a male speaker. The two thresholds are shown as dashed lines in Figures A9.9(*b*) and A9.10(*b*). In Figure A9.9, the voice bar before the release of the /d/ is clearly detected. Figure A9.10 shows both the voice bar of the initial /b/, and the voice bar associated with the final /d/, for the word

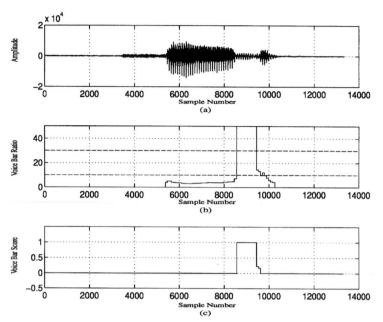

FIGURE A9.9 Voice bar detection for the word said. (*a*) Time-domain waveform. (*b*) Voice bar ratio and thresholds. (*c*) Voice bar score.

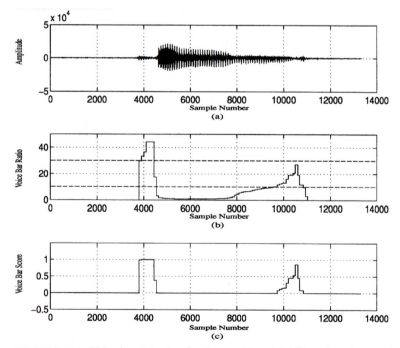

FIGURE A9.10 Voice bar detection for the word bond. (*a*) Time-domain waveform. (*b*) Voice bar ratio and thresholds. (*c*) Voice bar score.

"bond." Also note in Figures A9.9 and A9.10 that the voice bar associated with the final /d/ is much shorter in the word "bond" than in the word "said" due to the preceding nasal in "bond."

A9.5.5 Formant Tracking

Formant tracking is accomplished in a manner completely different from the typical detection process. The output of the algorithm is also different from the other feature detector outputs. Actually, the formant tracking algorithm is a "front-end" processor for the nasal detection algorithm, since the nasal detector does not use volume functions, but rather a ratio of the amplitudes of the first two formant frequencies to determine the nasal feature score.

Formant tracking is accomplished by an algorithm developed by McCandless (1974). Only a brief description is given, since the algorithm is documented elsewhere. Only the voiced portions of the speech signal are analyzed to estimate the formant tracks.

The algorithm attempts to find a best match between the peaks of the frequency response obtained from the filter produced from the LPC coefficients, and estimates for the first four formant frequencies. The amplitudes of the first four formant peaks are also estimated. Initially, the estimates for the four formant frequencies are set to values that are typical for the male voice (or the female voice, if female speech is being analyzed). For the male voice, the initial estimates are $F_1 = 320$ Hz, $F_2 = 1440$ Hz, $F_3 = 2760$ Hz, and $F_4 = 3200$ Hz. For the female voice, the initial estimates are $F_1 = 480$ Hz, $F_2 = 1760$ Hz, $F_3 = 3200$ Hz, and $F_4 = 3520$ Hz. The algorithm matches each peak of the frequency response of the LPC filter with the closest formant frequency estimate. The estimates for the formant frequencies are updated after each frame of speech is processed, provided that a match has been made.

In any given frame, if there is no match between the LPC filter peaks and the formant frequency estimates, the algorithm attempts to increase spectral resolution by iteratively evaluating the "frequency response" of the LPC filter on a circle in the z-plane with a radius less than one. This is done by evaluating the z-transform of the LPC filter with $z = re^{j\theta}$, where r denotes the radius of the circle. The initial radius value is unity, and is decreased by 0.004 during each iteration until a match is obtained, or until the radius is less than 0.88. This procedure is able to resolve two closely spaced poles if they are relatively close to the unit circle. If the radius is reduced to less than 0.88 and a match is still not found for all of the formant frequencies in the frame, the algorithm re-evaluates the matches it has made for the frame. During the re-evaluation, the algorithm either changes the matches it has made, or assigns a zero value to the formant frequency in question for that particular frame.

The final results are smoothed by checking each of the first three formant frequency tracks individually for any zero values (this is not a part of the McCandless algorithm). If the formant frequency is zero for either one or two (consecutive) frames, the frequency and amplitude are linearly interpolated and the zero values are removed.

The first three estimated formant frequency and estimated formant amplitude tracks for the voiced portion of the word "meat" spoken by a male speaker are shown in Figure A9.11. Although only the first two formant tracks are used in this system, the first three formants are retained for future expansion of the system.

A9.5.6 Nasal Detection

Nasal detection is accomplished by comparing the estimated amplitudes of the first two formants obtained in the McCandless formant tracker. A nasal ratio, N(i), is calculated for each frame as

$$N(i) = \frac{A2(i)}{A1(i)} \tag{A9.5.6.1}$$

where A1(i) is the estimated first formant amplitude, A2(i) is the estimated second formant amplitude, and i is the current frame index (Mermelstein, 1977). The nasal ratio is then smoothed by a fifth-order median filter. A nasal score, NS(i), within the continuous range [0, 1] is calculated for each frame by

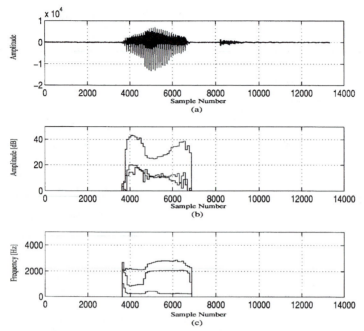

FIGURE A9.11 McCandless formant tracker for the word meat. (*a*) Time-domain waveform. (*b*) Amplitudes of the first three formants. (*c*) Center frequencies of the first three formants.

comparing the smoothed nasal ratio with two thresholds. The score is given by

$$
NS(i) = \begin{cases} 0, & \text{if } N(i) \geq T_{upper} \\ 1, & \text{if } N(i) < T_{lower} \\ \dfrac{T_{upper} - N(i)}{T_{upper} - T_{lower}}, & \text{if } T_{lower} \leq N(i) < T_{upper} \end{cases} \quad \text{(A9.5.6.2)}
$$

where $T_{upper} = 0.20$ and $T_{lower} = 0.05$. The thresholds are determined empirically.

Additional processing is achieved by applying pattern recognition rules to distinguish nasals from other segment types. First, if a frame has a voice bar score greater than 0.75, the nasal score for that frame is set to zero. This is because strong voice bars typically exhibit strong nasal scores. The opposite, however, is not true. The nasal score is then set to zero if the frame has a zero sonorant score. This is done to prevent non-sonorant frames from being classified as nasal. Finally, the nasal score for all frames in any continuous nasal segment that is less than 25 msec in length are set to zero.

Figure A9.12 shows nasal detection for the word "bond" spoken by a male speaker. The two thresholds are shown as dashed lines in Figure A9.12(*b*). Note that the nasal score rises slowly after the transition from the vowel to the nasal. This shows that the formant-amplitude-based algorithm is not as accurate as the other feature detection algorithms that are based upon the short-term frequency response. Still, the algorithm is able to correctly identify the nasal region.

A9.5.7 Semivowel Detection

Semivowel detection is based on a method developed by Espy-Wilson (1986). The algorithm deviates slightly from the standard detector, although it uses the volume functions from Equation (A9.4.2.1). A LFV function from Equation (A9.4.2.1) is calculated with $G = 1$, $A = 1$, and $B = 20$. The LFV is equivalent to a bandpass filter from 20 Hz to 391 Hz. A HFV function from Equation (A9.4.2.1) is calculated for $G = 1$, $A = 21$, and $B = 50$. The HFV is equivalent to a bandpass filter from 410 Hz

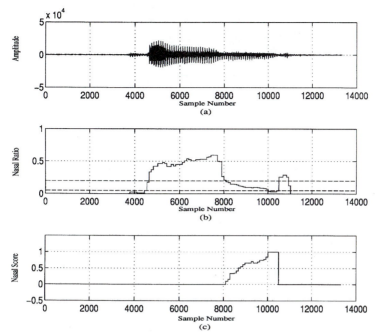

FIGURE A9.12 Nasal detection for the word bond. (*a*) Time-domain waveform. (*b*) Nasal ratio and thresholds. (*c*) Nasal score.

to 977 Hz. A murmur ratio, MUR(i), is calculated for each frame as

$$MUR(i) = \frac{LFV(i)}{HVF(i)} \qquad (A9.5.7.1)$$

The murmur ratio is then smoothed with a fifth-order median filter. A murmur score, MS(i), within the continuous range [0, 1] is calculated for each frame by comparing the smoothed murmur ratio with two thresholds. The score is given by

$$MS(i) = \begin{cases} 1, & \text{if } MUR(i) \geq T_{upper} \\ 0, & \text{if } MUR(i) < T_{lower} \\ \dfrac{MUR(i) - T_{lower}}{T_{upper} - T_{lower}}, & \text{if } T_{lower} \leq MUR(i) < T_{upper} \end{cases} \qquad (A9.5.7.2)$$

where $T_{upper} = 12$ and $T_{lower} = 4$. The thresholds are determined empirically. The semivowel score, SVS(i), is then calculated for each frame as

$$SVS(i) = (1 - MS(i))(1 - VBS(i))VCS(i) \qquad (A9.5.7.3)$$

where i is the frame index, VBS(i) is the voice bar score from Section A9.5.4, and VCS(i) is the voiced consonant score from Section A9.5.3. The value of SVS is limited to the range [0, 1]. If SVS is greater than one, it is set to unity for the frame. Equation (A9.5.7.3) shows that if a frame has a high voiced consonant score, a low murmur score, and a low voice bar score, it will have a high semivowel score.

Additional processing is done to smooth SVS. If the frame has a nasal score greater than 0.5, SVS is set to zero for that frame. This is because some strong nasals get labeled as semivowels. In addition, the semivowel scores for all of the frames in any continuous semivowel segment less than 30 msec in duration are set to zero. This eliminates mislabeling of short segments that are not semivowels.

Figure A9.13 shows semivowel detection for the word "wield" spoken by a male speaker. The algorithm correctly detects the leading /w/ as well as the /l/ near the end of the word immediately

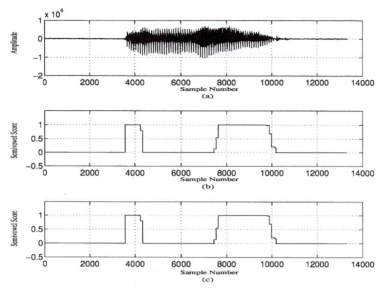

FIGURE A9.13   Semivowel detection for the word wield. (*a*) Time-domain waveform. (*b*) Semivowel score before post-processing. (*c*) Semivowel score after post-processing.

preceding the release of the plosive /d/. Note that listening reveals that the actual release in this example more closely resembles an unvoiced /t/ rather than a voiced /d/. Also note that the post-processing does not detect any nasal segments nor any semivowel segments less than 30 msec long. Therefore, the results for both before and after post-processing are the same.

A9.5.8   Voiced Fricative Detection

The voiced fricative detection algorithm deviates from the standard detector, although it does calculate feature scores from fixed thresholds. The first step in voiced fricative detection is to add preemphasis to the frequency response of the filter produced by the LPC coefficients. In other studies, the typical preemphasis method is to calculate the first difference of the sampled data waveform before calculating the LPC coefficients. In this system the first difference is not calculated before the LPC analysis, so the first difference function is approximated by a weighting function, W, in the frequency domain given by

$$W\left(e^{j\pi \frac{m}{256}}\right) = \frac{m}{256}, \quad \text{for } 0 \le m \le 255 \tag{A9.5.8.1}$$

The magnitude of the weighting function's frequency response is within 3 dB of the magnitude of the frequency response of a first-order differentiator for all frequencies in the filter passband. The preemphasized frequency response for frame i, $\hat{H}_i$, is

$$\hat{H}_i\left(e^{j\pi \frac{m}{256}}\right) = W\left(e^{j\pi \frac{m}{256}}\right) H_i\left(e^{j\pi \frac{m}{256}}\right), \quad \text{for } 0 \le m \le 255 \tag{A9.5.8.2}$$

where H_i is calculated from Equation (A9.4.2.2) for frame i with $G = 1$. The mean frequency of the preemphasized frequency response, MF(i), is then found for each frame as

$$MF(i) = \frac{1}{H_{\text{total}(i)}} \sum_{m=0}^{255} \left(\frac{m}{256} \frac{F_s}{2} \left|\hat{H}_i\left(e^{j\pi \frac{m}{256}}\right)\right|\right) \tag{A9.5.8.3}$$

where $F_s = 10$ kHz, and i is the frame index. $H_{\text{total}(i)}$ is given for frame i as

$$H_{\text{total}(i)} = \sum_{m=0}^{255} \left|\hat{H}_i\left(e^{j\pi \frac{m}{256}}\right)\right| \tag{A9.5.8.4}$$

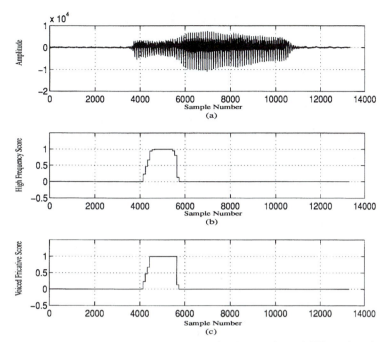

FIGURE A9.14 Voiced fricative detection for the word zoo. (*a*) Time-domain waveform. (*b*) High frequency score. (*c*) Voiced fricative score.

MF(i) is then smoothed by a third-order median filter. A high frequency score, HFS(i), is calculated for each frame as

$$
\text{HFS(i)} = \begin{cases} 1, & \text{if MF(i)} \geq T_{upper} \\ 0, & \text{if MF(i)} < T_{lower} \\ \dfrac{\text{MF(i)} - T_{lower}}{T_{upper} - T_{lower}}, & \text{if } T_{lower} \leq \text{MF(i)} < T_{upper} \end{cases} \tag{A9.5.8.5}
$$

where $T_{upper} = 3200$ and $T_{lower} = 2400$. The thresholds are determined empirically.

The voiced fricative score, VFS(i), is then calculated using HFS(i). If the frame is voiced and the sonorant score is zero, then VFS(i) = 1 for the frame. This is done because the frame is voiced and also has a relatively large amount of high-frequency energy (i.e., the frame is non-sonorant). If the frame is voiced and the sonorant score is 1, then VFS(i) = HFS(i) for the frame. In this case, the voiced fricative score depends solely upon the frame's high-frequency energy distribution. The final step is to set VFS(i) to zero for all frames in any voiced fricative segment less than 15 msec in duration. This is done to eliminate false detection of short segments.

Figure A9.14 shows the results of voiced fricative detection for the word "zoo" spoken by a male speaker. Examination of the spectrogram (not shown) reveals that there is little high-frequency energy at the beginning of the /z/. This explains why the algorithm does not classify the beginning of the /z/ as a voiced fricative.

A9.5.9 Unvoiced Stop and Fricative Detection

If a frame is classified as unvoiced, it is either an unvoiced stop or an unvoiced fricative. The algorithm used to distinguish between the two segment types differs from the standard feature detector, and uses both time-based and frequency-based parameters. First, the mean frequency is calculated for each frame from Equations (A9.5.8.1) through (A9.5.8.4). The mean frequency track is smoothed by a third-order median filter. The high frequency score is then calculated for each frame from

Equation (A9.5.8.5) with $T_{upper} = 3800$ and $T_{lower} = 2400$. The base-ten logarithm of the power, $P_{log\,10}$, is calculated from the initial LPC analysis results of each frame. Next, all adjacent unvoiced frames are grouped into segments. For example, the word "sit" has two unvoiced segments, /s/ and /t/, and each of these unvoiced segments is comprised of multiple, adjacent, unvoiced frames.

The slope of $P_{log\,10}$ of the initial twelve frames (60 msec) of each unvoiced segment is examined. If the segment is shorter than twelve frames, all of the frames are used. A first-order approximation (i.e., a straight line), $M_{seg}(j)$, of the slope of $P_{log\,10}$ is calculated for each segment j using the MATLAB function "polyfit." This is a least-squares fit. A segment slope score, $MS_{seg}(j)$, is computed for each segment from $M_{seg}(j)$ by

$$M_{seg}(j) = \begin{cases} 0, & \text{if } M_{seg}(j) \geq T_{upper} \\ 1, & \text{if } M_{seg}(j) < T_{lower} \\ \dfrac{T_{upper} - M_{seg}(j)}{T_{upper} - T_{lower}}, & \text{if } T_{lower} \leq M_{seg}(j) < T_{upper} \end{cases} \qquad (A9.5.9.1)$$

where $T_{upper} = 1.0$, $T_{lower} = -1.0$, and j is the index of the current segment. The "seg" subscript is included to draw attention to the fact that the slope score is calculated as a single value for the entire unvoiced segment. All of the frames in a given unvoiced segment are assigned the same MS_{seg} value. The frame slope score is denoted as MS(i). Thus, $MS(i) = MS_{seg}(j)$ for each frame i in segment j.

Calculation of the unvoiced stop score, USS(i), takes advantage of the fact that unvoiced stops are inherently shorter in duration than unvoiced fricatives (Cole and Cooper, 1975; Klatt, 1979; Umeda, 1977). The unvoiced stop score, USS(i), is given for each frame by

$$USS(i) = K_s MS(i) \qquad (A9.5.9.2)$$

where i is the current frame index, and K_s is given as

$$K_s = \begin{cases} G_s\left(1 + \dfrac{T_{stop} - L_j}{T_{stop}}\right), & L_j < T_{stop} \\ 1, & T_{stop} \leq L_j \leq T_{fric} \\ \left(1 + \dfrac{L_j - T_{fric}}{T_{fric}}\right)^{-1}, & L_j > T_{fric} \end{cases} \qquad (A9.5.9.3)$$

where $G_s = 8.0$, $T_{stop} = 50$ msec, $T_{fric} = 80$ msec, and L_j denotes the length of the unvoiced segment j (in milliseconds). The two thresholds and the gain, G_s, are all determined empirically. The term K_s acts as a duration-dependent scale factor that greatly amplifies the stop score for unvoiced segments less than 50 msec long. To a lesser degree, K_s also attenuates the stop score for unvoiced segments greater than 80 msec long. The unvoiced stop score, USS(i), is then limited to the range [0, 1]. If USS(i) is greater than one for a given frame, it is set to unity for the frame.

The final unvoiced fricative score, UFS(i), is calculated for each frame by

$$UFS(i) = HFS(i) \qquad (A9.5.9.4)$$

which is simply the high frequency score for the frame.

Figure A9.15 shows both the unvoiced stop and the unvoiced fricative scores for the word "pest" spoken by a male speaker. The /p/ is correctly detected as an unvoiced stop, the /s/ is correctly detected as an unvoiced fricative, and the /t/ has both a high unvoiced stop score and a high unvoiced fricative score. This is because of the high mean frequency of the /t/. Note in Figure A9.15(b), that the /t/ is incorrectly split into two different segments, which is an error caused by the V/U/S algorithm. Still, despite the V/U/S error, the /t/ has a greater unvoiced stop score than unvoiced fricative score, which is desirable.

A9.6 SPEECH SEGMENTATION

The algorithms described in the previous section focus primarily on the acoustic features associated with individual frames. However, in order to segment and label the speech into the segment categories defined in Section A9.2, the boundaries between the phoneme segments must be determined. This is

CHECK FOR 2 PARTS
(2CDs)

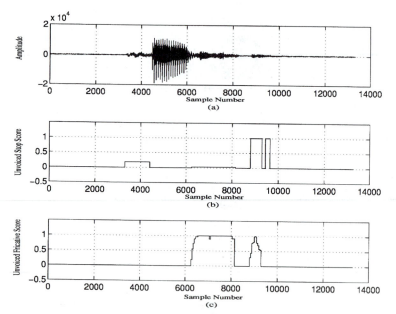

FIGURE A9.15   Unvoiced fricative and stop detection for the word pest. (*a*) Time-domain waveform. (*b*) Unvoiced stop score. (*c*) Unvoiced fricative score.

achieved with two algorithms that are described in the following subsections. The results from the two algorithms are then combined to determine the final segment boundaries and durations.

A9.6.1   Spectral-Based Boundary Detection and Segmentation

The first segmentation algorithm is based upon changes in the short-term frequency spectra of the speech signal. It uses an algorithm developed by Glass and Zue (1986) that measures the similarity between a current frame and its neighbors. To do so, the absolute value of the frequency response of the filter produced by the LPC coefficients from Equations (A9.4.2.2) and (A9.4.2.3) is calculated for each frame. A Euclidian distance measure, $D(x, y)$, is defined as

$$D(x, y) = \sum_{m=0}^{255} \left\| H_x \left(e^{j\pi \frac{m}{256}}\right) \right| - \left| H_y \left(e^{j\pi \frac{m}{256}}\right) \right\| \qquad (A9.6.1.1)$$

where x is the current frame index, y is a past or future frame index, and $H_x(e^{j\pi \frac{m}{256}})$ is the single-sided frequency response evaluated for frame x at the points $\exp(j\pi m/256)$ for $0 \leq m \leq 255$. From Glass and Zue (1986), the decision strategy is to associate the current frame x with past frames if

$$\max(D(x, y)) < \min(D(x, v)) \quad x - 4 \leq y \leq x - 2, x + 2 \leq v \leq x + 4 \qquad (A9.6.1.2)$$

and to associate the current frame x with future frames if

$$\min(D(x, y)) > \max(D(x, v)) \quad x - 4 \leq y \leq x - 2, x + 2 \leq v \leq x + 4 \qquad (A9.6.1.3)$$

No association (a "don't care" state) is made if neither of these conditions is met. After each frame is associated with one of the three states, a segment boundary is determined to occur whenever the current frame's association changes from the past to the future. The location of the boundary is at the first sample of the frame where the transition to the future occurs. Post processing is also done to remove any boundaries that occur in the middle of silent segments.

   Figure A9.16 shows the spectral-based boundary detection results for the word "wield" spoken by a male speaker. The algorithm marks boundaries at the beginning of the /w/, at the beginning of the relatively stationary portion of the vowel /i/, at the end of the transition from the vowel to the liquid /l/, and at the beginning of the release of the /d/. While all of these points are clearly seen from

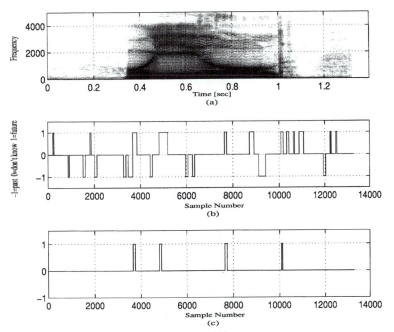

FIGURE A9.16   Spectral-based boundary detection for the word wield.
(*a*) Spectrogram. (*b*) Frame association. (*c*) Spectral boundaries.

Figure A9.16(*a*) as dividing lines between different parts of the word, the locations may not always agree with the results obtained by manual parsing. For example, in the transition from the /i/ to the /l/, manual parsing might put the location of the transition at the middle of the second formant transition region, instead of at the end of the transition. However, this lack of agreement between automatic and manual segmentation results can often be attributed to the lack of a universally accepted method to manually specify the "correct" transition point between two phonemes.

A9.6.2   V/U/S Boundary Detection

The second segmentation algorithm is based upon the voiced/unvoiced/silent (V/U/S) feature detection algorithm results. The raw V/U/S results are processed using the pattern recognition rules listed in Table A9.1. Boundaries are determined to occur wherever transitions in the V/U/S track occur. The boundary is marked at the first sample of the first frame of the new segment.

Figure A9.17 shows the V/U/S boundary detection results for the word "wield" spoken by a male speaker. Note that as discussed in Section A9.5.7, the release of the /d/ more closely resembles a /t/, and is classified as unvoiced.

A9.6.3   Final Segmentation

Results from both the spectral segmentation and the V/U/S segmentation algorithms are used in the final segmentation process. All boundaries from the V/U/S algorithm are marked as boundaries in the final result. Any boundary from the spectral-based boundary detection algorithm that occurs in the middle of a voiced segment (as determined from the V/U/S results) is also marked as a boundary in the final result, provided that the boundary occurs at a frame that is located greater than two frames away from any V-U, U-V, V-S, or S-V boundary. This "two-frame rule" keeps the two algorithms from marking the same phoneme boundary as two separate, but closely spaced boundaries.

Figure A9.18 shows the final segmentation results for the word "wield" as spoken by a male speaker. Note that the boundaries in both Figure A9.18(*b*) and A9.18(*c*) are added together to create the final segmentation results that are shown in a later figure. The two spectral boundaries from

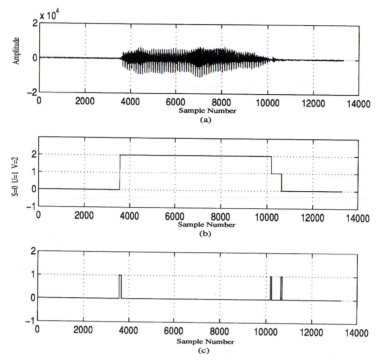

FIGURE A9.17   V/U/S boundary detection for the word wield. (*a*) Time-domain waveform. (*b*) V/U/S classification. (*c*) V/U/S boundaries.

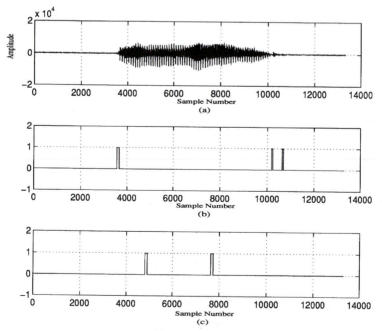

FIGURE A9.18   Final boundary detection for the word wield. (*a*) Time-domain waveform. (*b*) V/U/S boundaries. (*c*) Spectral boundaries.

Figure A9.16(*c*) at (approximately) sample numbers 3700 and 10,100 have been discarded in Figure A9.18(*c*) since they coincide with the V/U/S boundaries and are eliminated by the two-frame rule.

The inclusion of the spectral segment boundaries creates a type of "sub-segmentation" that further divides voiced segments into regions of smaller durations. Examples of this are the semivowel-vowel and the vowel-semivowel transitions shown in Figure A9.18. Here, the boundaries between the /w/ and the /i/, and the /i/ and the /l/, divide the voiced region into three distinct parts.

Spectral boundaries that occur in the middle of unvoiced segments are ignored. This is done to lessen mistakes in subsequent labeling, since the specific token words that are analyzed with this system do not contain double consonant patterns. Note, however, that double consonant patterns regularly exist in the English language. Therefore, in future work, the spectral boundaries that occur in unvoiced segments should be included in the final segmentation results.

A9.7 SEGMENT LABELING

Segment labeling is defined as the task of correctly assigning a label from one of the eight speech segment categories of Section A9.2 to each (unknown type) segment produced by the final segmentation algorithm of Section A9.6.3.

The labeling algorithm first examines the V/U/S results for each segment. If the segment is voiced, it can be labeled as a vowel, semivowel, nasal, voice bar, or voiced fricative. If the segment is unvoiced, it can be labeled as either an unvoiced stop or an unvoiced fricative. If the segment is silent, it can only be labeled as silent. For each segment, each of the possible feature scores is averaged over the duration of the segment. For example, if the unknown segment is unvoiced, then USS and UFS will both be averaged across the frames that comprise the segment. Since the segment in this example is unvoiced, the average scores for VWLS, NS, SVS, VBS, and VFS need not be calculated for the segment. Likewise, if the segment is voiced, the average scores for VWLS, NS, SVS, VBS, and VFS are all calculated, and the averages for USS and UFS are not. The average unvoiced stop score, USS_{mean}, is given by

$$USS_{mean}(j) = \frac{1}{b - a + 1} \sum_{i=a}^{b} USS(i) \qquad (A9.7.1)$$

where a is the index of the starting frame in segment j, and b is the index of the final frame in segment j. The averages for all of the other feature scores are calculated in the same manner.

Once the average scores are calculated for all of the unknown segments, a first choice label, $L1(j)$, and a second choice label, $L2(j)$, are selected for each segment j. The first choice label for each segment is the feature with the highest mean score. The second choice label for each segment is the feature with the second highest mean score. Reliability scores are also calculated for $L1(j)$ and $L2(j)$. The reliability score, $R1(j)$, for $L1(j)$ is defined as the mean score for the first choice label divided by the sum of all of the mean scores for that segment. For example, if a segment is voiced and $L1(j)$ is nasal, then $R1(j)$ is given by

$$R1(j) = \frac{NS_{mean}(j)}{VWLS_{mean}(j) + SVS_{mean}(j) + NS_{mean}(j) + VBS_{mean}(j) + VFS_{mean}(j)} \qquad (A9.7.2)$$

Likewise, if the segment is unvoiced, and $L1(j)$ is an unvoiced fricative, then $R1(j)$ is given by

$$R1(j) = \frac{UFS_{mean}(j)}{UFS_{mean}(j) + USS_{mean}(j)} \qquad (A9.7.3)$$

The reliability score, $R2(j)$, for $L2(j)$ is defined as the mean score for the second choice label divided by the sum of all the mean scores for that segment.

There are two different cases where $R1(j)$ and $R2(j)$ can be used to override $L1(j)$. In the first case, if for a given segment j, (1) the preceding segment is a vowel; and (2) the current segment is not a vowel; and (3) the current segment has a mean vowel score, $VWLS_{mean}$, greater than 0.45; and (4) the second choice for the segment, $L2(j)$, is a vowel; and (5) the reliability of the second

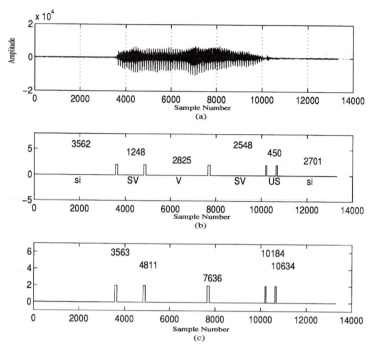

FIGURE A9.19 Final segmentation and labeling for the word wield. (*a*) Time-domain waveform. (*b*) Segment labels and segment durations (in number of samples). (*c*) Segment boundary points (boundary sample number).

choice, $R2(j)$, is greater than 0.35, then $L1(j)$ is changed to a vowel. This is done because the spectral sub-segmentation algorithm of Section A9.6.1 occasionally divides a single vowel into two or more parts. Often this is because the latter part of the vowel is actually starting a transition to a following non-vowel and is exhibiting coarticulation effects. The coarticulation effects may be strong enough to cause the final portion of the vowel to be classified as a non-vowel. Therefore, if the final vowel segment meets the conditions listed above, it can be assumed that coarticulation is taking place, and that the segment is actually a vowel.

The second case where $R1(j)$ and $R2(j)$ are used to possibly override $L1(j)$ is as follows. First, this rule is invoked only if the preceding rule was not invoked for the current frame. Then, if (1) the current segment is a vowel; and (2) the current segment has a mean vowel score, $VWLS_{mean}$, less than 0.50; and (3) the reliability of the second choice, $R2(j)$, is greater than 0.10, then $L1(j)$ is changed to the second choice, $L2(j)$. This is done because the average vowel score is less than 0.5, which implies from Section A9.5.3 that the average voiced consonant score (which is not calculated since it is not a segment category) is greater than 0.5. This indicates that the segment is actually some type of voiced consonant.

Figure A9.19 shows the final segmentation and labeling results for the word "wield" spoken by a male speaker. The symbol "si" denotes silence, the symbol "SV" denotes semivowel, the symbol "V" denotes vowel, and the symbol "US" denotes unvoiced stop. The durations of the individual segments are shown as the number of samples in A9.19*b*. The starting sample number of each new segment is shown in Figure A9.19(*c*).

A9.8 MANUAL MODIFICATION OF AUTOMATIC SEGMENTATION AND LABELING RESULTS

All speech recognition algorithms make mistakes. The number and nature of the mistakes depend upon many factors including the variability of human speakers, the choice of recognition categories,

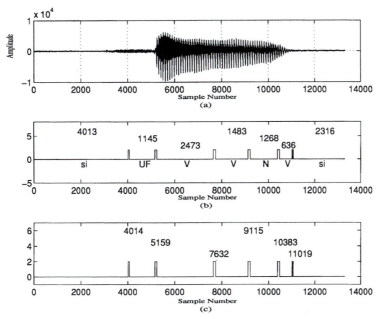

FIGURE A9.20 Final segmentation and labeling for the word foo. (*a*) Time-domain waveform. (*b*) Segment labels and segment duration (in number of samples). (*c*) Segment boundary points (boundary sample number).

and the recognition algorithms themselves. Since our primary goal is to modify the time sequence of speech, and not to create a totally new speech recognition scheme, the errors that occur in automatic segmentation and labeling are repaired manually before the time-modification algorithms are invoked.

This section describes the nature of the errors. The set of software programs along with a graphical user interface (GUI) created to help the user manually edit and correct the automatic segmentation and labeling results are described in Chapter 8.

A9.8.1 Description of Errors

A variety of errors can occur. Segmentation errors result when the algorithms pick either an incorrect number of segments, or incorrect locations for the segment boundaries. A labeling error results when the algorithms pick the wrong label for a segment. Figure A9.20 shows examples of these types of errors.

In Figure A9.20(*c*), the beginning of the unvoiced fricative is incorrectly detected as starting at sample number 4014. Examination of the time waveform shows that the actual beginning of the unvoiced fricative is closer to sample number 3300. This type of error typically occurs with weak, unvoiced fricatives such as /f/, since the energy level of the /f/ is not much greater than the energy level of the background noise. Strong, unvoiced fricatives such as /s/ and /ʃ/ do not exhibit this problem.

A second type of error is seen in Figure A9.20(*b*). The vowel is divided into four parts by the spectral segmentation algorithm. While this is not a problem in itself, it creates the possibility for labeling errors by requiring that the four parts of the same vowel be labeled separately.

A third type of error is also seen in Figure A9.20(*b*). An error occurs in the labeling stage for the third part of the vowel. The third part is labeled as a nasal (N), instead of a vowel (V). This occurs most often for long, voiced segments that are comprised of nasalized vowels and/or word-final vowels.

Figure A9.21 shows a different type of error caused by the spectral segmentation algorithm for the word "veal" spoken by a male speaker. In this case, the boundary between the /v/ and the /i/ is not detected. As a result, the combined /v/-/i/ segment is classified as a vowel, since the vowel score has the

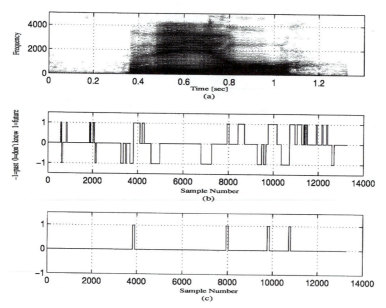

FIGURE A9.21 Spectral segmentation for the word veal. (*a*) Spectrogram. (*b*) Frame association. (*c*) Spectral boundaries.

greatest mean value for the segment. This typically occurs for the weak voiced fricatives, namely /v/ and /th/.

Figure A9.22 also shows the segmentation and labeling results for the word "veal" spoken by a male speaker. Note that the final portion of the word in Figure A9.22(*b*) is classified as an unvoiced stop. Examination of the spectrogram in Figure A9.21(*a*) shows that there is significant, unvoiced, low-frequency noise present at the end of the word. Listening reveals that the noise is caused by the speaker exhaling after completion of the word. Therefore, while unexpected, an unvoiced segment does exist at the end of the word, and it is correctly detected.

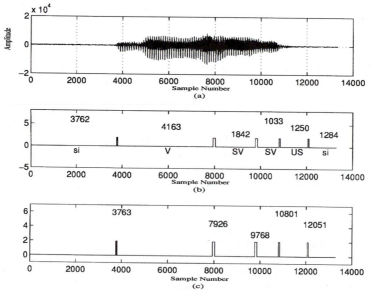

FIGURE A9.22 Final segmentation and labeling for the word veal. (*a*) Time-domain waveform. (*b*) Segment labels and segment durations (in number of samples). (*c*) Segment boundary points (boundary sample number).

A9.9   THE LINEAR PREDICTION CODING (LPC) SPEECH SYNTHESIZER

The all-pole synthesizer filter is expressed in the z-domain as

$$H_i(z) = \frac{1}{A_i(z)} \tag{A9.9.1}$$

where $A_i(z)$ for frame i of the speech signal is

$$A_i(z) = a_{0_i} + a_{1_i}z^{-1} + a_{2_i}z^{-2} + \cdots + a_{N_i}z^{-N} \tag{A9.9.2}$$

The vector $A_i(z)$ is calculated and stored for each analysis frame. The value $N = 13$ adequately models the human speech production system (Hu, 1993).

The error signal obtained during the LPC analysis is the residue signal, $r(n)$, which is used to excite the all-pole filter during synthesis. Let $R_i(n)$ be defined as the portion of the residue signal obtained during the analysis of frame i. For example, if frame 2 begins at sample number 101 and ends at sample number 150, then $R_2(n)$ is the 1 by 50 vector given by

$$R_2(n) = [r(101), r(102), r(103), \ldots, r(149), r(150)] \tag{A9.9.3}$$

The input to the LPC speech synthesizer for frame i can be described by the ordered pair (A_i, R_i), where A_i is a 1 by N vector of LPC coefficients, R_i is a 1 by M_i vector of the residue signal, and M_i is the length (in number of samples) of frame i. Note that M_i is not constant for a pitch-synchronous synthesizer, since the frame size usually changes from pitch period to pitch period.

A9.10   TIME MODIFICATION BASICS— FRAME SKIPPING AND FRAME DOUBLING

The basic time-modification method used here involves either elimination (for compression) or doubling (for expansion) of the (A_i, R_i) ordered pairs prior to being used by the LPC synthesizer. This is best illustrated by the following examples. To synthesize the original speech token without time modification, the ordered pairs (A_i, R_i) are sent to the LPC speech synthesizer for $i = (1, 2, 3, \ldots, L - 2, L - 1, L)$, where L is the total number of frames in the original speech signal. As a result, each frame is synthesized once. This is depicted in Figure A9.23(a). To synthesize the token at approximately twice the original speaking rate (one-half the original duration) the ordered pairs (A_i, R_i) are sent to the synthesizer for $i = (1, 3, 5, \ldots, L - 4, L - 2, L)$. This method skips every other frame during the synthesis process. This is depicted in Figure A9.23(b). Note that we said that the token is synthesized at approximately twice the original speaking rate because the pitch period (and therefore the duration) of each frame is not constant. To synthesize the token at one-half the original speaking rate (twice the original duration) the ordered pairs (A_i, R_i), where $i = (1, 1, 2, 2, 3, 3, \ldots, L - 2, L - 2, L - 1, L - 1, L, L)$, are sent to the synthesizer. This method synthesizes every frame twice, and is depicted in Figure A9.23(c). The three synthesized speech tokens created by these methods for the word "meat" are shown in Figure A9.24. Note that for each of these examples, the silent segment preceding the unmodified word (from sample number 0 to sample number 3300) is not modified. This is done only for demonstration purposes to preserve the alignment of the beginnings of the three synthesized speech tokens in the three graphs.

Although these examples are simple, they demonstrate the basic method of time modification of speech used in this system. This method involves the manipulation of the sequence of (A_i, R_i) ordered pairs used as inputs by the LPC speech synthesizer. Note however, that these examples do not demonstrate selective time modification, since they control the information that is removed or doubled in a simple manner. Selective time modification is accomplished by exercising greater control (than the previous examples) over the ordered pairs used for synthesis. For this system, the selection of the ordered pairs is based on multiple user-specified parameters that are based on the phonemic content of the speech token.

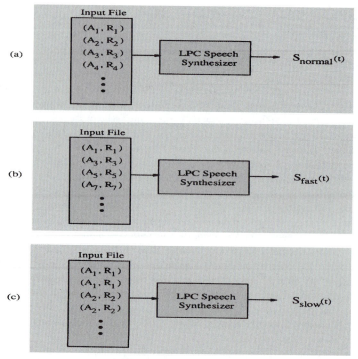

FIGURE A9.23 Examples of time modification using an LPC speech synthesizer. (*a*) Normal rate with no time modification. (*b*) Fast rate with approximately one-half the original duration. (*c*) Slow rate with twice the original duration.

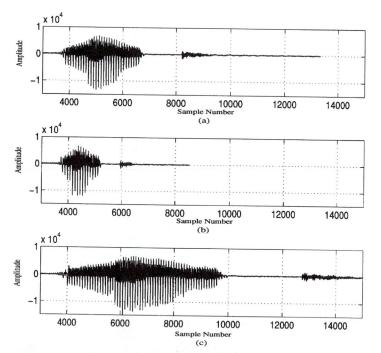

FIGURE A9.24 Examples of time-modified speech. (*a*) Normal rate with no time modification. (*b*) Fast rate with approximately one-half the original duration. (*c*) Slow rate with twice the original duration.

A9.11 USER-SPECIFIED TIME-MODIFICATION PARAMETERS

The user-specified time-modification parameters are based on the eight segment types defined previously. Each segment type (vowel, nasal, unvoiced fricative, etc.) has two, global, user-specified parameters. The first parameter is the duration scale factor (SF). It is expressed as a real number with a resolution of 0.01. There are a total of eight SF parameters, with one per segment type, that is, SF_{vowel}, SF_{nasal}, and so on. The SF parameter specifies the desired ratio of the final segment duration to the original segment duration. For example, if $SF_{vowel} = 0.33$, the duration of the vowel segment(s) in the time-modified word are approximately 33% of the duration of the corresponding vowel segment(s) in the original word. Again we say approximately, since the resolution of the algorithm is controlled by the frame size. The frame is the smallest unit that can be added or removed in the algorithm, and the duration of a discrete number of frames may not exactly equal 33% of the original duration.

The second global parameter is the minimum segment duration (MD), which specifies the minimum duration of the time-modified segment. It is expressed in milliseconds. There are a total of eight MD parameters, one per segment type; that is, MD_{vowel}, MD_{nasal}, and so on. Note that this parameter can override the desired final duration calculated from the segment's SF parameter. For example, if $SF_{vowel} = 0.00$, then the desired final duration of each vowel segment in the time-modified word is, initially, zero msec. However, if MD_{vowel} is greater than zero, the algorithm automatically adjusts SF_{vowel} so that the final duration of each vowel segment is as close as possible to MD_{vowel}. Note that the final duration may not be equal to MD_{vowel} since the resolution is equal to the duration of a single frame, as discussed previously.

The manual scale factor (MSF) is a third, user-specified parameter. It is not global in scope, and is not associated with one particular type of segment. It is expressed as a real number, with a resolution of 0.01. The MSF parameter is used to override the SF parameter for a given segment. A separate MSF value is specified for each segment in the speech token. For example, in the word "man," the initial /m/, the /a/, and the final /n/ each have a unique MSF parameter. The default MSF value is unity, and the MSF parameter must also be "activated" for each segment if it is to be used. The default state is "inactive." In addition, the MD parameter can override the desired final duration calculated from the MSF parameter in the same manner that the MD parameter can override the desired final duration calculated from the SF parameter.

The MSF parameter allows a single occurrence of a particular type of segment to be modified independently in words that have multiple occurrences of the segment type. An example of how the MSF parameter is used is as follows. Suppose, from the previous example, that the word "man" is being modified. If $SF_{nasal} = 0.50$, then both the initial and final nasals are modified during synthesis. If the user only wants the initial nasal to be modified, he or she activates the MSF parameter for segment 3 (the final /n/), and sets its value to 1.00. Since the activated MSF parameter for segment 3 overrides any global SF parameter, the final nasal /n/ has a scale factor of 1.00, which results in no time modification of the segment.

A9.12 MAPPING

Once the desired final durations are specified, the algorithm must determine the frames of a given segment that will be removed or doubled. In the simple examples of Section A9.10, every other frame was removed to achieve compression, and every frame was doubled to achieve expansion. However, these techniques do not offer the flexibility of modifying specific portions of a given segment (i.e., selective modification).

This algorithm uses a method that allows the user to assign a weighting function, or "map," to each segment. Each frame in the segment is assigned a weight between zero and one. During synthesis, the frames with the lowest weights are eliminated (if SF < 1.00), and the frames with the highest weights are doubled (if SF > 1.00). Obviously, if SF (or MSF) for the segment is 1.00, the weight is trivial, since frames are neither eliminated nor doubled.

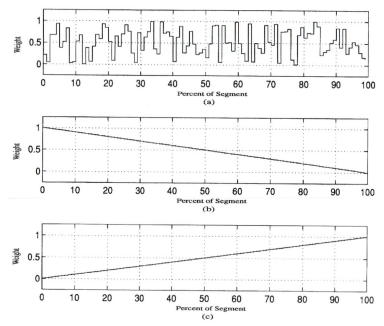

FIGURE A9.25   User selectable mapping functions. (*a*) random. (*b*) fixed_1. (*c*) fixed_2.

The user selects one of eight weighting maps for each segment. Five of the maps are fixed (the "fixed maps"), and the other three can be edited by the user (the "user maps"). The maps are shown in Figures A9.25 and A9.26. The random (fixed) map shown in Figure A9.25(*a*) arbitrarily assigns a random weight to each frame in the segment. The fixed_1 map in Figure A9.25(*b*) emphasizes the beginning of the segment. The fixed_2 map in Figure A9.25(*c*) emphasizes the end of the segment. The fixed_3 map in Figure A9.26(*a*) emphasizes the end points of the segment. The fixed_4 map shown in Figure A9.26(*b*) emphasizes the middle of the segment. One of the three user maps

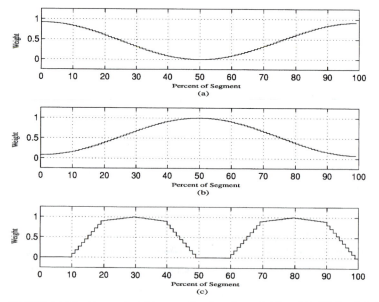

FIGURE A9.26   User selectable mapping functions. (*a*) fixed_3. (*b*) fixed_4. (*c*) user_1.

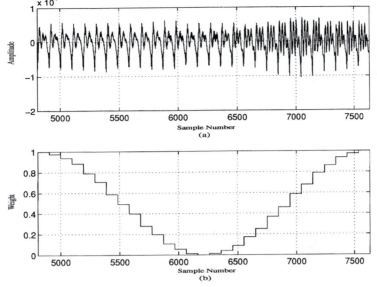

FIGURE A9.27   Interpolated map for the vowel segment of the word "wield."
(*a*) Time-domain waveform. (*b*) Interpolated weighting map (fixed_3 map).

(the user_1 map) is shown in Figure A9.26(*c*). The weighting function shown is arbitrary and can be edited by the user. The three user maps are identical in function. The difference between them is that they can each contain a different weighting curve. As a result, the remaining two user maps (user_2 and user_3) are not shown. The three user maps are saved for each speech token. As a result, each token has its own, unique, mapping functions. For example, the user_1 map for the word "meat" is not necessarily the same as the user_1 map for the word "sue."

Each map is stored as a 1 by 100 vector. Since few segments have exactly 100 frames, an interpolation process determines the actual weight for each frame in the segment. The process first maps the weighting map onto the segment of interest. This is done by creating a temporary map (of length 1 by N samples) by

$$\text{temp\_map(i)} = \text{map(j)}, \quad 1 \le i \le N, \quad j = \text{ceil}\left\{\frac{100i}{N}\right\} \tag{A9.12.1}$$

where map(j) is the original 1 by 100 weighting map, N is the length of the segment in samples, and ceil is a MATLAB function that rounds the argument up to the closest integer. Once the temporary map is calculated, the sample number that resides at the center of each frame is then determined. The weight for each frame of the interpolated map is the value of temp_map(i) for the center sample of each frame.

An example of map interpolation is shown in Figure A9.27 for the vowel segment of the word "wield" spoken by a male speaker. The fixed_3 map is shown after interpolation in Figure A9.27(*b*). Note that each pitch period (frame) in Figure A9.27(*a*) has exactly one corresponding weight in Figure A9.27(*b*).

A9.13   TIME MODIFICATION AND SYNTHESIS

The time modification and synthesis processes are outlined in Figure A9.28. The algorithm begins by creating a 1 by M vector, F_{save}, of frame indices where

$$F_{save} = [1, 2, 3, \ldots, M - 2, M - 1, M] \tag{A9.13.1}$$

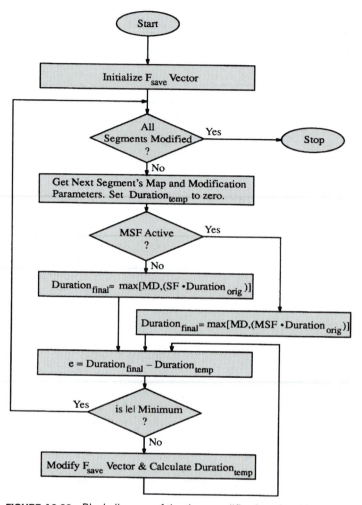

FIGURE A9.28 Block diagram of the time-modification algorithm.

and M is the total number of frames in the original, unmodified word. Once the scale factors and minimum durations are specified, the algorithm calculates the desired final duration for each segment. The process then enters an iterative loop for each segment. If frames are to be cut in a particular segment, then the algorithm examines the interpolated map for the segment, and removes the frame index from F_{save} that corresponds to the frame in the segment with the lowest weight. As a result, F_{save} becomes a 1 by M − 1 vector. If frames are to be added (instead of cut), the frame index with the highest weight is duplicated, and F_{save}, becomes a 1 by M + 1 vector. The duration of the total (new) number of frames in the segment is then calculated (this is denoted as $Duration_{temp}$ in Figure A9.28), and compared to the desired final duration. The algorithm continues in the loop, and removes (or adds) one frame index from F_{save} during each pass. Once the desired final duration is reached for the segment, the process exits the loop. The algorithm repeats the loop for each segment that is modified. The final result is F_{save}, which contains only the indices of the frames that are used to synthesize the time-modified speech.

Note that the algorithm cuts (or adds) the number of frames that causes the actual final duration for the segment to be as close as possible to the desired final duration. However, these two durations are not always the same. Tests show that the actual scale factor (calculated as the ratio of the duration of the modified segment to the duration of the corresponding unmodified segment) usually differs only by one or two percentage points from the desired scale factor. The tests also show that the actual

segment duration is within 2.5 msec of the desired segment duration for unvoiced segments, and 5.0 msec of the desired segment duration for voiced segments.

During synthesis, F_{save} controls the frames to be synthesized. Before synthesis, F_{save} is numerically sorted to guarantee that the elements occur in ascending order. The synthesizer then reads one frame index from F_{save} for each frame of speech that is synthesized. It uses the corresponding (A_i, R_i) ordered pair to obtain the excitation signal and filter coefficients.

The synthesizer architecture is a single, 13th-order, Direct-I type filter. The typical cascade of multiple, second-order sections is not used. Since all of the calculations are done in MATLAB, the filter architecture does not exhibit significant numerical precision problems (i.e., instability due to truncated coefficients).

A9.14 GLITCH PREVENTION

It is important to ensure that the synthesizer does not create artificial discontinuities, or "glitches," in the output signal. These glitches can occur at frame boundaries as the result of discontinuities in the F_{save} vector when frames are removed. The discontinuities can also result from frames being doubled, although the effects are not as severe as those caused by frame elimination. Frame removal can result in discontinuities in the final excitation (residue) signal, as well as large, abrupt changes in the frequency response of the LPC filter. In general, in order to prevent glitches, the contents of the 13 filter taps are analyzed at the frame boundaries. The contents of the filter taps are either preserved or modified, depending upon the specific case.

In the first case, if the indices of the frames on either side of the boundary are sequential with no missing frames, the filter contents are not modified. Thus, the final conditions of the filter after the last sample of the previous frame are the initial conditions of the filter for the first sample of the current frame. For example, if $F_{save} = [1, 2]$, then the final conditions for frame 1 are the initial conditions for frame 2.

In the second case, if the indices of the frames on either side of the boundary are not sequential and have doubled frames, the filter contents are not modified. Thus, the final conditions of the filter after the last sample of the previous frame are the initial conditions of the filter for the first sample of the current frame. For example, this applies if $F_{save} = [1, 1, 1]$. Note, however, that the excitation signal is modified according to a method developed by Hu (1993). This modification is done in two stages. In both stages, it is assumed that the previous frame index is the same as the current frame index. In the first stage, the amplitude of the excitation signal associated with the current frame is multiplied by a scale factor. This is done to simulate shimmer. The scale factor is constant over the duration of the frame, but changes randomly from frame to frame. The scale factor has a uniform distribution over the range [0.975, 1.025]. In the second stage, if the doubled frame is unvoiced, the excitation signal for the current frame is replaced by a zero-mean, white noise sequence that has the same length as the excitation signal. The amplitude of the white-noise excitation signal is scaled to have the same root-mean-square (RMS) value as the original excitation signal for the frame. This is done by Hu (1993) to prevent audible artifacts, such as "warble" or "phasing" effects that occur during time expansion of unvoiced segments. In theory, this is not required, since the excitation (i.e., the LPC residue) for unvoiced frames is a white-noise sequence. Therefore, the act of repeating the sequence should not create the artifacts described. However, due to the relatively short duration of an unvoiced frame (50 samples), the residue for a single frame typically does not exhibit a flat frequency spectrum. Therefore, the act of repeating the "non-white" sequence may indeed cause artifacts to be perceived. This is due to the possibility of periodicity existing in the resulting duplicated excitation sequences.

In the third case, if the indices of the frames on either side of the boundary are not sequential and one or more frames are missing, then a special rule is invoked. The filter calculates the initial conditions that would exist if the single, final missing frame was present, and uses these filter states as the initial conditions for the synthesized frame immediately following the missing (unsynthesized) frames. For example, if $F_{save} = [1, 2, 9]$, the filter calculates the final conditions that would be present if frame 8 were synthesized after frame 2, and uses these as the initial conditions for frame 9. The

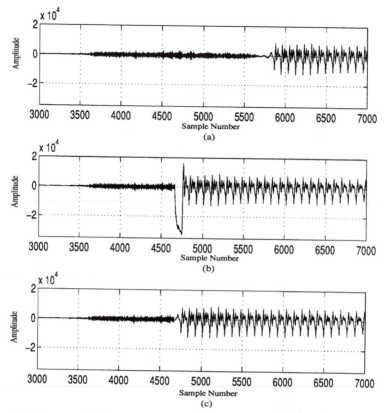

FIGURE A9.29   Glitch prevention for the unvoiced-voiced transition in the word sue. (*a*) Original unmodified waveform. (*b*) Synthesized with final 50% of /s/ removed and glitch prevention off. (*c*) synthesized with final 50% of /s/ removed and glitch prevention on.

motivation behind this rule is: The discontinuities are almost always a result of the energy that is stored in the digital filter, which can be greatly amplified by the new LPC coefficients immediately after the coefficients are updated. This occurs most often when frames are removed from an unvoiced-voiced transition region. Since the gain of the LPC filter is not normalized during the initial analysis to have a constant, average frequency response from frame to frame, the average gain of the LPC filter for voiced regions is much greater than the average gain of the filter during unvoiced regions. Thus, if no glitch prevention is performed, the stored energy in the digital filter during synthesis of an unvoiced frame is greatly amplified when the LPC coefficients are abruptly updated to those of a voiced frame.

An example of this is shown in Figure A9.29 for the transition from the /s/ to the /u/ in the word "sue" spoken by a male speaker. Figure A9.29(*a*) shows the original, unmodified word. Figure A9.29(*b*) shows the synthesized word where the final 50% of the /s/ is removed with no glitch prevention. Note that the filter creates a large glitch that dominates the output for several glottal cycles. Figure A9.29(*c*) shows the synthesized word where the final 50% of the /s/ is removed using the glitch prevention rule described above. There is no noticeable discontinuity at the transition, and the transition region closely resembles that of the original word (ignoring the position of the /s/ that is removed).

The theory described in this appendix has been implemented in the algorithms of the time modification toolbox in Chapter 8.

VOCAL FOLD
MODELS: THEORY

A10.1 INTRODUCTION

The larynx provides phonation for speech. When there is sufficient airflow through the glottis, sufficient air pressure drop across the glottis, and appropriate configurational and physiologic conditions for the laryngeal tissues, the vocal folds of the larynx vibrate (Ishizaka and Flanagan, 1972; Titze and Talkin, 1979a and 1979b; van den Berg and Tan, 1959). The human phonatory mechanism produces a vocal tract excitation of quasi-periodic pulses of air that are generated by the glottis as it valves the airflow from the trachea (Carhart, 1940).

The glottis is the three-dimensional airspace between the vocal folds. The "naturalness" of synthetic speech is closely related to the shape of the glottal flow, which mainly depends on the geometry of the glottis. Traditionally, the glottis is viewed and described primarily from the superior view because most visual imaging techniques are limited to this aspect. Radiographic imaging can provide an additional third dimension, but the detail is poor because of the rapid dynamic changes of the glottis during phonation. Ultrasonic imaging is also limited in its ability to resolve the details of the time-varying boundary between the tissue and air (Titze, 1989).

The vibratory pattern of the vocal folds is a major factor relating laryngeal function to sound production (Childers and Lee, 1991; Childers and Wu, 1990). Due to the relative inaccessibility of the larynx, many direct examinations of laryngeal function are precluded. One method that can integrate the various subsystems of the phonatory mechanism is a computer model of the vocal folds. Computer simulation models can be constructed with adequate relationships among the system parts to not only predict various relationships, but also some hypotheses can be advanced concerning phonatory mechanisms (Scherer, 1981). An understanding of phonatory mechanisms can benefit speech synthesis and analysis, linguistics, and clinical practice including the detection, diagnosis, and treatment of vocal disorders.

The problem of determining the internal structure and movement from speech or other measured data is central to speech analysis. One aspect of acoustic phonetics, for example, deals with inferring articulatory shapes and movement from the speech waveform (or its spectrum). Usually, the success of this inverse problem depends on the ability to accurately model the underlying biophysical processes. Constructing simple predictive models of phonatory acoustics, tissue mechanics, and glottal aerodynamics is difficult. Unlike the acoustic theory of wave propagation in nonuniform ducts, which for years has been the foundation of studies in resonance and articulation (Fant, 1960), the myoelastic–aerodynamic theory of phonation has only recently become quantitative. It has been recognized that an adequate mathematical description of the interaction between fluid flow and tissue movement in the larynx is considerably more complex than acoustic wave propagation in ducts.

Imaging and modeling the vocal folds is difficult, yet holds promise both to increase our scientific understanding of the speech process and to improve our potential for advancing the clinical procedures for treating disorders of the larynx. The vocal fold models toolbox in Chapter 9 provides two vocal fold vibratory models that yield a three-dimensional vibratory image of the vocal folds. This appendix provides the background and theory for the toolbox. Since a knowledge of the structure of the vocal folds as a mechanical vibrator is important for modeling, we first introduce the basic anatomy of the larynx, followed by the accepted theory of vocal fold vibration, and an overview of some existing vocal fold vibratory models.

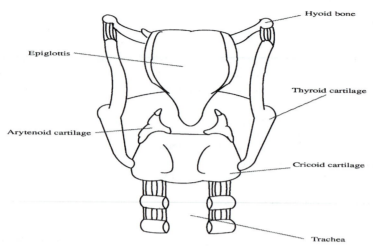

FIGURE A10.1   Posterior view of the larynx.

A10.2 BASIC ANATOMY OF THE LARYNX

A10.2.1 Skeletal Structure of the Larynx and Its Function

The larynx, shown as a sketch in Figure A10.1, is composed of nine cartilages, three paired (arytenoid, corniculate, and cuneiform) and three unpaired (thyroid, cricoid, and epiglottis), and one hyoid bone, as depicted in Figure A10.2 (Schneiderman, 1984; Shearer, 1979).

The hyoid bone can be considered as support for the tongue as well as part of the laryngeal system. The horseshoe-shaped hyoid serves as an attachment for many of the extrinsic laryngeal

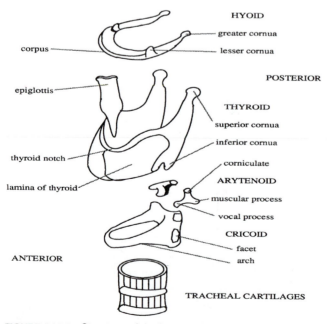

FIGURE A10.2   Structure of the larynx.

muscles. The thyroid is the largest cartilage of the larynx. Its superior cornua attaches indirectly to the corresponding major cornua of the hyoid, and its inferior cornua attaches to the posterior aspect of the cricoid arch. The cricoid cartilage is shaped like a signet ring with the anterior arch, a narrow convex ring, and posteriorly the lamina form the "signet." Two arytenoid cartilages rest on the superior border of the cricoid lamina. Each arytenoid approximates a pyramidal shape with a lateral projection that is called the muscular process and an anterior projection that is called the vocal process. Its superior aspect, or apex, curves slightly backward. The cricoarytenoid joint is a saddle joint that permits a rocking motion and a limited amount of gliding action (Zemlin, 1964 and 1988). The corniculate cartilages (cartilages of Santorini) are two pyramidal shaped nodules located on the apex of each arytenoid for protection of the arytenoid. The cuneiform cartilages (cartilages of Wrisberg) are rod shaped elastic cartilages found in the posterior portions of the aryepiglottic folds that give support to the membrane. The epiglottis is a single leaf-like structure bound by ligaments to the base of the tongue, walls of the pharynx, and thyroid cartilage. This structure acts to close off the laryngeal airway and deflect bulbi of food posteriorly into the esophagus during swallowing.

A10.2.2 Articulations and the Larynx

The cricothyroid and cricoarytenoid cartilages (the articulations are known by the same name) in the larynx are important for understanding the speech process. The cricothyroid articulations are located between the inferior horn of the thyroid and the lateral wall of the cricoid at the point where the arch and laminae meet (see Figures A10.1 and A10.2). Two types of movements are possible for this articulation. One is a gliding movement in a ventrodorsal (anterior to posterior) direction of either cartilage on the other. The second is a tipping (tilting) or rocking motion of either cartilage on the other around an axis that is parallel to a line joining the two eardrums when the head is facing forward (Figure A10.3). The combination of tipping and gliding is the anatomic basis of a stretching or tightening movement of the vocal folds.

Figure A10.3(*a*) shows a gliding movement of the thyroid on the cricoid. The solid line indicates a normal position; the dashed lines show the extremes of the glides. In this case, the thyroid moves on the cricoid as shown by the arrow. Figure A10.3(*b*) shows a tipping (tilting) movement of

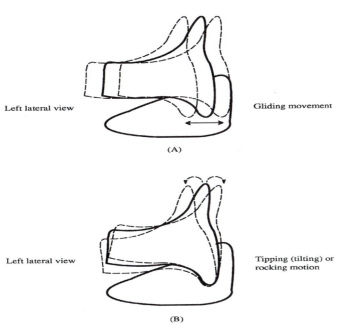

Left lateral view   Gliding movement

(A)

Left lateral view   Tipping (tilting) or rocking motion

(B)

FIGURE A10.3 Movements of the thyroid on the cricoid: cricothyroid articulations.

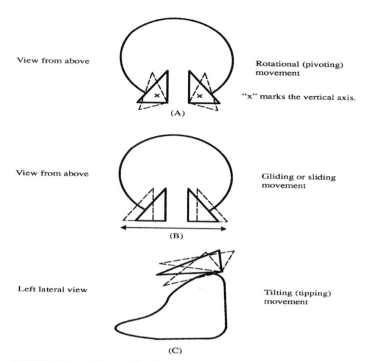

View from above

Rotational (pivoting)
movement

"x" marks the vertical axis.

(A)

View from above

Gliding or sliding
movement

(B)

Left lateral view

Tilting (tipping)
movement

(C)

FIGURE A10.4   Misconceived movements of the arytenoid (cricoarytenoid articulations).

the thyroid on the cricoid. The solid line indicates a normal position; the dashed lines show tipped positions. Note that the inferior horn does not glide on the cricoid.

The cricoarytenoid articulations are very complex. The two arytenoids are perched on the high back part (signet portion) of the cricoid (see Figure A10.1 and Figure A10.2). The mechanics of the cricoarytenoid joint control abduction and adduction of the vocal folds, and thereby facilitate respiration, protect the airway, and permit phonation and other functions of the larynx. The obscure position of this joint and the complex structure have led to at least three misconceptions of arytenoid motion as shown in the Figure A10.4 (von Leden and Moore, 1961). The first misconception is that there is a rotational movement around an axis that parallels the spinal column. The second misconception is that there is a gliding or sliding movement of the arytenoids on the superior border of the lamina of the cricoid. This gliding movement is considered to be back and forth along an axis that is parallel with a line drawn between the eardrums. The third misconception is that there is a tilting (tipping) movement around a horizontal axis that parallels a line drawn between the eardrums (van Riper and Irwin, 1958).

A monocular inspection of the larynx, for instance, leads the observer to the conclusion that the motions of the cricoarytenoid joint may be based on a vertical axis of rotation or on a linear lateral glide (von Leden and Moore, 1961). Based on anatomic, cinematographic, and mathematical studies, von Leden and Moore (1961) revised the three misconceived movements between each arytenoid and the cricoid as shown in Figure A10.5.

There is a rocking or rotating movement around the axis of the joint (the principal axis of rotation). The principal axis of rotation extends in a dorsomediocranial and ventrolaterocaudal direction, Figure A10.5(a). This rocking motion represents an example of internal rotation (similar to the rotation of the earth around its polar axis). The distance of the vocal process from the axis of rotation provides the leverage for the massive movement of the vocal folds during the opening and closing of the glottis. Incidentally, the same rocking movement lowers the vocal process in adduction and slightly shortens the vocal folds.

There is a linear gliding movement parallel to the principal axis of rotation. The direction of linear motion occurs along the longitudinal dimension of the cricoid facet; that is, dorsomediocranially

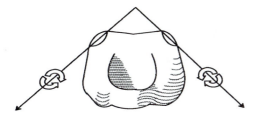

A rocking or rotating movement
around the principal axis of the joint

(A)

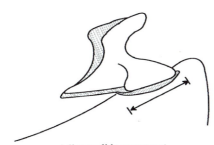

A linear glide movement
parallel to the principal axis of the joint

(B)

FIGURE A10.5 The two principal (correct) types of motion at the cricoidarytenoid joint.

and ventrolaterocaudally, Figure A10.5(*b*). This excursion is limited to approximately 2 mm, the extension of the cricoid facet beyond the longitudinal diameter of the arytenoid facet. In isolation, this motion tends to shorten or lengthen the vocal cords during vocal adjustments, with a small amount of lateral displacement.

A third, although very limited, type of rotating motion is around the secondary axis of rotation, which pivots outside the cricoarytenoid joint, near the attachment of the posterior cricoarytenoid ligaments into the cricoid lamina. This movement represents an illustration of external rotation (comparable to the motion of the earth around the sun). Zemlin (1964 and 1988) feels that the third type of movement is a very restricted and controversial rotary motion. Because of the nature of the joint, this motion is negligible and quite probably does not occur in the normal larynx. However, it is sometimes recognized and has been largely confirmed by means of a mathematical analysis by von Leden and Moore (1961).

Since the vocal folds are attached at the anterior to the thyroid and at the posterior to the arytenoids, and both the thyroid and the two arytenoids are attached to the thyroid with complex articulations, an almost infinite variety of positions of the vocal folds is possible. It is this complexity of movement that makes it so difficult to analyze the possible movements of the vocal folds. Yet it is this same complexity that enables us to make an incredible variety of sounds.

A10.2.3 Laryngeal Membranes

Extrinsic membranes and ligaments connect the laryngeal cartilages, hyoid bone, and trachea. The thyrohyoid membrane runs between the thyroid cartilage and the hyoid bone. The hyoepiglottic membrane runs from the epiglottis to the hyoid, and the cricotrachial ligament connects the cricoid to the first tracheal ring. The intrinsic membrane covers the median surface of the larynx, connecting the laryngeal cartilages together.

A10.2.4   Muscles of the Larynx

The laryngeal musculature is divided into extrinsic and intrinsic groups. Extrinsic muscles are those that support the larynx, and are also called strap muscles. Their function is to move the larynx as a whole. They have at least one attachment to a structure outside the larynx. The intrinsic muscles control phonation. They have both relatively fixed and movable muscle attachments within the larynx, and may be grouped according to the function of opening or closing the glottis.

The intrinsic muscles are described next (Moore, 1971; Schneiderman, 1984; Shearer, 1979). The posterior cricoarytenoid muscle is the abductor of the glottis, which is situated on the posterior surface of the cricoid cartilage (Figure A10.6). It rotates the arytenoid so that the muscular process is drawn backward and the vocal process outward. Thus, the vocal folds are moved laterally and the glottis is widened. The lateral fibers of this muscle help slide the arytenoids laterally.

There are four adductor muscles (oblique arytenoid, transverse arytenoid, lateral cricoarytenoid, and thyroarytenoid) that close the glottis. The oblique arytenoid (paired) originate on the posterior surface of the muscular process of one arytenoid (Figure A10.7). Together the two oblique arytenoids serve as a weak sphincter for the superior aperture of the larynx. The transverse arytenoid (unpaired) covers the entire posterior surfaces of the arytenoids, extending from the base to the summit (Figure A10.6). It approximates the arytenoid cartilages by sliding toward one another. This muscle closes the posterior part of the glottis. Complete glottal closure is often assisted by the lateral cricoarytenoid muscle.

The lateral cricoarytenoid (paired) are located interior to the thyroid cartilage in the lateral wall of larynx (Figure A10.8). They rotate the arytenoid so that the muscular process is drawn forward with the vocal process medialward. The thyroarytenoid (paired) form the substance of the vocal folds and include the thyrovocalis muscle and thyromuscularis fibers (Figure A10.9). They may influence the vibration of the vocal folds by drawing the arytenoids closer to the thyroid. The thyroarytenoids make the folds more flaccid. By varying the tension in the walls to which the vocal folds attach, the thyroarytenoids affect the mode of vibration of the folds. The cricothyroid muscle lies on the external surface of the larynx arising along the lower border and outer surface of the cricoid arch (Figure A10.10). This muscle increases the distance between the angle of the thyroid and the arytenoids, thus increasing the tension of the vocal folds.

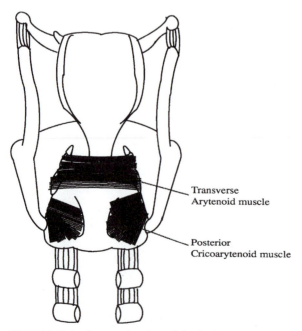

Transverse
Arytenoid muscle

Posterior
Cricoarytenoid muscle

FIGURE A10.6   A posterior view of the intrinsic muscles of the larynx.

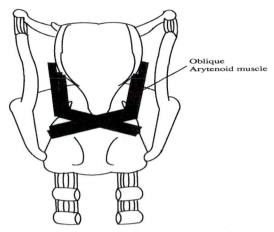

FIGURE A10.7 A posterior view of the intrinsic muscles of the larynx.

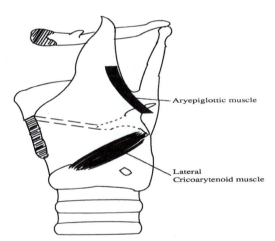

FIGURE A10.8 Intrinsic muscles of the larynx as viewed from the left with the left side of the thyroid cartilage removed.

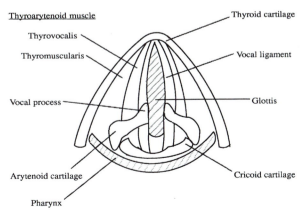

FIGURE A10.9 Intrinsic muscles of the larynx. A superior view of transverse section at the level of the vocal folds.

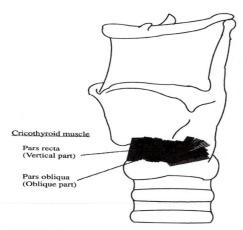

Cricothyroid muscle

Pars recta
(Vertical part)

Pars obliqua
(Oblique part)

FIGURE A10.10   A left lateral view of the intrinsic muscles of the larynx.

A10.2.5   Laryngeal Cavity

The laryngeal cavity is the upward continuation of the cavity of the trachea from the cricoid cartilage to the superior entrance bounded by the glossoepiglottic folds (Figure A10.11). The laryngeal cavity has three divisions. The superior subdivision or vestibule, extends from the superior aperture of the larynx to the ventricular or false folds. The middle subdivision, or ventricle, extends from the ventricular folds to the true vocal folds. The inferior division extends from the vocal folds to the trachea (Boone, 1971).

The larynx is a valving system. The ventricular folds (paired) run horizontally along the lateral wall of the laryngeal cavity. Each soft and somewhat flaccid fold contains the lower part of the quadralateral membrane with its ventricular ligament and mucous glands, which lubricate the vocal folds. The ventricular folds are more widely separated than the vocal folds, and form a valve that serves to keep air inside the lungs when excess intralung pressure is needed. The ventricle of Morgagni (which is a cave-like cavity having its opening between the ventricular fold and the vocal fold on the same side) aids the valving action of the ventricular folds. Laterally it has an upward extension. When the ventricular folds are approximated and the intrathoracic pressure is increased, the pressure inside the ventricle of Morgagni is increased, and, thus, the ventricular folds are pressed harder against each other, aiding the folds to resist even greater intrathoracic pressure.

The vocal folds or true vocal folds (paired) are parallel to and inferior to the ventricular folds. They extend from the posterior surface of the thyroid angle to the vocal processes of the arytenoids (Figure A10.9). Each band is attached along the lateral wall of the larynx and is composed of a

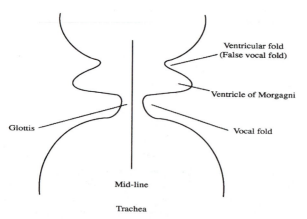

Ventricular fold
(False vocal fold)

Ventricle of Morgagni

Glottis

Vocal fold

Mid-line

Trachea

FIGURE A10.11   Schematic representation of the laryngeal cavity. A frontal view.

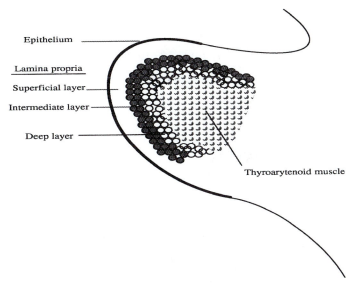

FIGURE A10.12 Schematic representation of a coronal section through the right vocal fold showing tissue layers of the vocal fold.

medial ligament (vocal ligament) and two lateral muscle groups (the thyromuscularis and thyrovocalis muscles). Its mucous membrane is thin and pale in color. Its median edge (vocal ligament) is pearly white. The vocal folds form a valve that prevent the entrance of air or other substances into the trachea and lungs.

The glottis is the opening between the vocal folds. The intermembranous portion of the glottis is the anterior section bounded by the vocal folds. The intercartilagious portion is the posterior section bounded laterally by the medial surfaces of the arytenoid cartilages and posteriorly by the transverse arytenoid muscle (Figure A10.9). The width of the glottis is determined by movements of the arytenoid cartilages. When the arytenoids slide toward each other or rotate so that their vocal processes are approximated, the glottis is narrowed. The opening is widest during inhalation and narrowest during phonation. Tilting the arytenoids, though it changes slightly the length of the glottis, is important, and affects the tension of the vocal folds. The length of the glottis may be altered by the rotating movements allowed by the cricothyroid joint.

A10.2.6 Vocal Fold Tissue Layers

Figure A10.12 shows a drawing of the layered structure of the right vocal fold in coronal section (Mackenzie, Lavar, and Hiller, 1983; Titze, 1994). The outermost layer is thin, 0.05 to 0.10 mm thick, made up of stratified squamous (layered and scalelike) epithelium (Hirano, 1977). The epithelium encapsulates softer, fluidlike tissue, somewhat like a balloon filled with water. The lamina propria, a layered system of nonmuscular tissues, is between the epithelium and muscle, and can conveniently be divided into three layers: superficial, intermediate, and deep. The superficial layer is approximately 0.5 mm thick in the middle of the vocal fold (Hirano, Kurita, and Nakashima, 1981) and consists primarily of loosely organized elastin fibers surrounded by interstitial fluids. Elastin fibers are made of a special type of protein structure that allows for ample elongation (like a rubber band). The intermediate layer is also made up primarily of elastin fibers, but they are more uniformly oriented in the anterior–posterior (longitudinal) direction. There are also some collagen fibers. The intermediate and deep layers of the lamina propria together are about 1 to 2 mm thick (Hirano, Kurita, and Nakashima, 1981). The deep layer is made up primarily of collagen fibers. These fibers have a protein structure that limits elongation. Like a cotton thread, they are nearly inextensible. The fibers in the deep layer also run parallel in the anterior–posterior direction. The thyroarytenoid muscle, approximately 7 to 8 mm thick, lies laterally to the lamina propria. This is the major portion of the vocal fold. In total, the epithelium, the three layers of the lamina propria, and the muscle constitute a five-layer scheme.

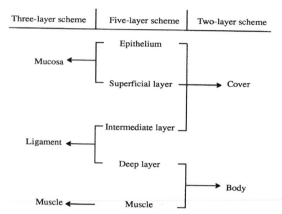

FIGURE A10.13 Three schemes used to label the layered structure of the vocal folds.

Different labeling schemes have been used to group the vocal fold soft-tissue layers, depending on the physiology to be described. In a three-layer scheme, the mucosa consists of the epithelium and the superficial layer of the lamina propria, the ligament consists of the intermediate and deep layers of the lamina propria, and muscle refers to the thyroarytenoid muscle (Hirano, 1974, 1975, 1977; Hirano and Sato, 1993). In a two-layer scheme, the body is equivalent to the deep layer of the lamina propria and the muscle, and the term "cover" is used to describe the combination of epithelium, superficial, and intermediate layers of the lamina propria (Hirano and Kakita, 1985). The three schemes are summarized in Figure A10.13.

A10.3 THEORY OF VOCAL FOLD VIBRATION

A10.3.1 Myoelastic–Aerodynamic Theory

The myoelastic–aerodynamic theory states that phonation occurs when the vocal folds are approximated by muscle contraction on the arytenoid cartilages. Air from the lungs increases in speed as it flows through the narrowed glottis. The increased air flow through the glottis results in a drop in pressure along the margin of the vocal folds. When tissue pressure inherent within the folds exceeds the pressure at the glottal margin, the folds are "sucked" closer together. This effect is the Bernoulli principle. This principle states that as a gas or liquid increases in velocity across a plane, there is a pressure drop along the plane (i.e., less pressure is exerted perpendicular to the flow). Applying this to the action of the vocal folds, the continuing effect of narrowing the glottis and increasing the air flow velocity eventually closes the glottis. At this point, the subglottal air pressure increases until it exceeds the tissue pressure holding the folds together, and the folds are moved apart. The entire cycle then repeats itself.

A10.3.2 Description of Vocal Fold Vibration

We can divide a single vibratory cycle into three distinct phases to describe the vocal fold vibration (Figure A10.14). The first is an opening phase, during which the vocal folds pull apart, increasing the area of the glottal opening. Second is a closing phase, during which the vocal folds come together, reducing the glottal area. Finally, there is a closed phase, during which the vocal folds are maximally closed. Note that in some vibratory modes as in a breathy voice, a distinct closed phase may not exist and the area of the glottal opening shows an almost sinusoidal variation with time.

Based on observations using excised larynges (Baer, 1981a, 1981b), ultra-high speed photography (Hildebrand, 1976; Moore, 1975; Timcke, von Lenden, and Moore, 1958), ultrasonography (Hamlet, 1981), and x-ray stroboscopy (Saito et al., 1981), the movements of the vocal folds during these three phases in normal chest (modal) voice can be described as follows.

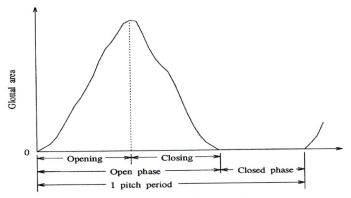

FIGURE A10.14 Divisions of the glottal area cycle.

During the opening phase, the vocal folds first separate inferiorly and the opening moves upward with a wave-like motion in the mucous membrane. Occasionally, the opening first appears on the superior surface as a small "chink" that opens up in a "zipper" like fashion (Baken, 1987; Childers et al., 1986; Childers et al., 1990). The closing phase begins with contact between the lower edges of the glottis. The closure then proceeds along the length of the lower edge and is followed by the mucosal layers coming together. The closed phase is not necessarily associated with an increasing amount of contact between the vocal folds. It is often observed (Baer, 1981a) that as the vocal folds come into contact in a vertical plane, they may be pulling apart at the same time in a different vertical plane.

A schematic representation of vocal fold vibration from Stevens and Klatt (1974) is shown in Figure A10.15. The sketches numbered 1 to 7 in part *a* represent schematized sections through the

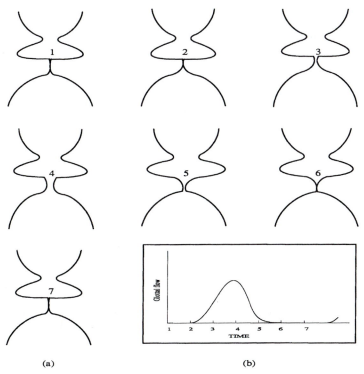

(a)   (b)

FIGURE A10.15 A schematic representation of vocal fold vibration.

larynx at various instants of time during a cycle of vocal fold vibration. A sketch of the waveform of the volume velocity through the glottis is given in *b*. The points indicated on the sketch correspond to the various sections in part *a* of the figure. These sketches are not based on actual measurements, but are derived from reports of stroboscopic tomography of the larynx, and observations from high speed and stroboscopic pictures of the vibrating larynx. See Childers et al. (1986) for other more detailed descriptions. The vocal fold model toolbox in Chapter 9 shows this motion.

A10.4 REVIEW OF VOCAL FOLD VIBRATORY MODELS

The dynamics of vocal folds have been extensively studied for several decades and a number of models of the vocal folds have been developed. These models include the one-mass model (Flanagan and Landgraf, 1968), the two-mass model (Flanagan and Ishizaka, 1978; Ishizaka and Flanagan, 1972, 1977), the multiple-mass model (Titze, 1973, 1974), the continuum model (Titze and Strong, 1975; Titze and Talkin, 1979a), the four-parameter model (Titze, 1984, 1989), and the body-cover model (three-mass model) (Story and Titze, 1995). One can classify the above models into four main categories as follows. Mechanical models include the one-mass model, two-mass model, and multiple-mass model; and the continuum model, ribbon model with four parameters, and the body-cover model with three-masses.

A10.4.1 Mechanical Models

We can classify the one-mass, two-mass, and multiple-mass models as mechanical models because these models represent the glottal source as lumped mechanical oscillators. In a lumped mechanical model, the subglottal system is represented by an air reservoir with pressure (P_s) that provides an air flow with the volume velocity (U_g). The vocal folds are mechanically modeled by an oscillatory system of masses, viscous damping, and springs (Titze and Strong, 1975; Titze and Talkin, 1979a).

A10.4.1.1 One-Mass Model
In the one-mass model, the vocal fold vibrations are modeled with a single mass-spring oscillator driven by airflow from the lungs as shown in Figure A10.16. The one-mass model has the following characteristics. It is simple with a low computational burden. There is source-tract interaction. The phase-difference between the motion of fold edges is disregarded and the glottal area and volume velocity can be simulated.

A10.4.1.2 Two-Mass Model
In the two-mass model, the vocal folds are divided in depth into an upper and a lower mass due to the anatomic and functional division between the mucosa and the vocalis. Each part consists of a simple mechanical oscillator having mass, spring, and damping (m, s, r) as in Figure A10.17. The springs represent the elastic properties of the folds. The damping represents dissipative forces such as viscosity and friction. There is also an interaction between the two masses, represented by a coupling stiffness S_c. The coupling stiffness represents the

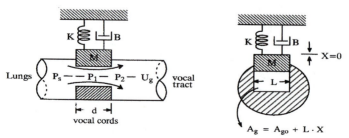

FIGURE A10.16 One-mass model of the vocal folds.

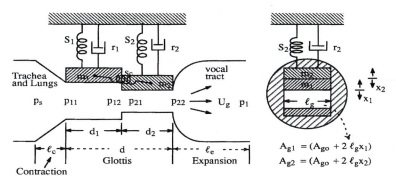

FIGURE A10.17 Two-mass model of the vocal folds.

fact that as one of the masses is displaced relative to the other, there is a force tending to restore the masses to their equilibrium position relative to one another. The two-mass model has the following characteristics: realistic simulation of glottal properties, phase-difference between the motion of fold edge is considered [the mucosal surface wave is not considered in Ishizaka and Flanagan (1972), but it is in the modified model of Koizumi et al. (1987)]. Natural speech can be produced with a reasonable computational burden.

A10.4.1.3 Multiple-Mass Model

Although the two-mass model is a milestone in quantifying vocal fold vibration, it models only the vocal folds as a minimal mechanical structure capable of responding to aerodynamic forces and sustaining oscillation. It is not capable of exhibiting various longitudinal vibratory modes observed in human phonation. Titze (1973), in an attempt to enlarge the horizontal degrees of freedom, proposed a 16-mass model composed of two rows of eight masses each (Figure A10.18). The top-row of masses represents primarily the mucosa and the bottom row represents primarily the vocal ligament and the vocalis muscle. The forces T_m and T_v represent the longitudinal tensions as determined by the balance of forces between the cricothyroid and thyroarytenoid muscles. Specifically, the spring constants for the upper and lower rows increase nonlinearly with elongation of the vocal folds. The 16-mass model has the following characteristics. It is complex, with a high computational burden. The mucosal surface wave can be simulated. It has the ability to regulate the model by parameters that have direct physiologic correlates. There is increased naturalness of the utterances, and the phonation is in at least two distinct registers.

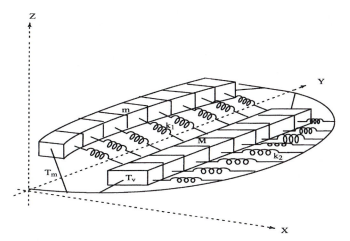

FIGURE A10.18 A 16-mass model of the vocal folds.

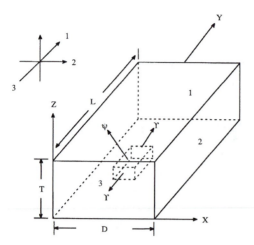

The origin of the coordinate system is centered at the vocal processes.

Within the rectangular parallelepiped representing the vocal fold.

Surfaces 1, 2, and 3 are fixed, and others are free.

ψ : Displacement vector of the differential element

Υ : Longitudinal stress of the differential element

FIGURE A10.19 A continuum model of the vocal folds.

A10.4.2 Continuum Model

One could extend the 16-mass model by increasing the number of masses and degrees of freedom. Titze and Strong (1975) went a step further and represented the vocal folds not as a coupled set of discrete masses but as a continuous deformable medium (Figure A10.19). The incompressibility of the vocal folds dictates a coupling between the horizontal and vertical motion. An important consequence of the incompressibility of the vocal folds is that the most easily excited vibratory mode appears to involve vertical phase differences, since this mode tends to preserve the volume of the vocal folds. The Titze and Strong (1975) study also showed that the layered structure of the vocal folds is ideally adapted to support vocal fold vibration. The longitudinal fibrous structure is more loose in the vertical direction than in the longitudinal direction. This allows vertical phase differences to occur (Chan, 1989).

The continuum model is highly informative concerning the relationship between the vocal fold structure and the vocal fold vibratory modes. However, the shape of the vocal folds in this model is restricted to a rectangular form. The tissue properties are uniform in the plane normal to the longitudinal direction for ease of manipulation (Titze, 1976). In addition, the model lacks a complete representation of the interaction between the aerodynamic air flow and the elastic vocal fold tissue because the normal modes of vocal fold vibration are derived based on an eigen-value analysis of the fold tissue.

A10.4.3 Ribbon Model

Vocal fold vibration occurs mainly in a thin layer of the nonmuscular tissue at the vocal fold surface. It is estimated that the effective depth of vibration into the vocal fold is on the order of 1 mm. Hence, one can think of the vibrating portion as a stretched ribbon that is fixed at the horizontal endpoints ($Y = 0$ at the posterior arytenoid part, $Y = L$ at the anterior thyroid part) but is free to bend and flex in the vertical dimension between those endpoints. The motion of the ribbon, therefore, can be described by a wave equation with appropriate boundary conditions, and its eigenfunction will give the approximate vibration patterns of vocal folds. Based on this concept, a kinematic four-parameter model (Figure A10.20) for the three-dimensional glottis was presented by Titze (1984, 1989). Titze's model can

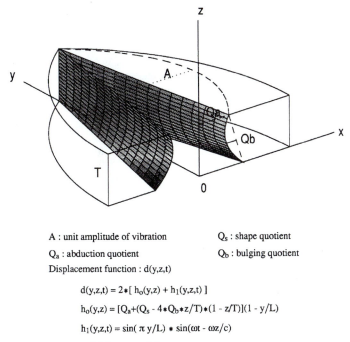

A : unit amplitude of vibration $\qquad$ Q_s : shape quotient

Q_a : abduction quotient $\qquad$ Q_b : bulging quotient

Displacement function : $d(y,z,t)$

$$d(y,z,t) = 2*[\, h_0(y,z) + h_1(y,z,t)\,]$$

$$h_0(y,z) = [Q_a + (Q_s - 4*Q_b*z/T)*(1 - z/T)](1 - y/L)$$

$$h_1(y,z,t) = \sin(\pi\, y/L) * \sin(\omega t - \omega z/c)$$

FIGURE A10.20 Ribbon model.

provide the glottal flow, glottal area, and vocal fold contact area waveforms. The static glottis is controlled by the abduction quotient (Q_a), the shape quotient (Q_s), and the bulging quotient (Q_b). The phase quotient (Q_p) and fundament frequency (F_o) control the dynamic glottis. The displacement function, $h_1(y, z, t)$, is used to calculate the glottal area and is a sinusoidal function.

A10.4.4 Body-Cover Model (Three-Mass Model)

The body-cover concept (Hirano, 1974) is generally used to describe the layered structure of vocal folds (Figures A10.12 and A10.13). It suggests that the vocal folds can be divided into two tissue layers with different mechanical properties. The body layer consists of muscle fibers and some tightly connected collagen fibers of the vocal ligament. The cover layer consists of pliable, noncontractile tissue that acts as a flexible sheath around the body layer. The cover typically is loosely connected to the body during vibration. The motion of the cover layer is usually observed as a surface wave that propagates from the bottom of the vocal folds to the top, thus experiencing movement in both the lateral and vertical directions. Self-sustained vocal fold oscillation is highly dependent on this surface-wave behavior (typically referred to as the vertical phase difference) and is the primary mechanism for transferring energy from the glottal flow to the tissue to fuel the vibration. The body layer is primarily involved in lateral motion. Based on his findings, Hirano (1974) suggests that the vocal folds should be treated as a double structured vibrator with stiffness parameters that should be based on the relative actions of the thyroarytenoid and cricothyroid muscles. Thus, the resultant vibration of the vocal folds is composed of the coupled oscillations of the body and cover layers.

In the two-mass model, the lower mass is made thicker (vertical dimension in the coronal plane) and more massive than the upper element in an attempt to include the effects of the body layer. But, because a provision does not exist for coupled oscillation of both layers, the two-mass model is essentially a "cover" model rather than a "body-cover" model. In order to more realistically represent the body-cover vocal fold structure, Story and Titze (1995) extended the two-mass model into a three-mass model (Figure A10.21).

The three-mass model consists of two "cover" masses coupled laterally to a "body" mass by nonlinear springs and viscous damping elements. The body mass, which represents muscle tissue, is

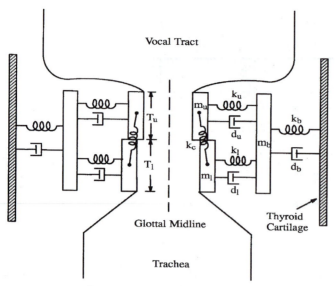

FIGURE A10.21   Body-cover model of the vocal folds.

further coupled laterally to a rigid wall (assumed to represent the thyroid cartilage) by a nonlinear spring and a damping element. The two cover springs are intended to represent the elastic properties of the epithelium and the lamina propria, while the body spring simulates the tension produced by contraction of the thyroarytenoid muscle. Thus, contractions of the cricothyroid and thyroarytenoid muscles are incorporated in the values used for the stiffness parameters of the body and cover springs. The two cover masses are coupled to each other through a linear spring, which can represent vertical mucosal wave propagation.

A10.5 THE TWO-MASS VOCAL FOLD MODEL: THEORY

This section describes the purpose for and the implementation of the two-mass model (Ishizaka and Flanagan, 1972). This is followed by the development of an acoustic model, and some examples.

A10.5.1 Purpose

The two-mass model provides sufficient theory to investigate variations of and the relationship among kinematic parameters of a vocal fold model as well as selected physiologic parameters; for example, pre-phonatory shape of the glottis, fold tension, and lung pressure. The kinematic parameters include fundamental frequency, vertical phase difference, vertical equilibrium shape, vertical amplitude profile, and maximum excursion of the vocal folds during vibration. There are limitations to the model. For example, variations of the vocal fold dynamics along the length of the glottis cannot be simulated because the two-mass model can only account for vertical dissimilarities.

A10.5.2 Acoustic Model for the Implementation of Two-Mass Model

An acoustic model is required to implement the two-mass model (Ishizaka and Flanagan, 1972). The acoustic model is a simple articulatory speech synthesizer, which does not include a noise source model or a nasal tract model. Usually, an articulatory speech synthesizer contains an articulatory model and an acoustic model. The articulatory model maps the positions of the key articulators, such as the jaw, tongue, lips, and velum, to the cross-sectional area function of the vocal tract. The

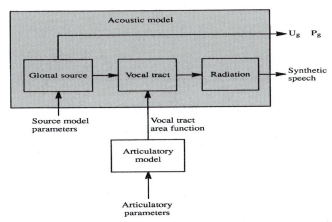

FIGURE A10.22 Acoustic model for the two-mass vocal fold model.

acoustic model is a set of ordinary differential equations (acoustic equations) that describe the acoustic properties of the vocal system. To obtain synthetic speech, one can solve the acoustic equations using numerical methods. The basic structure of the acoustic model for the implementation of the two-mass model is shown in Figure A10.22.

A10.5.2.1 Acoustic Model of the Glottal Source
The schematic diagram of the two-mass model (Ishizaka and Flanagan, 1972) appears in Figure A10.17. The motions of the masses in the two-mass model are governed by the aerodynamic forces that activate the larynx as well as the myoelastic forces of the springs and the dampers. The basic equations of motion are

$$m_1 \frac{d^2x_1}{dt^2} + r_1 \frac{dx_1}{dt} + k_1x_1 + k_c(x_1 - x_2) + F_1 = 0 \qquad (A10.5.2.1.1)$$

$$m_2 \frac{d^2x_2}{dt^2} + r_2 \frac{dx_2}{dt} + k_2x_2 + k_c(x_2 - x_1) + F_2 = 0 \qquad (A10.5.2.1.2)$$

where x_i represents the lateral displacement of the two masses, F_i represents the aerodynamic forces exerted on each mass, r_i represents the viscous loss (resistance), $i = 1$ for the lower mass, $i = 2$ for the upper mass, $r_i = 2\zeta_i\sqrt{m_ik_i}$, ζ is the damping ratio, m is the mass, and k is the spring constant. In the model, the springs are given a nonlinear characteristic to conform to the stiffness as measured from excised human vocal folds (Ishizaka and Flanagan, 1972). During closure of the glottis, there is a contact force that results in additional deformation. The spring restoration forces are represented as

$$f_{si} = k_ix_i(1 + \eta_{ki}x_i), \quad \text{for } i = 1, 2 \qquad (A10.5.2.1.3)$$

$$f_{hi} = h_i(x_i + x_{oi})(1 + \eta_{hi}(x_i + x_{oi})), \quad \text{for } (x_i + x_{oi}) < 0 \qquad (A10.5.2.1.4)$$

where $i = 1$ for the lower mass, $i = 2$ for the upper mass, k and η_k are the linear and nonlinear stiffness parameters of the spring, and h and η_h are additional linear and nonlinear stiffness parameters of the spring during the closed glottal phase.

The physiologic constants of the two-mass model are set to values suggested by Ishizaka and Flanagan (1972) and are summarized in the Figure A10.23. Note that the lower mass (the vocalis–ligament combination) is five times as massive and five times as thick as the upper mass (the mucous membrane). For simulation of normal voices, the parameter values in Figure A10.23 are held constant, except for control of the fundamental frequency. Ishizaka and Flanagan (1972) proposed a tension parameter, Q, to control the fundamental frequency. The fundamental frequency in the range of 120 to 220 Hz varies almost linearly with the tension parameter Q. For other vocal registers such as falsetto and vocal fry, the parameter values in Figure A10.23 are no longer valid. Similar problems occur when attempts are made to simulate the vibratory pattern of abnormal vocal folds. Therefore, the parameters in the two-mass model must be changed according to laryngeal adjustments for other voicing modes.

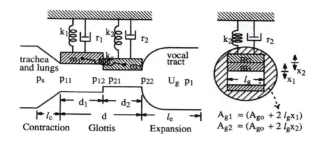

$m_1 = 0.125$ g	lower mass
$m_2 = 0.025$ g	upper mass
$d_1 = 0.25$ cm	thickness of m_1
$d_2 = 0.05$ cm	thickness of m_2
$l_g = 1.5$ cm	effective length of vocal folds
$k_2 = 8000$ dyne/cm	linear stiffness of spring 2
$k_1 = 80000$ dyne/cm	linear stiffness of spring 1
$k_c = 25000$ dyne/cm	stiffness of coupled spring
$\eta_{k1} = \eta_{k2} = 100$ dyne/cm	nonlinear stiffness of spring 1 and 2
$\eta_{h1} = \eta_{h2} = 500$ dyne/cm	nonlinear stiffness of contact springs
$h_1 = 3 \cdot k_1$	linear stiffness of contact spring 1
$h_2 = 3 \cdot k_2$	linear stiffness of contact spring 2
$\zeta_1 = 0.1$	damping ratio for open phase
$\zeta_2 = 0.6$	damping ratio for open phase
$\zeta_1 = 1.1$	damping ratio for closed phase
$\zeta_2 = 1.6$	damping ratio for closed phase

FIGURE A10.23 The physiologic constants for the two-mass model.

The acoustic impedance elements of the glottal orifice constitute an equivalent circuit of the glottis shown in the Figure A10.24, where R_c represents the abrupt contraction at the inlet to the glottis; R_{v1} and R_{v2} represent viscous losses at the lower-fold edge, upper-fold edge, respectively; R_{12} represents the change in kinetic energy per volume of fluid at the junction between masses m_1 and m_2; R_e represents the expansion of the glottal outlet; L_c, L_{g1}, and L_{g2} represent the inertances of the air masses (Ishizaka and Flanagan, 1972).

A10.5.2.2 Acoustic Model of the Vocal Tract
The vocal tract is a three-dimensional lossy cavity composed of nonuniform cross-sections and nonrigid walls (Sondhi, 1974; Sondhi, 1986). Although the appropriate Navier–Stokes equations with the boundary conditions for nonrigid walls describe the acoustic properties of the vocal tract, a large number of calculations are required to solve such equations and neither the shape of the vocal tract nor the physical properties of the walls are known with sufficient accuracy to establish a reliable model. These limitations suggest the need for a simplified version of the acoustic model of the vocal tract. One simplification is to assume plane wave propagation in the vocal tract. This is reasonable since, first, the soft tissue along the vocal tract prevents radial propagation of the sound wave, and, second, the average lateral (cross-sectional) dimension of the vocal tract is about 2 cm, which is smaller than the wavelength of a sound wave at 4 kHz, which is $\lambda = c/f = 34,300/4000 = 8.6$ cm. Strictly speaking, this assumption is valid only for frequencies below 4 kHz. But for speech, where 5 kHz is considered to be an appropriate bandwidth, the plane wave propagation assumption is quite adequate.

With the assumption of plane wave propagation in the vocal tract, then only the cross-sectional area and the perimeter along the length of the vocal tract determine the acoustic characteristics of the vocal tract. Thus, the acoustic equations can be described in one dimension instead of three, which is a significant simplification. The area function of the vocal tract is then approximated by a sufficiently small number of successive sections with each section having a constant cross-sectional area.

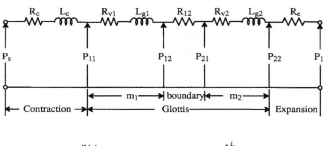

$$R_c = 1.37 \frac{\varrho}{2} \frac{|U_g|}{A_{g1}^2} \qquad L_c = \int_0^{l_c} \frac{\varrho}{A_c(x)} dx$$

$$R_{v1} = 12 \frac{\mu l_g^2 d_1}{A_{g1}^3} \qquad L_{g1} = \frac{\varrho d_1}{A_{g1}}$$

$$R_{12} = \frac{\varrho}{2} \left(\frac{1}{A_{g2}^2} - \frac{1}{A_{g1}^2} \right) |U_g| \qquad R_e = -\frac{\varrho}{2} \frac{2}{A_{g2} A_1} \left(1 - \frac{A_{g2}}{A_1} \right) |U_g|$$

$$R_{v2} = 12 \frac{\mu l_g^2 d_2}{A_{g2}^3} \qquad L_{g2} = \frac{\varrho d_2}{A_{g2}}$$

U_g = glottal flow P_s = lung pressure

ϱ = air density μ = shear viscosity coefficient

l_c = the length of contraction area A_g = glottal area

A_c = cross-sectional area of the contraction region

A_1 = cross-sectional area of the 1st vocal tract element

FIGURE A10.24 Equivalent circuit for the glottis.

For each section of the vocal tract, the acoustic model is derived as follows. Portnoff (1973) has shown that sound waves in the lossless tube satisfy the following equations.

$$-\frac{\partial p}{\partial x} = \rho \frac{\partial (u/A)}{\partial t} \qquad (A10.5.2.2.1)$$

$$-\frac{\partial u}{\partial x} = \frac{1}{\rho c^2} \frac{\partial (pA)}{\partial t} + \frac{\partial A}{\partial t} \qquad (A10.5.2.2.2)$$

where $p = p(x, t)$ is the sound pressure, $u = u(x, t)$ is the volume velocity, ρ is the density of air, c is the velocity of sound, and $A = A(x, t)$ is the area function of the tube. Applying Equations (A10.5.2.2.1) and (A10.5.2.2.2) to the section specified by the cross-sectional area A yields

$$-\frac{\partial p}{\partial x} = \frac{\rho}{A} \frac{\partial (u)}{\partial t} \qquad (A10.5.2.2.3)$$

$$-\frac{\partial u}{\partial x} = \frac{A}{\rho c^2} \frac{\partial (p)}{\partial t} \qquad (A10.5.2.2.4)$$

Based on the similarity between these equations and the equations for lossless, uniform electrical transmission lines, the tube with length l can be represented by an inductance, $L = \frac{\rho l}{A}$, followed by a shunt capacitance, $C = \frac{Al}{\rho c^2}$. The similarity between the acoustic wave propagation in a cylindrical tube and the propagation of an electrical wave along a transmission line are summarized in Chapter 5.

The effects of the vibration of the vocal tract wall can be added to the above model. The pressure variations inside the vocal tract will cause the cross-sectional area to change, since it exerts a force on the tract's elastic walls. Assuming that the walls are subject to local reactions (i.e., the motion of one portion of the wall is dependent only upon the acoustic pressure on that portion and independent of the motion of any other part of the wall), the area $A(x, t)$ will be a function of the pressure $p(x, t)$. Since the pressure variations are very small, the resulting variation in the cross-sectional area can be treated as a small perturbation,

$$A(x, t) = A_0 + \Delta A = A_0 + yS_0 \qquad (A10.5.2.2.5)$$

where A_0 is the nominal area, ΔA is a small perturbation, S_0 is the circumference of the tube, and y is the displacement of the yielding walls due to the sound pressure inside the tube. The wall vibration is modeled as a mass–compliance–viscosity mechanical model and is governed by Newton's law. Let m, b, and k represent the mass, the mechanical resistance, and the stiffness of the wall per unit length of the tube, respectively. According to Newton's law

$$m\frac{\partial^2 y}{\partial t^2} + b\frac{\partial y}{\partial t} + ky = pS_0 \qquad (A10.5.2.2.6)$$

Define the volume velocity generated by the wall vibration as

$$u_w = \frac{\partial(yS_0\ell)}{\partial t} \qquad (A10.5.2.2.7)$$

These two equations can be combined to obtain

$$p = \frac{m}{S_0^2\ell}\frac{\partial u_m}{\partial t} + \frac{b}{S_0^2\ell}u_m + \frac{k}{S_0^2\ell}\int u_m \, dt \qquad (A10.5.2.2.8)$$

The equivalent circuit for this equation is an RLC series circuit where

$$L_w = \frac{m}{S_0^2\ell}$$

$$R_w = \frac{b}{S_0^2\ell}$$

and

$$C_w = \frac{S_0^2\ell}{k}$$

are the components of the wall vibration impedance.

The wall impedance can be included in every elemental section of the vocal tract as a distributed element (Flanagan, 1972a; Flanagan and Ishizaka, 1976; Flanagan et al., 1975, 1980; Ishizaka et al., 1975; Maeda, 1982a) or inserted as a lumped shunt element, one for the pharynx and one at the level of the cheeks (Badin and Fant, 1984; Lin, 1990; Wakita and Fant, 1978). As Wakita and Fant (1978) indicated, the lumped wall impedance, which is independent of the vocal tract configurations may not give satisfactory results.

Figure A10.25 presents data concerning the wall mass, viscosity, and compliance found in the literature. In some cases, the compliance is not used since it has no effect on the resonances of the model (Wakita and Fant, 1978). Maeda (1982a) pointed out that the total mass of the walls may vary unrealistically if the yielding wall parameters are specified in terms of a unit surface area. Thus, the per unit length specification was used in his vocal tract simulation (Hsieh, 1994).

The effects of viscous friction and thermal conduction at the wall sites are much less pronounced than those for the wall vibration. Flanagan (1972) considered these losses in detail and showed that the effects of viscous friction can be accounted for by including a frequency-dependent resistor, R, in series with the inductor, L. The effects of heat conduction through the vocal tract wall can be accounted for by adding a frequency-dependent resistor, $\frac{1}{G}$, in parallel with the capacitor, C. The resistor, R,

	b (gm/sec)	m (gm)	k (dyne/cm)
Ishizaka et al. (1975), at cheek	1060	1.5	33,300
Ishizaka et al. (1975), at neck	2320	2.4	49,100
Flanagan et al. (1975)	1600	1.5	–
Maeda (1982)	1400	1.5	30,000
Lin (1990a)	1600	1.4	–

FIGURE A10.25 Per unit area yielding wall parameters.

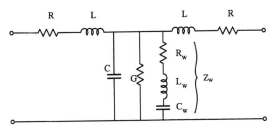

FIGURE A10.26   Equivalent circuit of an analog transmission line element of the vocal tract.

is significant in time domain simulations; when a constriction occurs, the resistance becomes very large and the air flow is blocked. As a result, a section of the vocal tract may be represented by a finite number of transmission line elements whose structure is given in Figure A10.26. The definitions of the circuit components are given in Figure A10.27. The series resistor R is used to represent the acoustic loss due to viscous drag in which the energy loss is proportional to the square of the volume velocity. The shunt conductance G represents the loss due to heat conduction, which is proportional to the pressure squared. The shunt impedance is the acoustic equivalent mechanical impedance of the

$$R = \frac{S\sqrt{\varrho\mu\omega}}{2\sqrt{2}A^2}l \qquad\qquad \text{; Series resistance}$$

$$L = \frac{\varrho}{2A}l \qquad\qquad \text{; Series inductance}$$

$$C = \frac{A}{\varrho c^2}l \qquad\qquad \text{; Shunt capacitance}$$

$$G = \frac{(\eta - 1)S}{\varrho c^2}\sqrt{\frac{\lambda\omega}{2\xi\varrho}}l \qquad\qquad \text{; Shunt conductance}$$

$$R_w = \frac{b}{S^2 l} \qquad\qquad \text{; Resistance in wall impedance}$$

$$L_w = \frac{m}{S^2 l} \qquad\qquad \text{; Inductance in wall impedance}$$

$$C_w = \frac{S^2 l}{k} \qquad\qquad \text{; Capacitance in wall impedance}$$

where

$S = 2S_A\sqrt{A\pi}$: circumference of element.

S_A: section shape factor, for a circular cross-section, $S_A=1$;
        for an elliptic cross-section, $S_A=2$.

l : length of elemental tube.

A : cross-sectional area of element.

ϱ : density of air, 1.14×10^{-3} gm/cm^3 (moist air at body temperature, $37°$C).

c : sound velocity, 3.53×10^4 cm/sec (moist air at body temperature, $37°$C).

μ : viscosity, 1.86×10^{-4} dyne-sec/cm^2 ($20°$C, 0.76 m.Hg).

λ : coefficient of heat conduction of air, 0.055×10^{-3} cal/cm-sec-deg ($0°$C).

η : adiabatic gas constant, 1.4.

ξ : specific heat, 0.24 cal/gm-degree ($0°$C, 1 atmos.).

ω : radian frequency.

FIGURE A10.27   Physical definitions of the components in Figure A10.26.

yielding wall. This wall impedance, which represents a mass–compliance–viscosity loss of the soft tissue, has three components, R_w, L_w, and C_w. The acoustic model of the vocal tract is established by concatenating these element models. No standardized model of the vocal tract has been established to date and the choices for the component values vary among researchers. The values representing the best choices remain to be determined (Wakita and Fant, 1978).

A10.5.3 Acoustic Model of the Radiation

Acoustic energy escapes from the vocal tract via the lips. From the transmission-line analogs, the lips are treated as a radiation impedance that loads the vocal tract. The radiation impedance contains a resistive part that represents acoustic energy loss and a reactance part that represents the mass inertia of air at the lips (Fant, 1960). Radiation from a spherical baffle is one model for the radiation impedance that is represented by nonlinear functions (Morse, 1948; Morse and Ingard, 1968). Stevens et al. (1953) made approximations and represented the radiation impedance by a resistive load with three other frequency-dependent components. Fant made another approximation and modeled the impedance by two frequency-dependent components, one being resistive and the other inductive (Fant, 1960; Wakita and Fant, 1978).

Another simplified radiation model is to assume that the radiating surface is set in a plane baffle of infinite extent. In this case, the radiation impedance is formed by a first order Bessel function and a Struve function (Flanagan, 1972; Rayleigh, 1945; Wakita and Fant, 1978). Flanagan (1972) provided a good approximation to this complicated representation by a parallel connection of a resistance and an inductance. The most important feature of Flanagan's model (1972) is that both circuit components are frequency independent. Figure A10.28 illustrates the Stevens et al. (1953) model and Flanagan (1972) model.

Comparisons between models have been made by researchers (Badin and Fant, 1984; Lin, 1990; Wakita and Fant, 1978). The Stevens et al. (1953) model yields the most accurate result. However, the Flanagan (1972) model is usually preferred for time-domain synthesis (Flanagan and Ishizaka, 1976; Flanagan et al., 1975, 1980; Maeda, 1982a) and is used in our acoustic model of radiation.

A10.5.4 Example of a Synthetic Vowel

One example of a synthesized vowel, /AA/ as in father, is presented. The vocal tract area function is adopted from the optimized results of Hsieh (1994), and is similar to the data in Appendix 5. The synthetic glottal area, glottal flow, speech waveforms, and the wideband spectrograms for vowel /AA/ are shown in the Figure A10.29.

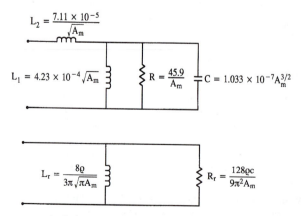

A_m is the area of the mouth opening

FIGURE A10.28 Radiation models. (*a*) Model by Stevens et al. (1953). (*b*) Flanagan model (1972).

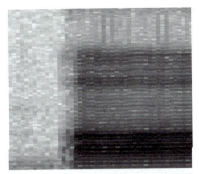

FIGURE A10.29 Synthesized glottal area, glottal flow, speech waveform, and wideband spectrogram for the vowel /AA/.

A time-domain comparison of the synthetic waveforms with the results of Ishizaka and Flanagan (1972) and a frequency-domain comparison of the formant frequencies and bandwidths, measured from the wideband spectrograms, with the vocal tract frequency responses from the optimized results of Hsieh (1994), are satisfactory. Thus, the two-mass vocal fold model and the synthesizer are appropriate tools for modeling the vibratory motion of the vocal folds.

In the implementation of the two-mass model in Chapter 9, the two masses are shown as two semirigid, connected panels or ribbons rather than as two blocks as seen in Figure A10.17. This convention is used since it is similar in form to the ribbon model, which is discussed next.

A10.6 THE RIBBON MODEL

A brief introduction to the ribbon model is given at the beginning of this appendix. The ribbon model (model I) is based on Titze (1984). One modification to the model includes the design of a flexible three-dimensional model of the vocal fold vibrations.

A10.6.1 Basic Objectives of the Model

Due to the relative inaccessibility of the larynx, imaging and modeling of the vocal folds are difficult, as described previously. Thus, one aim of the model is to characterize the vocal fold vibratory characteristics in terms of a three-dimensional glottal configuration with spatially varying tissue properties. Another aim of the model is to estimate several glottographic waveforms. The three glottographic waveforms obtained with the model are the projected glottal area, the vocal fold contact area, and the electroglottographic waveform.

A10.6.2 Assumptions for the Model

Several assumptions are made to simplify the mathematical expressions in the vocal fold vibratory model. The vibratory movement of vocal folds is confined to the lateral direction. Although motion of the vocal fold tissue does occur in other directions, especially after collision between the two folds, the projected area of the glottis is primarily determined by the vocal fold lateral motion. An algorithm accounts for changes in lateral contact area due to vertical movement when the folds collide. Each point on the medial vocal fold surface vibrates in a sinusoidal fashion; that is, as a harmonic oscillator (Titze, 1984). This is a first-order approximation to the true displacement function of vocal folds. Despite its simplicity, the model based on this approximation can simulate most features of the glottographic waveforms. On the other hand, the vibratory pattern of the vocal folds for abnormal voices can be quite irregular and other surface vibratory functions need to be examined. Uniform travelling speed is assumed for all travelling wave phenomena in the proposed model. Thus, the time delay in movement of different points on the vocal fold surface is proportional to their location in the direction of wave propagation.

A10.6.3 Choice of Model Parameters

Since each glottographic waveform reflects an average measurement of the vocal fold motion, it is natural to formulate the glottographic waveforms in terms of vocal fold kinematics. The choice of model parameters is based on their importance in developing and adjusting the self-oscillation of the vocal folds and their appropriateness in describing the kinematics of the vocal folds. Model building is essentially a systematic coordination of theoretical and empirical elements of knowledge into a joint construct. Empirical observations of the vocal fold vibrations suggest that specific information about the glottal configuration and vocal fold kinematics should be contained in the parameter set. Ultra-high speed photography of the vibrating folds (Childers and Krishnamurthy, 1985; Childers et al., 1986; Childers et al., 1990; Hirano et al., 1983; Moore and von Leden, 1958) and the results from previous vocal fold models (Ishizaka and Flanagan, 1972; Titze and Talkin, 1979a) have demonstrated that five factors are important in determining the vibratory patterns. First, there must be proper abduction and adduction of the vocal folds. It is well known that proper abduction must be achieved before vibration. Different levels of abduction may result in different vibratory modes (Ishizaka and Flanagan, 1972). Second, the vertical pre-phonatory shape of the glottis is important. Titze (1988) has shown that the vertical shape may be of significance in register control in human phonation and in determining the waveshape of the glottographic waveforms. Third, vertical phasing is important. The mucosal wave is also known as vertical phasing. A lag between the movements of the upper and lower portions of the vocal folds has been observed during phonation, except for falsetto. This phenomenon is more obvious during low pitch phonation. Photography of the vibration of the vocal folds in the frontal planes confirms the wave-like nature of the vibration (Moore and von Leden, 1958). This lag reflects the degree of coupling along the depth of the vocal folds. Fourth, there is also longitudinal phasing. A lag between the movements of the anterior and the posterior portions of the vocal folds is known as longitudinal phasing. During the opening phase, the vocal folds first separate along their posterior section and the separation progresses toward the anterior (Childers et al., 1986; Childers et al., 1990; Hirano et al., 1983). Fifth, the profiles of maximum excursion of the vibration along the length and the depth of the glottis is another factor. The maximum displacement from its equilibrium position is called the amplitude of oscillation or maximum excursion. The maximum excursion of the vocal folds and equilibrium positions may vary according to the glottal configuration.

The model parameters of the ribbon model is summarized in Figure A10.30. Note that there is an adjustment phasing in the displacement function, which is described in the following paragraph.

The maximum excursion profile reflects the degree of compliance of the vocal fold tissue along each dimension. Along its length, the vocal folds appear to be most pliable near the midpoint of the membranous portion. The posterior portion of the vocal fold appears to be slightly more pliant than the anterior portion. At the anterior and the posterior ends, the tissues appear to be firm. The largest amplitude of lateral excursion, however, occurs not at the midpoint but at a place posterior to the midpoint (Hirano et al., 1983). This can be accounted for, at least in part, by the fact that the anterior end of the vocal folds can be considered fixed, while the posterior end is movable. In order to compensate for the fact that the largest amplitude of lateral excursion occurs not at the midpoint, we chose one more parameter, the adjustment phasing ϕ_m. The value of ϕ_m is set to $\frac{\pi}{4}$ in our model. However, the user can change this value.

A10.6.4 Displacement Function for the Ribbon Model

The configurational adjustments of the ribbon model can be described geometrically as illustrated in the Figure A10.30(a). Let ξ_{01} and ξ_{02} be the pre-phonatory displacements (from the glottal midline) of the inferior and superior edges of the vocal folds at the level of the vocal process. Assume that the pre-phonatory glottal shape decreases linearly toward zero at the anterior commissure and that the vertical shape of the glottis is trapezoidal. Thus, we can define the pre-phonatory glottal configuration by the equation

$$\xi_0 = \left\{ \xi_{01} - (\xi_{01} - \xi_{02})\frac{z}{T} \right\} \left(1 - \frac{y}{L} \right) \tag{A10.6.4.1}$$

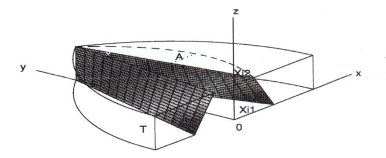

Pre-phonatory Glottal Configuration :

$$\xi_0(y,z) = \left\{\xi_{01} - (\xi_{01} - \xi_{02})\frac{z}{T}\right\}\left(1 - \frac{y}{L}\right)$$

where $\xi_{01} = $ Xi1; $\quad \xi_{02} = $ Xi2; $\quad A = \xi_m$

L is the length of the glottis; T is vocal fold thickness

(a)

Displacement Function :

$$\xi(y,z,t) = \xi_0(y,z) + \xi_1(y,z,t) = \xi_m\left(1 - \frac{y}{L}\right)\left(Q_a + Q_s - Q_s\frac{z}{T}\right) + \xi_1(y,z,t)$$

where $\quad \xi_1(y,z,t) = \xi_m \sin\left(\frac{\pi y}{L}\right) \sin\left(2\pi ft - \phi_v\frac{z}{T} - \phi_h\frac{y}{L} - \phi_m\right)$

Model Parameters :

abduction quotient $= Q_a = \dfrac{\xi_{02}}{\xi_m}$ $\qquad$ shape quotient $= Q_s = \dfrac{(\xi_{01} - \xi_{02})}{\xi_m}$

ξ_m is the maximum excursion amplitude $\qquad \phi_m = \frac{\pi}{4}$ is the adjustment phasing

$\phi_v = $ vpd is the vertical phase delay $\qquad \phi_h = $ hpd is the horizontal phase delay

(b)

FIGURE A10.30 The ribbon model. (*a*) Configuration of the glottis and vocal folds. (*b*) Displacement function and model parameters.

where T is the vocal fold thickness, L is the length of the glottis, and y and z are spatial dimensions as indicated.

Making the further assumption that the displacement at the anterior and posterior boundaries is zero (i.e., fixed), and that the displacement between these boundary points is sinusoidal (Titze, 1976; Titze and Story, 1975), we approximate the dynamic displacement (from the pre-phonatory position) to be

$$\xi_1(y,z,t) = \xi_m \sin\left(2\pi ft - \phi_v\frac{z}{T} - \phi_h\frac{y}{L} - \phi_m\right) \qquad (A10.6.4.2)$$

where ξ_m is the common amplitude for the upper and lower edges of the vocal folds, ϕ_v is the vertical phase delay between the folds, ϕ_h is the longitudinal phasing, $\phi_m = \frac{\pi}{4}$ is the adjustment phasing, f is the fundament frequency of vibration, and t is time. Left-right vocal fold symmetry is also assumed.

Two configurational parameters are defined as follows.

$$\text{abduction quotient} = Q_a = \frac{\xi_{02}}{\xi_m} \qquad (A10.6.4.3)$$

$$\text{shape quotient} = Q_s = \frac{\xi_{01} - \xi_{02}}{\xi_m} \qquad (A10.6.4.4)$$

Combining the pre-phonatory displacement Equation (A10.6.4.1) with the dynamic displacement

Equation (A10.6.4.2) and substituting Q_a and Q_s, we have the following expression

$$\xi(y, z, t) = \xi_m\left(1 - \frac{y}{L}\right)\left(Q_a + Q_s - Q_s\frac{z}{T}\right)$$

$$+ \xi_m \sin\left(2\pi ft - \phi_v\frac{z}{T} - \phi_h\frac{y}{L} - \phi_m\right) \quad (A10.6.4.5)$$

where $\xi(y, z, t)$ is the total displacement of each vocal fold from the midline described in terms of the configurational parameters. All negative values of $\xi(y, z, t)$ are set to zero for the glottal area computations. The displacement function, model parameters, and the ribbon model are summarized in Figure A10.30.

The normalization of the pre-phonatory glottal widths ξ_{01} and ξ_{02} with respect to the amplitude of vibration ξ_m, is not only convenient, but meaningful, since it reminds us of the DC/AC ratios of glottal area and flow used in glottal leakage assessment. In other words, the relative width of the glottal chink with respect to the maximum glottal width is more important than the absolute width of the chink (Titze, 1984).

A10.7 RELATING THE VOCAL FOLD VIBRATORY MOTION TO THE GLOTTOGRAPHIC WAVEFORMS

Three glottographic waveforms, electroglottographic, photoglottographic, and inverse-filtered glottal waveforms, have been related to glottal characteristics; that is, lateral contact area (Childers, Smith, and Moore, 1984) and glottal volume velocity (Wong, Markel, and Gray, 1979), respectively. In order to establish the relation between the glottographic waveforms and the vibratory patterns of the vocal folds, we first express these glottal characteristics in terms of the glottal displacement. Formulas are then derived for the simulation of the glottographic waveforms. The comparisons between the synthetic glottographic waveforms and the measured waveforms allow an evaluation of the vibratory model and the transduction formula.

A10.7.1 Derivation of the Projected Glottal Area

The formulation of the projected glottal area follows the work by Titze (1984). The projected glottal area is the area outlined by the glottis when seen from above the glottis. The glottal area is believed to be the primary descriptor of the glottal excitation. The glottal area becomes zero when the glottis is completely closed along its length.

Consider the length L of the glottis to be divided into a series of differential lengths dy. A differential projected glottal area can then be written as

$$dA_g(t) = 2\xi_{min}(y, t)\,dy \quad (A10.7.1.1)$$

where ξ_{min} is the minimum positive value of ξ in a differential vertical duct (negative values are set to zero). ξ_m can, in principle, be found by differentiating $\xi(y, z, t)$ with respect to z and setting the result equal to zero. This produces a contour of values along the length of the vocal folds at the vertical positions z_{min}. The total projected glottal area is then

$$A_g(t) = \int_0^L 2\xi_{min}(y, t)\,dy \quad (A10.7.1.2)$$

In practice, this glottal area function is determined numerically. In discrete form, Equation (A10.7.1.2) becomes

$$A_g(n) = \frac{L}{M}\sum_{j=1}^{M} 2\xi_{min,j,n} \quad (A10.7.1.3)$$

for the nth time step and M finite ducts along the length. The minimum value in each of the vertical ducts is found by simple comparison of N discretized values of vertical displacements within the duct $(z = k\Delta z, k = 1, N)$.

The locations of z_{min} may not be the same along the length. For example, if there are large phase differences in the tissue movement, the possibility that z_{min} is near the top anteriorly and near the bottom posterily (or vice versa) cannot be excluded. The minimum glottal area that is relevant for glottal airflow may not be the same as the projected glottal area. For a computation of the glottal volume velocity, it may be more appropriate to define the minimum glottal area as

$$A_g(n) = \min_k \left(\frac{L}{M} \sum_{j=1}^{M} 2\xi_{j,k,n} \right) \tag{A10.7.1.4}$$

where $\min_k$ indicates that the minimum value of k glottal areas stacked vertically is selected as the minimum glottal area.

A10.7.2   Derivation of the Vocal Fold Contact Area

The vocal fold contact area is the lateral area of contact between the folds when they come together. To compute the total contact area, an infinitesimal amount of contact area dy dz is added whenever the glottal width goes to zero at any coordinate (y, z) on the glottal mid-plane. The total contact area is then the summation of the partial contact area along the length and the depth of the vocal fold as

$$A_c = \sum \sum c(y, z) \, dy \, dz \tag{A10.7.2.1}$$

where $c(y, z) = 1$ for $\xi(y, z) \leq 0$ or $c(y, z) = 0$ for $\xi(y, z) > 0$.

In the above formulation, the vocal folds are allowed to overlap. We may consider the vocal fold tissue to be incompressible and every movement of the vocal folds results in their deformation in another direction. During vocal fold collision, the vocal folds press against each other and cause a change in the thickness of the vocal folds. For a better approximation to the collision process, the thickness at the collision surface should be dynamically adjusted according to the degree of overlap.

The approach for calculating the varying thickness during the collision is to use the incompressibility of the vocal tissue. The incompressibility property requires that the total volume of the vocal folds remains the same. This implies that

$$(x(t) - x_0) \, d_0 = (x_c(t) - x_0) \, d(t) \tag{A10.7.2.2}$$

where $x(t)$ is the lateral displacement of a unit area, x_0 is the position of the boundary, d_0 is the nominal unit contact area if there is no overlap, $x_c(t)$ is the location where contact occurs, and $d(t)$ the adjusted contact area of the vocal folds. One immediate result of this approximation is that the electroglottographic (EGG) waveform in the vicinity of the EGG minimum is rounded rather than flat. This result is a consequence of the change in contact area after the first collision.

A10.7.3   The Simulation of EGG Waveform

Electroglottography (EGG) is described in Chapter 4. It measures the electrical impedance of the tissue in the vicinity of the larynx. It is generally accepted that the EGG reflects the changes in the lateral area of contact (Childers and Krishnamurthy, 1985; Childers et al., 1986; Childers et al., 1990; Kitzing, 1986). Phonation alters the impedance, most likely due to changes in the current paths within the larynx. These changes occur when vocal fold motion alters the glottal configuration. A closed glottis creates a relatively small impedance compared to an open glottis. However, the vocal fold contact area has never been measured *in vivo*. Titze (1989) used excised canine larynges to examine the relation between EGG signals and the corresponding dynamic vocal fold contact area. His results did not refute the hypothesis of a linear relation between contact area and the EGG signal. Thus, it is assumed that a change in electrical impedance measured by the EGG is inversely proportional to the changes in lateral contact area. The lateral contact area and EGG signal are then related as follows.

$$EGG(t) = \frac{k_1}{c(t) + k_2} \tag{A10.7.3.1}$$

where $c(t)$ is the contact area, k_2 represents the shunt impedance of the adjacent tissues, and k_1 is a scaling constant.

A10.8 SELECTING VALUES FOR THE RIBBON MODEL PARAMETERS

The choice of model parameter values is based on empirical data, with each parameter setting being made separately. In fact, parameters such as vertical phase difference, vertical shape of the glottis, and fundamental frequency of vibration are correlated and should not be assigned arbitrarily. Knowledge of such correlations would help us exclude physically impossible combinations of parameters. In order to have a more systematic method for parameter settings, one could explore combinations and ranges of the parameter values using the data and model described earlier in this appendix. The approach is to systematically vary the control of the input, for example, lung pressure, vocal fold tension (or abduction of the glottis), and determine the corresponding changes in kinematics of the vocal folds. To obtain optimum model parameters by this method is difficult. The optimum model parameters can be extracted from the measured glottographic waveforms through a waveform-matching, model-fitting-data, optimization procedure. A simulated annealing algorithm is used to estimate the model parameters and the vocal fold configurations that minimize the error between the measured and the model-generated glottal area and EGG waveforms (Wu, 1996).

The ribbon model basically describes the kinematics of vocal fold vibration. For example, the lag between movements of the anterior and the posterior portions of the vocal folds is taken into account, and an adjustment phasing provides a method to compensate for the fact that the largest amplitude of lateral excursion does not occur at the midpoint of vocal fold length. A formulation to relate the lateral contact area to the EGG is also included in our model for the simulation of the EGG waveform.

A graphical user interface and animation are available in Chapter 9 for the two-mass and ribbon vocal fold models. The animation of vocal fold vibration can be viewed on the computer screen. These tools provide the user a view of the effects and interrelations between the model parameters and vocal fold vibrations.

A comparison between a measured and a model-derived glottal area waveform is shown in the Figure A10.31. The percentage error distance between model and data is 0.05. This example illustrates that the model does capture the vibratory patterns of the vocal folds quite well.

Additional data and models are available in Wu (1996). A movie of one vibratory cycle of the motion of actual vocal folds is also available on the CDs accompanying this text. See the README file on the CDs.

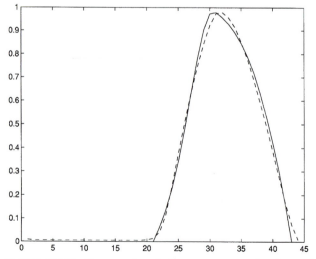

FIGURE A10.31 A comparison between a measured and a model-derived projected glottal for the ribbon model. Model is dashedline. Measured data is solidline.

ARTICULATORY SPEECH SYNTHESIS: THEORY

A11.1 INTRODUCTION

Both formant and LP synthesis models represent acoustic domain models of speech, with no interaction between the glottal flow excitation and the vocal-tract filter. Articulatory synthesis is the production of speech sounds using a model of the vocal tract, which directly or indirectly simulates the movements of the speech articulators. This method provides a means for gaining an understanding of speech production and for studying phonetics. In this model coarticulation effects arise naturally, and in principle it should be possible to deal with glottal source properties, including interaction between the vocal tract and the vocal folds, as well as the contribution of the subglottal system, and the effects of the nasal tract and sinus cavities.

Articulatory synthesis usually consists of two separate components as shown in Figure A11.1. In the articulatory model, the vocal tract is divided into numerous small sections and the corresponding cross-sectional areas are used as parameters to represent the vocal tract characteristics. In the acoustic model, each cross-sectional area is approximated by an electrical analog transmission line. To simulate the movement of the vocal tract, the area functions change with time. Each sound is represented by a target configuration of the vocal tract. The movement of the articulators of the vocal tract is specified by a change in the vocal tract configuration. This is done by specifying changes in the articulators from frame to frame over the speech file that is to be synthesized.

Presently, the complexity of articulatory synthesis is partially due to the analysis procedure, which usually requires an "articulatory-to-acoustic inverse transformation" from the speech signal, that is, speech inverse filtering. The complexity of the relationship between articulatory gestures and the acoustic signal makes it difficult to automatically generate the details of articulatory control needed to produce a synthetic copy of a given sample of human speech. Despite such drawbacks, articulatory speech synthesis has several advantages.

- The model has a direct relation to the human speech production process. Consequently, it is conjectured that articulatory synthesis may lead to a simpler and more elegant synthesis by rule; for example, text-to-speech applications (Parthasarathy and Coker, 1990, 1992) and articulation-based speech recognition systems (Erler and Deng, 1993).

- The articulatory parameters in the human voice production system vary slowly. Consequently, researchers have suggested that these parameters are potential candidates for efficient coding; for example, low bit-rate speech communication (Flanagan et al., 1980).

- To the extent that we can accurately obtain the speech gestures (articulatory movements or trajectories), articulatory synthesizers may be valuable for research scientists and physicians, since such synthesizers can be used to study linguistic theories, to provide a feedback mechanism for teaching speech production, and to explore the effects of vocal tract surgical techniques on speech production prior to surgical intervention (Childers, 1991); and they hold out the ultimate promise of high quality, natural-sounding speech with a simple control scheme (Klatt, 1987).

A properly constructed articulatory synthesizer is capable of reproducing all the naturally relevant effects for the generation of fricatives and plosives, modeling coarticulation transitions as well as source-tract interaction in a manner that resembles the physical process that occurs in real speech

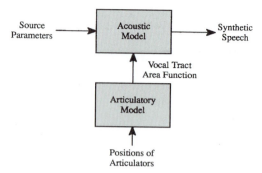

FIGURE A11.1 A model of articulatory speech synthesis.

production. Articulatory synthesizers will continue to be of great importance for research purposes, and to provide insights into various acoustic features of human speech. Thus, an articulatory synthesizer may provide both an efficient description of natural speech and a means for synthesizing natural-sounding speech. However, a major problem with the articulatory synthesizer is the lack of a means to derive articulatory configurations from the speech signal using speech inverse filtering. The software toolbox provided in Chapter 10 provides one means for speech inverse filtering. In addition, the toolbox displays the following characteristics:

- The articulator movements (gestures) for synthesizing words or sentences.
- The cross-sectional area and acoustic transfer function of the vocal tract.
- The pressure and volume–velocity waveforms at selected points in the vocal tract.
- The excitation source waveform and power spectral density.
- The synthesized speech waveform.

The optimization procedure is based on the simulated annealing (SA) algorithm. Using this method, the articulatory parameters are optimized to minimize the error distance between the natural (target) and the model-generated first four formants. The excitation source model for the articulatory synthesizer is a combination of a glottal source model (a modified LF-model), a subglottal, and a glottal area model. The toolbox can modify and calculate the vocal tract and the nasal tract models, which in turn are used to calculate the acoustic transfer function at selected points in the vocal tract. The toolbox is able to insert a noise source model at the center of a vocal tract constriction, or downstream from the constriction, or distributed along a specified spatial interval of the constriction. In addition, the toolbox can simulate viscous, heat conduction, and yielding wall losses in the vocal and nasal tracts. This model also includes the effects of the sinus cavities.

The speech synthesis process can vary the following parameters.

- The synthesis sampling frequency.
- The degree of nasalization, including the velopharyngeal port opening.
- The excitation source parameters, including subglottal coupling.
- The type and location of excitation.

A11.2 ARTICULATORY SPEECH SYNTHESIS

The articulatory speech synthesizer is based on a model of the physiology of the human speech production process. As shown in Figure A11.1, the articulatory synthesizer has two components. The articulatory model represents the articulatory positions, which are converted into vocal tract cross-sectional area functions. The acoustic model, which includes subglottal coupling, source-tract

interaction, the vocal tract, the nasal tract with sinus cavities, and acoustic radiation, simulates speech production and propagation as well as the physics of the physiological-to-acoustic transformation. This appendix presents the implementation of the articulatory model and the realization of the acoustic model. Our articulatory model is based on Mermelstein's model (1973), which is implemented in MATLAB. The time-domain approach is used to implement the acoustic model, since it offers the ability to simulate the dynamic properties of the vocal system as well as a method to improve the quality of the synthesized speech. Methods for estimating articulatory data from acoustic measurements are reviewed and described later in this appendix.

A11.2.1 Review of Articulatory Models

According to the acoustic theory of speech production, the human vocal tract can be modeled as an acoustic tube with nonuniform and time-varying cross-sections. This model modulates the excitation source to produce various sounds. The acoustic tube can be adjusted to various shapes by adjusting the articulatory parameters. These articulatory parameters are expressed in the form of a vector and specify the positions of the tongue body, tongue tip, jaw, lips, hyoid, and velum. Articulatory models are well known in the literature and can be classified into two major types: parametric area models and midsagittal distance models.

A11.2.1.1 *Parametric Area Models*   The parametric area models do not represent articulatory positions directly but rather, concentrate on modeling the area function as a function of distance along the tract subject to certain constraints (Atal et al., 1978; Fant, 1960; Fant and Lin, 1991; Flanagan et al., 1980; Lin, 1990; Stevens, 1998; Stevens and House, 1955; Yu, 1993). A common feature of these models is a specification of the minimum constriction area and its axial location. The area of the vocal tract is usually represented by a continuous function such as a hyperbola, a parabola, or a sinusoid (Lin, 1990). Consonant articulations have generally not been implemented. Figure A11.2 shows one example of a parametric area model.

A11.2.1.2 *Midsagittal Distance Models*   The midsagittal distance models are usually based on a representation of the midsagittal plane as seen from an x-ray image. They describe the speech articulator movements in a midsagittal plane and require the specification of the positions of the articulators (Levinson and Schmidt, 1983; Mermelstein, 1973; Prado, 1991; Sondhi and Schroeter, 1986) or a means to control the movements of the articulators by rules (Coker, 1976; Parthasarathy and Coker, 1990, 1992). The output is an estimate of the vocal tract cross-sectional area. Visualization and articulatory state interpretation are the major advantages of these models. Figure A11.3 shows an example of a midsagittal distance model.

A11.2.2 Implementation of the Articulatory Model

Articulatory models are used to transform articulatory parameters to a vector representation of the vocal tract cross-sectional area, which in turn are transformed to acoustic characteristics within the vocal tract. Our articulatory model is a modified version of the Mermelstein's model (1973),

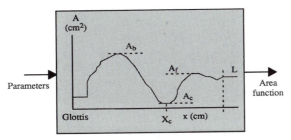

FIGURE A11.2   A parametric area model.

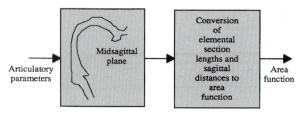

FIGURE A11.3 A midsagittal distance model.

which is implemented in MATLAB. Mermelstein's model (1973) achieves a match between x-ray tracings and a midsagittal vocal tract outline. However, there is insufficient information for a robust representation of the lower part of the pharynx and for the region between the tongue tip and the jaw. Our approach modifies the lower part of the pharynx and optimizes this region whenever necessary. This is discussed in more detail later in this appendix. We have also modified the hyoid and tongue-tip-to-jaw regions.

A11.2.2.1 Articulatory Parameters and Midsagittal Vocal Tract
Outline In the articulatory model, a set of variables is used to specify the inferior outline of the vocal tract (Figure A11.4). These variables, referred to as articulatory parameters, are:

- Tongue body center: This is represented with an arc (DL-B) of a circle with a moving center and fixed radius. The tongue body center, denoted as tongc, has polar coordinates (sc, thetaj + thetab) with respect to the fixed point F. However, the rectangular coordinates (tbodyx, tbodyy) are used for display and optimization.

- Tongue tip: The tongue tip is represented by the rectangular coordinates (tipx, tipy) of point T. Arcs B-T and T-PF, specify the tongue blade outline. Since the location of point B varies with the tongue-body center (tongc) and the jaw angle (jaw), the tongue blade movements depend on the tongue body and jaw positions.

- Jaw: The point JAW with polar coordinates (sj, thetaj) are used to represent the jaw location. The distance sj is kept constant for most phonemes. The parameter jaw is used to denote the

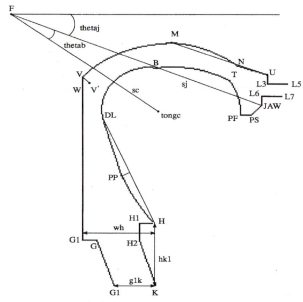

FIGURE A11.4 Articulatory model parameters.

angle thetaj. Note that the concave jaw is approximated by a polyline, a connected sequence of line segments (PF-PS-JAW-L6).

- Lips: The lips are represented by points L5 (upper) and L7 (lower). With respect to the point JAW, the coordinates of the lower lip are represented by (lipp, lipo), which specify the lip protrusion and lip opening, respectively. The use of lipp and lipo as separate variables allows lip closure, lip separation, or rounded lips. The upper lip L5 has the same coordinate values with respect to point U.

- Hyoid: The hyoid is specified by the parameter hyoid, the distance from point PP to the line segment H-DL. The point PP is on the normal bisector of the line segment H-DL, which is tangent to the tongue body arc outline at point DL. The line segment DL-PP and arc PP-H as well as the tongue body determine the anterior shape of the pharynx. The point H represents the intersection of the anterior edge of the epiglottis with the top edge of the hyoid bone. The point K represents an estimate of the anterior extremity of the larynx.

- The superior outline of the vocal tract is represented by the position of the upper teeth, U, the hard palate curve U-N-M, the highest point on the maxilla M, the soft palate arc M-V, the velum position V, the back wall of pharynx position W, and the highest point of the periarytenoid G. On the hard palate curve, the point N is located on the line segment M-U such that the distance M-N is twice the distance N-U. Circular arcs M-V and M-N are drawn with centers on a vertical line through M. The posterior–superior outline is generally considered fixed except for the soft palate curve near the velum point V. To specify the opening area of the velopharyngeal port, we treat the velum as an articulatory parameter.

- Velum: The state of the velum is represented by the position V of the tip of the uvula moving along a line segment (V-V'). The velar opening area is assumed proportional to the distance between the point V and the most elevated point of the velum. This distance is specified by the variable velum.

A11.2.2.2 Determination of the Vocal Tract Section Lengths and Cross-Sectional Areas

The vocal tract cross-sectional area function is determined by the areas of the sections with projections on the X-Y plane that form the sagittal grids of the vocal tract, as shown in Figure A11.5. These grid lines vary with the positions of the articulators (they are fixed in Mermelstein's model); that is, the interval between two adjacent parallel sagittal grid lines

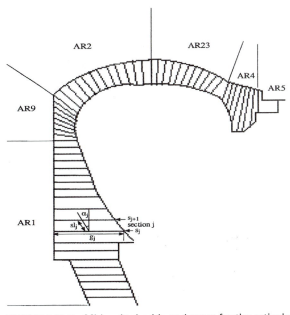

FIGURE A11.5   Midsagittal grids and areas for the articulatory model.

(regions AR1 and AR5) or the angle between two adjacent radial sagittal grid lines (rest regions) is not fixed. A total of 60 sections, 59 sections for the vocal tract plus one section (fixed length and area) for the outlet of the glottis, are used in our model. This feature provides more reliable estimates of the sagittal distances and cross-sectional areas.

The distance between the midpoints of two consecutive sagittal lines, s_j and s_{j+1}, represents the length of section j, sl_j (see Figure A11.5). The sagittal distance g_j of section j is defined as the grid line segment length between posterior–superior and anterior–inferior outlines. The sagittal distances are converted to cross-sectional areas by an empiric function based on previously published data (Mermelstein, 1973). In general, the cross-sectional area function is formulated as

$$A_j = F(j, g_j) \cos(\alpha_j) \tag{A11.2.2.2.1}$$

where j = 2 (vocal tract inlet), ..., 60 (lips). F is an empiric function and has a different formula for the pharyngeal region, oral region, and labial region, and α_j is the deviation angle of the direction of wave propagation from the normal to the jth grid line (Guo and Milenkovic, 1993; Mermelstein, 1973; Rubin et al., 1981).

In the pharyngeal region (AR1 and AR9 in Figure A11.5), the empiric function is

$$F(j, g_j) = \pi g_j b_j \tag{A11.2.2.2.2}$$

where g_j is one axis and $b_j = g_j + \Delta g$, where Δg belongs to the interval [1.5 3], is another axis of the ellipse, since we approximate each pharyngeal section as an elliptic cylinder. The b_j is proportionally increased as the grid line moves toward the larynx. In the soft palate region (AR2 in Figure A11.5), the empiric function has the form

$$F(j, g_j) = (2.0)g_j^{1.5} \tag{A11.2.2.2.3}$$

In the hard palate region (AR23 in Figure A11.5), the empiric function is given by

$$F(j, g_j) = (1.6)g_j^{1.5} \tag{A11.2.2.2.4}$$

For the labial region (AR5 in Figure A11.5), the empiric function is

$$F(j, g_j) = g_j[2.0 + 1.5(\text{lipo} - \text{lipp})] \tag{A11.2.2.2.5}$$

For the other region (AR4 in Figure A11.5), the empiric function is

$$\begin{aligned} F(j, g_j) &= (1.5)g_j, & \text{for } g_j < 0.5 \\ &= (0.75) + 3(g_j - 0.5), & \text{for } 0.5 < g_j < 2 \\ &= 5.25 + 5(g_j - 2), & \text{for } g_j > 2 \end{aligned} \tag{A11.2.2.2.6}$$

A11.2.2.3 Calculation of Formant Frequencies from the Vocal Tract Cross-Sectional Areas

The calculation of formant frequencies from a given vocal tract cross-sectional area function has been well established in the acoustic theory of speech production (Atal et al., 1978; Badin and Fant, 1984; Fant, 1960; Fant, 1985; Fant and Lin, 1991; Lin, 1990, 1992; Stevens, 1998; Wakita and Fant, 1978). By computing the acoustic transfer function of a given vocal tract configuration, we can decompose the formant frequencies from the denominator of the acoustic transfer function. Refer to Appendix A11-B for the detailed acoustic transfer function calculations.

Let an all-pole acoustic transfer function be

$$H(s) = \frac{1}{H_p(s)} \tag{A11.2.2.3.1}$$

The denominator $H_p(s)$ is normally a complex number

$$H_p(s) = N_b(s) + jN_a(s) \tag{A11.2.2.3.2}$$

For a lossless vocal tract $N_a(s)$ is zero. When the losses are small, $N_a(s)$ is small compared with $N_b(s)$. Consequently, the roots of the complex function $H_p(s)$ should be located in the neighborhood of the roots of $N_b(s)$. Based on this assumption, a two-step approach was proposed by Fant (1960) and was referred to as the N_b method (Lin 1990, 1992). Figure A11.6 illustrates the flow chart of the method.

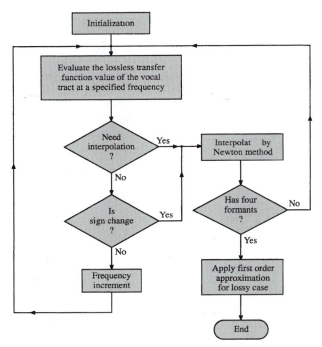

FIGURE A11.6 Flow chart of the N_b method for the decomposition of formants.

The first step of the N_b method is to search for the roots of $N_b(s) = 0$. At a given frequency f_n, the value $N_b(j2\pi f_n)$ is computed. The frequency is next increased (a few hundred hertz) and the value $N_b(j2\pi f_{n+1})$ is computed at the new frequency f_{n+1}. If the polarity changes, $(N_b(j2\pi f_n))$ $(N_b(j2\pi f_{n+1})) < 0$, within this interval, then a root is present. Newton's approximation or other methods can be used to determine the root within this interval. Let f_0 be the estimated root frequency found by setting $N_b(s) = 0$. The second step is to account for the finite $N_a(s)$ by means of a first-order approximation term for $H_p(s)$ in the vicinity of $(j2\pi f_0)$.

$$H_p(s) \cong H_p(j2\pi f_0) + (s - j2\pi f_0)(H_p'(j2\pi f_0)) \qquad \text{(A11.2.2.3.3)}$$

where

$$H_p'(j2\pi f_0) = \frac{d(H_p(s))}{ds}\bigg|_{s=j2\pi f_0} = N_a'(j2\pi f_0) - j(N_b'(j2\pi f_0)) \qquad \text{(A11.2.2.3.4)}$$

Set $H_p(s) = 0$ and let the roots be denoted as

$$s_n = \sigma_n + j(2\pi f_0 + \Delta\omega_n) \qquad \text{(A11.2.2.3.5)}$$

From the above equations, we have

$$\sigma_n = \frac{N_a N_b'}{(N_a')^2 + (N_b')^2} \quad \text{and} \quad \Delta\omega_n = -\sigma_n \frac{N_a'}{N_b'} \qquad \text{(A11.2.2.3.6)}$$

The final pole frequency and bandwidth is given by

$$f_n = f_0 + \frac{\Delta\omega_n}{2\pi} \quad \text{and} \quad B_n = -\frac{\sigma_n}{\pi} \qquad \text{(A11.2.2.3.7)}$$

By repeating the two-step procedure, one can terminate the search when the first four formants have been found or the incremental frequency is over 5 kHz.

In summary, the N_b method samples $N_b(s)$ at specific frequency increments, checking for changes in polarity of the function. Then a linear interpolation, such as Newton's method, is used to

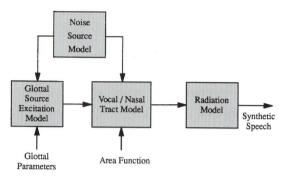

FIGURE A11.7   An acoustic model for the articulatory speech synthesizer.

obtain the frequency root. To determine the final frequencies of H(s), the derivatives N'_a and N'_b are approximated by finite differences.

A11.2.2.4   Estimate the Vocal Tract Cross-Sectional Area from the Formant Frequencies   One of the functions of the articulatory model is to compute the articulatory information (in particular, the vocal tract cross-sectional area) from the acoustic information (the first four formant frequencies) that are obtained from the speech signal. In general, an optimization scheme is used to solve this speech inverse problem. The optimization scheme varies the articulatory parameters iteratively to achieve a match between the model-generated first four formants and the target (desired) first four formants. Later in this appendix we describe a simulated annealing optimization scheme for this purpose.

A11.2.3   Acoustic Models

Basically, the acoustic model of the human vocal system embodies several submodels, as shown in Figure A11.7. Both the vocal tract and nasal tract models simulate the sound propagation in these tracts. The excitation source model represents and generates the voiced excitation waveforms for the vocal tract. The turbulent air flow at a constriction for fricatives or plosives is generated by the noise source model. The radiation model simulates the acoustic energy radiating from the lips and the nostrils.

A11.2.3.1   Vocal Tract Models   The vocal tract is a bent (or curved), three-dimensional acoustic tube with a slowly time-varying shape; it has soft wall vibrations, viscous friction and heat conduction losses, and varying boundaries at both the lips and glottis. There is a nasal side branch, beginning at the top of the pharynx, of fixed dimensions but variable coupling. The nasal tract is discussed later.

Research has demonstrated that the Navier–Stokes description of fluid flow has the feasibility for realistically characterizing the nonlinearities involved in voiced sound generation by the vocal cords, voiceless-fricative generation from turbulent flow at constrictions, and resonance and radiation effects conditioned by sound propagation in a nonuniform, lossy, soft wall human vocal tract (Hegerl and Höge, 1991; Iijima et al., 1992; Thomas, 1986). However, the results have been limited by the extensive computational requirements for solving the time-dependent, turbulent Navier–Stokes equations on a dense time-space grid for realistic geometric configurations. This limitation suggests the need for a simplified version of the acoustic model of the vocal tract.

For a bent (curved) vocal tract with variable cross-sectional area, the computation of its resonances (or its acoustic transfer function) is difficult. Fortunately, Sondhi (1986) has shown that the shift in the resonance frequencies below 4 kHz is in the range of 2% to 8% for typical dimensions of the vocal tract when it is straightened out. Thus, the vocal tract can be represented as a straight tube of varying cross-sectional area, as shown in Figure A11.8(*a*), with fixed shape (circular or elliptic) without a loss in accuracy. The next assumption is that sound propagation is planar along the axis of

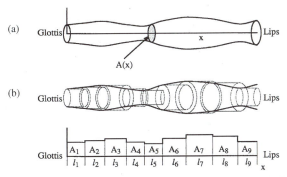

FIGURE A11.8 Vocal tract approximation. (*a*) Ideal vocal tract with variable cross-sectional area. (*b*) A step-wise approximation using concatenated acoustic tubes.

the tube. There are two reasons that make this assumption reasonable. First, the soft tissue along the vocal tract prevents radial propagation of the sound wave. Second, the average lateral (cross-sectional) dimension of the vocal tract is about 2 cm, which is much smaller than the wavelength of sound at 4 kHz, which is $\lambda = c/f = 34,300/4000 \cong 8.6$ cm. Strictly speaking, this assumption is valid only for frequencies below 4 kHz. But for speech, where 5 kHz is considered to be an appropriate bandwidth, the planar propagation assumption remains adequate. By neglecting the losses due to friction, heat conduction, and yielding wall vibration, a pair of equations characterizing the wave propagation in the vocal tract can be derived. In general, the solutions to such a pair of equations can only be obtained numerically. Thus, a further approximation is needed. A more tractable approach is to represent the vocal tract as a number of contiguous cylindrical sections, as depicted in Figure A11.8(*b*). If the number of concatenated sections is large, these short-length elemental sections provide a stepwise approximation of the continuous area function. We can expect that at resonant frequencies, the concatenated tubes are indistinguishable from the continuous ones. The uniform elementary cylindrical section is easy to treat. Once the lossless uniform tube has been analyzed, the effects of losses in the vocal tract can be accounted for.

A11.2.3.1.1 Sound Propagation.

The linear plane wave propagation in the vocal tract is governed by the law of continuity and Newton's force law (Morse and Ingard, 1968)

$$\frac{1}{\rho c^2} \frac{\partial p(x, t)}{\partial t} = -\nabla \vec{v}(x, y, z, t) \tag{A11.2.3.1.1.1}$$

$$\rho \frac{\partial \vec{v}(x, y, z, t)}{\partial t} = -\nabla \cdot p(x, t) \tag{A11.2.3.1.1.2}$$

where ∇ represents the gradient, $\nabla \cdot$ is the divergence, $p(x, t)$ is the variation in sound pressure in the tube at position x and time t, $\vec{v}(x, y, z, t)$ is the particle velocity vector within the vocal tract, c is the velocity of sound, and ρ is the density of air in the tube.

If plane wave propagation is assumed, then all particles at a given displacement x and a specific time t will have the same velocity independent of location (y, z) within the cross-sectional area $A(x, t)$. Since the velocity vector points in a single direction, we can drop the vector notation in Equations (A11.2.3.1.1.1) and (A11.2.3.1.1.2). Define the volume velocity flow at position x and time t as

$$u(x, t) = A(x, t)v(x, t) \tag{A11.2.3.1.1.3}$$

Applying the plane wave propagation assumption and substituting the volume velocity definition from Equation (A11.2.3.1.1.3) into Equations (A11.2.3.1.1.1) and (A11.2.3.1.1.2), we have the following equations (Deller et al., 1993; Rabiner and Schafer, 1978).

$$-\frac{\partial u(x, t)}{\partial x} = \frac{1}{\rho c^2} \frac{\partial (p(x, t)A(x, t))}{\partial t} + \frac{\partial A(x, t)}{\partial t} \tag{A11.2.3.1.1.4}$$

$$-\frac{\partial p(x, t)}{\partial x} = \rho \frac{\partial (u(x, t)/A(x, t))}{\partial t} \tag{A11.2.3.1.1.5}$$

There is no closed form solution except for the simple configurations. If, however, the cross-sectional area $A(x, t)$ and associated boundary conditions are specified, numerical solutions can be obtained. One method of simplifying the last pair of equations is to assume the vocal tract consists of concatenated uniform lossless sections, as depicted in Figure A11.8(b).

A11.2.3.1.2   Uniform Lossless Section.   Assume that the vocal tract is composed of S_N uniform elemental sections and each section has cross-sectional area A_k and length l_k, where $1 \leq k \leq S_N$. This corresponds to spatial sampling, with l_k being the sampling interval for the kth section. Consider the ith section, of length l_i, with constant cross-sectional area A_i. Define $x_{i-1} = \sum_{k=1}^{i-1} \ell_k$, which represents the vocal tract length from the glottis to section i. Then the coupled differential Equations (A11.2.3.1.1.4) and (A11.2.3.1.1.5) for this elemental section become

$$-\frac{\partial u_i(x, t)}{\partial x} = \frac{A_i}{\rho c^2} \frac{\partial (p_i(x, t))}{\partial t} \tag{A11.2.3.1.2.1}$$

$$-\frac{\partial p_i(x, t)}{\partial x} = \frac{\rho}{A_i} \frac{\partial (u_i(x, t))}{\partial t} \tag{A11.2.3.1.2.2}$$

where $u_i(x, t)$ and $p_i(x, t)$ are the volume velocity and pressure, respectively, along the x axis with $x_{i-1} \leq x \leq x_i$. The solutions to Equations (A11.2.3.1.2.1) and (A11.2.3.1.2.2) have the form (Deller et al., 1993; Flanagan, 1972; Rabiner and Schafer, 1978)

$$u_i(x, t) = u_i^+\left(t - \frac{x}{c}\right) - u_i^-\left(t + \frac{x}{c}\right) \tag{A11.2.3.1.2.3}$$

$$p_i(x, t) = \frac{\rho c}{A_i}\left[u_i^+\left(t - \frac{x}{c}\right) + u_i^-\left(t + \frac{x}{c}\right)\right] \tag{A11.2.3.1.2.4}$$

where $u_i^+(t - \frac{x}{c})$ and $u_i^-(t + \frac{x}{c})$ indicate forward (transmitted) and backward (reflected) traveling waves, respectively. The boundary conditions at both ends of each section determine the relationship between the traveling waves in adjacent sections. They are derived from the physical principle that pressure and volume velocity must be continuous in both time and space everywhere within the tract.

A11.2.3.1.3   Approaches for Vocal Tract Simulation.   Based on the above analysis, there are two approaches used for vocal tract simulation.

Wave propagation approach: This approach is based on the analytical solutions of Equations (A11.2.3.1.2.3) and (A11.2.3.1.2.4) for a lossless elemental uniform tube section. The pressure at any point within the section is considered to be made up of two components, a forward wave and a backward wave. At the junction of two cylindrical sections with different cross-sectional areas and lengths (see Figure A11.9), each wave has a forward propagation and backward reflection. Define

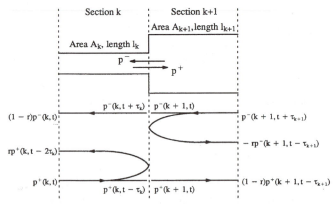

FIGURE A11.9   Reflection relationships at the junction of two lossless sections.

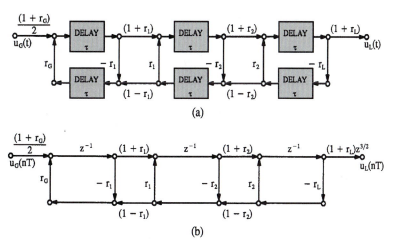

FIGURE A11.10 Discrete time lossless vocal tract. (*a*) Signal flow graph for lossless tube model. (*b*) Equivalent discrete time system.

the reflection coefficient r and the propagation delay τ, then apply the continuity conditions at each junction, and account for the losses at the glottis and lips as boundary conditions. This results in the signal flow graph and the equivalent discrete-time system, as depicted in Figure A11.10. See Rabiner and Schafer (1978) for detailed derivations.

This approach was first used by Kelly and Lochbaum (1962) for speech synthesis and has been called the Kelly–Lochbaum model or lattice structure. A more elegant realization is given by means of wave digital filters (WDF) (Fettweis and Meerkötter, 1975; Liljencrants, 1985; Meyer and Strube, 1984; Meyer et al., 1989; Meyer et al., 1991; Strube, 1982). The WDF has been implemented in special hardware for real-time synthesis (Meyer et al., 1989). Neglecting losses and using a fixed vocal tract length are the major drawbacks of this approach. By varying the sampling rate, the dynamic variation of vocal tract length can be simulated (Wright and Owens, 1993). Some progress has been made in incorporating losses (Liljencrants, 1985; Meyer et al., 1989).

Transmission-line approach: A transmission-line analog of the vocal tract (or equivalent electrical circuit model) is based on the similarity between the acoustic wave propagation in a cylindrical tube and the propagation of an electrical wave along a transmission line. The derivation from the basic equations of acoustic wave propagation to an equivalent electrical quadripole representation is well known (Fant, 1960; Flanagan, 1972; Linggard, 1985). The analogs are summarized in Chapter 5, Figure 5.1. For the circuit in the lower part of Figure 5.1, the series resistor R is used to represent the acoustic loss due to viscous drag, where the energy loss is proportional to the square of the volume velocity. The shunt conductance G represents the loss due to heat conduction, which is proportional to pressure squared. The shunt impedance Z_w is the acoustic equivalent mechanical impedance of the yielding wall. This wall impedance, which represents a mass–compliance–viscosity loss of the soft tissue, has three components, R_w, L_w, and C_w. Figure A10.27 in Appendix 10 provides some physical definitions for all the circuit components in Figure 5.1. Note that both R and G are a function of frequency. We describe the wall impedance in the next subsection.

A11.2.3.1.4 Wall Impedance.

The pressure variation inside the vocal tract causes the cross-sectional area to change, since it exerts a varying force on the tract's elastic walls. Assume that the walls are locally reacting and the resulting cross-sectional area variation is small, that is,

$$A(x, t) = A_0(x, t) + \Delta A(x, t) = A_0(x, t) + y(x, t)S_0(x, t) \qquad (A11.2.3.1.4.1)$$

where $A_0(x, t)$ is the nominal area, $\Delta A(x, t)$ is a small variation, $y(x, t)$ is the yielding amount of the walls, and $S_0(x, t)$ is the circumference of the tract (Maeda, 1982a; Rabiner and Schafer, 1978). The wall vibration is modeled as a mass–compliance–viscosity mechanical model. The pressure variation

is governed by the following differential equation

$$S_0(x, t)p(x, t) = m\frac{\partial^2 y(x, t)}{\partial t^2} + b\frac{\partial y(x, t)}{\partial t} + ky(x, t) \qquad (A11.2.3.1.4.2)$$

where m, b, and k are the mass, viscosity, and compliance, respectively, of the wall per unit length of the tract (Maeda, 1982a). Define the volume velocity generated by the wall vibration as

$$u_w(x, t) = \frac{\partial(y(x, t)S_0(x, t)\ell)}{\partial t} \qquad (A11.2.3.1.4.3)$$

where ℓ is the length of tract. By substituting Equation (A11.2.3.1.4.3) into Equation (A11.2.3.1.4.2), the wall vibration can be rewritten as

$$p(x, t) = \frac{m}{S_0^2(x, t)\ell}\frac{\partial^2 u_w(x, t)}{\partial t^2} + \frac{b}{S_0^2(x, t)\ell}\frac{\partial u_w(x, t)}{\partial t} + \frac{k}{S_0^2(x, t)\ell}u_w(x, t) \qquad (A11.2.3.1.4.4)$$

For an elemental uniform section, Equation (A11.2.3.1.4.4) is simplified to

$$\begin{aligned}
p(x, t) &= \frac{m}{S_0^2\ell}\frac{\partial^2 u_w(x, t)}{\partial t^2} + \frac{b}{S_0^2\ell}\frac{\partial u_w(x, t)}{\partial t} + \frac{k}{S_0^2\ell}u_w(x, t) \\
&= L_w\frac{\partial^2 u_w(x, t)}{\partial t^2} + R_w\frac{\partial u_w(x, t)}{\partial t} + \frac{1}{C_w}u_w(x, t)
\end{aligned} \qquad (A11.2.3.1.4.5)$$

where the circuit components, L_w, R_w, and C_w representing the wall vibration impedance, have the apparent defintions.

The wall impedance can be either included in every elemental section of the vocal tract as a distributed element (Flanagan, 1972a; Flanagan and Ishizaka, 1976; Flanagan et al., 1975, 1980; Ishizaka et al., 1975; Maeda, 1982a) or inserted as a lumped shunt element, one in the pharynx and one at the level of the cheek (Badin and Fant, 1984; Lin, 1990; Wakita and Fant, 1978). As Wakita and Fant (1978) indicated, the lumped wall impedance, which is independent of the vocal tract configurations may not give satisfactory results. The distributed wall impedance is used in our software.

Figure A10.25 in Appendix 10 presents data concerning the wall mass, viscosity, and compliance found in the literature. In some cases, the compliance is not used since it has virtually no effect on the resonances of the model (Wakita and Fant, 1978). The data measured by Ishizaka and colleagues (1975) are used in our software. As Maeda (1982a) pointed out, the total mass of the walls may vary unrealistically if the yielding wall parameters are specified in terms of a unit surface area. Thus, the per unit length specification was used in Maeda's vocal tract simulation. We follow Maeda's specification.

A11.2.3.2 Nasal Tract and Sinus Cavities

The nasal tract constitutes a side branch of the vocal tract. The velopharyngeal port controls the coupling between these two tracts for producing certain sounds. A general rule is that when the opening area is smaller than 20 mm^2 there is no apparent nasality. A wider opening produces nasal resonance, and speech is perceived as nasal when the area approaches 50 mm^2 (Borden and Harris, 1980). In our articulatory model, the opening area of the velopharyngeal port is simulated by lowering the velum along a line segment, as mentioned in Section A11.2.2.1.

The nasal tract has two channels at the nostrils. Usually, an acoustically approximated single tract is used owing to its quasi-symmetrical profile. The minor errors due to this approximation have been analyzed by Lin (1990). Figure A11.11 shows the area function of the nasal tract used by Maeda (1982b), where the nasal tract is assumed to be 11 cm long and consists of 11 elemental uniform sections. Generally, the nasal tract has a fixed structure except for the first few sections, indicated by a dashed line in Figure A11.11, where the area varies with the velopharyngeal port opening. Maeda (1982b) used linear interpolation to interpolate the areas (the second and third sections) between the coupling section (the first section) and the first fixed section (the fourth section).

From sweep frequency measurements of the acoustic transfer function (Lindqvist-Gauffin and Sundberg, 1976) and simulation studies (Fant, 1985; Lin, 1990; Maeda, 1982b) of the nasal tract, it has been found that the nasal sinuses should be considered as a part of the acoustic system. In a model of speech production, Lindqvist-Gauffin and Sundberg (1976) indicated that at least two

TABLE A11.1   Data on the Sinuses

	R_{sin} (dyne · sec/cm^5)	L_{sin} (g/cm^4)	C_{sin} (cm^4 · sec^2/g)	F (Hz)	B (Hz)
Maxillary (Lin, 1990)	1.1	3.42×10^{-3}	29.7×10^{-6}	500	51.2
Frontalis (Lin, 1990)	7.2	11.4×10^{-3}	1.13×10^{-6}	1,400	100.5
Maxillary (Sondhi and Schroeter, 1987)	1.0	5.94×10^{-3}	15.8×10^{-6}	519.5	26.8

shunting cavities, the sinus maxillares and the sinus frontales, must be added to improve the nasal quality. Since the opening area, which couples the sinus cavities to the nasal tract, is rather small, the sinus cavities can be regarded as Helmholtz resonators. According to the Lindqvist-Gauffin and Sundberg (1976) study, a reasonable estimate of the resonant frequencies would be 200 to 800 Hz for the maxillary sinuses and 500 to 2000 Hz for the frontal sinuses. The effect of the sinus resonance on the acoustic system is modeled as a shunt circuit element (Fant, 1985), as shown in Figure A11.12, and the resonance can be tuned to the required frequency. Fant (1985) inserted the sinus maxillares and frontales at positions 6 cm and 8 cm from the nostrils, respectively. The two sinuses are tuned to resonate at 500 Hz and at 1400 Hz respectively. However, Maeda (1982b) inserted only the sinus maxillares at a position 4 cm from the nostrils and showed that the quality of all nasalized vowels was satisfactory. Table A11.1 lists data for the shunt circuit components used in the literature (Lin, 1990; Sondhi and Schroeter, 1987).

To investigate the effects of the nasal tract and sinus cavities, our software system provides the user with a method to vary the nasal tract structure (area and length), to assign the number of coupled sinus cavities, and to change the circuit component values and the coupling position of each sinus.

A11.2.3.3   Radiation Models of Lips and Nostrils   Acoustic energy escapes

from the vocal tract via the lips. From the transmission-line analogs, the lips are treated as a radiation impedance that loads the vocal tract. The radiation impedance contains a resistive part that represents acoustic energy loss and a reactance part that represents the mass inertia of air at the lips (Fant, 1960; Stevens, 1998). Radiation from a spherical baffle is one model for the radiation impedance that is represented by nonlinear functions (Morse, 1948; Morse and Ingard, 1968). Stevens et al. (1953) made approximations and represented the radiation impedance by a resistive load and three other frequency-dependent components. Fant made another approximation and modeled the impedance by two frequency-dependent components, one being resistive and the other inductive (Fant, 1960; Wakita and Fant, 1978).

Another simplified radiation model is to assume that the radiating surface is set in a plane baffle of infinite extent. In this case, the radiation impedance is formed by a first order Bessel function and Struve function (Flanagan, 1972a; Rayleigh, 1945; Wakita and Fant, 1978). Flanagan (1972) provided a good approximation to this complicated representation by a parallel connection of a resistance and an

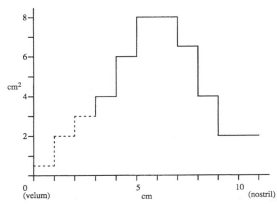

FIGURE A11.11   The area function for the nasal tract (after Maeda, 1982b).

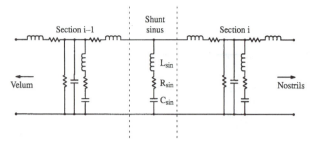

FIGURE A11.12 An extra shunt inserted between two nasal sections to model the sinus.

inductance. The most important feature of Flanagan's model (1972a) is that both circuit components are frequency independent. Figure A10.28 in Appendix 10 illustrates the Stevens et al. (1953) model and the Flanagan (1972a) model.

Comparisons between models have been made by researchers (Badin and Fant, 1984; Lin, 1990; Wakita and Fant, 1978). The Stevens et al. (1953) model yields the most accurate result. However, the Flanagan (1972a) model is usually preferred for time-domain synthesis (Flanagan and Ishizaka, 1976; Flanagan et al., 1975, 1980; Maeda, 1982a) and is used in our synthesis model. The same radiation model is used for the nostrils.

The relationship between the volume velocity at the lips or nostrils and the radiated pressure at a distance d cm from the lips or nostrils is given by (Fant, 1960)

$$\frac{p_r(\omega)}{u_r(\omega)} = \frac{\rho\omega}{4\pi d} K_T(\omega) \qquad \text{(A11.2.3.3.1)}$$

The factor $K_T(\omega)$ is a function to provide a smooth high frequency emphasis. Due to a lack of experimental verification, $K_T(\omega)$ is generally set to unity and the relationship is essentially a differentiation (Badin and Fant, 1984).

A11.2.3.4 Excitation Source Models

Basically, there are two kinds of speech sounds. One is voiced, which involves quasi-periodic vibrations of the vocal folds. The other is unvoiced, which involves the generation of turbulence noise by the rapid flow of air past a narrow constriction. In the case of voiceless speech, the excitation waveform appears somewhat like a random noise source, which we will discuss in section A11.2.3.6. For voiced speech, the excitation source is a quasi-periodic pulse train located at the glottis.

A11.2.3.4.1 Excitation at the Glottis.

In the case of voiced speech, the conventional LPC methods use only an impulse train as the excitation, which does not generate natural sounds (Childers and Wu, 1990). It is well-known that the "naturalness" of synthetic speech is closely related to the shape of the glottal pulse (Childers and Lee, 1991; Childers and Wu, 1990; Holmes, 1973; Klatt and Klatt, 1990; Rosenberg, 1971). We do not yet have a complete understanding of the phonatory behavior of the vocal folds. Thus, we lack an efficient model for the voice source. However, several models capable of describing the major characteristics of the glottal flow have been proposed. They can be classified into two major categories: interactive and noninteractive models (Fujisaki and Ljungqvist, 1986).

In the interactive models, there are two approaches to generate the glottal volume velocity. For the method known as the nonphysical approach, the glottal flow is calculated by modeling the glottal area (Allen and Strong, 1985; Ananthapadmanabha and Fant, 1982; Pinto et al., 1989; Titze, 1984) or conductance (Rothenberg, 1981) function and by incorporating the various impedances of the acoustic system into the model. For the method known as the physical approach, structural modeling of the mechanical vibration of the vocal cords (Flanagan and Landgraf, 1968; Ishisaka and Flanagan, 1972; Titze, 1973) and a kinematic model for the three-dimension glottis (Titze, 1989) have been attempted. The need to know the details of the physical characteristics of the various parts of the vocal cords is the major drawback of the interactive models. Furthermore, the computational burden for such models is high. We have, however, implemented one mechanical vibratory model in Chapter 9 with the theory presented in Appendix 10.

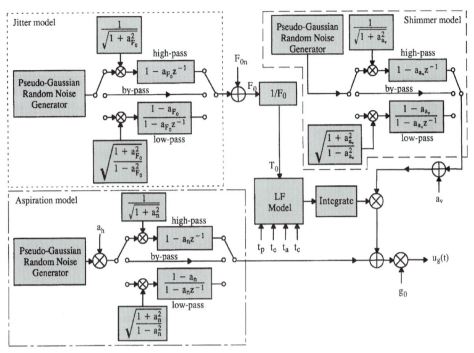

FIGURE A11.13 A simplified excitation model (Lalwani and Childers, 1991a).

In contrast, the noninteractive models directly parameterize the glottal flow or flow derivative function. If the parameters are sufficient to represent the glottal waveform, it may be possible to reconstruct the waveform from a given set of parameters. Therefore, the parameterization provides a method for generating, classifying, and storing a large number of glottal waveforms for various voicing conditions. A number of non–interactive models exist in the literature (Ananthapadmanabha, 1982; Fant, 1979; Fant et al., 1985; Fujisaki and Ljungqvist, 1986; Klatt and Klatt, 1990; Rosenberg, 1971). The Liljencrants–Fant (LF) model (Fant et al., 1985) is often used because: (1) it is preferred by listeners when they evaluate synthesized speech (Childers, 1991; Eggen, 1992); and (2) it has been shown to be superior to other models of the same complexity (Fant et al., 1985; Fujisaki and Ljungqvist, 1986). The LF model requires four parameters for modeling the differential glottal waveform (see Appendix 7). For additional details of the LF model refer to Fant (1986, 1988, 1993), Fant and Lin (1988, 1989), and Lin (1990). Another advantage of the LF model is that parameters of the model can be measured or estimated from the inverse filtered speech and the EGG signal or from the inverse filtered speech signal only (Childers and Lee, 1991; Lee, 1988).

It is well known that "jitter," the aperiodicity of the fundamental frequency of voicing (Horii, 1979), and "shimmer," the period-to-period random fluctuations in glottal-pulse amplitude (Horii, 1980), also contribute to a natural sounding voice. Klatt and Klatt (1990) included a slow quasi-random drift called "flutter" into the voicing source model to simulate jitter (pitch perturbation) but did not include a shimmer model. Lalwani and Childers (1991a) proposed a unified glottal excitation model that includes the pitch perturbation model with rate of perturbation control and the aspiration noise model with amplitude modulation into the LF model. This unified model has the capability to include the "shimmer," that is, the amplitude perturbation model. We use a simplified version of this unified model as a noninteractive glottal excitation model in our software. Figure A11.13 illustrates the block diagram of such a simplified excitation model.

A11.2.3.4.2 Excitation in the Vocal Tract. A speaker routinely phonates using the vocal folds. Unfortunately, over 1.5 million nonspeaking persons in the United States, excluding some deaf individuals (Klatt, 1987), cannot phonate using the vocal folds. Although aerodynamic-mechanical devices can aid the vocally handicapped, they produce no sound until pulmonary air is

diverted through them (Hilgers and Schouwenburg, 1990). In 1980, an electrically driven intraoral artificial larynx was invented (Lowry, 1981). This new speech prosthesis consists of a small speaker (with battery) and a resonator horn, which are joined to a dental plate and placed in the oral cavity of the subject (Myrick and Yantorno, 1993). With such a device the subject can produce intelligible speech, although the quality is still inferior. Such a device, called UltraVoice, is available through Health Concepts, Inc. (Malvern) in Pennsylvania.

Since this electrical-driven speech prosthesis must be placed in the vocal tract, it is reasonable to expect that the driving point acoustic transfer function is different from that seen by the glottis. Thus, the excitation signal must have a different waveform from the glottal pulse to produce the same speech sounds. Myrick and Yantorno (1993) presented the vocal tract frequency response when the excitation is located at the sixth section of a ten-section vocal tract by using the Kelly–Lochbaum model (1962) and the lossless transmission-line analog model (Flanagan, 1972a). C. Wu (1996) provides a detailed derivation of a new form of the discrete time acoustic equations with the excitation located within the vocal tract.

Our software system provides several advanced features for the user to investigate the properties when the excitation is located in the vocal tract. They are:

- The excitation can be located at any section of a sixty-section vocal tract with soft-wall vibration, thermal, and heat conduction losses.

- The nasal tract, with or without sinus cavities, can be coupled into the vocal tract by varying the opening area of the velopharyngeal port.

- The subglottal system can be coupled to the vocal tract when the glottis is opened.

A11.2.3.5 *Glottal Impedance and Subglottal Models*   According to the classical formulation of the acoustic theory of speech production (Fant, 1960; Flanagan, 1972a), the voicing source is characterized as a current source. This assumes that the glottal waveform depends very little on the shape or impedance of the vocal tract. Similarly, the vocal tract is modeled by a time-invariant linear filter since the glottal impedance is assumed much higher than the vocal tract impedance. However, recent work by several researchers has shown that there does exist a certain degree of dependency of the glottal flow on the load of the vocal tract and the subglottal cavities. A number of major interaction consequences have been identified. They are:

- First formant, F_1, ripple in the source waveform: One may often observe a "hump" in the rising portion of the glottal volume velocity waveform obtained by using inverse filtering (Childers and Wu, 1990).

- Nonlinear interaction between F_0 and F_1: The pharyngeal pressure standing waves may cause a nonlinear increase in the glottal source strength whenever F_1 is near an integral multiple of F_0 (Ananthapadmanabha and Fant, 1982).

- Truncation of the F_1 damped sinusoid: The time-varying glottal impedance affects the vocal tract transfer function primarily by increasing the first-formant bandwidth, which leads to a truncation of the F_1 damped sinusoid when the glottis is open (Ananthapadmanabha and Fant, 1982).

- Pulse-skewing: The inertive loading by the subglottal and supraglottal acoustic systems results in a skewing to the right of the glottal pulse (Rothenberg, 1981).

Although speech can be synthesized by using a simple noninteractive excitation model, it seems to be essential that a glottal excitation model used in the articulatory speech synthesizer should reproduce the variations in the acoustic features of the excitation more naturally (Sondhi and Schroeter, 1987). Interactive physical models are hard to implement since one needs the physiologic characteristics of the vocal folds, and, furthermore, most models are computationally inefficient. On the other hand, interactive nonphysical models are attractive for researchers. Klatt and Klatt (1990) included the ability to change the first-formant bandwidth pitch synchronously to simulate the interaction between source and vocal tract in their formant synthesizer. For the articulatory synthesizer, a prescribed glottal area time function is usually used for source–tract interaction.

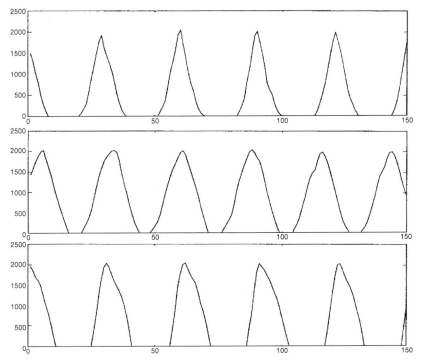

FIGURE A11.14   Three glottal area waveforms measured from ultra-high speed films.

The glottal area is defined as the opening between the vocal folds. It is time-varying during voiced phonation and quasi-steady for voiceless phonation. One possible method to obtain the time-varying glottal area is from ultra-high speed films (Childers and Krishnamurthy, 1985; Childers and Larar, 1984; Childers et al., 1984; Childers et al., 1990; Moore and Childers, 1983). The area function measured in this manner is the projected area, that is, the minimum area of the glottis. Figure A11.14 shows three measured glottal area waveforms from ultra-high speed films, where one can see that the projected glottal area tends to have a roughly triangular shape that is slightly skewed to the left. The sharp peak of the glottal area function usually results in the excitation at the apex being exaggerated (Lin, 1990). On the other hand, Cranen and Boves (1987) have derived the glottal area from a vertically uniform glottis by using the two-mass model (Ishizaka and Flanagan, 1972). The glottal area function derived in this manner is called the effective glottal area function and does not coincide with the projected glottal area. However, some simple functions such as a triangle, a sine, and a raised cosine are used to model the glottal area (Ananthapadmanabha and Fant, 1982). Figure A11.15 illustrates the glottal area waveforms modeled by triangular, sine, and raised-cosine functions. Our software program implementation provides these three functions as options for modeling the glottal area. In addition, for the triangular and raised-cosine functions, the opening and closing durations can be specified to model the glottal area skewing.

The time-varying glottal impedance is determined by the time-varying glottal area function. It contains a resistance and an inductance. Assume that the glottis is modeled as a rectangular slit with A_g, l_g, and d as the area, length, and thickness, respectively. Then the glottal inductance is given by

$$L_g = \frac{\rho d}{A_g} \qquad \text{(A11.2.3.5.1)}$$

The resistance of the glottis, according to an experiment by van den Berg and co-workers (1957), is formulated by

$$R_g = \frac{12\mu d\ell_g^2}{A_g^3} + k_g \frac{\rho u_g}{2A_g^2} \qquad \text{(A11.2.3.5.2)}$$

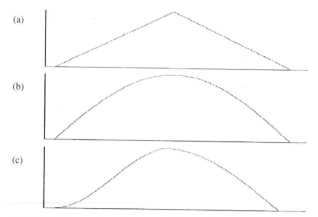

FIGURE A11.15 Glottal area models. (*a*) Triangular. (*b*) Sine. (*c*) Raised cosine.

where μ is the viscosity of air, k_g is a coefficient, ρ is the density of air in the tube, and u_g is glottal volume velocity. Some typical values of k_g used in the literature are 0.875 (van den Berg et al., 1957), 0.9 (Stevens, 1971), 1.1 (Ananthapadmanabha and Fant, 1982), and 1.38 (Maeda, 1982a).

The subglottal system, which includes the tracheal tube and lungs is usually omitted in vocal tract simulation, since its effect on speech spectra is assumed to be minor, except for unvoiced sounds, where the glottis is large (Ishizaka et al., 1976). However, when the glottal opening is large, which means the glottal impedance is no longer very high, the coupling between the subglottal system and the vocal tract is not negligible. From measurements of laryngectomized subjects, Ishizaka et al. (1976) measured the acoustic input impedance of the subglottal system. Ananthapadmanabha and Fant (1982) used Ishizaka et al. (1976) experimental data and represented the subglottal system as three cascaded RLC resonance modules called the Foster-chain circuit. Figure A11.16 shows the circuit and the corresponding component values. The subglottal formants were located at 640, 1335, and 2110 Hz, with the corresponding bandwidths of 246, 155, and 140 Hz. The effects of the Foster-chain subglottal model on the vocal tract formants and bandwidths have been analyzed (Ananthapadmanabha and Fant, 1982; Badin and Fant, 1984; Lin, 1990). The summary is that the acoustic effect of the subglottal system is small, except for unvoiced sounds, where the glottal opening is fairly large.

Combining the simplified Lalwani and Childers excitation source model (1991a), Ananthapadmanabha and Fant glottal area model (1982), and Foster-chain subglottal model, we use an interactive excitation model, shown in Figure A11.17.

A11.2.3.6 *Noise Source Models*

When there is a flow of air through a constriction or past an obstruction, turbulence is created (Shadle, 1991; Stevens, 1971, 1993a, 1993b). The random velocity fluctuations in the flow can act as a source of sound called turbulence. Three types of consonants produced in this manner are fricatives, stops (plosives), and affricates. Fricatives are generated with the turbulent flow excitation located in the region of a constriction in the vocal tract. Plosives are produced by making a complete closure of the tract, building up pressure, and abruptly releasing it. The stop release is frequently followed by a turbulence noise excitation. Affricates are

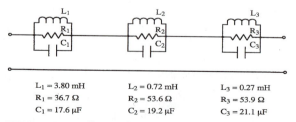

$L_1 = 3.80 \text{ mH}$ $L_2 = 0.72 \text{ mH}$ $L_3 = 0.27 \text{ mH}$
$R_1 = 36.7 \ \Omega$ $R_2 = 53.6 \ \Omega$ $R_3 = 53.9 \ \Omega$
$C_1 = 17.6 \ \mu\text{F}$ $C_2 = 19.2 \ \mu\text{F}$ $C_3 = 21.1 \ \mu\text{F}$

FIGURE A11.16 Foster-chain circuit model of the subglottal system.

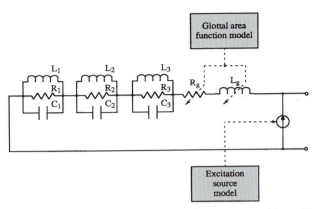

FIGURE A11.17 An interactive excitation source model used in our software.

dynamic sounds that can be modeled as the concatenation of a stop and a fricative. A special phoneme, denoted as HH or /h/, called aspirate, is produced with turbulent flow through the glottis. Refer to Broad (1977a), Borden and Harris (1980), and Stevens (1998) for more details on the generation of the unvoiced speech sounds.

Under the plane wave assumption, the sound pressure of turbulent flow can be taken as proportional to the square of the volume velocity of the airflow and inversely proportional to the constriction area, (Stevens, 1971). The turbulence noise source may be located at the center of, or immediately downstream from, the constriction region or possibly, at a combination of these places, or spatially distributed along the constriction region (Fant, 1960; Flanagan and Cherry, 1968; Lin, 1990; Stevens, 1971, 1993a, 1993b, 1998). The spectrum of the turbulence noise is broadly distributed over a wide range of frequencies (2 to 8 kHz), with some accentuation in the mid-audio range (Childers and Lee, 1991; Stevens, 1971, 1993a, 1993b, 1998).

Basically, the noise source model defines the characteristics of the noise source as a function of the airflow through the constriction and of the constriction cross-sectional area. Meyer-Eppler (1953) (Broad, 1977b) found that the rms sound pressure, P_{rms}, of the noise could be expressed as

$$P_{rms} = A_c \left(R_e^2 - R_{ec}^2 \right) \qquad \text{(A11.2.3.6.1)}$$

where R_e is the Reynolds number and R_{ec} is the critical Reynolds number. Fant (1960) adopted a serial noise pressure source and reformulated the P_{rms} as a function of the pressure drop through the constriction and the effective width of the constriction. Lin (1990) extended Fant's model to include the frictional and turbulent losses inside the constriction. Both Fant and Lin tried to reconstruct the fricative spectra from area functions by using the acoustic transfer function. Some fricatives have been modeled quite successfully and some are unsatisfactory (Badin, 1989, 1991). Klatt (1980) used a random number generator, a spectrum-shaping filter, and an amplitude modulator to model the turbulent flow for the formant synthesizer. The spectrum-shaping filter was designed to simulate the spectral characteristics of the turbulent flow. A first order IIR filter was used to obtain the volume velocity due to a random pressure source. Childers and Lee (1991) have used a FIR filter to model highpass-filtered turbulence noise. Cook (1991, 1993) used a four-pole filter to model the spectral properties of the noise source.

By including a latent random pressure source, P_n, and an inherent constriction loss, R_n, in each elemental section of the vocal tract, Flanagan and Cherry (1968) could automatically introduce the turbulent flow excitation at any section. The P_n source was produced from Gaussian noise, which was bandpass-filtered from 500 to 4000 Hz, and the flow, U_n, was lowpass-filtered to 500 Hz before it modulated the P_n noise source. Figure A11.18 illustrates the schematic diagram. Such a turbulence noise model has been used in several studies (Flanagan and Ishizaka, 1976; Flanagan et al., 1975, 1980). However, as Sondhi and Schroeter (1987) pointed out, the Flanagan and Cherry (1968) model did not produce satisfactory unvoiced sounds due to the high "back" cavity impedance. Sondhi and Schroeter (1986, 1987) modified the above distributed and series pressure noise source model into

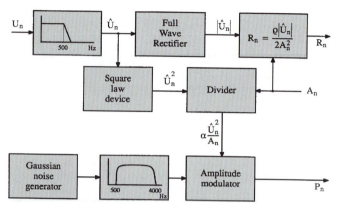

FIGURE A11.18 Schematic diagram of Flanagan and Cherry (1968) turbulence source generation model. A_n is the constriction area.

a parallel flow source $U_n = P_n/R_n$, which was located downstream from the constriction. The P_n is given by

$$P_n = (\text{turbg})(\text{rand}) \left(R_e^2 - R_{ec}^2\right), \quad \text{for } R_e > R_{ec}$$
$$= 0, \qquad\qquad\qquad\qquad \text{for } R_e < R_{ec}$$
(A11.2.3.6.2)

where turbg is empirically determined as the turbulence gain, and rand is a random number uniformly distributed between -0.5 and 0.5. A first-order IIR filter with cutoff frequency 2000 Hz is used to lowpass the flow. Figure A11.19(a) and (b) shows the equivalent circuits of the serial and parallel turbulence sources, respectively.

We adopt the turbulence noise source model from Sondhi and Schroeter (1986, 1987). However, our model allows the user to place the turbulence noise source at the center of, or immediately downstream or upstream from, the constriction region, or spatially distributed along the constriction region. The turbulence gain and critical Reynolds number can also be specified.

We considered an acoustic model of the human vocal system. Next, we construct a transmission-line circuit model for the vocal system. Figure A11.20 illustrates the model structure of the vocal system for the proposed articulatory synthesizer. Based on this structure, the acoustic transfer function for different characteristics of the vocal system can be evaluated. The main purpose of this model structure is for deriving the acoustic equations for synthesizing speech. Refer to Appendices A11–B and A11–C for acoustic transfer function calculations and the derivation of the acoustic equations, respectively.

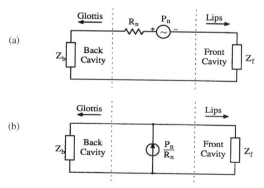

FIGURE A11.19 Equivalent circuits for the turbulence source. (a) Serial (after Flanagan and Cherry, 1968). (b) Parallel (after Sondhi and Schroeter, 1986).

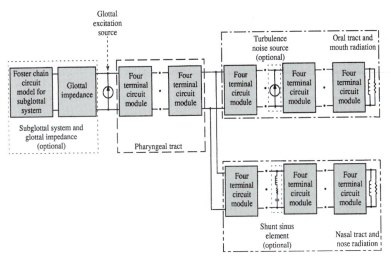

FIGURE A11.20   A model structure for the vocal system for the articulatory speech synthesizer.

A11.2.4   Analysis of Various Vocal System Characteristics

In this section, we analyze the effects of various vocal system characteristics. This analysis provides the basis for selecting the appropriate vocal system model structure and component values for the articulatory synthesizer. Five American vowels and diphthongs (/AA, IH, IY, OW, UW/) are investigated under different vocal characteristics. The vocal tract cross-sectional areas for these vowels and diphthongs are given in Appendix 5, while the methods for calculating the acoustic transfer function are given in Appendix A11-B.

A11.2.4.1   Frequency-Dependent Components
As mentioned in Section A11.2.3.1.3, the series resistance R and the shunt conductance G in the equivalent circuit representation of a lossy section are frequency dependent. In the time-domain approach for the articulatory synthesizer, the frequency-dependent components have to be simulated at a fixed frequency. The effects of using a fixed frequency for these two components on the acoustic transfer function are not well documented. Wakita and Fant (1978) illustrated the effects on formant frequencies and bandwidths for the five Russian vowels given in Appendix 5, when the frequency was fixed at 1 kHz. They determined that the formant frequencies are scarcely affected, but that the bandwidths are affected rather appreciably. We have investigated the acoustic transfer functions for five vowels for three fixed frequencies: 1 kHz, 2.5 kHz, and 4 kHz. The acoustic transfer functions were calculated with the glottis closed and no nasal tract coupling. The formants were not affected appreciably, which agrees with Wakita and Fant's (1978) result. For formant frequencies below 1.5 kHz, the formant bandwidths for the fixed frequency cases were wider than the frequency-dependent case. For formant frequencies above 1.5 kHz, the 1 kHz case had the narrowest formant bandwidths, which is narrower than the frequency-dependent case. As a trade-off, we have set the frequency at 2.5 kHz for the frequency-dependent components.

A11.2.4.2   Number of Vocal Tract Sections
As described in Section A11.2.3.1.2, the vocal tract can be approximated by a concatenation of uniform elemental sections. The number of elemental sections, S_N, should be large enough so that the acoustic characteristics of the concatenated tubes is indistinguishable from a continuous tube. In this section, we examine the influence of spatial sampling on the acoustic transfer function of the vocal tract. Figure A11.21 shows the acoustic transfer functions of different numbers of vocal tract sections for five vowels. The acoustic transfer functions are calculated with the glottis closed and no nasal tract coupling. It can be seen that the formant frequencies generally shift upward as the spatial sampling interval increases. The curves for 60 sections (solid line) are nearly identical to the results obtained for the last section with the frequency set at

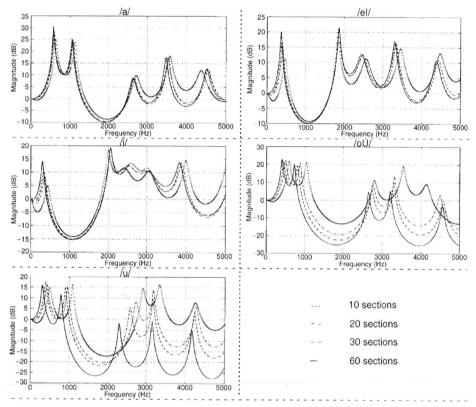

FIGURE A11.21   The effect of the spatial sampling interval on the acoustic transfer function for five vowels.

2.5 kHz for the frequency-dependent components. The exception is for the second and third formants for the vowel /IY/, where the formants shift downward. The vowels have been labeled using the IPA symbols. The corresponding upper case ARPAbet symbols are as follows: /a/ = /AA/ (as in Bach), /eI/ = /IH/ (as in bit), /i/ = /IY/ (as in beet), /oU/ = /OW/ (as in boat), /u/ = /UW/ (as in boot). From Figure A11.21, we can see that the spatial sampling interval, that is, the number of elemental sections, has a more significant effect on the acoustic transfer functions of vowels /u/ = /UW/ and /oU/ = /OW/ than others. However, a ten-section cross-sectional area function is not sufficient to represent the acoustic characteristics of the vocal tract.

A11.2.4.3 Nasal Tract System   To study the acoustic properties of the nasal tract, the acoustic transfer function of the nasal tract for different opening areas of the velopharyngeal port are calculated. Figure A11.22 shows the resonant characteristics of the nasal tract for various velopharyngeal port opening areas. Basically, the nasal cavity has three resonant frequencies. The first resonance is not affected by the velopharyngeal port opening area. However, the second and third resonances shift downward as the velopharyngeal port opening area increases.

The effect of the extra sinus cavities on the acoustic transfer function of the nasal tract can be examined from Figure A11.23. The maxillary sinus and the frontal sinus are located at 4 cm and 8 cm, respectively, from the nostrils. The two sinus cavities are tuned to 500 Hz and 1400 Hz, respectively. The frontal sinus has a limited effect on the nasal tract acoustic transfer function. It has the effect of a zero-pole pair in the vicinity of its resonance frequency. For this reason, Maeda (1982b) ignored this sinus without losing the essentials of the nasal tract. The first resonance of the sinus nasal tract is shifted downward with a lower peak level as a result of the maxillary sinus coupling. The maxillary sinus also brings about a pole-zero pair in the vicinity of its resonance frequency.

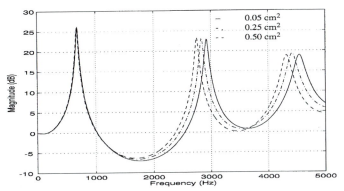

FIGURE A11.22   The effect of the velopharyngeal port opening on the nasal tract acoustic transfer function.

Lowering the velum creates a side passage for the air flow through the nasal cavity, giving rise to complex modifications of the acoustic characteristics of the sound. Figure A11.24 shows how this mechanism affects the acoustic transfer function of the vocal system. The velopharyngeal opening area is 0.5 cm². For the nasal tract with the sinus cavity, only the maxillary sinus is included and is tuned to 500 Hz. It is well known that the parallel branching of the nasal tract at the velum causes antiresonances in the vocal tract acoustic transfer function. The antiresonances are at the vicinities of 1 kHz and 3 kHz and can be seen clearly for some vowels. The effect of an extra sinus on the acoustic transfer functions for vowels /a/ = /AA/ and /eI/ = /IH/ is more significant than for vowels /i/ = /TY/, /u/ = /UW/, and /oU/ = /OW/. This result supports Maeda's statement (1982b) that the high vowels, such as /TY/ and /UW/, are more nasalized than the middle and low vowels, such as /AH/ and /AA/, even when the nasal sinus is not included.

A11.2.4.4   Glottal Impedance and Subglottal System   The influence of the glottal impedance and the subglottal system on the acoustic transfer function of the vocal system can be examined from Figure A11.25. When the glottal area is small; that is, the glottal impedance is relatively high, the influence is insignificant. For a large glottis, the increased loading of the vocal tract causes an increase in the bandwidths and to some extent in the formant frequencies. It is obvious that the influence of the subglottal system depends on the glottal impedance. When the glottal area is small, the influence of the subglottal resonances is small, and vice versa.

A11.2.4.5   Excitation in the Vocal Tract   Placing the excitation source in the vocal tract results in a very complicated system, as mentioned in Section A11.2.3.4.2. This section

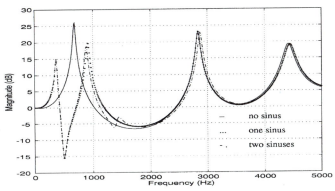

FIGURE A11.23   The effect of extra sinus cavities on the nasal tract acoustic transfer function.

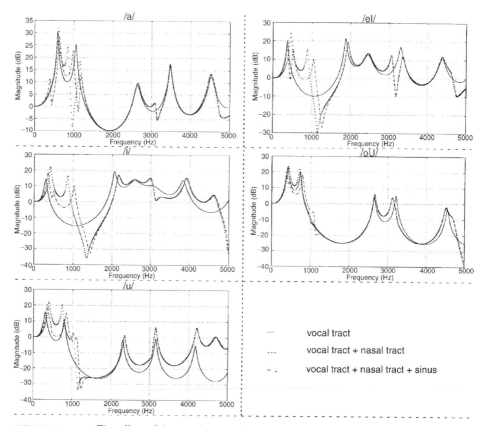

FIGURE A11.24   The effect of the nasal tract and an extra sinus on the acoustic transfer function for five vowels.

examines the effects of relocating the excitation on the acoustic transfer function (see Figure A11.26). It is found that the vocal tract resonant frequencies are relatively unaffected when the excitation is placed forward from the pharynx to the front oral cavity, even though this introduces antiresonances into the acoustic transfer function. Another feature is that the number of antiresonances increases when the excitation is placed forward from the pharynx to the front oral cavity. The excitation source waveform can be obtained by deconvolving the speech signal from the modified acoustic transfer function. It is expected that such an excitation waveform differs from the glottal waveform, which generally contains no zeros in its transfer function. Different vowels have different antiresonances, due to different vocal tract shapes. This makes the modeling of the excitation waveform inside the vocal tract difficult. One possible way to generate the excitation waveform inside the vocal tract is to prefilter the glottal pulse with the inverse filter of the modified acoustic transfer function.

A11.2.5   Articulatory Speech Synthesis

Basically, there are three approaches used in articulatory speech synthesis. The wave digital filter approach (Fettweis and Meerkötter, 1975; Lawson and Mirzai, 1990) extends the Kelly–Lochbaum model (1962). This approach is based on forward and backward traveling waves in a lossless acoustic tube (Meyer et al., 1989; Rubin et al., 1981; Strube, 1982; Titze, 1973) and can produce real-time synthesis (Meyer et al., 1989). It usually omits many acoustic effects, such as proper handling of existing losses in the tract, realistic modeling of the glottis, and appropriate modeling of source–tract interaction. Also the vocal tract length cannot be varied easily since the length of each section is fixed and related to the sampling frequency (Wakita, 1973). Recently, some progress has been made in

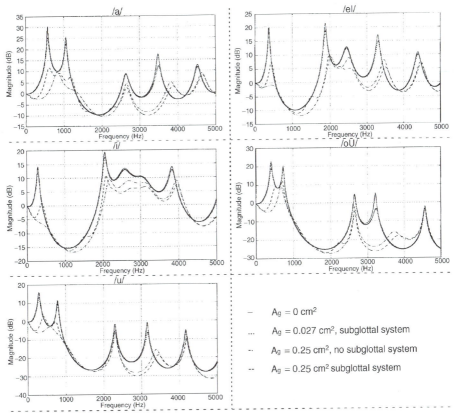

FIGURE A11.25 The effect of the glottal impedance and the subglottal system on the acoustic transfer function for five vowels.

modeling voiceless excitation, damping, and the glottal excitation (Meyer et al., 1989). The dynamic variation of vocal tract length can be simulated by varying the sampling rate (Wright and Owens, 1993).

The second approach uses a hybrid time–frequency domain method, which models the highly nonlinear glottal characteristics in the time domain and the linear tract with frequency-dependent losses and wall vibration characteristics in the frequency domain (Allen and Strong, 1985; Sondhi and Schroeter, 1986, 1987). The tract filter function and glottal source excitation function are interfaced by an inverse Fourier transformation and digital convolution. The problems with this approach are that it is incapable of producing the dynamic transitions of certain phonemes; for example, plosives, and it needs additional care to cope with the interaction between voiced and voiceless sources (Lin, 1990). In addition, it does not calculate the pressure and volume velocity.

The third approach is to model the human vocal system as a large set of linear or nonlinear difference equations to be solved in each sampling interval to give samples of the pressure and volume velocity at each point in the transmission-line circuit (Flanagan and Cherry, 1968; Flanagan and Ishizaka, 1976; Flanagan and Landgraf, 1968; Flanagan et al., 1975, 1980). The values of pressure and volume velocity at one time instant are used to determine the losses for the next time interval. This approach has been referred to as the time-domain approach (Sondhi and Schroeter, 1987). Figure A11.27 shows the schematic diagram of this approach.

In the time-domain approach, a very high sampling rate is usually required to avoid frequency-warping distortion (Wakita and Fant, 1978). In addition, the frequency-dependent components are simulated at a fixed frequency (see Section A11.2.4.1). Natural-sounding speech, however, can be generated. Several advantages have made the time-domain approach popular, although its computation

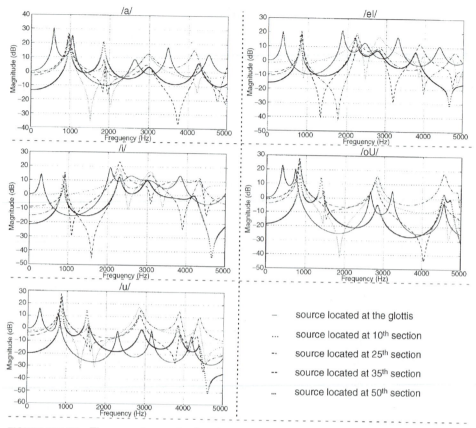

FIGURE A11.26   The acoustic transfer function for the five vowels when the source is located at different positions within the vocal tract.

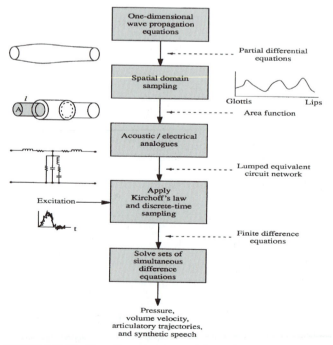

FIGURE A11.27   Time-Domain approach for articulatory speech synthesis.

is lengthy. These advantages are that the aerodynamic interaction is inherently included, the pressure and volume velocity at any point can be computed, and the dynamic articulatory gestures can be obtained when combined with the articulatory model. In our software, the time-domain approach is used for realization of the acoustic model.

Maeda (1982a) simplified Flanagan's model by replacing the mechanical vibratory model of the vocal cords with glottal area control parameters, discarding noise sources within the vocal tract, and omitting the effects of the nasal sinuses. These simplifications made the synthesis procedure much faster. Bocchieri (1983) and Bocchieri and Childers (1984) introduced other simplifications, such as reducing the number of noise sources and modeling the vocal tract variation by drawing a sequence of midsagittal vocal tract outlines on a graphic terminal. Based on Maeda's (1982a) work, Childers and Ding (1991) implemented an articulatory speech synthesizer by using a discrete circuit model that converts the acoustic equations into linear algebraic equations.

We rederived the acoustic equations (see Appendix A11–C) of the vocal system to include the subglottal system, the glottal impedance, the turbulence noise source, and the sinus cavities. Table A11.2 compares several articulatory synthesizers with our implementation (Figure A11.20).

A11.2.5.1 Realization

Figure A11.28 shows a software block diagram of the model constructed for time-domain articulatory synthesis. Two options are provided for the interpolation of the vocal tract configuration: vocal tract cross-sectional area and articulatory parameters. If the articulatory parameters are interpolated, the articulatory model is used to transform the parameters to the vocal tract cross-sectional area. The number of vocal tract sections is 60. The vocal tract cross-sectional area is transformed to the equivalent RLC-network. On the other hand, the excitation parameters are interpolated and the excitation waveform is generated, according to the interpolated parameters, as the source input to the circuit network. The nasal sinus cavities and/or the subglottal system can be included optionally in the circuit network. By applying Kirchoff's and Ohm's laws and the trapezoidal algorithm, the discrete-time acoustic matrix equations are formed (see Appendix A11–C for the details). The pressure at the midpoint of each section and volume velocity at the junction of adjacent sections are calculated using the elimination procedure and a backward substitution. The

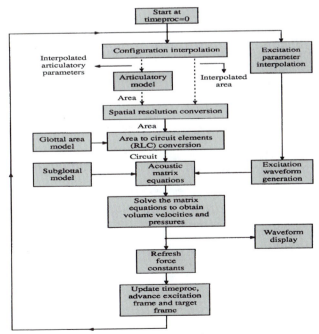

FIGURE A11.28 Block diagram of the articulatory speech synthesizer implemented in our software.

TABLE A11.2 Comparison of Several Articulatory Speech Synthesizers

	Flanagan	Maeda	Bocchieri and Childers	Childers and Ding	Prado	Proposed Synthesizer
Model of excitation	Self-oscillating two-mass	A slit	Self-oscillating two-mass	Passive one-mass	Parametric two-mass	LF
Jitter and shimmer models included	No	No	No	No	No	Yes
Noise source at the glottis	No	No	Yes	Yes	Yes	Yes
Noise source in the vocal tract	Every section	No	At constriction	At constriction	At constriction	Center of, or downstream or upstream from, or distributed along the constriction
Excitation in the vocal tract	No	No	No	No	No	Yes
Method for changing model parameters	Recompile source code	Recompile source code	Recompile source code	Through parameter files	Through parameter files	On screen by moving slider

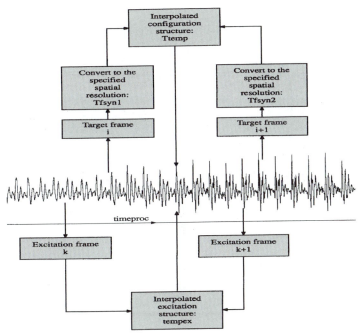

FIGURE A11.29   The timing sequence and interpolation of target frames and excitation frames.

synthetic speech is the backward difference between the sum of the volume velocities at the nostrils and lips at the current time instant and the sum of the volume velocities at the nostrils and lips at the previous time instant. The synthesis procedure is repeated by refreshing the force constants (see Appendix A11–C), updating the time instant, and advancing the target and/or excitation frames until the time epoch is reached.

A sketch of the configuration and excitation parameter interpolations is illustrated in Figure A11.29. Only two target frames and two excitation frames are shown in this figure. Either the vocal tract cross-sectional areas or the articulatory parameters are interpolated between the current target frame i and the next target frame i + 1 during the synthesis of speech. Assume that the target frame structure is pointed to by pointers Tfsyn1 and Tfsyn2, respectively. A temporary pointer, Ttemp, is used to point to the interpolated structure configuration. Similarly, a temporary pointer, tempex, is used to point to the interpolated excitation structure. Both data structures are used to generate speech.

A11.2.5.2   *Interpolation Function*   Only one interpolation function, namely linear, is provided to interpolate the vocal tract configuration: (1) vocal tract cross-sectional area; or (2) articulatory parameters. For any one articulatory parameter or any one vocal tract cross-sectional area, linear interpolation can be described through the following function.

$$y = \alpha + \beta t \qquad \qquad \text{(A11.2.5.2.1)}$$

where y is the interpolated articulatory parameter or vocal tract cross-sectional area, t is time, and α and β are interpolation parameters. The two interpolation parameters are determined by the two target frames, i and i + 1.

A11.3   SPEECH INVERSE FILTERING

The recovery of articulatory movements from the speech signal, sometimes referred to as the speech inverse filtering problem, is difficult due to the nonuniqueness of the solution. This problem has been

the subject of research for several applications, including articulatory synthesis, speech recognition, low-bit-rate speech coding, and text-to-speech synthesis. Here, we offer a new solution using the simulated annealing algorithm, which is a "constrained multidimensional nonlinear optimization problem." The coordinates of the jaw, tongue body, tongue tip, lips, velum, and hyoid compose the multidimensional articulatory vector. A comparison between the model-derived and the target-frame first four formant frequencies forms the cost function. There are two constraints: (1) the articulatory-to-acoustic transformation function; and (2) the boundary conditions for the articulatory parameters. The optimum articulatory vector is obtained by finding the minimum cost function. Once the optimum articulatory vector is determined, the articulatory model determines the vocal tract cross-sectional area function, which in turn is used by the articulatory speech synthesizer.

A11.3.1   Review of the Derivations of the Vocal Tract Area Function

Geometric data concerning the vocal tract is essential to understanding articulation, and is a key factor in speech production. The acoustical theory of speech production (Fant, 1960) views the vocal tract as an acoustical tube with a varying cross-sectional area. The success of articulatory modeling depends to a large extent on the accuracy with which the vocal tract cross-sectional area function, A(x), can be specified for a particular utterance. Measurement of the vocal tract geometry is difficult. Basically, there are two methods for obtaining the vocal tract cross-sectional area function: (1) direct measurements from images such as x-rays; and (2) estimating the area function from acoustic data.

A11.3.1.1   Direct Measurements   Direct measurements of the vocal tract have been made from lateral x-ray images (e.g., Chiba and Kajiyama, 1941; Fant, 1960; Johansson et al., 1983). Unfortunately, these direct measurements and their evaluations are laborious. In addition, the exposure to x-ray for utterances of long durations is a problem owing to dosage limitations. Magnetic resonance imaging (MRI) (Baer et al., 1991), which is free from the disadvantages associated with x-ray methods, might appear to be the best available method to collect the necessary data. The drawback, however, is that the subject may fatigue since the imaging process requires a long time. Additional drawbacks stem from the fact that the resolution of air–tissue boundaries may depend on the thickness of the tissue section, and the calcified structures contain little mobile hydrogen and, thus, may be indistinguishable from the airway.

A11.3.1.2   Estimation from Acoustic Data   Several researchers have proposed analytical methods to derive the vocal tract cross-sectional area function, A(x), from acoustic data. Two approaches are based on LPC and the tube impulse response, respectively. The LPC approach is based on the fact that the filtering process of the lossless nonuniform acoustic tube model of the vocal tract is identical to that of the optimal inverse filter model with proper boundary conditions at the glottis and the lips (Atal and Hanauer, 1971; Wakita, 1973, 1979; Wakita and Gray, 1975; Gobl, 1988). The reflection coefficients are extracted by inverse filtering the speech signal. Then, the vocal tract cross-sectional area function is obtained from the set of reflection coefficients. The main problem with this approach is the articulatory compensation or the "ventriloquist effect;" that is, the fact that different vocal tract shapes can produce the same formant frequencies (Atal et al., 1978; Bonder, 1983; Charpentier, 1984; Mermelstein, 1967; Schroeder, 1967; Schroeder and Strube, 1979). For the tube impulse response approach, the basic concept is that if the transfer function of the vocal tract is known, then the A(x) can be derived uniquely (Gopinath and Sondhi, 1970; Milenkovic, 1984, 1987; Paige and Zue, 1970; Schroeder, 1967; Sondhi, 1979; Sondhi and Resnick, 1983). However, finding the transfer function of the vocal tract involving the use of impedance tubes with externally generated excitation does not allow the subject to phonate sounds.

To finesse some of the difficulties of the analytical methods, the "sorting" and "codebook" methods perform sampling of the articulatory parameters from the articulatory model and establish tables of vocal tract shapes and related acoustical representations. For the sorting method, reference tables are established by covering the articulatory space with a uniform or nonuniform grid and storing

the acoustic values computed at every vertex of the grid (Atal et al., 1978; Charpentier, 1984; Cook, 1991). These tables can be used to look up the effective vocal-tract geometric representations that have similar acoustic features. Some refinements, such as singular value decomposition and local region linearization (Atal et al., 1978), have been used to solve the ambiguous geometric subspace. On the other hand, the codebook method samples the articulatory space randomly and prunes it to retain only the reasonable shapes in the codebook. This method provides the basis for the vector quantization of the articulatory space (Larar et al., 1988). In 1990, Schroeter et al. (1990) made some improvements for the generation of codebooks by using a dynamic programming search. The codebooks are accessed through evaluating a weighted cepstral distortion measure as given by Meyer et al. (1991). There are several drawbacks with this numerical approach: cumbersome computations, sensitivity to the source excitation, mapping ambiguities, and acoustic modeling limitations (Schroeter and Sondhi, 1994).

A recent approach is to apply an artificial neural network (ANN) model to the speech inverse filtering, since it is a promising approach to implement codebooks (Shirai, 1993). The ANN model is trained with a large set of acoustic parameter patterns. Then, a test pattern of acoustic parameters is used to search the codebook to retrieve a corresponding articulatory pattern parameter set (Båvegård and Högberg, 1992, 1993; Papcun et al., 1992; Rahim et al., 1993; Xue et al., 1990). However, the learning of a large set of training patterns to span the articulatory space is still a challenge for the ANN model (Xue et al., 1990). In addition, as Schroeter and Sondhi (1994) pointed out, no clear advantage has so far been shown for ANN compared to other approaches.

The feedback methods try to optimize the articulatory parameters that are adjusted until the synthetic speech features differ minimally from the actual speech features. The selected speech features can be formants (Prado, 1991; Prado et al., 1992), spectral (Flanagan et al., 1980; Guo and Milenkovic, 1993; Gupta and Schroeter, 1993; Levinson and Schmidt, 1983; Parthasarathy and Coker, 1990, 1992; Riegelberger and Krishnamurthy, 1993; Heike, 1979), or others. The optimization can be done on a phoneme-by-phoneme basis (Parthasarathy and Coker, 1990, 1992) or on a frame-by-frame basis (Flanagan et al., 1980; Gupta and Schroeter, 1993; Levinson and Schmidt, 1983; Prado, 1991; Prado et al., 1992). Several search algorithms have been used, such as the Hooke and Jeeves algorithm (Flanagan et al., 1980; Gupta and Schroeter, 1993; Parthasarathy and Coker, 1990, 1992), the optimal gradient algorithm (Levinson and Schmidt, 1983), and combinations of the modified Fletcher–Reeves method and linear successive approximation (Prado, 1991; Prado et al., 1992). The problem of local minima related to the nonlinearity in the speech inverse filtering is a major impediment of this method.

Advances in computer technology have allowed the solution of optimization problems that require large numbers of complicated function evaluations to be computed on relatively inexpensive machines in a reasonable time. Thus, stochastic methods, such as genetic algorithms that serve as search procedures based on the mechanics of natural selection and natural genetics (Goldberg, 1989), can be applied to the speech inverse filtering problem. Some preliminary results have been obtained by McGowan (1994). Articulatory trajectories of an articulatory model were recovered by means of a genetic algorithm from the first three formant frequencies using a task-dynamic model (Saltzman, 1986; Saltzman and Kelso, 1987; Saltzman and Munhall, 1989) of speech articulation. Tests on synthesized utterances show that the method can recover the major aspects of an original trajectory, but it has trouble in obtaining the precise timing of events. An additional difficulty for the genetic algorithm, as Goffe et al. (1994) experienced, stems from the fact that it needs further development to become more usable for continuous function problems, since it has difficulty with a relatively flat surface.

In general, finding the global minimum value of a cost function with many degrees of freedom is difficult, since the cost function tends to have many local minima. A procedure for solving such optimization problems should sample values of the cost function in a manner that has a high probability of finding a near-optimal solution and should also lend itself to efficient implementation. Over the past few years, simulated annealing has emerged as a viable technique that meets these criteria. Simulated annealing is based on models found in nature; for example, some processes in thermodynamics (Kirkpatrick et al., 1983; Metropolis et al., 1953) can be modeled using a stochastic optimization method. Simulated annealing explores a function's entire surface and tries to optimize the function while moving uphill and downhill. Thus, this technique is largely independent of the starting values, which is often a critical factor in conventional optimization algorithms. Simulated annealing also makes less stringent assumptions regarding the function than do conventional algorithms. For example, the function need not be continuous since the method does not require the calculation of derivatives.

Because of these relaxed assumptions, it can more easily deal with functions that have ridges and plateaus. In addition, it can be applied to optimize a "black box" system for which one only needs to define the state (the parameter space) and to compute the corresponding energy (cost function value). Finally, functions that are not defined for some parameter values can also be optimized by the simulated annealing method (Bohachevsky et al., 1986; Corana et al., 1987; Goffe et al., 1992, 1994; Vanderbilt and Louie, 1984).

Based on the above reviews and discussions, we selected the simulated annealing algorithm for optimizing the nonlinear acoustic-to-articulatory transformation; that is, speech inverse filtering.

A11.3.2 Simulated Annealing Algorithms

Simulated annealing was first derived from statistical mechanics, where the thermodynamic properties of a large system in thermal equilibrium at a given temperature were studied (Metropolis et al., 1953). A description of the physical annealing process inspired this algorithm. In this situation, a solid metal is to be melted at a high temperature. After slow cooling (annealing), the molten metal arrives at a low energy state, since careful cooling brings the material to a highly ordered, crystalline state. Inherent random fluctuations in energy allow the annealing system to escape local energy minima to achieve the global minimum. However, if the material is cooled very quickly (or quenched), it might not escape local energy minima and when fully cooled it may contain more energy than annealed metal. Simulated annealing attempts to minimize an analogue of energy in an annealing process to find the global minimum. Kirkpatrick et al. (1983) were the first to propose and demonstrate the application of simulated annealing techniques to problems of combinatorial optimization, specifically to the problems of wire routing and component placement in VLSI design. Both Vanderbilt and Louie (1984) and Bohachevsky et al. (1986) modified simulated annealing for continuous variable problems. However, the Corana et al. (1987) implementation of simulated annealing for continuous variable problems appears to offer the best combination of ease of use and robustness, so it is used for our optimization process.

A11.3.2.1 Origin of the Algorithm As far back to 1953, Metropolis et al. (1953) proposed a method for computing the equilibrium distribution of a set of particles in a "heat bath" using a computer simulation method. For the system in thermal equilibrium at a given temperature T, they assumed that the probability $\pi_T(c)$ that the system is in a given configuration c depends upon the energy E(c) of the configuration and follows the Boltzmann distribution.

$$\pi_T(c) = \frac{e^{-\frac{E(c)}{kT}}}{\sum_{s \in C} e^{-\frac{E(s)}{kT}}} \tag{A11.3.2.1.1}$$

where k is Boltzmann's constant and C is the set of all possible configurations. The configuration of the system is identified with the set of spatial positions of the particles. A stochastic relaxation technique was developed to simulate the behavior of the system. Suppose that the system is in configuration C_t at time t. A candidate configuration C_n for the system at time $t + 1$ is generated randomly. The criterion for selecting or rejecting configuration C_n as a new configuration (state) depends on the difference of energies between configuration C_n and configuration C_t. Define p, the ratio of the probability of being in C_n to the probability of being in C_t, as

$$p = \frac{\pi_T(C_n)}{\pi_T(C_t)} = e^{-\frac{E(C_n) - E(C_t)}{kT}} \tag{A11.3.2.1.2}$$

Then, apply a criterion, which has come to be known as the Metropolis criterion or algorithm, to decide the acceptance of C_n. The Metropolis criterion can be stated as follows. If $p > 1$, that is, the energy of C_n is strictly less than the energy of C_t, then configuration C_n is automatically accepted as the new configuration for time $t + 1$. If $p \leq 1$, that is, the energy of C_n is greater than or equal to that of C_t, then configuration C_n is accepted as the new configuration with probability p. So a move to a state of higher energy is accepted in a limited way. By repeating this process for a large enough number of moves, that is, as $t \rightarrow \infty$, regardless of the starting configuration, it can be shown that the

distribution of configurations generated converges to the Boltzmann distribution (Geman and Geman, 1984).

A11.3.2.2 The Cooling Schedule

A fundamental question arises in statistical mechanics concerning the system in the limit as it approaches a low temperature, for example, whether cooling produces crystalline or glassy solids in a metallurgic process. To achieve ground state (a low-energy crystalline configuration), simply lowering the temperature is not sufficient. Rather, a cooling schedule must be followed, where the temperature of the system is elevated, and then gradually lowered, spending enough time at each temperature to guarantee that thermodynamic equilibrium has been reached. If insufficient time is spent at each temperature, especially at a lower temperature, then the probability of achieving a low-energy crystalline state is greatly reduced.

The application of the annealing process to optimization problems involves several steps. First, one must identify the analogues of the physical concepts in the optimization problem. The energy function becomes the cost function. The configuration of particles becomes the combination of independent variable values. The rearrangement of particles becomes the iterative improvement of function values by changing variable values. Finding a low-energy configuration is a near-optimal solution, and the temperature becomes the control parameter for the process. Second, one must have a way of generating the candidate states. Usually, states are generated with a probability density function $g(x)$ that has a gaussian-like peak. Third, one must have a way of selecting the new state. A state acceptance probability allows occasional hill-climbing as well as descents. The acceptance probability is based on the chances of obtaining a new state relative to a previous state. Two acceptance probability equations have been used successfully, the Boltzmann machine and the Metropolis algorithm, which are given as follows:

Boltzman machine

$$p(\Delta E) = \frac{1}{\left(1 + e^{-\frac{\Delta E}{T}}\right)}, \quad \text{for all } \Delta E \tag{A11.3.2.2.1}$$

Metropolis algorithm

$$\begin{aligned} p(\Delta E) &= 1.0, \quad \text{for } \Delta E \leq 0 \\ &= e^{-\frac{\Delta E}{T}}, \quad \text{for } \Delta E > 0 \end{aligned} \tag{A11.3.2.2.2}$$

where $\Delta E = E(\text{new state}) - E(\text{current state})$ is the energy gap between the new state and the current state. The Boltzmann machine is known to better approximate the physical metaphor, but is more computationally expensive (Davis and Ritter, 1987). The Metropolis algorithm is used in our implementation. Fourth, one must specify a cooling schedule consisting of:

- An initial value of the control parameter; that is, the initial artificial temperature T.
- A decrement function for decreasing the value of the control parameter; that is, the cooling rate.
- A final value of the control parameter or a stop criterion.
- A finite number of moves for each downward control parameter value, that is, the amount of time spent at each temperature.

Such an analogy was first suggested by Kirkpatrick et al. (1983). They linked the algorithm with combinatorial optimization, specifically to the problems of wire routing and component placement in VLSI design. The rapid increase in inexpensive computing power has lead to several applications of the simulated annealing algorithm, including computer and circuit design (Vecchi and Kirkpatrick, 1983), image restoration and segmentation (Carnevali et al., 1985; Geman and Geman, 1984), the traveling salesman problem (Bonomi and Lutton, 1984), artificial intelligence (Hinton and Sejnowski, 1983), digital filter design (Benvenuto, et al., 1992; Pitas, 1994), and vector quantization (Rose et al., 1992). Because of the success of the simulated annealing in combinatorial optimization problems, its potential has been investigated for solving continuous function minimization problems. Vanderbilt and Louie

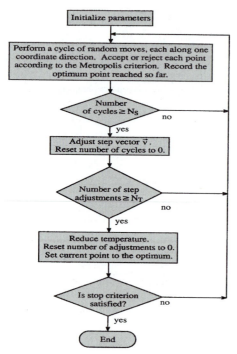

FIGURE A11.30 The simulated annealing algorithm (after Corana et al., 1987).

(1984) first modified simulated annealing by using a covariance matrix for controlling the transition probability. Bohachevsky et al. (1986) presented a simple method that was easy to implement, wherein the length of a generation step is constant. However, the Corana et al. (1987) implementation of simulated annealing for continuous variable problems appears to offer the best combination of ease of use and robustness, and has been used in econometric problems (Goffe et al., 1992, 1994).

A11.3.2.3 The Simulated Annealing Algorithm

The Corana et al. (1987) algorithm is schematically shown in Figure A11.30. While a detailed description of the algorithm can be found there, we briefly describe it as follows. Let $\vec{x}$ be an M-dimensional vector with components $[x_1, x_2, \ldots, x_M]$. Let $\varepsilon(\vec{x})$ be the cost function and $lb_j \leq x_j \leq ub_j, j = 1, \ldots, M$; be the M variables with corresponding boundaries. The algorithm proceeds iteratively as follows. First, a cost-function evaluation is made at the initial point $\vec{x}$ and its value ε is recorded. Next, a new candidate point, $\vec{x}_n$, is generated by varying element i of $\vec{x}$, namely

$$x_{ni} = x_i + r(v_i) \tag{A11.3.2.3.1}$$

The variable r is a uniformly distributed random number from $[-1, 1]$ and v_i is element i of $\vec{v}$, the step length vector of $\vec{x}$. The new function value ε_n is then computed. If ε_n is less than ε, $\vec{x}_n$ is accepted, $\vec{x}$ is set to $\vec{x}_n$, ε is set to ε_n, and the search path moves downhill. If this is the smallest ε at this point, it and $\vec{x}$ are recorded, since this is the best current value. If ε_n is greater than or equal to ε, the Metropolis criterion (Metropolis et al., 1953) determines acceptance. Compute the value p as follows.

$$p = e^{\frac{\varepsilon - \varepsilon_n}{T}} \tag{A11.3.2.3.2}$$

In this equation, T represents the current temperature. Generate a uniformly distributed random number from $[0, 1]$. Decide the action based on the result of value comparison between p and p_u. If p_u is less than p, the new point is accepted, $\vec{x}$ is updated with $\vec{x}_n$, and the search path moves uphill. Otherwise, $\vec{x}_n$ is rejected; that is, no move is made. Thus, the process repeats from the new point (candidate point is accepted) or from the current position (candidate point is rejected). From Equation

(A11.3.2.3.2), we can see that the probability of an uphill move decreases when the temperature is lower and the difference in the function's value is larger.

After N_S steps through all elements of $\vec{x}$, the step length vector $\vec{v}$ is adjusted so that 50% of all moves are accepted. The goal is to make the algorithm follow the cost function (Corana et al., 1987). A greater percentage of accepted points means that the candidate points are too close to the current point. Thus, the step length vector $\vec{v}$ is enlarged. For a given temperature, this step adjustment increases the number of rejections and decreases the percentage of acceptances. On the other hand, a higher percentage of rejected points means that the candidate points are too far from the current point. A reduced step length decreases the rejection rate.

After N_T times through the above loops (corresponding to thermal equilibrium), the temperature, T, is reduced. The temperature is updated according to the following equation.

$$T_n = (r_T)T \tag{A11.3.2.3.3}$$

The reduction coefficient r_T has a value between 0 and 1. The starting point at the new temperature T_n is the optimum point obtained at the last temperature T. This makes the search path start at the most favorable point. Since the temperature characterizes the degree of "excitation" of the system, a lower temperature decreases the number of uphill moves, so the number of rejections increases and the step size reduces. The lower temperature (smaller step size) makes the search space shrink and focus on the most promising area, that is, concentrates most of the search in a smaller subset of low energy points.

The terminating criterion checks if there have been no significant moves for the last N_ε temperatures. Assume that the optimum value obtained at temperature T_k is ε_k^*. Let ε_{opt} be the current optimum value at the temperature T_{k+1}. If

$$\left| \varepsilon_k^* - \varepsilon_{k-m}^* \right| \leq \eta, \quad \text{where } m = 1, \ldots, N_\varepsilon$$
$$\left| \varepsilon_k^* - \varepsilon_{opt} \right| \leq \eta, \tag{A11.3.2.3.4}$$

then stop the search. Note that η is a specified small constant. This check makes sure that the global or near-global minimum is reached. Another stop criterion is that the total number of cost-function evaluations exceeds a specified constant N_{tot}.

In summary, the simulated annealing algorithm starts at some high temperature specified by the user. A sequence of points is then generated until an equilibrium is approached. During this random walk process, the step length vector is periodically adjusted to better follow the cost function behavior. After thermal equilibrium, the temperature is reduced and a new sequence of moves is made starting from the current optimum point, until thermal equilibrium is reached again, and so forth. The process is terminated at a low temperature such that no more useful moves can be made, according to the stopping criterion.

A11.3.3 Speech Inverse Filtering Strategy and Procedure

In general, the relationship between the shape of the vocal tract and its acoustic output can be represented by a multidimensional function of a multidimensional argument

$$\vec{y} = f(\vec{x}) \tag{A11.3.3.1}$$

where $\vec{x}$ is a vector formed by the coordinates of the articulators, $\vec{y}$ is a vector formed by the corresponding acoustic features, and f is the function relating these vectors. Given an acoustic measurement $\vec{y}_d$, the problem is to find an articulatory state $\vec{x}_0$ such that $f(\vec{x}_0)$ is the best match to $\vec{y}_d$. In other words, with the optimization approach, $\vec{x}_0$ is the solution to the nonlinear optimization problem

$$\vec{x}_0 = \text{minimal arguement of } \|f(\vec{x}) - \vec{y}_d\| \tag{A11.3.3.2}$$

where $\| \cdot \|$ is a norm on the acoustic space.

A11.3.3.1 *Strategy* Speech inverse filtering is a "constrained multidimensional nonlinear optimization problem." As we defined in Section A11.2.2.1, the coordinates of the tongue body

(tbodyx, tbodyy), tongue tip (tipx, tipy), lips (lipp, lipo), jaw (jaw), and hyoid (hyoid) compose the multidimensional articulatory vector; that is

$$\vec{x} = [\text{thodyx, tbodyy, tipx, tipy, lipp, lipo, jaw, hyoid}] \qquad (A11.3.3.1.1)$$

Note that $\vec{x}$ is an eight-dimensional vector. Usually, the velum is set at different default positions for nasal, non-nasal, or nasalized phonemes, but it can be optimized for some phonemes. The dimensions of the lower pharynx are also allowed to be optimized whenever this is necessary.

We designate the articulatory vector as

$$\vec{x} = [x_1, x_2, \ldots, x_M] \qquad (A11.3.3.1.2)$$

where the value of M represents the number of dimensions of the articulatory domain to be optimized. As mentioned in the previous paragraph, M has a value of 8. For nasal and nasalized sounds, we may include the velum as an additional articulatory parameter; that is, M is set to 9. For middle vowels, some back vowels, and semivowels, three more parameters, which are anterior–posterior movements of K and H, g1k, and wh, and the height between K and H, hkl (refer to Figure A11.4), are included; that is, M is set to 11. Finally, one more parameter, the velum, can be included, that is, M = 12.

The acoustic vector is composed of the first four formant frequencies; that is, $\vec{y} = [F_1, F_2, F_3, F_4]$. The cost function (error distance) is derived from a comparison of the first four formant frequencies of the articulatory model and the first four formant frequencies determined from speech analysis (the target formants). A percentage of the weighted least-absolute-value (l_1-norm) error distance is defined as

$$\varepsilon(\vec{F}_m(\vec{x})) = \sum_{i=1}^{4} \frac{W_i |F_{mi}(\vec{x}) - F_{ti}|}{F_{ti}} \% \qquad (A11.3.3.1.3)$$

where F_{mi} is the ith model-derived formant, which is a function of the articulatory vector, F_{ti} is the ith target-frame formant estimated from the analysis of speech signal, and W_i is the assigned weight.

The constraints, which include the articulatory-to-acoustic transformation function f (Equation A11.3.3.1) and the boundary conditions of the articulatory parameters, are described as follows

$$\vec{y}_m = f(\vec{x}) = f([x_1, x_2, \ldots, x_M]) = \vec{F}_m(\vec{x}) = [F_{m1}(\vec{x}), F_{m2}(\vec{x}), F_{m3}(\vec{x}), F_{m4}(\vec{x})] \qquad (A11.3.3.1.4)$$

where $lb_j \leq x_j \leq ub_j, j = 1, \ldots, M$ are the lower and upper bounds of the articulatory parameters, and the subscript m represents the model-derived parameter values.

The object of the optimization process is to find the optimal articulatory vector that generates the acoustic vector (model-derived) as close to the desired (target-frame) as possible. The ideal minimum value of $\varepsilon(\vec{F}_m(\vec{x}))$ is 0%, but some approximations used in the articulatory model (see Section A11.2.2) make this value hard to obtain. The first approximation is related to the articulatory model. A non-robust representation of the lower part of the pharynx and the tongue tip-to-jaw region may cause some deviations of the midsagittal vocal tract outline. The second, and more significant deviation, is the uncertainty of the sagittal distance to cross-sectional area transformations. Different empirical transformation formulas can be found in the literature (Heinz and Stevens, 1964; Mermelstein, 1973; Sundberg et al., 1987). The final approximation is the area to formant frequency conversion. We have determined that an error criterion requiring the final value of error distance function to be less than 1% is adequate.

A11.3.3.2 Procedure

To extract the articulatory trajectories from a speech sentence, the first step is to obtain a smoothed formant trajectory from the speech signal. A feature is available in our software for doing this. Then N target frames are selected. The target frame selection is based on the results of the speech analysis, which includes the formant trajectory, the location of the word endpoints, and the estimated phoneme boundaries of the speech signal. An example of selecting (marking) selected target frames of the formant tracks obtained from a speech file is shown in Chapter 10. Next, the speech inverse filtering procedure is applied to each target frame to obtain the optimum articulatory parameters.

Figure A11.31 shows the block diagram of the speech inverse filtering procedure, which is performed frame-by-frame. For each target frame, an initial value of the error distance function (cost function) is evaluated from the initial articulatory vector. The evaluation of the error distance function

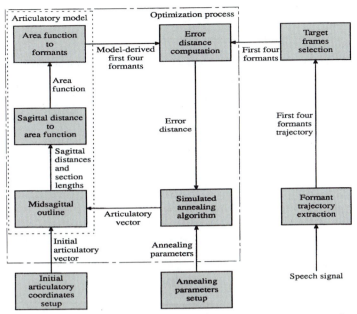

FIGURE A11.31 Block diagram of the speech inverse filtering procedure.

includes the computations of the sagittal distances and the section lengths, the calculations of the vocal tract cross-sectional area and the acoustic transfer function, the decomposition of the first four formants from the acoustic transfer function, and the calculation of the error distance. Then, the simulated annealing algorithm controls the movement of the search path. Each movement requires the generation of a next candidate point, the error distance function evaluation for the candidate point, and the decision to move. After a number of steps, the temperature is lowered and a new search begins. The process stops if the near-global minimum is reached or the maximum allowed number of function evaluations is exceeded. The speech inverse filtering procedure terminates when all target frames are optimized. The articulatory parameters and the vocal tract cross-sectional areas of all the optimized N target frames can be saved as a disk file for later use or can be directly passed to the articulatory synthesizer for synthesis.

Table A11.3 lists the components of the data structure. The first four formant frequencies, the frame starting time, and the frame duration are the initial components. After the minimum error distance is obtained, the first section area of the nasal tract, the optimal articulatory coordinates, the section lengths, and the cross-sectional areas are stored as the content of the target-frame data structure.

TABLE A11.3   Components of the Target Frame Structure

Data Type	Component	Description
Short	tfsetflag	Flag for optimization
Double	ttime	Frame starting time
Double	tdur	Frame duration
Float	ntla	Velopharyngeal port area
Structure pointer	areaf	Area function
Structure pointer	vtlen	Section lengths
Structure pointer	shape	Articulatory coordinates
Structure pointer	tfptr	Target formants
Structure pointer	next	
Structure pointer	previous	Both used for double-linked list

TABLE A11.4   Default Annealing Parameter Values

Annealing Parameters	Default Values
T = artificial temperature (as control parameter)	0.1–0.2 degrees
r_T = temperature reduction coefficient	0.85
N_S = number of steps to adjust the step length vector	20
N_T = number of adjustments at each temperature	5
N_ε = number of successive temperatures to test for stopping	4
η = termination criterion	0.005
N_{tot} = total number of function evaluations	5001
v_i = step length; where i = 1, 2, ..., M	3.0

The annealing parameters that control the simulated annealing algorithm include the initial temperature T, the temperature reduction coefficient r_T, the number of steps to adjust the step length vector N_S, the number of step adjustments at each temperature N_T, the number of successive temperature reductions to test for termination N_ε, a small constant used for the termination criterion η, and the maximum number of function evaluations N_{tot}. The analogues between the annealing process and the articulatory problem can be identified as follows. First, the percentage of the weighted least-absolute-value (l_1-norm) error distance, Equation (A11.3.3.1.3), corresponds to the energy of the material. The articulatory vector, Equation (A11.3.3.1.1), corresponds to the configuration of particles. The change of articulatory parameters corresponds to the rearrangement of particles. Finding a near-optimal articulatory vector corresponds to finding a low-energy configuration. The temperature of the annealing process, T, becomes the control parameter for the speech inverse filtering process. Second, the Metropolis algorithm corresponds to the random fluctuations in energy. Third, the temperature reduction coefficient r_T corresponds to the cooling rate. Fourth, the finite number of moves at each downward control temperature value, $(N_S)(N_T)$, corresponds to the amount of time spent at each temperature.

Reasonable values of the parameters (Table A11.4) are used as defaults for the optimization process. However, a guideline of the optimization process is given in Appendix A11–D along with two examples.

A11.3.4   Results

A11.3.4.1   *Optimization of Vowels*   Appendix A11–A presents the articulatory and acoustic characteristics for typical American vowels. The midsagittal vocal tract outline and the corresponding vocal tract cross-sectional area function are obtained from sustained vowel phonations by using the simulated annealing algorithm. Appendix A11–A, shows that the simulated annealing optimization algorithm works well, since most of the error distances are less than 0.5%. From these results, we arrive at the following observations.

- Different vowels are characterized by a different set of resonant frequencies (formants), thus, a different vocal tract shape. For example, front vowels (/IY, IH, EH, AE/) are characterized by a large difference between F_1 and F_2, which in turn corresponds to a large back cavity; middle vowels (/AA, ER, AH /) and back vowels (/UW, UH, OW, AO/) have a small difference between F_1 and F_2, which in turn corresponds to a narrow back cavity.

- The three middle vowels (/AA, ER, AH/) and the low, back vowel (/AO/) have similar vocal tract shapes, except for the retroflexed vowel, /ER/, which has a more significant tongue curl-up, which results in a distinctly low F_3.

- For all middle vowels (/AA, ER, AH/) and some other vowels (/AE, AO, OW/), the three articulatory parameters of the lower pharynx; that is, wh, g1k, and hk1, must be optimized.

- To investigate the complexity of the error distance function for each vowel, we used the same initial vocal tract configuration for the optimization process for all vowels. Experience with the optimization process indicates that the error distance function for the high front vowels

(/IY, IH /) and the high, back, rounded vowels (/UW, UH /) converges faster than for other vowels. The middle vowel (/AA /) and the low, back vowel (/AO/) have the slowest convergent rate because these two vowels have a more complex error distance function. For the middle vowel (/ER/), the curl of the tongue causes a distinctly low F_3, which may result in a slow convergence of the optimization process. To articulate the low, back vowel (/AO/), the tongue needs to be lowered and the jaw needs to be opened to widen the oral cavity. Also, the tongue must move back to narrow the pharyngeal cavity. These articulations may make the error distance function complex and slow the convergence of the optimization process. Notice that a more complex error distance function needs a higher initial temperature and more evaluations, since it can have more local minima to escape. See Appendix A11–D for a guideline of the optimization process.

A11.3.4.2 *Optimization for a Sentence*   In Appendix A11–D, the simulated annealing algorithm is applied to perform the speech inverse filtering for the same sentence spoken by two male subjects. The following are some general observations regarding sentence optimization.

- There are two semivowels (/W, Y/) in the speech token analyzed. According to the phonetic classifications, these two semivowels have been categorized as glides. The formant tracks for the sentence, "We were away a year ago," show that the formants, especially the second and third formants, glide up or down to the next vowel. The /W/ glides (frames 1, 6, and 10 of subject A, and frames 1, 4, and 8 of subject B in Appendix A11–D) have three places of articulation in common: the protruded lips, high tongue tip, and high back tongue. However, some of these have a significant upward tongue tip curl. As frame 16 of subject A and frame 13 of subject B show, the tongue blade of the semivowel /Y/ approximates the palate and has been called a palatal glide. In summary, the vocal tract shape for glides "glide" to the next vowel with a fast movement of the tongue and lips.

- Both subjects have quite similar vocal tract shapes for vowel /IY/ and for vowel /ER/, respectively. The lip opening decreases during the transition from vowel /IY/ to semivowel /W/ (frame sequences 3-4-5-6 of subject A, and 2-3-4 of subject B) in order to have protruded lips for /W/.

- The diphthong /EI/ ending with vocal tract shape for /IH/ entails tongue movement forward up from the /EI/ (frame sequences 11-12-13-14 of subject A, and 9-10-11 of subject B). The diphthong /OU/ ending with the vocal tract shaped for /UH/ entails tongue movement back and up, concurrent with lip protrusion (frame sequences 24-25-26 of subject A, and 20-21-22 of subject B).

- The voiced stop /G/ is usually classified as a velar consonant. From frame 22 of subject A and frame 18 of subject B, a classification as the palatal-velar consonant is more correct (Borden and Harris, 1980, p. 117).

- The simulated annealing algorithm performs well. On the average, over 87% of the total frames have an error distance less than 0.1%.

A11.3.4.3 *Remarks*   The above results illustrate the usefulness of the simulated annealing algorithm, which is efficient and flexible in dealing with the problems that are inherent to the acoustic-to-articulatory transformation. However, the selection of parameters for the annealing schedule is difficult, since we know little about the relation between the argument domain (articulatory vector) and the technology (the algorithm). The guideline in Appendix A11–D and the default annealing parameter values in Table A11.4 are considered a good procedure. The evaluation of the error distance function is the most computationally intensive part of the program. On the average, 2000 computations per minute are needed to obtain relatively quick results.

A11.4 SYNTHESIS OF UNVOICED SPEECH

The speech synthesis we have discussed so far has focused on voiced speech. In this section, we discuss the synthesis of unvoiced speech using the articulatory speech synthesizer. The acoustics and

aerodynamics involved in the production of unvoiced speech are far from being completely understood. It is not sufficient to determine the segments of unvoiced speech by estimating the vocal tract transfer function and the waveshape at the glottis. Fricatives are generated at a constriction within the vocal tract, while the glottis is the source for aspiration /HH/. Still, there are many unvoiced speech sounds that are generated by both types of constrictions: one at the glottis and another in the vocal tract (Stevens, 1971). Here we describe some factors relevant to the production of unvoiced speech.

A11.4.1 Introduction

When there is a flow of air through a constriction or past an obstruction, turbulence is created (Shadle, 1991; Stevens, 1971, 1993a, 1993b). Unvoiced speech can be divided into three groups, that is, unvoiced fricatives, unvoiced stops, and affricates. Aspiration noise can be regarded as unvoiced speech as well. An unvoiced fricative is generated by exciting the vocal tract with a steady air flow, which becomes turbulent near the constriction where the velocity of the airflow increases due to the reduced cross-sectional area of the constriction. The location of the constriction determines which sound is produced. The constriction separates the vocal tract into two cavities: the front cavity and the back cavity. The unvoiced speech sound is generated from the front cavity, and the back cavity traps energy and thereby introduces antiresonances into the vocal output. An unvoiced stop consonant is produced by forming a complete closure in the vocal tract and then releasing that closure. The closure is formed by a particular articulator; that is, the lips, the tongue blade, or the tongue body, and then released by moving the articulator rapidly. The initial rapid increase of the cross-sectional area at the constriction gives rise to a transient, and there is a brief burst of turbulence noise following the transient. An unvoiced affricate is a dynamic sound that can be modeled as a concatenation of the unvoiced stop and the fricative consonant. Therefore, this sound has the characteristics of both stop and fricative consonants. An aspiration is produced by turbulent air flow at the glottis with little or no vibration of the vocal folds. Because the vocal tract shape is in the position for the following vowel during the production of the aspiration noise, the characteristics of the aspiration /HH/ are similar and dependent on the vowel that follows the aspiration noise. Since the noise source for aspiration is located at the glottis, there is no front and back cavity in aspiration.

Among these four unvoiced sound groups, the fricative and aspiration are generated by a relatively steady flow of air. Therefore, it is much easier to analyze and synthesize fricatives and aspirations than other unvoiced consonant categories. To understand factors that control the unvoiced speech generation process, we describe the models for generation of unvoiced consonants, which include a turbulence noise generator and the vocal tract structure.

A11.4.2 Unvoiced Speech Production

A11.4.2.1 *Models for the Generation of Unvoiced Consonants*

A11.4.2.1.1   **Fricative Consonants.**   A schematized model of the vocal tract for an unvoiced fricative consonant is shown in Figure A11.32. A_g and A_c are the cross-sectional areas of the glottis and the constriction, respectively. Likewise, L_g and L_c are the length of the glottis and the

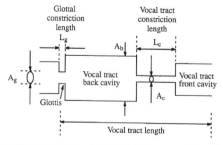

FIGURE A11.32   A model of the vocal tract for a fricative consonant.

vocal tract constriction, respectively. Unvoiced fricatives are produced by a turbulence noise source near a constriction either at the glottis or above the glottis depending on the cross-sectional areas, A_g and A_c. The location of the constriction, that is, the place of the articulation, determines the sound produced. If $A_g > A_c$, the supraglottal constriction plays a major role in the generation of fricative speech. The constriction is formed by the tongue body and/or tongue tip. The length of the constriction, L_c, is typically a few centimeters long. Aspiration noise is generated when $A_g < A_c$. The air flow is independent of the supraglottal constriction. The length of the constriction formed by the glottis, L_g, may be only a few millimeters, which is the average width of the glottis.

As shown in Figure A11.32, when the noise source is located in the vocal tract, the constriction separates the vocal tract into two cavities: the front cavity and the back cavity. Speech is radiated from the front cavity, while the back cavity serves to trap energy, and, thereby, introduces antiresonances into the vocal output. It is reported that the back cavity resonances may have negligible effect on the fricative spectrum if the constriction is sufficiently long and narrow (Heinz and Stevens, 1961). Therefore, the most prominent feature of the fricative spectrum is determined by the turbulence noise source at the constriction and by the resonances of the oral cavity in front of the constriction; that is, the front cavity resonance frequencies.

The constriction in the vocal tract may be abrupt or gradual depending on the articulatory position. It has been found experimentally that the pressure drop across the constriction is proportional to the square of the particle velocity, $(V_c)^2$, and the pressure drop across the constriction Δ_p is approximated by the Bernoulli equation

$$\Delta p = k\rho \frac{V_c^2}{2} = k\rho \frac{U^2}{2A^2} \qquad \text{(A11.4.2.1.1.1)}$$

where k is a constant, ρ is the density of the air, V_c is the flow velocity in the constriction, U is the volume velocity in the constriction, and A is the cross-sectional area of the constriction. The constant k is dependent on the ratio of the cross-sectional area A_b and A_c. It is also dependent on the rate of contraction and expansion of the constriction. Stevens found that a value of 0.9 was a reasonable value for constrictions for normal speakers (Stevens, 1971, 1998).

A11.4.2.1.2 Stop Consonants.

For an unvoiced stop consonant, the vocal tract is also modeled as a tube with two constrictions; one at the glottis and one formed by an articulator within the vocal tract. The principal driving source for the flow is the subglottal pressure. At the release of an unvoiced stop consonant, the initial intraoral pressure is equal to the subglottal pressure. Both the glottal constriction area A_g and the supraglottal constriction area A_c change with time as the supraglottal constriction is formed and released. As pressure builds up above the glottis during closure for a consonant, outward forces are exerted on the vocal-fold surfaces, causing a passive increase in the glottal area. The rate of change of cross-sectional area of the supraglottal constriction near the consonantal closure and release can be estimated from cineradiographic, photographic, and airflow data (Childers 1977). It is reasonable to assume the glottal area to be constant and the supraglottal constriction area to increase with a trajectory of the form

$$A_c = (1 - e^{-\frac{t}{\tau}}) \qquad \text{(A11.4.2.1.2.1)}$$

The airflow through the constriction and through the glottis in the few tens of milliseconds following the release gives rise to a sequence of four types of sources: transient, fricative, aspiration, and voicing (Stevens, 1993b). Immediately following the release, there is a brief transient as the air that has been compressed in the vocal tract discharges through the opening constriction. Following this initial transient, the rapid airflow through the constriction gives rise to turbulence and hence to a sound source immediately downstream from the constriction. This source is identified as fricative noise. The turbulence noise source is usually represented as a sound-pressure source near an obstacle downstream from the constriction. Rapid airflow through the glottis also gives rise to turbulence noise, called aspiration noise, with source characteristics similar to those for fricative noise. Vocal-fold vibration begins simultaneously with or immediately following the aspiration noise.

A11.4.2.1.3 Affricate Consonants.

The release mechanism for an affricate consonant is different from that for a simple stop consonant. The constriction that is formed by the major articulator to produce an affricate has two parts that can be manipulated independently, an anterior section

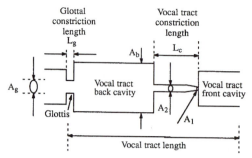

FIGURE A11.33 A model of the vocal tract for an affricate consonant.

and a posterior section. The configuration of the vocal tract can be schematized as in Figure A11.33, where the vocal tract posterior to the constriction is represented as a relatively wide uniform tube. The consonantal constriction has a narrowing near its anterior end, with area A_1 and a longer section with area A_2 behind this narrower section. The cross-sectional area of the glottal opening is A_g. The articulator used to produce an affricate can be the lower lip, the tongue blade, or the tongue body. The acoustic pattern includes an initial transient and an interval over which the fricative noise undergoes changes in amplitude and spectrum as the tongue-tip constriction increases in size.

There is a diverse sequence of acoustic and aerodynamic events at the release of an affricate consonant. During the closure interval, there is a pressure buildup behind the constriction and A_1 is set to zero. A consequence of this increased intraoral pressure is the creation of a force on the articulatory structure that forms the constriction, and it causes a passive downward displacement of the surface. The release occurs in two stages: first the area A_1 is increased, and this initial release is followed by a slower increase of the area A_2 of the longer constriction. At the initial release of the closure, the rate of movement of the articulatory structure that forms anterior area A_1, and hence the rate of increase of A_1, is expected to be rapid, since the movement is enhanced by forces due to pressure behind the constriction. This rate of increase of cross-sectional area is taken to be 50 cm²/sec during the initial 1 to 2 msec (Stevens, 1993a, 1998).

Following the initial release at the anterior end of the constriction, the force on the surface of the articulator behind the constriction A_1 is reduced, and there may be an initial inward movement of the surface. After that, the posterior area A_2 is maintained for a few tens of milliseconds and then released. The constriction during the later part of the release sequence (after about 50 msec) is much like the constriction for a fricative consonant. Finally, as this longer constriction is released, there may be a time interval in which turbulence noise is generated at the glottal constriction A_g before glottal vibration commences.

For the transient source after the initial 0.1 to 0.2 msec, the initial rate of increase of the airflow can be estimated roughly by assuming that the flow U is proportional to the area A of the opening, and is given by

$$U \cong A \sqrt{\frac{2P_m}{\rho}} \qquad (A11.4.2.1.3.1)$$

where P_m is the intraoral pressure and ρ is the density of the air. After the initial transient source, turbulence noise will be generated as a consequence of airflow through the palatoalveolar constriction impinging on the lower incisors. The effective amplitude of the turbulence noise source is assumed to be proportional to $U^3 A^{-2.5}$, where U is the volume velocity through the constriction and A is the cross-sectional area of the constriction (Stevens, 1971, 1998).

Figure A11.34 gives an estimate of the time course of the change in cross-sectional area for the front portion of the affricate constriction, represented by A_1 in the schematized configuration in Figure A11.33. Also shown in Figure A11.34 is the estimated cross-sectional area of the palatal constriction following the initial release. This is the equivalent of the area A_2 in the schematized configuration in Figure A11.33. The time course of this cross-sectional area over the time interval beyond 50 msec in Figure A11.34, together with the change in cross-sectional area of the glottal opening A_g, are similar to the trajectories that are normally assumed for the release of a fricative consonant (Stevens et al., 1992; Stevens, 1998). In summary, the requirement for an affricate is that the constriction formed by

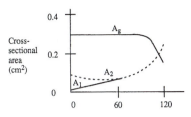

Time from initial release (ms)

FIGURE A11.34 Time course of the change in cross-sectional area for an affricate consonant (after Stevens et al., 1993a).

the relevant articulator can be divided into an anterior portion, which executes the initial rapid release from closure, and a longer posterior portion, which is shaped to produce fricative noise corresponding to the fricative portion of the sequence.

A11.4.2.2 Model for Turbulence Noise Generation

A11.4.2.2.1 Characteristics of Equivalent Noise Source. The configuration of the vocal tract for the generation of an unvoiced turbulence noise source can be modeled as in Figure A11.35. This model was originally suggested by Fant (1960) who used it for calculations for various fricative and stop consonants. The source is assumed to be independent of the constriction area A_c. The impedance, Z_2, is seen looking upstream from the sound pressure source, P_s, and is frequency dependent. The impedance looking downstream from the pressure source is Z_1. According to this model, the front and back cavity resonance frequencies are determined by Z_1 and Z_2, respectively. The volume velocity U_0 at the mouth flows through the radiation impedance Z_r. The transfer function $\frac{U_0}{P_s}$, that is, the transfer function from the pressure source to the volume velocity at the lips can be calculated as

$$\frac{U_0}{P_s} = \frac{1}{Z_1 + Z_2}$$

(A11.4.2.2.1.1)

Therefore, due to the frequency dependent impedance Z_2, the spectrum of the turbulence noise is characterized by poles at the natural frequencies for which $Z_1 + Z_2 = 0$, and by zeros at the frequencies for which $Z_2 = \infty$.

It has been found experimentally that the spectrum of the series pressure source in Figure A11.35 is relatively flat over a frequency range of two or three octaves centered on 0.2 V/D, where V is the velocity and D is a characteristic dimension (Stevens, 1971). More specifically, the center frequency can be represented as

$$f_c = 0.2 \frac{U}{A^{\frac{3}{2}}}$$

(A11.4.2.2.1.2)

where U is the volume velocity of the air flow near the constriction and A is the cross-sectional area of the constriction. For typical values of the volume velocities and constriction sizes encountered in turbulence noise generated speech, the center frequency is in the range of 500 to 3000 Hz. The lower end of this range is for aspiration noise /HH/ and the higher end is for fricative noise.

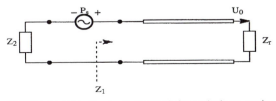

FIGURE A11.35 Equivalent circuit for turbulence noise generation.

In summary, the most important two components in this model are the turbulence noise source, P_s, at the constriction and the impedance Z_1 looking downstream from the pressure source that determines the front cavity resonance frequencies. As the position of the constriction moves forward from the glottis to the lips, the pole of the transfer function moves higher in frequency, for example, from about 500 Hz for aspiration noise /HH/ to about 4000 Hz for fricative noise /S/ (Stevens, 1971; 1998).

A11.4.2.2.2 Noise Source Model.

For convenience we repeat some of the results from Section A11.2.3.6. The airflow in a narrow tube generates turbulence noise, and the intensity and spectrum of the noise are decided by the Reynolds number R_e. If $R_e > R_{ec}$, where R_{ec} is the critical Reynolds number, then a noise with high intensity is generated, which is a turbulence noise. The Reynolds number is defined as

$$R_e = \frac{\rho \upsilon h}{\mu}$$

where ρ is air density, υ is linear velocity of air flow, h is characteristic dimension of the constriction, and μ is viscosity of air (Sorokin, 1994).

Basically, the noise source model defines the characteristics of the noise source as a function of the airflow through the constriction and of the constriction cross-sectional area. Broad (1977b) found that the rms sound pressure of the noise could be expressed as

$$P_{rms} = A_c \left(R_e^2 - R_{ec}^2 \right) \tag{A11.4.2.2.2.1}$$

where R_e is the Reynolds number and R_{ec} is the critical Reynolds number. Fant (1960) adopted a serial noise pressure source and reformulated the P_{rms} as a function of the pressure drop through the constriction and the effective width of the constriction. Under the plane wave assumption, the sound pressure of turbulent flow can be taken as proportional to the square of the volume velocity of the airflow and inversely proportional to the constriction area, A_c (Stevens, 1971). The location of the turbulence noise source may be located at the center of, or immediately downstream from, the constriction region, or possibly at a combination of these places, or spatially distributed along the constriction region (Fant, 1960; Flanagan and Cherry, 1968; Lin, 1990; Stevens, 1971, 1993a, 1993b).

The spectrum of the turbulence noise is broadly distributed over a wide range of frequencies (2 to 8 kHz) with some accentuation in the mid-audio range (Childers and Lee, 1991; Stevens, 1971, 1993a, 1993b). Klatt (1980) used a random number generator, a spectrum-shaping filter, and an amplitude modulator to model the turbulent flow for the formant synthesizer. The spectrum-shaping filter was designed to simulate the spectral characteristics of the turbulent flow. A first order IIR filter was used to obtain the volume velocity due to a random pressure source. Childers and Lee (1991) have used a FIR filter to model highpass-filtered turbulence noise. Cook (1991, 1993) used a four-pole filter to model the spectral properties of the noise source.

By including a latent random pressure source, P_n, and an inherent constriction loss, R_n, in each elemental section of the vocal tract, Flanagan and Cherry (1968) could introduce the turbulent flow excitation at any section. However, as Sondhi and Schroeter (1987) pointed out, the Flanagan and Cherry (1968) model did not produce satisfactory unvoiced sounds due to the high "back" cavity impedance. Sondhi and Schroeter (1986, 1987), thus, modified the model to a parallel flow source $U_n = \frac{P_n}{R_n}$, which was located downstream from the constriction. The P_n is given by

$$\begin{aligned} P_n &= (\text{turbg})(\text{rand}) \left(R_e^2 - R_{ec}^2 \right), \quad &\text{for } R_e > R_{ec} \\ &= 0, \quad &\text{for } R_e < R_{ec} \end{aligned} \tag{A11.4.2.2.2.2}$$

where turbg is empirically determined as the turbulence gain and rand is a random number uniformly distributed between -0.5 and 0.5. The source resistance R_n is

$$R_n = \frac{\rho |\bar{U}|}{2A^2} \tag{A11.4.2.2.2.3}$$

where $\bar{U}$ is a digitally low-pass filtered version of the volume velocity U at the constriction and is

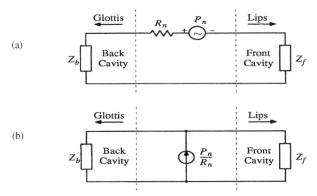

FIGURE A11.36 Equivalent circuits for the turbulence source. (*a*) Serial (after Flanagan and Cherry, 1968). (*b*) Parallel (after Sondhi and Schroeter, 1986).

defined as

$$\bar{U}(n) = \bar{U}(n-1) + [U(n) - \bar{U}(n-1)]2\pi f_g T \tag{A11.4.2.2.2.4}$$

The value of the cutoff frequency f_g is not critical. Flanagan and co-workers (1975) used 500 Hz in order to ensure stability. A first-order IIR filter with a cutoff frequency of 2000 Hz was used to lowpass filter the flow. Figure A11.36(*a*) and (*b*) show the equivalent circuits of the serial and parallel turbulence sources, respectively.

Based on the previous discussion, we adopt the turbulence noise source model from Sondhi and Schroeter (1986, 1987) into the transmission-line circuit model of the vocal system. This model allows the user to place the turbulence noise source at the center of, or immediately downstream or upstream from, the constriction region, or spatially distributed along the constriction region. The turbulence gain and critical Reynolds number can also be specified.

A11.4.3 Synthesis Results

Examples of several speech tokens were synthesized using the articulatory speech synthesis toolbox given in Chapter 10. The speech tokens consisted of the affricate /CH/, fricative /S/, and the sentence, "Should we chase those cowboys?"

A11.4.3.1 *Example 1—Synthesis of the Affricate /CH/* Affricates and fricatives are produced by the formation of a narrow supraglottal constriction and the generation of turbulence at the vicinity of this constriction. The experimental sections for fricatives are taken at the interval of maximum intensity, which occurs near the middle or towards the end of the sound. Obtaining articulatory data for unvoiced consonants from speech is only in a preliminary stage (Sorokin, 1994; Stevens, 1993a, 1993b, 1998). Thus, a vocal tract cross-sectional area (Figure A11.37) estimated from the x-ray photograph of Badin (1991) was adopted for the affricate /CH/ in this experiment. The actual width of a very narrow constriction cannot always be estimated accurately from sagittal x-ray photographs. In addition, the perpendicular dimensions are generally inaccessible to measurements. The simulations were carried out on the assumption that the minimum cross-sectional area was 0.16 cm². Errors in the estimation of these dimensions do not severely affect the calculations, since the length of a narrow constriction is more crucial than the cross-sectional area.

Modeling the complex turbulent phenomena is another challenging problem. In this experiment, the affricate /CH/ was synthesized with the turbulence noise source located at (1) the center of; (2) immediately downstream; (3) upstream from the constriction region; and (4) spatially distributed along the constriction region. The model presented in Figure A11.36(*b*) was used to simulate the turbulence noise source. The turbulence gain and critical Reynolds number were specified at 0.000002 and 2700, respectively. The glottal volume velocity was assumed to be a DC source and set at 1000 cm³/sec.

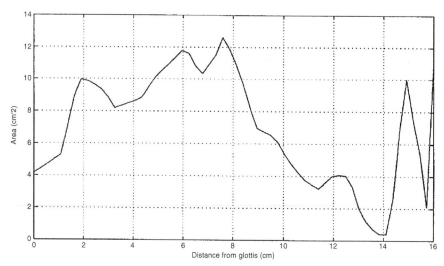

FIGURE A11.37 Vocal tract cross-sectional area for the affricate /CH/.

In Figure A11.38, the transfer function of /CH/ is displayed for various locations of the noise source. Placing the excitation source in the vocal tract introduces antiresonances into the acoustic transfer function. The vocal tract resonant frequencies are relatively unaffected when the excitation is placed forward from the pharynx toward the front oral cavity. Another feature is that the number of antiresonances increases as the excitation is moved forward from the pharynx to the front oral cavity. Figure A11.39 gives a comparison of the power spectral density of the synthesized and original speech for the affricate /CH/. A 6th-order LPC model was used to analyze a segment of the speech signal. Note that the spectral characteristics of the noise source located at the center of and upstream from the constriction region are similar but differ slightly from the other two locations. The resonant peak near 3700 Hz is due to the resonance of the small, short front cavity.

The resonance is less prominent when the turbulence noise source is distributed along the constriction region. The resonant frequency and amplitude for the downstream case (2000 Hz) are lower than those for original speech case (2200 Hz). The difference may be due to differences in the dimensions of the vocal tract for the location of the constriction and the source characteristics of the turbulence noise for the two speakers. The synthesized and original speech are similar in the low frequency region. This means that the back cavity has little effect on the synthesized speech.

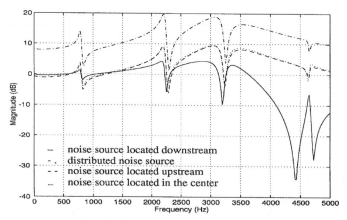

FIGURE A11.38 Transfer function for the affricate /CH/ for various locations of the noise source.

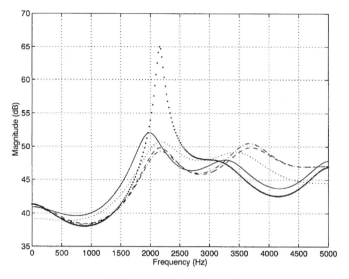

—   noise source located downstream
- .   noise source located in the center
- -   noise source located upstream
....   distributed noise source
...   real speech

FIGURE A11.39   Power spectral density of the synthesized and original affricate /CH/.

A11.4.3.2   Example 2—Effect of the Back Cavity on the Synthesized Fricative /S/

For some fricative sounds generated with a supraglottal constriction, the effect of the back cavity resonance on the speech spectrum can be neglected if the constriction is narrow and long (Heinz and Stevens, 1961). Here, we confirm that the back cavity resonance does not greatly effect the synthetic fricative sound as long as the supraglottal constriction is sufficiently narrow and long. The shape of the area function corresponding to the back cavity is changed as specified in the left column of Figure A11.40. The front cavity and constriction of the three area functions remain the same; that is, the cross-sectional area of the supraglottal constriction is 0.16 cm^2, the front cavity length is 2.02 cm, and the constriction length is 2.02 cm.

The synthesized speech is similar to the original fricative /S/. Since the vocal tract configuration for /S/ has a distinct front cavity and back cavity, it is not difficult to observe the effect of the back cavity resonance on the speech spectrum. Synthesized fricatives generated with two different back cavity shapes are compared with a synthesized fricative using data for an area function measured from an x-ray (Fant, 1960) in Figure A11.40.

As expected, the formant structure of the three spectra are quite similar, since the back cavity resonance is decoupled from the constriction and the front cavity. Consequently, there is little influence on the overall speech spectrum. These results are in agreement with the fricative speech production theory; that is, the back cavity resonance does not play an important role in generating fricative sounds.

A11.4.3.3   Example 3—Synthesis of the Sentence, "Should We Chase Those Cowboys?"

The sentence, "Should we chase those cowboys?" consists of voiced and unvoiced sounds. For the voiced portion of the original speech, speech inverse filtering was performed with the simulated annealing algorithm to obtain the vocal tract cross-sectional area. Based on the pitch contour and LF parameters obtained in the analysis phase, the excitation waveform model was constructed. Unvoiced consonants have traditionally proven difficult to model and synthesize because of the complex nature of their production mechanisms and the lack of sufficient articulatory and aerodynamic data for these sounds. The dynamic, time-varying characteristics and the complex

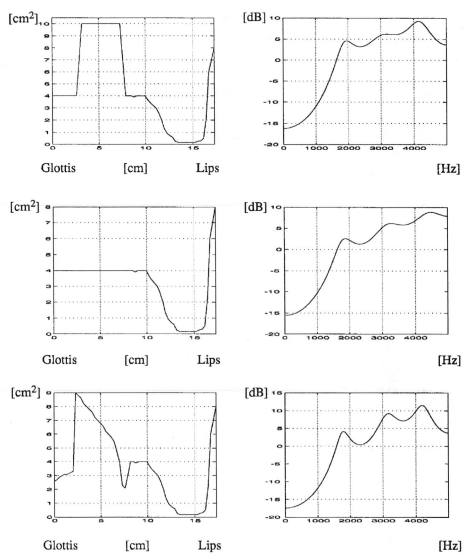

FIGURE A11.40 Area function and LPC spectrum for a synthesized /S/ for different back cavity shapes.

nature of their production have made plosives the most difficult phonemes to model and synthesize. Thus, only the fricative-like portion of unvoiced sounds and the corresponding cross-sectional areas are considered here.

The experimental sections for fricatives and affricates are taken at the interval of maximum intensity, which occurs near the middle or end of the sound. The experimental sections for stops are also taken at the interval of maximum intensity, which generally falls at the beginning of the unvoiced sound interval identified as the stop burst. The unvoiced simulations also assume a minimum cross-sectional area of 0.16 cm^2. The model presented in Figure A11.36(b) was used to simulate the turbulence noise source. The turbulence gain and critical Reynolds number were specified at 0.000002 and 2700, respectively. The glottal volume velocity was assumed to be a DC source and set at 1000 cm^3/sec. The original and synthetic speech signals and wideband spectrograms are shown in Figure A11.41,

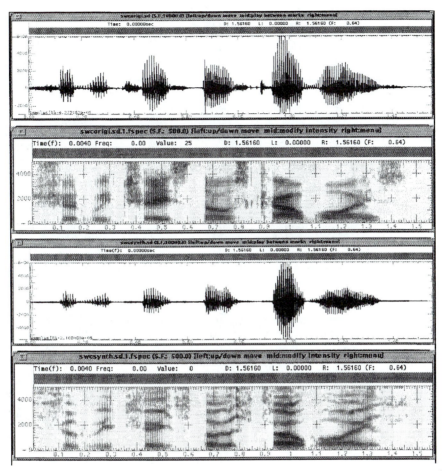

FIGURE A11.41   The original (upper) and synthesized (lower) speech and spectrograms for the sentence, "Should we chase those cowboys?"

and are reasonably similar. The synthesized speech is highly intelligible, but is not as natural as the original.

A11.4.4 Discussion

From the examples described previously (and others), it is concluded that the articulatory speech synthesis toolbox in Chapter 10 is able to produce a good quality synthetic speech. This indicates that the articulatory synthesis tool effectively identifies and simulates the human vocal system. From the results of these experiments, we summarize the production of unvoiced speech as follows.

- Schroeter and Sondhi (1994) concluded that the synthesis of fricatives in the articulatory synthesizer is not yet satisfactory. However, from our experiments, the turbulence noise source location was found to have important acoustic consequences. The downstream case is able to generate spectral characteristics close to the original speech. The major problem for synthesis of unvoiced sounds lies in the estimation of the relevant parameters from the acoustic speech signal, such as inferring articulatory and source information.

- Our results confirmed that the back cavity resonance does not have much effect on the synthetic fricative sound as long as the supraglottal constriction is sufficiently narrow and long. For the

fricative /S/, the supraglottal constrictions, which are configured by the tongue tip and tooth ridge, are narrow and long enough to decouple the back cavity resonance with the front cavity.

- Analysis and synthesis of voiced sounds provide us with an understanding of the production of phonetic information and vocal characteristics. Numerous algorithms have been proposed and many of them give successful results. For unvoiced consonants, the acoustics and aerodynamics involved in the production of unvoiced speech are far from being completely understood because of the complex nature of their production mechanisms and lack of sufficient articulatory and aerodynamic data. The experimental sections for unvoiced sounds in our analysis and synthesis are taken at their intervals of maximum intensity. Since many portions of speech are voiced sounds, the analysis and synthesis based on voiced sounds and experimental sections of unvoiced sounds are enough to generate an intelligible synthetic speech.

A11.5 SUMMARY

This appendix focuses on four main areas of the articulatory synthesizer model: the articulatory model, the acoustic model, the analysis of various vocal system characteristics, and the articulatory speech synthesizer. After a brief review of the articulatory model, we defined the articulatory parameters and described the articulatory model in some detail. Our articulatory model represents the vocal tract by 60 sections to provide more reliable estimates of the cross-sectional areas. We have made an attempt to cover the acoustic model of the entire vocal system, which includes the vocal tract, the nasal tract, the sinuses, the glottal impedance, the subglottal tract, the glottal excitation source, and the turbulence noise source. A transmission-line circuit model of the vocal system is provided. A model for unvoiced speech production is included along with several examples. Also included in this appendix is an analysis of several characteristics of the vocal system that is based on the calculation of the acoustic transfer function, which is described in Appendix A11–B. Such an analysis provides a basis for choosing appropriate parameters for the articulatory synthesizer. The effect of relocating the excitation source on the acoustic transfer function is also described. For more details see C. Wu (1996). Finally, the strategy of the implementation of the articulatory synthesizer is presented after a review of the articulatory synthesis approaches. The time-domain approach is implemented to provide the ability to investigate the dynamic properties of the vocal system. A derivation of the discrete time acoustic equations is given in Appendix A11–C. One interpolation function, linear, is used to interpolate the vocal tract cross-sectional area or articulatory parameters.

We described the simulated annealing optimization algorithm in detail after reviewing the derivations of the vocal tract cross-sectional area. The simulated annealing algorithm is based on the Corana et al. (1987) approach. The articulatory vector defines the set of parameters to be optimized. The cost function is a percentage of the weighted least-absolute-value error distance. It defines a comparison of the first four formant frequencies between the model-generated and the target-frame (from speech analysis). A 1% error criterion gives satisfactory results. Once the optimum articulatory vector is obtained, the articulatory model determines the vocal tract cross-sectional area function, which in turn is used by the articulatory speech synthesizer. Results and discussion of speech inverse filtering for twelve typical American vowels and two sentences are presented in Appendix A11–A and Appendix A11–D, respectively. Default annealing parameters that control the simulated annealing algorithm are also given in Table A11.4. The simulated annealing algorithm is efficient and flexible in dealing with the problems that are inherent to speech inverse filtering. The author is grateful for the results provided by Y. F. Hsieh (1994) and C. "John" Wu (1996).

EXAMPLES OF FEATURES FOR SOME TYPICAL VOWELS

A11–A.1 INTRODUCTION

Appendix 5 presents twelve vowels that illustrate the positions of the jaw, tongue, and lips; that is, the approximated vocal tract configuration in the form of a midsagittal outline (upper left of figure), along with the synthesized acoustic waveform (upper right), the vocal tract cross-sectional area function (lower left), and the vocal tract frequency response (lower right). These results were obtained using the articulatory speech synthesizer, which minimized the error between the vowel target formants and the articulatory model formants. The vowels and the error are listed here. Consult Appendix 5 for the other data.

- Vowel IY as in beet, error 0.01432%
- Vowel IH as in bit, 0.00632%
- Vowel EY as in bait, error 0.03832%
- Vowel EH as in bet, error 0.01204%
- Vowel AE as in bat, error 0.06795%
- Vowel UW as boot, error 0.02982%
- Vowel UH as in book, error 0.01867%
- Vowel OW as boat, error 0.19218%
- Vowel AO as in bought, error 0.05938%
- Vowel AA as in Bach, error 0.065087%
- Vowel AH as in but, error 0.21787%
- Vowel ER as in bird, error 0.57565%

ACOUSTIC TRANSFER FUNCTION CALCULATION

The formant frequencies are required by the simulated annealing algorithm to calculate the articulatory-to-acoustic inverse transform. The formant frequencies need to be decomposed from the acoustic transfer function or from the input impedance, both of which are calculated from the vocal tract cross-sectional area function. This appendix presents a derivation of the acoustic transfer function for various structures using a circuit network representation and transmission matrix theory. In Appendix 11, Section A11.2.3.1.3, it was mentioned that an elemental tube with uniform area A and length l can be analogous to the transmission line circuit model. The circuit in Figure 5.1 and the definitions given in Figure A10.27 form the basis for the following derivations. The circuit in Figure 5.1 can be rearranged as a four terminal T network, as shown in Figure A11–B.1. This network representation facilitates the acoustic transfer function calculation, which is defined as the ratio of the total output volume velocity from the vocal tract and the nasal tract to the excitation volume velocity source input. Define the propagation constant as $\gamma = \sqrt{zy}$, where $z = R + j\omega L$, $y = G + j\omega C + \frac{1}{Z_w}$, and $Z_w = R_w + j\omega L_w + \frac{1}{j\omega C_w}$. Let the characteristic impedance be

$$Z_a = Z_0 \tanh\left(\frac{\gamma \ell}{2}\right)$$

and

$$Z_b = \frac{Z_0}{\sinh(\lambda \ell)}$$

See Fant (1960) and Flanagan (1972a) for more details.

It is well known that the input-output characteristics of a four-terminal network are described by a matrix equation of the form

$$\begin{bmatrix} P_i(\omega) \\ U_i(\omega) \end{bmatrix} = [T(\omega)] \begin{bmatrix} P_o(\omega) \\ U_o(\omega) \end{bmatrix}$$
(A11–B.1)

where $P_i(\omega)$ and $U_i(\omega)$ are the sound pressure and volume velocity, respectively, at the input of the network; $P_o(\omega)$ and $U_o(\omega)$ are the corresponding outputs of the network; and $T(\omega)$ is the transmission matrix (also called ABCD matrix or chain matrix) of the four-terminal network. From Ohm's law and the current loop law, the transmission matrix of Figure A11–B.1 is

$$T(\omega) = \begin{bmatrix} A & B \\ C & D \end{bmatrix} = \begin{bmatrix} 1 + \frac{Z_a}{Z_b} & 2Z_a + \frac{Z_a^2}{Z_b} \\ \frac{1}{Z_b} & 1 + \frac{Z_a}{Z_b} \end{bmatrix}$$
(A11–B.2)

Now, consider two elemental sections with a shunt impedance Z_s in between and with radiation impedance Z_r as load. We represent the radiation impedance as $Z_r = R_r + j\omega L_r$, where $R_r = \frac{128\varrho c}{9\pi^2 A_m}$, $L_r = \frac{8\varrho}{3\pi\sqrt{\pi A_m}}$, and A_m is the lip or nostril opening area. The network representation is constructed in Figure A11–B.2. Note that the shunt impedance Z_s can be considered as the modelling circuit for the nasal sinus (see Appendix A11). Assume that the transmission matrices of section $i-1$ and section i

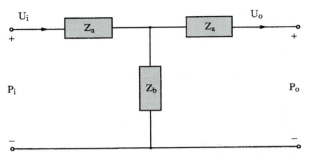

FIGURE A11–B.1 Four-terminal T-network for a uniform elemental tube.

are $T_{i-1}(\omega)$ and $T_i(\omega)$ respectively. Forming a dummy four-terminal T-network as in Figure A11–B.1 with $Z_a = 0$ and $Z_b = Z_s$ for the shunt element Z_s and applying Equation (A11–B.2), the transmission matrix is

$$T_s(\omega) = \begin{bmatrix} A_s & B_s \\ C_s & D_s \end{bmatrix} = \begin{bmatrix} 1 & 0 \\ \frac{1}{Z_s} & 1 \end{bmatrix} \qquad \text{(A11–B.3)}$$

From the theory of cascade circuit networks, the overall transmission matrix for Figure A11–B.2 is the product of the individual transmission matrices for the sections. Thus the relation between pressure and volume velocity at the input P_{i-1}, U_{i-1} and at the radiation P_r, U_r can be written as

$$\begin{bmatrix} P_{i-1} \\ U_{i-1} \end{bmatrix} = T_{i-1}(\omega) \cdot T_s(\omega) \cdot T_i(\omega) \begin{bmatrix} P_r \\ U_r \end{bmatrix} = \begin{bmatrix} A_f & B_f \\ C_f & D_f \end{bmatrix} \begin{bmatrix} P_r \\ U_r \end{bmatrix} \qquad \text{(A11–B.4)}$$

From this matrix equation and Ohm's law, $P_r = Z_r U_r$, the transfer function $H(\omega)$ from U_{i-1} to U_r is

$$H(\omega) = \frac{U_r(\omega)}{U_{i-1}(\omega)} = \frac{1}{C_f Z_r + D_f} \qquad \text{(A11–B.5)}$$

and the input impedance Z_{in} is

$$Z_{in}(\omega) = \frac{P_{i-1}(\omega)}{U_{i-1}(\omega)} = \frac{A_f Z_r + B_f}{C_f Z_r + D_f} \qquad \text{(A11–B.6)}$$

By use of the four-terminal network representation and the transmission matrix, we can calculate the acoustic transfer function of the vocal tract system for human speech production. Since the excitation source can be at the glottis or inside the vocal tract, there are four cases.

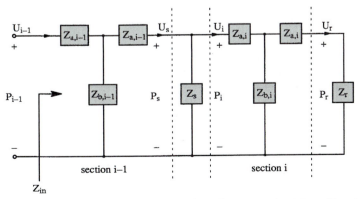

FIGURE A11–B.2 Network representation of a two-section tube with a shunt element in between and a radiation load.

Case A: Excitation Source Located at the Glottis.   Let Z_{or} and Z_{nr} be the radiation impedances of the vocal tract and the nasal tract, respectively. Let Z_{sub} as the impedance of the glottis and the subglottal system looking backward from the glottis. Use $A_p B_p C_p D_p$, $A_o B_o C_o D_o$, and $A_n B_n C_n D_n$ to represent the transmission matrices of the pharyngeal tube, oral tract, and nasal tract, respectively. Figure A11–B.3(a) shows the equivalent network representation of the vocal system.

To calculate the acoustic transfer function $H(\omega)$ from U_g to $U_r = U_{or} + U_{nr}$, we first compute the input impedance of the oral tract and the input impedance of the nasal tract, both of which are seen downstream from the bifurcation point. From Equation (A11–B.6), we have the input impedance of oral tract as

$$Z_o = \frac{A_o Z_{or} + B_o}{C_o Z_{or} + D_o} \tag{A11–B.7}$$

and the nasal tract input impedance as

$$Z_n = \frac{A_n Z_{nr} + B_n}{C_n Z_{nr} + D_n} \tag{A11–B.8}$$

The second step is to calculate the acoustic transfer function $H_v(\omega)$ of the vocal tract from U_g to U_{or}. Forming an equivalent network for Figure A11–B.3(a) (see Figure A11–B.3(b)) and applying the analysis results of Figure A11–B.2 (refer to Equation A11–B.5), we have

$$H_v = \frac{Z_{sub}}{Z_p + Z_{sub}} \cdot \frac{1}{C_{vv} Z_{or} + D_{vv}} \tag{A11–B.9}$$

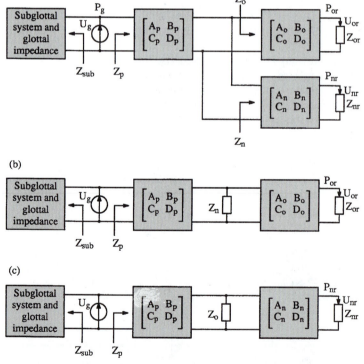

FIGURE A11–B.3   Case A with excitation source located at the glottis. (*a*) Network representation of vocal system. (*b*) Network representation for calculating $H_v(\omega)$. (*c*) Network representation for calculating $H_n(\omega)$.

and the corresponding transmission matrix as

$$\begin{bmatrix} A_{vv} & B_{vv} \\ C_{vv} & D_{vv} \end{bmatrix} = \begin{bmatrix} A_p & B_p \\ C_p & D_p \end{bmatrix} \begin{bmatrix} 1 & 0 \\ \frac{1}{Z_n} & 1 \end{bmatrix} \begin{bmatrix} A_o & B_o \\ C_o & D_o \end{bmatrix}$$ (A11–B.10)

Similarly, from Equation (A11–B.7), the input impedance of the vocal tract seen downstream from the glottis is

$$Z_p = \frac{A_{vv} Z_{or} + B_{vv}}{C_{vv} Z_{or} + D_{vv}}$$ (A11–B.11)

Applying the same procedure to the Figure A11–B.3(c), the acoustic transfer function $H_n(\omega)$ from U_g to the output of the nasal tract U_{nr} is

$$H_n = \frac{Z_{sub}}{Z_p + Z_{sub}} \cdot \frac{1}{C_{nn} Z_{nr} + D_{nn}}$$ (A11–B.12)

and the corresponding transmission matrix is

$$\begin{bmatrix} A_{nn} & B_{nn} \\ C_{nn} & D_{nn} \end{bmatrix} = \begin{bmatrix} A_p & B_p \\ C_p & D_p \end{bmatrix} \begin{bmatrix} 1 & 0 \\ \frac{1}{Z_o} & 1 \end{bmatrix} \begin{bmatrix} A_n & B_n \\ C_n & D_n \end{bmatrix}$$ (A11–B.13)

The vocal system acoustic transfer function is $H(\omega) = H_v(\omega) + H_n(\omega)$.

Case B: Excitation Source Located in the Pharyngeal Tube.   Let $A_{pl} B_{pl} C_{pl} D_{pl}$ represent the transmission matrix of the lower part of the pharynx with the input at the excitation location and output is at the glottis, as shown in Figure A11–B.4. The results obtained in Case A can be applied to this case. Replace the Z_{sub} in Equation (A11–B.9) and (A11–B.12) with the Z'_{sub}, where the Z'_{sub} is calculated as follows.

$$Z'_{sub} = \frac{A_{pl} Z_{sub} + B_{pl}}{C_{pl} Z_{sub} + D_{pl}}$$ (A11–B.14)

Then, the sum of Equations (A11–B.9) and (A11–B.12) yields the final acoustic transfer function.

Case C: Excitation Source Located at the Bifurcation of the Vocal Tract and the Nasal Tract.   As shown in Figure A11–B.5, the network representation is the same as Figure A11–B.3 except the excitation source is located at the bifurcation of the vocal tract and the nasal tract. First, the input impedances Z_o and Z_n are calculated as in Equations (A11–B.7) and (A11–B.8). Then $H_v(\omega)$ and $H_n(\omega)$ are obtained as follows.

$$H_v = \frac{Z'_{sub} Z_n}{Z'_{sub} Z_o + Z_o Z_n + Z_n Z'_{sub}} \cdot \frac{1}{C_o Z_{or} + D_o}$$ (A11–B.15)

$$H_n = \frac{Z'_{sub} Z_o}{Z'_{sub} Z_o + Z_o Z_n + Z_n Z'_{sub}} \cdot \frac{1}{C_n Z_{nr} + D_n}$$ (A11–B.16)

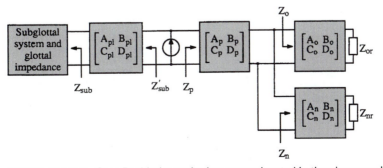

FIGURE A11–B.4   Case B with the excitation source located in the pharyngeal tube.

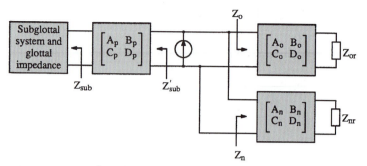

FIGURE A11–B.5   Case C with the excitation source located at the bifurcation point.

where the impedance Z'_{sub} is

$$Z'_{sub} = \frac{A_p Z_{sub} + B_p}{C_p Z_{sub} + D_p} \qquad (A11–B.17)$$

The sum of $H_v(\omega)$ and $H_n(\omega)$ yields the acoustic transfer function for the vocal system.

Case D: Excitation Source Located in the Oral Tract.

As shown in Figure A11–B.6(a), we use $A_{ob}B_{ob}C_{ob}D_{ob}$ and $A_{of}B_{of}C_{of}D_{of}$ to represent the transmission matrix of the back and front parts of oral tract, respectively. From Equations (A11–B.7), (A11–B.8), and (A11–B.17), we

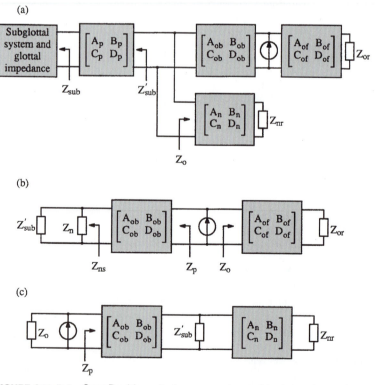

FIGURE A11–B.6   Case D with excitation source located in the oral tract. (*a*) Network representation of vocal system. (*b*) Network representation for calculating $H_v(\omega)$. (*c*) Network representation for calculating $H_n(\omega)$.

calculate the impedances Z_o, Z_n, and Z'_{sub} as follows.

$$Z_o = \frac{A_{of}Z_{or} + B_{of}}{C_{of}Z_{or} + D_{of}} \qquad (A11\text{–}B.18)$$

$$Z_n = \frac{A_n Z_{nr} + B_n}{C_n Z_{nr} + D_n} \qquad (A11\text{–}B.19)$$

$$Z'_{sub} = \frac{A_p Z_{sub} + B_p}{C_p Z_{sub} + D_p} \qquad (A11\text{–}B.20)$$

If we define Z_{ns} (see Figure A11–B.6(b)) as

$$Z_{ns} = \frac{Z_n Z'_{sub}}{Z_n + Z'_{sub}} \qquad (A11\text{–}B.21)$$

then the impedance Z_p can be calculated as

$$Z_p = \frac{A_{ob}Z_{ns} + B_{ob}}{C_{ob}Z_{ns} + D_{ob}} \qquad (A11\text{–}B.22)$$

Now, apply Equation (A11–B.9) to obtain $H_v(\omega)$ as follows.

$$H_v = \frac{Z_p}{Z_p + Z_o} \cdot \frac{1}{C_{of}Z_{or} + D_{of}} \qquad (A11\text{–}B.23)$$

For $H_n(\omega)$, we form the network as shown in Figure A11–B.6(c) and apply Equation (A11–B.12) to yield

$$H_n = \frac{Z_o}{Z_p + Z_o} \cdot \frac{1}{C_{on}Z_{nr} + D_{on}} \qquad (A11\text{–}B.24)$$

where the corresponding transmission matrix is

$$\begin{bmatrix} A_{on} & B_{on} \\ C_{on} & D_{on} \end{bmatrix} = \begin{bmatrix} A_{ob} & B_{ob} \\ C_{ob} & D_{ob} \end{bmatrix} \begin{bmatrix} 1 & 0 \\ \frac{1}{Z'_{sub}} & 1 \end{bmatrix} \begin{bmatrix} A_n & B_n \\ C_n & D_n \end{bmatrix} \qquad (A11\text{–}B.25)$$

Then, the final acoustic transfer function is $H_v(\omega) + H_n(\omega)$.

We have considered the acoustic transfer function of the soft-wall, lossy vocal system. For some cases, calculation of the acoustic transfer function of a lossless or lossy, hard-wall vocal system is adequate and also easier. For a rigid and lossy vocal system, the wall impedance Z_w of each section is removed, causing the impedance Z_w to be infinity. In this case, the admittance y becomes $y = G + j\omega C$. For a rigid and lossless vocal system, this means that no dissipation elements and no wall impedance are present for each section. Then the propagation constant γ becomes an imaginary quantity.

In the above derivations, we assume that impedance Z_{sub} is composed of the Foster-chain circuit model for the subglottal system with glottal impedance $Z_g = R_g + j\omega L_g$, where R_g and L_g are defined in Appendix A11. Since the glottal area is a time-varying function, we assume that the glottal area and glottal volume velocity are constants when we evaluate the acoustic transfer function. In Case A, if the subglottal system and the glottal impedance are not included, then we let the impedance $Z_{sub} \rightarrow \infty$, which corresponds to an open circuit, that is, the block representing the subglottal system and the glottal impedance in Figure A11–B.3 are disconnected. For other cases, two glottal conditions are considered. If the glottis is open, the effects of the subglottal system and the glottal impedance are evaluated. Otherwise, if the glottis is closed, we let $Z_{sub} = 0$, that is, Z_{sub} is a short circuit. The velopharyngeal port opening area, which couples the nasal tract to the vocal system, also affects the acoustic transfer function. Our software system allows the user to change the velopharyngeal port opening area, the number of coupled sinus cavities, and the location of the excitation source when evaluating the acoustic transfer function.

DERIVATION OF DISCRETE TIME ACOUSTIC EQUATIONS

As covered in Section A11.2.3.1, the vocal tract tube can be described by two coupled partial differential acoustic equations. These two acoustic equations are functions of both time and space. Approximating the vocal tract as a sequence of elemental sections corresponds to digitizing the vocal tract in space, that is, spatial sampling. For each elemental section, the transmission-line analog approach is applied to form the equivalent circuit model, as seen in Figure 5.1. Connecting the equivalent circuit of each section together in combination with the equivalent circuit models of the other parts of the vocal system (subglottal system, glottis, and nasal sinus cavities), a lumped circuit network representation of the vocal system can be formed, as shown in Figure A11.18; the time-domain approach, the Kirchoff's and Ohm's laws are applied to the circuit network to obtain sets of differential equations. These differential equations, which correspond to the equivalent acoustic equations, which govern the generation and the propagation of acoustic waves inside the vocal system, are transformed into discrete-time representations. This appendix provides a detailed derivation of the discrete-time acoustic equations, that is, the difference matrix equations. The discretization scheme is similar to the work of Maeda (1982a). Our model, however, provides more features, such as the subglottal system, nasal sinus cavities, and turbulence noise source.

Consider the transmission-line circuit model, shown in Figure A11-C.1, of the ith section of the vocal tract. Define the volume velocity (current) at the input of the ith section, that is, at x_{i-1} in space, as $U_i(t) = u_i(x_{i-1}, t)$, where $x_{i-1} = \sum_{k=1}^{i-1} l_k$ is the vocal tract length from the glottis to the ith section. Similarly, define the central pressure (voltage) of the ith section as $P_i(t) = p_i(x_i - \frac{l_i}{2}, t)$, where l_i is the section length of section i. From Kirchoff's and Ohm's laws, we have the following differential equations.

$$P_{i-1}(t) - P_i(t) = \frac{d}{dt}\{[L_{i-1}(t) + L_i(t)]U_i(t)\} + [R_{i-1}(t) + R_i(t)]U_i(t)$$

$$U_i(t) - U_{i+1}(t) = u_{i1}(t) + u_{i2}(t) + u_{i3}(t)$$

$$= G_i(t)P_i(t) + \frac{d}{dt}\{C_i(t)P_i(t)\} + u_{i3}(t) \qquad \text{(A11-C.1)}$$

$$P_i(t) = R_{w,i}(t)u_{i3}(t) + \frac{d}{dt}\{L_{w,i}(t)u_{i3}(t)\} + \int_0^t \left\{\frac{u_{i3}(\tau)}{C_{w,i}(\tau)}\right\} d\tau$$

There are three terms in Equation (A11-C.1); simple, differential, and integral terms. Define the general forms of these three terms as follows.

$$y_1(t) \equiv c_1(t)x(t)$$

$$y_2(t) \equiv \frac{d}{dt}\{c_2(t)x(t)\} \qquad \text{(A11-C.2)}$$

$$y_3(t) \equiv \int_0^t \{c_3(\tau)x(\tau)\} d\tau$$

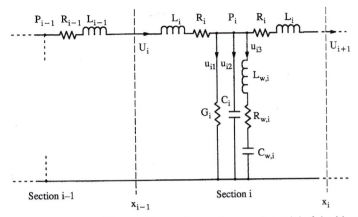

FIGURE A11–C.1 A lumped transmission-line circuit model of the ith section.

where $c_i(t)$ $(i = 1, 2,$ or $3)$ is a coefficient that represents the time-varying circuit component, and $x(t)$ can be $P_i(t)$, $U_i(t)$, or $u_{i3}(t)$. Let $y_i(n) = y_i(t = nT)$, $c_i(n) = c_i(t = nT)$, and $x(n) = x(t = nT)$ represent the sampled values of $y_i(t)$, $c_i(t)$, and $x(t)$, respectively, at $t = nT$, where $n = 0, 1, 2, \ldots,$ and T denotes the sampling time interval. For the first term, it is obvious that

$$y_1(n) = c_1(n)x(n) \tag{A11–C.3}$$

For the differential term, we first integrate both sides

$$\int_{(n-1)T}^{nT} y_2(t)\, dt = \int_{(n-1)T}^{nT} \frac{d}{dt}\{c_2(t)x(t)\}\, dt = \int_{(n-1)T}^{nT} d\{c_2(t)x(t)\} \tag{A11–C.4}$$

The trapezoidal rule is applied to the left hand side of Equation (A11–C.4) to yield

$$\int_{(n-1)T}^{nT} y_2(t)\, dt \equiv \frac{T}{2}[y_2(n) + y_2(n-1)] \tag{A11–C.5}$$

Then the discrete-time form of Equation (A11–C.4) is

$$\frac{T}{2}[y_2(n) + y_2(n-1)] = c_2(n)x(n) - c_2(n-1)x(n-1) \tag{A11–C.6}$$

Equation (A11–C.6) can be rewritten in recursive form as

$$y_2(n) = \frac{2}{T}c_2(n)x(n) - Q(n-1) \tag{A11–C.7}$$

where

$$Q(n-1) = \frac{4}{T}c_2(n-1)x(n-1) - Q(n-2)$$

Similarly, the integral term is approximated by

$$y_3(n) = \frac{T}{2}c_3(n)x(n) + V(n-1) \tag{A11–C.8}$$

where

$$V(n-1) = Tc_3(n-1)x(n-1) + V(n-2)$$

Now applying the rules of Equations (A11–C.3), (A11–C.7), and (A11–C.8) to the set of differential Equations (A11–C.1), the transformed equations are

$$P_{i-1}(n) - P_i(n) = \left\{R_{i-1}(n) + R_i(n) + \frac{2}{T}[L_{i-1}(n) + L_i(n)]\right\} U_i(n) - Q_{L_{i-1}L_i}(n-1)$$

$$U_i(n) - U_{i+1}(n) = \left\{G_i(n) + \frac{2}{T}C_i(n)\right\} P_i(n) - Q_{C_i}(n-1) + u_{i3}(n) \tag{A11–C.9}$$

$$P_i(n) = \left\{R_{w,i}(n) + \frac{2}{T}L_{w,i}(n) + \frac{T}{2}\frac{1}{C_{w,i}(n)}\right\} u_{i3}(n) + V_{C_{w,i}}(n-1) - Q_{L_{w,i}}(n-1)$$

where

$$Q_{L_{i-1}L_i}(n-1) = \frac{4}{T}[L_{i-1}(n-1) + L_i(n-1)]U_i(n-1) - Q_{L_{i-1}L_i}(n-2)$$

$$Q_{C_i}(n-1) = \frac{4}{T}C_i(n-1)P_i(n-1) - Q_{C_i}(n-2)$$

$$V_{C_{w,i}}(n-1) = \frac{T}{C_{w,i}(n-1)}u_{i3}(n-1) + V_{C_{w,i}}(n-2)$$

$$Q_{L_{w,i}}(n-1) = \frac{4}{T}L_{w,i}(n-1)u_{i3}(n-1) - Q_{L_{w,i}}(n-2)$$

Define

$$a_i(n) = \frac{2}{T}L_i(n) + R_i(n)$$

$$Y_{w,i}(n) = \frac{1}{\frac{2}{T}L_{w,i}(n) + R_{w,i}(n) + \frac{T}{2}\frac{1}{C_{w,i}(n)}}$$

$$b_i(n) = \frac{1}{\frac{2}{T}C_i(n) + G_i(n) + Y_{w,i}(n)}$$

$$V_i(n-1) = Q_{C_i}(n-1) - Y_{w,i}(n)[Q_{L_{w,i}}(n-1) - V_{C_{w,i}}(n-1)] \qquad \text{(A11–C.10)}$$

$$u_{i3}(n) = Y_{w,i}(n)[P_i(n) + Q_{L_{w,i}}(n-1) - V_{C_{w,i}}(n-1)]$$

$$H_i(n) = a_{i-1}(n) + a_i(n) + b_{i-1}(n) + b_i(n)$$

$$F_i(n) \equiv b_{i-1}(n)V_{i-1}(n-1) + Q_{L_{i-1}L_i}(n-1) - b_i(n)V_i(n-1)$$

After some manipulations, we obtain the difference equations for the ith section, as follows.

$$F_i(n) = -b_{i-1}(n)U_{i-1}(n) + H_i(n)U_i(n) - b_i(n)U_{i+1}(n)$$
$$P_i(n) = b_i(n)[U_i(n) - U_{i+1}(n) + V_i(n-1)] \qquad \text{(A11–C.11)}$$

Next, consider a shunt element, which models the nasal sinus, inserted in between two sections, i and i + 1. Figure A11–C.2 is the corresponding circuit model. The set of differential equations is

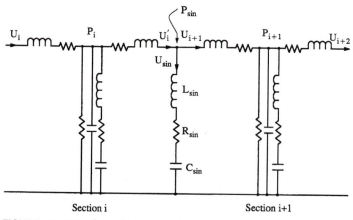

FIGURE A11–C.2   A shunt element of the nasal sinus inserted between two sections.

given by

$$P_i(t) - P_{sin}(t) = \frac{d}{dt}\left[L_i(t)U_i'(t)\right] + R_i(t)U_i'(t)$$

$$P_{sin}(t) - P_{i+1}(t) = \frac{d}{dt}[L_{i+1}(t)U_{i+1}] + R_{i+1}(t)U_{i+1}(t) \qquad (A11–C.12)$$

$$P_{sin}(t) = R_{sin}U_{sin}(t) + L_{sin}\frac{dU_{sin}(t)}{dt} + \frac{1}{C_{sin}}\int_0^t U_{sin}(\tau)\,d\tau$$

Applying the rules of Equations (A11–C.3), (A11–C.7), and (A11–C.8) to Equation set (A11–C.12), the corresponding difference equations are

$$P_i(n) - P_{sin}(n) = a_i(n)U_i'(n) - Q_{L_i}'(n-1)$$

$$P_{sin}(n) - P_{i+1}(n) = a_{i+1}(n)U_{i+1}(n) - Q_{L_{i+1}}(n-1) \qquad (A11–C.13)$$

$$P_{sin}(n) = b_{sin}\left[U_i'(n) - U_{i+1}(n)\right] - Q_{L_{sin}}(n-1) + V_{C_{sin}}(n-1)$$

where

$$a_i(n) = \frac{2}{T}L_i(n) + R_i(n)$$

$$U_{sin}(n) = U_i'(n) - U_{i+1}(n)$$

$$Q_{L_i}'(n-1) = \frac{4}{T}L_i(n-1)U_i'(n-1) - Q_{L_i}'(n-2)$$

$$Q_{L_{i+1}}(n-1) = \frac{4}{T}L_{i+1}(n-1)U_{i+1}(n-1) - Q_{L_{i+1}}(n-2)$$

$$Q_{L_{sin}}(n-1) = \frac{4}{T}L_{sin}U_{sin}(n-1) - Q_{L_{sin}}(n-2)$$

$$V_{C_{sin}}(n-1) = \frac{4}{C_{sin}}U_{sin}(n-1) + V_{C_{sin}}(n-2)$$

$$b_{sin} = \left[\frac{2}{T}L_{sin} + R_{sin} + \frac{T}{2C_{sin}}\right]$$

After rearranging items, the difference equations governing the shunt element can be rewritten as

$$F_{sin}(n) = -b_i(n)U_i(n) + H_{sf}(n)U_i'(n) - b_{sin}U_{i+1}(n)$$

$$F_{i+1}(n) = -b_{sin}U_i'(n) + H_{sb}(n)U_{i+1}(n) - b_{i+1}(n)U_{i+2}(n) \qquad (A11–C.14)$$

$$P_{sin}(n) = b_{sin}\left[U_i'(n) - U_{i+1}(n)\right] - Q_{L_{sin}}(n-1) + V_{C_{sin}}(n-1)$$

$$P_i(n) = b_i(n)\left[U_i(n) - U_i'(n) + V_i(n-1)\right]$$

where

$$F_{sin}(n) \equiv Q_{L_{sin}}(n-1) + Q_{L_i}'(n-1) + b_i(n)V_i(n-1) - V_{C_{sin}}(n-1)$$

$$F_{i+1}(n) \equiv Q_{L_{i+1}}(n-1) - Q_{L_{sin}}(n-1) - b_{i+1}(n)V_{i+1}(n-1) + V_{C_{sin}}(n-1)$$

$$H_{sf}(n) = a_i(n) + b_i(n) + b_{sin}$$

$$H_{sb}(n) = a_{i+1}(n) + b_{i+1}(n) + b_{sin}$$

Now, we consider the circuit representation at the bifurcation point of the vocal tract and the nasal tract. From Figure A11–C.3, the differential equations are given by

$$P_i(t) - P_{NC}(t) = \frac{d}{dt}[L_i(t)U_{NC}(t)] + R_i(t)U_{NC}(t)$$

$$P_{NC}(t) - P_{i+1}(t) = \frac{d}{dt}[L_{i+1}(t)U_{i+1}] + R_{i+1}(t)U_{i+1}(t)$$

$$P_{NC}(t) - P_{N1}(t) = \frac{d}{dt}[L_{N1}(t)U_{N1}] + R_{N1}(t)U_{N1}(t) \qquad (A11–C.15)$$

$$U_i(t) - U_{NC}(t) = u_{i1}(t) + u_{i2}(t) + u_{i3}(t)$$

$$= G_i(t)P_i(t) + \frac{d}{dt}\{G_i(t)P_i(t)\} + u_{i3}(t)$$

$$U_{NC}(t) = U_{i+1}(t) + U_{N1}(t)$$

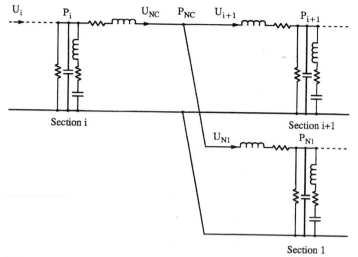

FIGURE A11–C.3   The circuit model of the bifurcation point.

After applying the discretization rules, previous derivation results, and some simple manipulations, one obtains the following difference equations

$$F_i(n) = -b_{i-1}(n)U_{i-1}(n) + H_i(n)U_i(n) - b_i(n)U_{NC}(n)$$

$$F_{NC}(n) = -b_i(n)U_i(n) + H_{NC}(n)U_{NC}(n) + P_{NC}(n)$$

$$F_{i+1}(n) = -P_{NC}(n) + H_{i+1}(n)U_{i+1}(n) - b_{i+1}(n)U_{i+2}(n)$$

$$P_i(n) = b_i(n)\,[U_i(n) - U_{NC}(n) + V_i(n-1)] \qquad\qquad \text{(A11–C.16)}$$

$$F_{N1}(n) = -P_{NC}(n) + H_{N1}(n)U_{N1}(n) - b_{N1}(n)U_{N2}(n)$$

$$U_{NC}(n) = U_{i+1}(n) + U_{N1}(n)$$

where

$$F_i(n) \equiv Q_{L_{i-1}L_i}(n-1) + b_{i-1}(n)V_{i-1}(n-1) - b_i(n)V_i(n-1)$$

$$F_{NC}(n) \equiv Q'_{L_i}(n-1) + b_i(n)V_i(n-1)$$

$$F_{i+1}(n) \equiv Q_{L_{i+1}}(n-1) - b_{i+1}(n)V_{i+1}(n-1)$$

$$F_{N1}(n) \equiv Q_{L_{N1}}(n-1) - b_{N1}(n)V_{N1}(n-1)$$

$$H_{NC}(n) = a_i(n) + b_i(n)$$

$$H_{i+1}(n) = a_{i+1}(n) + b_{i+1}(n)$$

$$H_{N1}(n) = a_{N1}(n) + b_{N1}(n)$$

$$Q'_{L_i}(n-1) = \frac{4}{T}L_i(n-1)U_{NC}(n-1) - Q'_{L_i}(n-2)$$

Next, we consider the radiation part. Figure A11–C.4 shows the circuit model of the mouth section and radiation. The differential equations for this circuit are

$$P_i(t) - P_r(t) = \frac{d}{dt}\,[L_i(t)U_r(t)] + R_i(t)U_r(t)$$

$$U_r(t) = \frac{P_r(t)}{R_r(t)} + \int_0^t \frac{P_r(\tau)}{L_r(\tau)}\,d\tau \qquad\qquad \text{(A11–C.17)}$$

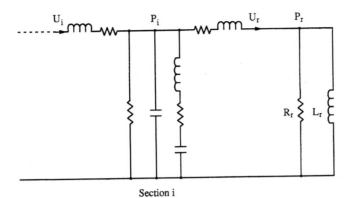

Section i

FIGURE A11–C.4 The circuit model of radiation.

The corresponding difference equations are

$$F_i(n) = -b_{i-1}(n)U_{i-1}(n) + H_i(n)U_i(n) - b_i(n)U_r(n)$$

$$F_r(n) = -b_i(n)U_i(n) + H_r(n)U_r(n)$$

$$P_i(n) = b_i(n)\left[U_i(n) - U_r(n) + V_i(n-1)\right]$$

$$P_r(n) = b_r(n)\left[U_r(n) - V_{L_r}(n-1)\right]$$

(A11–C.18)

where

$$F_i(n) = Q_{L_{i-1}L_i}(n-1) + b_{i-1}(n)V_{i-1}(n-1) - b_i(n)V_i(n-1)$$

$$F_r(n) = Q'_{L_i}(n-1) + b_i(n)V_i(n-1) + b_r(n)V_{L_r}(n-1)$$

$$H_r(n) = a_i(n) + b_i(n) + b_r(n)$$

$$b_r(n) = \frac{1}{\frac{1}{R_r(n)} + \frac{T}{2}\frac{1}{L_r(n)}}$$

$$Q'_{L_i}(n-1) = \frac{4}{T}L_i(n-1)U_r(n-1) - Q_{L_i}(n-2)$$

$$V_{L_r}(n-1) = \frac{T}{L_r(n-1)}P_r(n-1) + V_{L_r}(n-2)$$

Finally, we consider the excitation input. We have two types: one is without the glottal impedance and the subglottal system; one is with the glottal impedance and the subglottal system. We first derive the difference equations for no glottal impedance and no subglottal system. From Figure A11–C.5, it is easy to obtain the difference equations from previous derivation results.

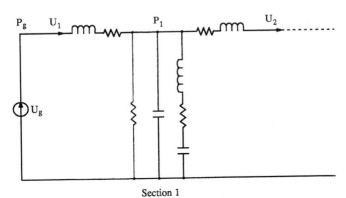

Section 1

FIGURE A11–C.5 The circuit model of the first section of vocal tract with excitation.

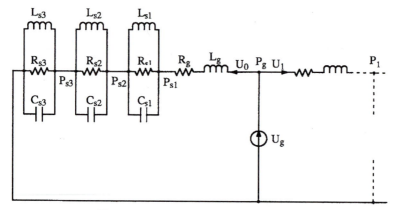

FIGURE A11–C.6 The circuit model of subglottal system and glottal impedance with excitation.

They are

$$F_1(n) = -P_g(n) - b_1(n)U_2(n)$$
$$U_1(n) = U_g(n) \tag{A11–C.19}$$
$$P_1(n) = b_1(n)[U_1(n) - U_2(n) + V_1(n-1)]$$

where

$$F_1(n) \equiv Q_{L_i}(n-1) + b_1(n)V_1(n-1) - [a_1(n) + b_1(n)]U_1(n)$$

For the case of excitation source with the glottal impedance and the subglottal system, we form the circuit model in Figure A11–C.6. The corresponding differential equations are given by

$$U_0(t) = \frac{[P_{s1}(t) - P_{s2}(t)]}{R_{s1}} + \frac{1}{L_{s1}} \int_0^t [P_{s1}(\tau) - P_{s2}(\tau)]\,d\tau + C_{s1}\frac{d}{dt}[P_{s1}(t) - P_{s2}(t)]$$

$$= \frac{[P_{s2}(t) - P_{s3}(t)]}{R_{s2}} + \frac{1}{L_{s2}} \int_0^t [P_{s2}(\tau) - P_{s3}(\tau)]\,d\tau + C_{s2}\frac{d}{dt}[P_{s2}(t) - P_{s3}(t)]$$

$$= \frac{P_{s3}(t)}{R_{s3}} + \frac{1}{L_{s3}} \int_0^t P_{s3}(\tau)\,d\tau + C_{s3}\frac{dP_{s3}(t)}{dt} \tag{A11–C.20}$$

$$P_g(t) - P_{s1}(t) = R_g(t)U_0(t) + \frac{d}{dt}[L_g(t)U_0(t)]$$
$$U_g(t) = U_0(t) + U_1(t)$$

Applying the discretization rules to above equations and solving $P_{si}(n)$, $i = 1, 2, 3$, in order, then the difference equations can be written as

$$F_s(n) = P_g(n) + X_{sub}(n)U_1(n)$$
$$U_g(n) = U_1(n) + U_0(n) \tag{A11–C.21}$$
$$F_1(n) = -P_g(n) + H_1(n)U_1(n) - b_1(n)U_2(n)$$

where

$$F_s(n) \equiv V_{s1}(n-1) + V_{s2}(n-1) + V_{s3}(n-1) + X_{sub}(n)U_g(n) - Q_{L_s}(n-1)$$
$$F_1(n) \equiv Q_{L_1}(n-1) - b_1(n)V_1(n-1)$$
$$V_{si}(n-1) = b_{si}\left[C_{si}Q_{C_{si}}(n-1) - \frac{V_{L_{si}}(n-1)}{L_{si}}\right], \quad i = 1, 2, 3$$

$$b_{si}(n) = \frac{1}{\frac{1}{R_{si}(n)} + \frac{T}{2}\frac{1}{L_{si}(n)} + \frac{2}{T}C_{si}}, \quad i = 1, 2, 3$$

$$Q_{C_{si}}(n-1) = \frac{4}{T}[P_{si}(n-1) - P_{si+1}(n-1)] - Q_{C_{si}}(n-2), \quad i = 1, 2$$

$$Q_{C_{s3}}(n-1) = \frac{4}{T}P_{s3}(n-1) - Q_{C_{s3}}(n-2)$$

$$V_{L_{si}}(n-1) = T[P_{si}(n-1) - P_{si+1}(n-1)] + V_{L_{si}}(n-2), \quad i = 1, 2$$

$$V_{L_{s3}}(n-1) = TP_{s3}(n-1) + V_{L_{s3}}(n-2)$$

$$Q_{C_{si}}(n-1) = \frac{4}{T}[P_{si}(n-1) - P_{si+1}(n-1)] - Q_{C_{si}}(n-2), \quad i = 1, 2$$

$$Q_{C_{si}}(n-1) = \frac{4}{T}P_{s3}(n-1) - Q_{C_{s3}}(n-2)$$

$$P_{si}(n) = \sum_{k=i}^{1} b_{sk}U_0(n) + \sum_{k=i}^{1} V_{sk}(n-1), \quad i = 1, 2, 3$$

$$P_g(n) = V_{s1}(n-1) + V_{s2}(n-1) + V_{s3}(n-1) + X_{sub}(n)U_0(n) - Q_{L_g}(n-1)$$

$$Q_{L_g}(n-1) = \frac{4}{T}[L_g(n-1)U_0(n-1)] - Q_{L_g}(n-2)$$

$$X_{sub}(n) = a_g(n) + b_{s1} + b_{s2} + b_{s3}$$

$$a_g(n) = R_g(n) + \frac{2}{T}L_g(n)$$

We have derived the difference equations for different parts of the vocal system. Note that in order to set all recursive equations, $Q(n-1)$ and $V(n-1)$, in motion, the initial rest conditions of the vocal system need to be assumed; that is, $Q(0) = 0$ and $V(0) = 0$ for all sections. To write out the matrix equations for the entire vocal system, we assume that

1. The vocal tract has a number of sections denoted as NING.
2. The nasal tract has a number of sections denoted as NTN.
3. The bifurcation point is located at the downstream of a vocal tract section denoted as NTS.
4. The excitation source is located at the input to first section of vocal tract.
5. One sinus is inserted at the downstream of a nasal-tract section denoted as NS.

For the vocal system that has no glottal impedance and no subglottal system but has one nasal sinus coupling, the three sets of simultaneous difference equations or the matrix equations for the pharyngeal, oral, and nasal tracts, respectively, are

$$U_1(n) = U_g(n)$$
$$k = 0: \quad F_1(n) = Q_{L_1}(n-1) - b_1(n)V_1(n-1) - H_1(n)U_1(n)$$
$$= -P_g(n) - b_1(n)U_2(n)$$
$$k = 1: \quad F_2(n) = b_1(n)V_1(n-1) + Q_{L_1L_2}(n-1) - b_2(n)V_2(n-1) + b_1(n)U_1(n)$$
$$= H_2(n)U_2(n) - b_2(n)U_3(n)$$
$$2 \le k < NTS: \quad F_{k+1}(n) = b_k(n)V_k(n-1) + Q_{L_kL_{k+1}}(n-1) - b_{k+1}(n)V_{k+1}(n-1)$$
$$= -b_k(n)U_k(n) + H_{k+1}(n)U_{k+1}(n) - b_{k+1}(n)U_{k+2}(n)$$
$$\text{if } k + 2 = NTS + 1 \Rightarrow U_{k+2}(n) = U_{NC}(n)$$
$$k = NTS: \quad F_{NC}(n) = Q'_{L_{NTS}}(n-1) + b_{NTS}(n)V_{NTS}(n-1)$$
$$= -b_{NTS}(n)U_{NTS}(n) + H_{NC}(n)U_{NC}(n) + P_{NC}(n)$$
$$P_i(n) = b_i(n)[U_i(n) - U_{i+1}(n) + V_i(n)], \quad \text{where } i = 1, \ldots, NTS$$

if

$$i = NTS,$$

then

$$U_{NTS+1}(n) = U_{NC}(n)$$

$$
\begin{bmatrix}
F_1 \\
F_2 \\
\cdot \\
\cdot \\
\cdot \\
F_{NTS} \\
F_{NC}
\end{bmatrix}
=
\begin{bmatrix}
-1 & -b_1 & 0 & \cdots & 0 & 0 & 0 \\
0 & H_2 & -b_2 & \cdots & 0 & 0 & 0 \\
\cdot & \cdot & \cdot & & \cdot & \cdot & \cdot \\
\cdot & \cdot & \cdot & & \cdot & \cdot & \cdot \\
\cdot & \cdot & \cdot & & \cdot & \cdot & \cdot \\
0 & 0 & 0 & \cdots & H_{NTS} & -b_{NTS} & 0 \\
0 & 0 & 0 & \cdots & -b_{NTS} & H_{NC} & 1
\end{bmatrix}
\begin{bmatrix}
P_g \\
U_2 \\
\cdot \\
\cdot \\
\cdot \\
U_{NC} \\
P_{NC}
\end{bmatrix}
\qquad (A11–C.22)
$$

$$
\begin{aligned}
F_{NTS+1}(n) &\equiv Q_{L_{NTS+1}}(n-1) - b_{NTS+1}(n)V_{NTS+1}(n-1) \\
&= -P_{NC}(n) + H_{NTS+1}(n)U_{NTS+1}(n) - b_{NTS+1}(n)U_{NTS+2}(n) \\
F_i(n) &\equiv Q_{L_{i-1}L_i}(n-1) + b_{i-1}(n)V_{i-1}(n-1) - b_i(n)V_i(n-1) \\
&= -b_{i-1}(n)U_{i-1}(n) + H_i(n)U_i(n) - b_i(n)U_{i+1}(n)
\end{aligned}
$$

where

$$i = NTS + 2, \ldots, NING, \quad \text{if } i = NING$$

then

$$
\begin{aligned}
U_{NING+1}(n) &= U_r(n) \\
F_r(n) &\equiv b_{NING}(n)V_{NING}(n-1) + b_r(n)V_{L_r}(n-1) + Q'_{NING}(n-1) \\
&= -b_{NING}(n)U_{NING}(n) + H_r(n)U_r(n) \\
P_i(n) &= b_i(n)[U_i(n) - U_{i+1}(n) + V_i(n-1)], \quad \text{where } i = NTS + 1, \ldots, NING - 1 \\
P_{NING}(n) &= b_{NING}(n)[U_{NING}(n) - U_r(n) + V_{NING}(n-1)] \\
P_r(n) &= b_r(n)[U_r(n) - V_{L_r}(n-1)]
\end{aligned}
$$

$$
\begin{bmatrix}
F_{NTS+1} \\
F_{NTS+2} \\
F_{NTS+3} \\
\cdot \\
\cdot \\
\cdot \\
F_{NING} \\
F_r
\end{bmatrix}
=
\begin{bmatrix}
-1 & H_{NTS+1} & -b_{NTS+1} & 0 & \cdots & 0 & 0 & 0 \\
0 & -b_{NTS+1} & H_{NTS+2} & -b_{NTS+2} & \cdots & 0 & 0 & 0 \\
0 & 0 & -b_{NTS+2} & H_{NTS+3} & \cdots & 0 & 0 & 0 \\
\cdot & \cdot & \cdot & \cdot & & \cdot & \cdot & \cdot \\
\cdot & \cdot & \cdot & \cdot & & \cdot & \cdot & \cdot \\
\cdot & \cdot & \cdot & \cdot & & \cdot & \cdot & \cdot \\
0 & 0 & 0 & 0 & \cdots & -b_{NING-1} & H_{NING} & -b_{NING} \\
0 & 0 & 0 & 0 & \cdots & 0 & -b_{NING} & H_r
\end{bmatrix}
\begin{bmatrix}
P_{NC} \\
U_{NTS+1} \\
U_{NTS+2} \\
\cdot \\
\cdot \\
\cdot \\
U_{NING} \\
U_r
\end{bmatrix}
$$

$$(A11–C.23)$$

$$
\begin{aligned}
F_{N1}(n) &= Q_{L_{N1}}(n-1) - b_{N1}(n)V_{N1}(n-1) \\
&= -P_{NC}(n) + H_{N1}(n)U_{N1}(n) - b_{N1}(n)U_{N2}(n) \\
F_{Ni}(n) &= Q_{L_{Ni-1}L_{Ni}}(n-1) + b_{Ni-1}(n)V_{Ni-1}(n-1) - b_{Ni}(n)V_{Ni}(n-1) \\
&= -b_{Ni-1}(n)U_{Ni-1}(n) + H_{Ni}(n)U_{Ni}(n) - b_{Ni}(n)U_{Ni+1}(n) \\
& i = 2, \ldots, N_s, N_{s+2}, \ldots, NTN;
\end{aligned}
$$

if

$$i = N_s$$

then

$$
\begin{aligned}
U_{NN_s+1}(n) &= U'_{NN_s}(n); \quad \text{if } i = NTN, \text{ then } U_{NNTN+1}(n) = U_{Nr}(n) \\
F_{sin}(n) &= Q_{L_{sin}}(n-1) + Q'_{L_{NN_s}}(n-1) + b_{NN_s}(n)V_{NN_s}(n-1) - V_{C_{sin}}(n-1) \\
&= -b_{NN_s}(n)U_{NN_s}(n) + H_{sf}(n)U'_{NN_s}(n) - b_{sin}U_{NN_s+1}(n)
\end{aligned}
$$

$$F_{NNs+1}(n) = Q_{L_{NNs+1}}(n-1) - Q_{L_{sin}}(n-1) + V_{C_{sin}}(n-1) - b_{NNs+1}(n)V_{NNs+1}(n-1)$$
$$= -b_{sin}U'_{NNs}(n) + H_{sb}(n)U_{NNs+1}(n) - b_{NNs+1}(n)U_{NNs+2}(n)$$
$$F_{Nr}(n) = b_{NNTN}(n)V_{NNTN}(n-1) + b_{Nr}(n)V_{L_{Nr}}(n-1) + Q'_{L_{NNTN}}(n-1)$$
$$= -b_{NNTN}(n)U_{NNTN}(n) + H_{Nr}(n)U_{Nr}(n)$$
$$P_{Ni}(n) = b_{Ni}(n)[U_{Ni}(n) - U_{Ni+1}(n) + V_{Ni}(n-1)],$$

where

$$i = 1, 2, \ldots, NTN; \quad \text{if } i = Ns, \text{ then } U_{NNs+1}(n) = U'_{NNs}(n)$$

if

$$i = NTN, \quad \text{then } U_{NNTN+1}(n) = U_{Nr}(n)$$
$$P_{sin}(n) = b_{sin}\left[U'_{NNs}(n) - U_{NNs+1}(n)\right] - Q_{L_{sin}}(n-1) + V_{C_{sin}}(n-1)$$
$$P_{Nr}(n) = b_{Nr}(n)[U_{Nr}(n) - V_{L_{Nr}}(n-1)]$$

$$
\begin{bmatrix}
F_{N1} \\
F_{N2} \\
F_{N3} \\
\vdots \\
F_{sin} \\
F_{NNs+1} \\
\vdots \\
F_{NNTN} \\
F_{Nr}
\end{bmatrix}
=
\begin{bmatrix}
-1 & H_{N1} & -b_{N1} & \cdots & 0 & 0 & 0 & \cdots & 0 & 0 \\
0 & -b_{N1} & H_{N2} & \cdots & 0 & 0 & 0 & \cdots & 0 & 0 \\
0 & 0 & -b_{N2} & \cdots & 0 & 0 & 0 & \cdots & 0 & 0 \\
\vdots & \vdots & \vdots & & \vdots & \vdots & \vdots & & \vdots & \vdots \\
0 & 0 & 0 & \cdots & -b_{NNs} & H_{sf} & -b_{sin} & \cdots & 0 & 0 \\
0 & 0 & 0 & \cdots & 0 & -b_{sin} & H_{sb} & \cdots & 0 & 0 \\
\vdots & \vdots & \vdots & & \vdots & \vdots & \vdots & & \vdots & \vdots \\
0 & 0 & 0 & \cdots & 0 & 0 & 0 & \cdots & H_{NNTN} & -b_{NNTN} \\
0 & 0 & 0 & \cdots & 0 & 0 & 0 & \cdots & -b_{NNTN} & H_{Nr}
\end{bmatrix}
\begin{bmatrix}
P_{NC} \\
U_{N1} \\
U_{N2} \\
\vdots \\
U'_{NNs} \\
U_{NNs+1} \\
\vdots \\
U_{NNTN} \\
U_{Nr}
\end{bmatrix}
$$

$$(A11-C.24)$$

If there is no nasal sinus cavity, the matrix Equation (A11–C.24) reduces to the following.

$$F_{N1}(n) = Q_{L_{N1}}(n-1) - b_{N1}(n)V_{N1}(n-1)$$
$$= -P_{NC}(n) + H_{N1}(n)U_{N1}(n) - b_{N1}(n)U_{N2}(n)$$
$$F_{Ni}(n) = Q_{L_{Ni-1}L_{Ni}}(n-1) + b_{Ni-1}(n)V_{Ni-1}(n-1) - b_{Ni}(n)V_{Ni}(n-1)$$
$$= -b_{Ni-1}(n)U_{Ni-1}(n) + H_{Ni}(n)U_{Ni}(n) - b_{Ni}(n)U_{Ni+1}(n)$$

where

$$i = 2, \ldots, NTN; \quad \text{if } i = NTN, \text{ then } U_{NNTN+1}(n) = U_{Nr}(n)$$
$$F_{Nr}(n) = b_{NNTN}(n)V_{NNTN}(n-1) + b_{Nr}(n)V_{L_{Nr}}(n-1) + Q'_{L_{NNTN}}(n-1)$$
$$= -b_{NNTN}(n)U_{NNTN}(n) + H_{Nr}(n)U_{Nr}(n)$$
$$P_{Ni}(n) = b_{Ni}(n)[U_{Ni}(n) - U_{Ni+1}(n) + V_{Ni}(n-1)],$$

where

$$i = 1, \ldots, NTN; \quad \text{if } i = NTN, \text{ then } U_{NNTN+1}(n) = U_{Nr}(n)$$
$$P_{Nr}(n) = b_{Nr}(n)[U_{Nr}(n) - V_{L_{Nr}}(n-1)]$$

$$
\begin{bmatrix}
F_{N1} \\
F_{N2} \\
F_{N3} \\
\cdot \\
\cdot \\
\cdot \\
F_{NNTN} \\
F_{Nr}
\end{bmatrix}
=
\begin{bmatrix}
-1 & H_{N1} & -b_{N1} & 0 & \cdots & 0 & 0 & 0 \\
0 & -b_{N1} & H_{N2} & -b_{N2} & \cdots & 0 & 0 & 0 \\
0 & 0 & -b_{N2} & H_{N3} & \cdots & 0 & 0 & 0 \\
\cdot & \cdot & \cdot & \cdot & & \cdot & \cdot & \cdot \\
\cdot & \cdot & \cdot & \cdot & & \cdot & \cdot & \cdot \\
\cdot & \cdot & \cdot & \cdot & & \cdot & \cdot & \cdot \\
0 & 0 & 0 & 0 & \cdots & -b_{NNTN-1} & H_{NNTN} & -b_{NNTN} \\
0 & 0 & 0 & 0 & \cdots & 0 & -b_{NNTN} & H_{Nr}
\end{bmatrix}
\begin{bmatrix}
P_{NC} \\
U_{N1} \\
U_{N2} \\
\cdot \\
\cdot \\
\cdot \\
U_{NNTN} \\
U_{Nr}
\end{bmatrix}
$$

$$(A11-C.25)$$

If the glottal impedance and the subglottal system are included, the matrix Equation (A11–C.22) extends to the following.

$$
\begin{bmatrix} F_s \\ F_1 \\ F_2 \\ \cdot \\ \cdot \\ \cdot \\ F_{NTS} \\ F_{NC} \end{bmatrix}
=
\begin{bmatrix}
1 & X_{sub} & 0 & \cdots & 0 & 0 & 0 \\
-1 & H_1 & -b_1 & \cdots & 0 & 0 & 0 \\
0 & -b_1 & H_2 & \cdots & 0 & 0 & 0 \\
\cdot & \cdot & \cdot & & \cdot & \cdot & \cdot \\
\cdot & \cdot & \cdot & & \cdot & \cdot & \cdot \\
\cdot & \cdot & \cdot & & \cdot & \cdot & \cdot \\
0 & 0 & 0 & \cdots & H_{NTS} & -b_{NTS} & 0 \\
0 & 0 & 0 & \cdots & -b_{NTS} & H_{NC} & 1
\end{bmatrix}
\begin{bmatrix} P_g \\ U_1 \\ U_2 \\ \cdot \\ \cdot \\ \cdot \\ U_{NC} \\ P_{NC} \end{bmatrix}
\qquad \text{(A11–C.26)}
$$

From the derived discrete-time acoustic matrix equations, we can form four different vocal system model structures, which are:

1. (A11–C.22), (A11–C.23), and (A11–C.24) for the vocal system model with nasal sinus but no glottal impedance and no subglottal system.

2. (A11–C.22), (A11–C.23), and (A11–C.25) for the vocal system model with no nasal sinus and no glottal impedance and no subglottal system.

3. (A11–C.26), (A11–C.23), and (A11–C.24) for the vocal system model with nasal sinus and with the glottal impedance and the subglottal system.

4. (A11–C.26), (A11–C.23), and (A11–C.25) for the vocal system model with no nasal sinus but with the glottal impedance and the subglottal system.

There are three matrix equations for each structure. Each matrix equation can be written as $y = A \cdot x$, where A is a non-square band diagonal sparse matrix of coefficients, y is a column vector of force constants, and x is the unknown column vector. The elements in the sparse coefficient matrix and the force constants inside the vocal system are defined in the previous paragraphs. In addition to the three matrix equations, we need the boundary condition at the nasal coupling point, $U_{NC}(n) = U_{i+1}(n) + U_{Ni}(n)$. For Cases 3 and 4, we need one more boundary condition at the glottis, $U_g(n) = U_1(n) + U_0(n)$. It may also be noted that the three matrix equations for each case are coupled by the term $P_{NC}(n)$ and boundary condition(s). As Maeda (1982a) pointed out, if we eliminate $P_{NC}(n)$ and $U_{NC}(n)$ analytically, the formulation results in an unstable system. Therefore, $P_{NC}(n)$ and $U_{NC}(n)$ are included in the unknown column vectors in order for stable solutions.

There are many methods that can be used to solve these sparse linear system equations. Since the coefficient matrices are band diagonal sparse matrices, an efficient elimination procedure followed by a substitution procedure can be used. Once the values of $U_i(n)$ and $U_{Ni}(n)$ for all i, $P_{NC}(n)$, and $U_{NC}(n)$ are solved, the pressure $P_i(n)$ and $P_{Ni}(n)$ for all i, and the volume velocity $u_{i3}(n)$ and $u_{Ni3}(n)$ for all i can be computed. Then, the force constants and coefficient matrices are updated for the next recursion.

COMMENTS ON USAGE OF SIMULATED ANNEALING PROCEDURE

In Appendix 11, we covered the details of speech inverse filtering using the simulated annealing optimization algorithm. The annealing parameters control the performance of the optimization process. The various acoustic characteristics of the speech signal and target-frame selections affect the optimization process. The default values of the annealing parameters are not the best values for all cases. Although we do not know the combinations of the annealing parameters that perform well for the optimization process for different target frames, the following guidelines provide some rules for adjusting the appropriate annealing parameters.

Step 1: Set the desired nasalization extent and set the number of dimensions of the articulatory vector at the appropriate dimensions, for example, $M = 8$ for front vowels; $M = 9$ for nasalized front vowels; $M = 11$ for middle, back vowels, and semivowels; and $M = 12$ for nasalized vowels (see the descriptions in Section A11.3.3.1). Start the optimization process with the default initial articulatory vector and the default annealing parameters. If the error distance is less than 1% after the process stops, go to Step 5. If not, go to Step 2.

Step 2: Check if the current vocal tract shape (or cross-sectional area) is reasonable. If the shape is not reasonable go to Step 3. Otherwise, record the error distance as ε_p and the current final temperature as T_p. Then set the initial temperature $T = \text{floor}[T_p]$. Start the optimization process again. If the new error distance is less than ε_p, then this step is repeated until the error criterion is met. If not go to Step 4.

Step 3: Examine the vocal tract shape (articulatory vector) for the initial settings and all subsequent values. Several adjustments of the annealing parameters can be used. The following order of adjustments are recommended: raise the initial temperature, increase the value of the reduction factor, increase the total number of evaluations, and change the other annealing parameters as desired. Then begin the process and apply Step 2.

Step 4: Examine the vocal tract shape as described in Step 3. Increase the number of dimensions of the articulatory vector from $M = 8$ to $M = 11$ and start the process. Apply Step 2.

Step 5: Check the vocal tract shape with x-ray tracings or schematic vocal tract profiles as published in the literature or compare with vocal tract profiles as shown in Appendix 5. If the vocal tract outline is similar to those published, then the optimization process is done. If not, this may mean that the "ventriloquist effect" has occurred. One can adjust the settings of the nine articulatory sliders in the **Articulatory Position Settings** popup window so that the initial configuration becomes closer to the true outline. Then go to Step 1 to start the optimization process again.

Step 6: If the above steps fail to reduce the error criterion to a satisfactory value, then return to the target-frame selection phase and reselect the current target frame. Start the optimization process from the Step 1.

The above guidelines are illustrated in the following examples for the sentence, "We were away a year ago," spoken by two male subjects, A and B. Figures A11–D.1 and A11–D.2 show the optimized target frames for subjects A and B, respectively. See Appendix A11–A for the vowel results.

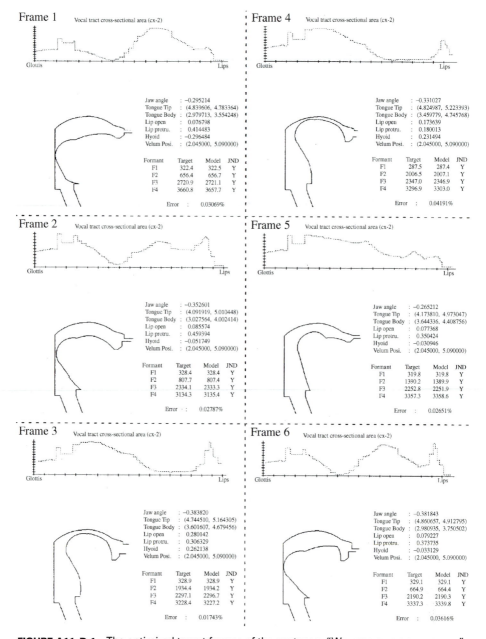

FIGURE A11–D.1   The optimized target frames of the sentence, "We were away a year ago," spoken by male subject A.

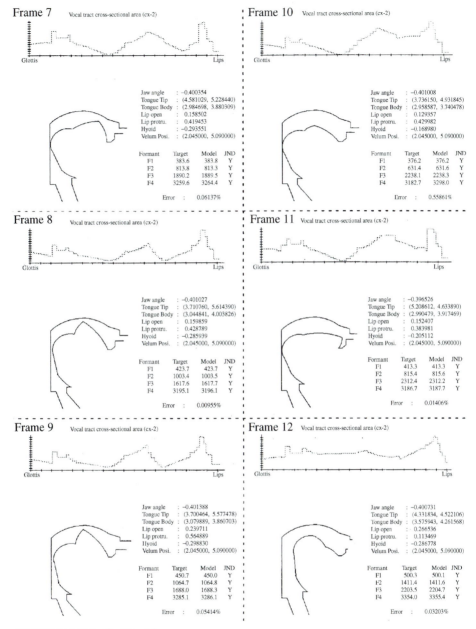

Frame 7 Vocal tract cross-sectional area (cx-2)

Jaw angle	:	−0.400354
Tongue Tip	:	(4.581029, 5.228440)
Tongue Body	:	(2.984698, 3.880309)
Lip open	:	0.158502
Lip protru.	:	0.419453
Hyoid	:	−0.293551
Velum Posi.	:	(2.045000, 5.090000)

Formant	Target	Model	JND
F1	383.6	383.8	Y
F2	813.8	813.3	Y
F3	1890.2	1889.5	Y
F4	3259.6	3264.4	Y

Error : 0.06137%

Frame 10 Vocal tract cross-sectional area (cx-2)

Jaw angle	:	−0.401008
Tongue Tip	:	(3.736150, 4.931845)
Tongue Body	:	(2.958587, 3.740478)
Lip open	:	0.129357
Lip protru.	:	0.429982
Hyoid	:	−0.168980
Velum Posi.	:	(2.045000, 5.090000)

Formant	Target	Model	JND
F1	376.2	376.2	Y
F2	631.4	631.6	Y
F3	2238.1	2238.3	Y
F4	3182.7	3298.0	Y

Error : 0.55861%

Frame 8 Vocal tract cross-sectional area (cx-2)

Jaw angle	:	−0.401027
Tongue Tip	:	(3.710760, 5.614390)
Tongue Body	:	(3.044841, 4.003826)
Lip open	:	0.159859
Lip protru.	:	0.428789
Hyoid	:	−0.285939
Velum Posi.	:	(2.045000, 5.090000)

Formant	Target	Model	JND
F1	423.7	423.7	Y
F2	1003.4	1003.5	Y
F3	1617.6	1617.7	Y
F4	3195.1	3196.1	Y

Error : 0.00955%

Frame 11 Vocal tract cross-sectional area (cx-2)

Jaw angle	:	−0.396526
Tongue Tip	:	(5.208612, 4.633890)
Tongue Body	:	(2.990479, 3.917469)
Lip open	:	0.152407
Lip protru.	:	0.383981
Hyoid	:	−0.205112
Velum Posi.	:	(2.045000, 5.090000)

Formant	Target	Model	JND
F1	413.3	413.3	Y
F2	815.4	815.6	Y
F3	2312.4	2312.2	Y
F4	3186.7	3187.7	Y

Error : 0.01406%

Frame 9 Vocal tract cross-sectional area (cx-2)

Jaw angle	:	−0.401388
Tongue Tip	:	(3.700464, 5.577478)
Tongue Body	:	(3.079889, 3.860703)
Lip open	:	0.239711
Lip protru.	:	0.564889
Hyoid	:	−0.298830
Velum Posi.	:	(2.045000, 5.090000)

Formant	Target	Model	JND
F1	450.7	450.0	Y
F2	1064.7	1064.8	Y
F3	1688.0	1688.3	Y
F4	3285.1	3286.1	Y

Error : 0.05414%

Frame 12 Vocal tract cross-sectional area (cx-2)

Jaw angle	:	−0.400731
Tongue Tip	:	(4.331834, 4.522106)
Tongue Body	:	(3.575943, 4.261568)
Lip open	:	0.266536
Lip protru.	:	0.113469
Hyoid	:	−0.286778
Velum Posi.	:	(2.045000, 5.090000)

Formant	Target	Model	JND
F1	500.3	500.1	Y
F2	1411.4	1411.6	Y
F3	2203.5	2204.7	Y
F4	3354.0	3355.4	Y

Error : 0.03203%

FIGURE A11–D.1 Continued.

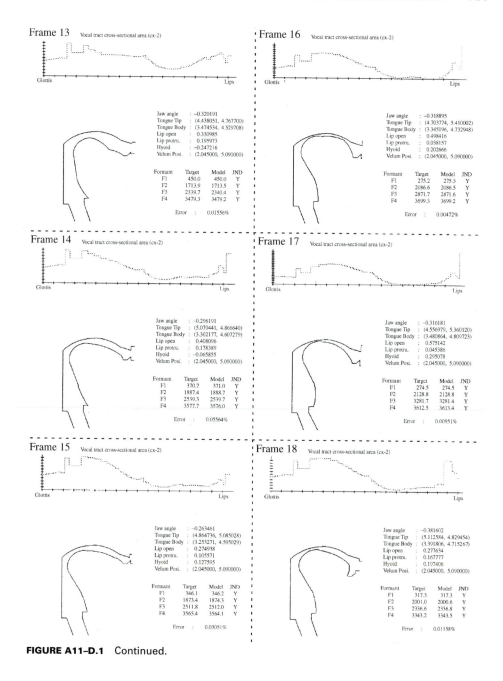

FIGURE A11–D.1   Continued.

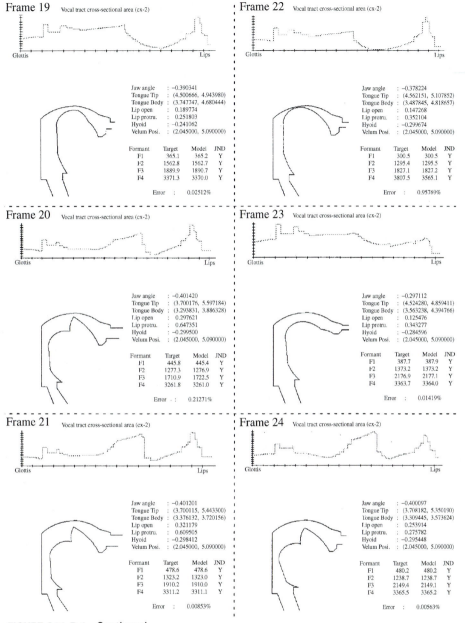

FIGURE A11–D.1 Continued.

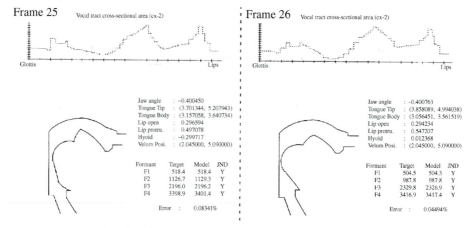

FIGURE A11–D.1   Continued.

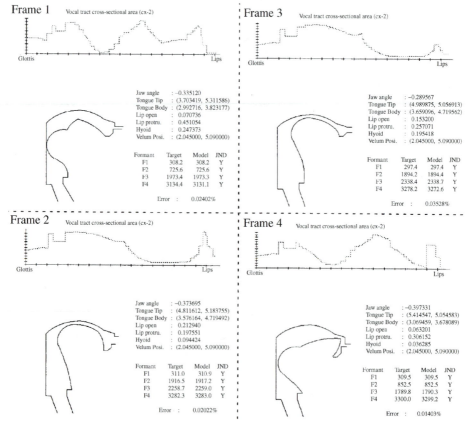

FIGURE A11–D.2   The optimized target frames of the sentence, "We were away a year ago," spoken by male subject B.

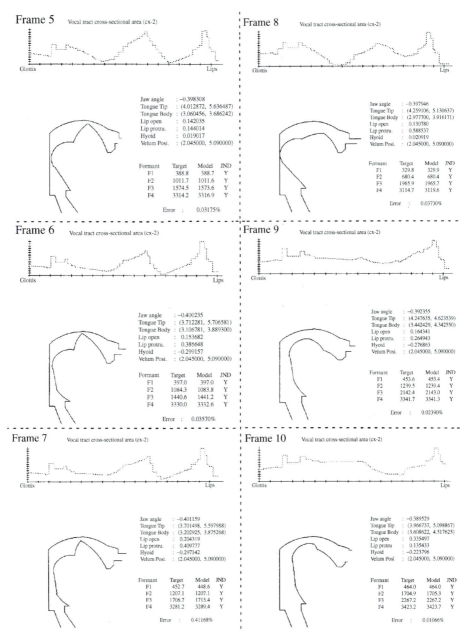

FIGURE A11–D.2 Continued.

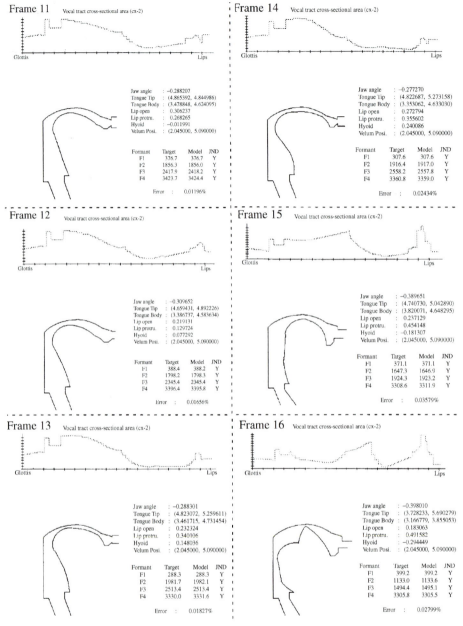

FIGURE A11–D.2 Continued.

descriptive terminology for the phonetic description of voice quality (Laver, 1980) and to classify voice qualities (Murry and Singh, 1980; Singh and Murry, 1978), but more work is needed in this area, particularly at the interface between speech technology and speech science.

The term quality has been used in different contexts. A phonetician might use quality to describe articulatory differences, as when comparing the vowels in different words. A speech pathologist might speak of laryngeal quality (hoarse, harsh, breathy) or of resonance (vocal tract) and nasal quality. A singer may use quality to express differences in vocal registers, which are related to laryngeal vibratory characteristics. Researchers have attempted to assess speech quality in terms of loudness (Rothauser et al., 1968, 1971). In quality assessment tasks, the intelligibility of individual phonemes may be a major factor listeners attend to (Pisoni et al., 1985). Quality may be determined from a listener's appraisal of a speech stimulus, using comparisons between a test stimulus and a reference utterance of known attributes such as "breathy," "crisp," "rough," and so on. (Colton and Estill, 1981; Logan et al., 1989; Rothauser et al., 1971). A listener's assessment of synthetic speech may be achieved by comparing synthetic speech tokens with tokens of natural speech.

A12.4 ASSESSING SYNTHETIC SPEECH

From a technical point of view, one prefers to have an objective method for assessing intelligibility, naturalness, or quality because the results obtained by such methods are presumably reproducible. Although the acoustic correlates of quality and naturalness are poorly understood, several attempts have been made to design an objective evaluation procedure using factors derived from the speech signal (Barnwell, 1979; Eskenazi, 1988; Eskenasi et al., 1990; Itoh et al., 1984; Jayant and Noll, 1984; Juang et al., 1982; Juang, 1984; Kitawaki, 1992; Koljonen and Karjalainen, 1984; Makhoul et al., 1976; Naik, 1984; Nakatsui and Mermelstein, 1982; Nocerino et al., 1985; Quackenbush et al., 1988; Steeneken and Houtgast, 1980; Viswanathan et al., 1984). A distance metric or distortion measure, based on a set of signal features, is the foundation for these objective measures. Two speech tokens are compared using identical feature sets extracted from each token. The quality is determined by measuring the distance between these feature sets with a distance metric. By selecting a perceptually consistent set of features, one hopes to achieve a high degree of correlation between objective distance measures and listener (or subjective) ratings of the same tokens (Makhoul et al., 1976, 1985; Markel and Gray, 1976). Typical distance measures use spectral data, employing a logarithmic magnitude scale, to assess the difference between two spectra (a reference sample and a test sample). Distance measures have used cepstral coefficients and a log likelihood ratio of spectra. The most frequently used spectral distance measure is perhaps the Itakura–Saito distance (Juang, 1984; Markel and Gray, 1976; Nocerino et al., 1985; Rabiner and Schafer, 1978). Many intelligibility experiments have been concerned with the evaluation of system distortions of natural speech; for example, (1) the effects of signal-to-noise ratios; (2) transmission bandwidth restrictions; (3) clipping and filtering; and (4) other signal distortions (Jayant and Noll, 1984; Licklider and Pollack, 1948; Quackenbush et al., 1988). Such studies, even though they may use distance or distortion measures, usually do not examine intelligibility issues as they relate to speech analysis–synthesis systems. One problem with a distance or distortion measure is that, on occasion, a decrease in the magnitude of the measure of a few decibels may be perceived by the ear quite easily but not so at other times (Makhoul et al., 1985). Consequently, distance measures do not always correlate well with listener evaluations of speech (natural or synthetic).

For synthesized speech the study of intelligibility has typically examined the relative contributions of various analysis and synthesis parameters. Wong and Markel (1977) evaluated the intelligibility of the linear predictive vocoder by systematically varying the parameters of predictor order, frame rate, and frame size. Their evaluation was accomplished using a subjective diagnostic rhyme listening test. They found that voiced speech segments were rated consistently as more intelligible than unvoiced segments. If the predictor coefficients for unvoiced segments were calculated more frequently (e.g., every 11.25 msec instead of every 22.5 msec), then the intelligibility of unvoiced segments improved but was still less than that for voiced segments.

Both the naturalness and intelligibility of synthesized sentences were evaluated by Carlson et al. (1979). They used MIT's text-to-speech system and Klatt's (1980) formant synthesizer. Their study

focused on durational factors in speech synthesis; for example, vowel and consonant duration, and showed that good durational structure for a sentence is important both for naturalness and intelligibility. Other studies have compared the intelligibility of natural speech to the speech of text-to-speech synthesizers (Hoover et al., 1987; Logan et al., 1989; Luce et al., 1983; Pisoni et al., 1985; Pisoni and Hunnicut, 1980). Generally, the error rates for synthetic speech were higher than those obtained with natural speech. One must remember that the intelligibility and quality of speech production by text-to-speech systems is greatly affected by the phonetic and linguistic rules employed by the system.

Context affects the intelligibility of speech (Mack and Gold, 1985; Pisoni, 1982; Pisoni and Hunnicut, 1980). Word recognition using nonsense sentences was less for synthetic speech than for natural speech, presumably because of the information provided by the meaning of the sentence.

Cooper (1987) notes from the work of Nooteboom (1983) and Waterworth (1983) that the judicious placement of pauses at appropriate grammatical positions in sentences may increase the intelligibility of the speech through improved listener recall. This technique seems to work as well for computer generated telephone numbers, as for waveform coders. But one need not manipulate characteristics of sentences of synthetic speech to improve a listener's assessment performance. For example, a listener's evaluation scores will become better as familiarity with (or learning to understand) synthetic speech increases (Schwab et al., 1985).

Manipulating the formants in a synthesizer can affect the intelligibility of the synthesized speech. Suppressing the magnitude of the second formant degraded intelligibility much more so than suppressing the first formant (Agrawal and Lin, 1975). This is in agreement with Thomas (1968), who attributed the high intelligibility of clipped speech to the presence of a strong (relatively unaffected) second formant.

A speaker's vocal characteristics apparently influence the quality of reproduction of the speaker's voice. Using a linear predictive coding (LPC) synthesizer Kahn and Garst (1983) found that male voices were more intelligible than female voices and that the presence of nasality or whisper in the speaker's voice degraded the intelligibility of the synthesized voice for both males and females.

Smith et al. (1981) found that the intelligibility of eigenparameter encoded speech was not improved over LPC speech and that synthetic speech quality was dependent on the eigenparameter quantization technique used. These investigators also found, as did Kahn and Garst (1983), that the LPC synthesizer does not model all voices equally well. They concluded that the LPC model needs to be improved if the synthetic speech is to approach that of natural speech in both intelligibility and quality, a point noted by Atal and David (1979) and Wong (1980).

A12.5 METHODS FOR ASSESSING INTELLIGIBILITY AND QUALITY

From the previous review, it appears that speech naturalness, quality, and intelligibility cannot be characterized as orthogonal coordinates in a three-dimensional space. As we presently use these terms, factors that affect one concept may affect the other two. So a definition of speech quality is needed.

One simple definition of speech quality has been suggested by Rothauser et al. (1968, 1969, 1971). They described speech quality in terms of only four factors: intelligibility, preference, loudness, and speaker recognizability. With the exception of loudness, none of these factors is directly related to a single measurement procedure. We suggest focusing on the factor "preference," a term that describes the average attitude of a listener toward a speech signal while he or she compares it with another speech signal. Preference tests can provide an answer to the question: Which one of two speech tokens to be compared is preferred by an average listener? According to Rothauser et al. (1968, 1969, 1971), the aspect of preference with respect to the overall speech quality becomes dominant when all of the following conditions are fulfilled, which appears to occur in practical situations.

1. The intelligibility of the speech signal is sufficiently high, so that it loses its importance as a prime quantitative speech quality factor and design criterion.

2. The level of the speech signal is maintained at an optimum loudness, eliminating the influence of loudness as a quality factor for the speech signals to be evaluated. Optimum loudness is

specified as the average speech level at which the single listener or a listener group prefers the sound level of the speech presented for the task to be performed.

3. The recognizability of the speaker is of no interest to the listener for the listening task. This is usually the case when the listener is expected to gain no further information from the speech signal than he or she might get from reading the written text.

Under these circumstances, preference may be said to represent speech quality for all practical purposes. Rothauser et al. (1969) suggested that the listener's preference may be expressed as the proportion of the listening group in percent that prefers the speech test signal to the speech reference signal as a source of information. The ratings are typically plotted as a reference scale or expressed as the number of times a signal was preferred or as a percentage of times the signal was preferred. The listener should be capable of discriminating speech quality and be able to express his or her preference in a consistent way. The degree to which the listeners' responses are consistent may be judged by correlational techniques, and used to eliminate listeners whose responses produce highly inconsistent data (Rothauser et al., 1969). Listeners can consider quality as equivalent to naturalness under the conditions discussed in the list presented in this section.

A12.6 LISTENING TESTS

Following Rothauser et al. (1968, 1969, 1971), we suggest that listeners hear two successive speech tokens, which can be sentences or other speech tokens. The listeners can be told the content of the sentences (tokens) that they will hear. The listeners should be given instructions. For example, that they are to indicate their preference for the sentence (token) that sounded the most natural, where naturalness is defined as "human sounding," or a similar preference task.

The speech tokens are usually presented in forced choice pairs. The subjective evaluation will yield a rating based on perceptually defined quantities rather than signal characteristics that can be measured. By presenting the synthesized speech material in a systematic manner, one hopes to identify the synthesis conditions that are "preferred" by a group of listeners. To ensure that the listeners can make the required discriminations in a consistent manner, each listener's performance should be graded. The order of presentation for each pair should be randomized for the complete experiment. For example, the listener might hear the speech token sequence A followed by B. Later, the sequence B followed by A would be presented. For the listener's scoring to be included in a study, a listener is to be consistent at the 75% level or better for all paired comparisons. For example, if only eight presentations of four different paired signals are made (A-B, C-D, E-F, G-H, B-A, D-C, F-E, H-G), then the listener could have at most only one disagreement among the paired signals.

One example of the presentation format for a listening test is a paired comparison in which three synthesized sentences for a particular speaker are compared with each other and with the original sentence, excluding same sentence comparisons. This gives six possible pairs. Each pair is presented twice in each listening session. The pair is also presented twice in reverse order. The order of presentation is randomized. Thus, a total of 24 pairs of sentences are presented for each speaker type. The listeners are told the content of the three sentences that they will hear.

A master listening tape might be generated or a computer stored equivalent. A tone of 500 Hz might be used to cue the listener that the speech material is to follow. The tone is followed by each pair of sentences (A, B), repeated twice; that is, (A, B), (A, B). There is a 1-second interval between the tone and the first A and the subsequent B. This is followed by a 2-second interval before the second A, followed by a 1-second interval before the second B. A 4-second interval is then provided for the listeners to make and score their choice before the next presentation occurrs. No ties in choice are allowed. The results of the listening test are rated usually as a percentage, representing the number of times each stimulus is preferred as more natural sounding when compared with all other stimuli. For example, a rating of 60% means that a particular sentence is preferred as sounding more natural 60% of the time. A rating of 50% would mean that the two sentences being compared would be equally preferable.

Another rating, referred to as the relative preference rating, represents the number of times sentence A is preferred to sentence B, expressed as a percentage of their joint occurrences in both

forward (A, B) and reverse (B, A) order. For example, a relative preference rating of 60% for sentence A compared with sentence B means that out of all the joint occurrences of the two sentences, A is preferred to B 60% of the time.

A12.7 SOME FACTORS THAT AFFECT THE QUALITY OF SYNTHETIC SPEECH

Several factors that affect the quality of synthetic speech produced by analysis–synthesis are: (1) formant locations and bandwidths (or number of poles and their positions); and (2) the excitation waveshape, which may include source–tract interaction. These factors are affected by the analysis techniques employed by the investigator. Furthermore, the manner in which these factors vary dynamically over time must be tracked by the analysis procedure (Childers and Wu, 1990).

Factors that affect LPC speech are modeling errors, inaccuracies in analysis, pitch measurement and modeling errors, errors in voicing detection, and LPC parameter quantization (Childers and Wu, 1990). "Buzziness" in LPC synthesized speech can be due discontinuities between speech segments introduced by segment-by-segment changes that occur in LPC parameters (Childers and Wu, 1990; Kuwabara, 1984). To cure this problem, Kuwabara (1984) suggests smoothing the synthesized speech segments pitch synchronously. The "warbling" in some LPC synthetic speech is often caused by improper measurement and modeling of the gain parameter between speech frames (Childers and Hu, 1994).

Formant synthetic speech can be improved by changing the formant bandwidths with time to account for source–tract interaction and/or by adjusting the source waveform to reflect the affect of formant interaction with the source (Childers, 1995; Childers and Ahn, 1995; Childers and Lee, 1991; Childers and Wong, 1994; Childers et al., 1989; Pinto et al., 1989).

Source excitation characteristics are important for speech synthesis; for example, jitter and shimmer, excitation waveform shape, and source–tract interaction (Childers, 1995; Childers and Ahn, 1995; Childers and Ding, 1991; Childers and Hu, 1994; Childers and Lee, 1991; Childers and Wong, 1994; Childers and Wu, 1990; Childers et al., 1989; Eskenasi et al., 1990; Lalwani and Childers, 1991a and 1991b; Pinto et al., 1989).

A12.8 SUMMARY

The purpose of this appendix is to acquaint the reader with some aspects of the assessment of speech intelligibility and quality and to present one definition for speech quality. In addition, an outline of an example listening test and its evaluation is given. For a discussion of similar factors for speech coding see Kitawaki (1992), which provides a description of the assessment of speech quality for waveform coding and analysis–synthesis. The Kitawaki paper discusses several assessment methods, subjective and objective methods of assessment of coded speech, distortion measures, speaker dependence on quality, and other factors. Several other references related to speech coding include Atal, Cuperman, and Gersho (1991, 1993), Beker and Piper (1985), Papamichalis (1987), Parsons (1986), Furui (1989), Furui and Sondhi (1992), Saito (1992), Singhal and Atal (1989), Tetschner (1993), Witten (1982).

LIST OF TOOLBOXES

Chapter Introduced	Name of toolbox Folder
Chapter 1	display_speech_1
Chapter 2	analysis
Chapter 4	display_speech_egg
Chapter 6	formant
Chapter 7	vocos and display_speech
Chapter 8	time
Chapter 9	vocal_fold
Chapter 10	formant_track and artm

INDEX